Biology and Evolution of Ferns and Lycophytes

With their team of contemporary scholars, the editors present a thorough coverage of fundamental topics necessary for obtaining an up-to-date understanding of the biology of ferns and lycophytes. The book is organized into major topics that build from the individual and its biochemistry and structure, to genetics and populations, to interactions among individuals and the conservation of species, and concludes with perspectives on evolutionary history and classification. Each chapter is organized to review past work, explore current questions, and suggest productive directions for continued discoveries about these fascinating plants. Written for advanced undergraduates, graduates and academic researchers, *Biology and Evolution of Ferns and Lycophytes* fills a major gap in the literature on vascular land plants by providing a modern overview of the biology and evolution of this important group of organisms.

Tom A. Ranker received his Ph.D. in Botany from the University of Kansas. He is President of the International Association of Pteridologists (2005–2011) and is a Past President of the American Fern Society. From 1990 to 2007, Dr. Ranker was the Curator of Botany at the Museum of Natural History and a faculty member in the Department of Ecology and Evolutionary Biology, University of Colorado at Boulder. In 2008 he joined the faculty of the Department of Botany, University of Hawai'i at Manoa, Honolulu. In 2007 Dr. Ranker was elected as a Fellow of AAAS.

Christopher H. Haufler received his Ph.D. in Botany from Indiana University, and then held post-doctoral positions at Harvard University and at the Missouri Botanical Garden before becoming a faculty member at the University of Kansas in 1979, where he has been ever since. He is past President of the Botanical Society of America, the American Society of Plant Taxonomists, and the American Fern Society. He has received a Centennial Award from the Botanical Society of America and a W. T. Kemper Fellowship for Teaching Excellence.

Biology and Evolution of Ferns and Lycophytes

Edited by

TOM A. RANKER
University of Hawai'i at Manoa
Honolulu, HI, USA

CHRISTOPHER H. HAUFLER
University of Kansas
Lawrence, KS, USA

CAMBRIDGE UNIVERSITY PRESS
Cambridge, New York, Melbourne, Madrid, Cape Town, Singapore, São Paulo, Delhi

Cambridge University Press
The Edinburgh Building, Cambridge CB2 8RU, UK

Published in the United States of America by Cambridge University Press, New York

www.cambridge.org
Information on this title: www.cambridge.org/9780521874113

© Cambridge University Press 2008

This publication is in copyright. Subject to statutory exception
and to the provisions of relevant collective licensing agreements,
no reproduction of any part may take place without
the written permission of Cambridge University Press.

First published 2008

Printed in the United Kingdom at the University Press, Cambridge

A catalog record for this publication is available from the British Library

ISBN 978-0-521-87411-3 hardback
ISBN 978-0-521-69689-0 paperback

Cambridge University Press has no responsibility for the persistence or
accuracy of URLs for external or third-party internet websites referred to
in this publication, and does not guarantee that any content on such
websites is, or will remain, accurate or appropriate.

Dedication

To Genie, Marsha, and our Parents, with Love

Contents

List of contributors   *page* xii
Preface   xv
Acknowledgments   xviii

Part I Development and morphogenesis

1 Photoresponses in fern gametophytes   3
   MASAMITSU WADA
   1.1   Introduction   3
   1.2   Spore germination   7
   1.3   Cell growth   9
   1.4   Phototropism and polarotropism   12
   1.5   Cell division   16
   1.6   Apical cell bulging   19
   1.7   Chloroplast movement   20
   1.8   Nuclear movement   30
   1.9   Reproductive organs   31
   1.10 Photoreceptors for photomorphogenesis   31
   1.11 Concluding remarks   39
         References   40

2 Alternation of generations   49
   ELIZABETH SHEFFIELD
   2.1   Introduction   49
   2.2   "The" fern life cycle   51
   2.3   Historical summary   52
   2.4   Variations on a theme   55
   2.5   Sporogenesis   61
         References   68

3 Meristem organization and organ diversity 75
RYOKO IMAICHI

   3.1 Introduction 75
   3.2 Stem 76
   3.3 Leaf 82
   3.4 Shoot branching, dichotomous versus monopodial 86
   3.5 Roots 90
   3.6 Psilotalean rhizomes (subterranean axes) 94
   3.7 Root-producing organs, rhizophores and rhizomorphs 95
   3.8 Summary and future goals 97
   References 98

Part II Genetics and reproduction

4 Population genetics 107
TOM A. RANKER AND JENNIFER M. O. GEIGER

   4.1 Introduction 107
   4.2 Population genetics and reproductive biology 108
   4.3 Genetic structure of populations 121
   4.4 Gene flow and divergence 122
   4.5 Population genetics of dispersal and colonization 123
   4.6 Summary and future prospects 124
   References 125

5 Antheridiogens 134
JAKOB J. SCHNELLER

   5.1 Introduction 134
   5.2 History of discovery 134
   5.3 General effect of antheridiogen 135
   5.4 Occurrence of different antheridiogens and their chemical structure 138
   5.5 Experiments under laboratory conditions 140
   5.6 Dark germination: a further influence of antheridiogen 143
   5.7 Antheridiogen in nature 145
   5.8 Biological and evolutionary implications of the antheridiogen system 149
   5.9 Future goals 150
   References 151

6 Structure and evolution of fern plastid genomes 159
PAUL G. WOLF AND JESSIE M. ROPER

   6.1 Introduction 159
   6.2 The golden age of fern chloroplast genomics 161

6.3 The age of complete plastid genome sequences 162
6.4 PCR mapping of fern plastid genomes 164
6.5 Conclusions and prospects 171
References 171

7 Evolution of the nuclear genome of ferns and lycophytes 175
TAKUYA NAKAZATO, MICHAEL S. BARKER, LOREN H. RIESEBERG, AND GERALD J. GASTONY

7.1 Introduction 175
7.2 Historical summary 177
7.3 Review of critical recent advances 180
7.4 Synthesis of current perspectives 188
7.5 Future goals and directions 189
References 193

Part III Ecology

8 Phenology and habitat specificity of tropical ferns 201
KLAUS MEHLTRETER

8.1 Introduction 201
8.2 Historical summary 201
8.3 Review of critical recent advances 203
8.4 Synthesis of current perspectives 215
8.5 Future goals and directions 216
8.6 Importance of long-term studies 216
References 217

9 Gametophyte ecology 222
DONALD R. FARRAR, CYNTHIA DASSLER, JAMES E. WATKINS, JR., AND CHANDA SKELTON

9.1 Introduction 222
9.2 Ecomorphology 225
9.3 Ecophysiology 233
9.4 Ecovalidation 242
9.5 Summary 250
References 251

10 Conservation biology 257
NAOMI N. ARCAND AND TOM A. RANKER

10.1 Introduction 257
10.2 Conservation importance of ferns and lycophytes 258
10.3 Threats to global fern and lycophyte diversity 260
10.4 Life cycle challenges for conservation 264

10.5 *Ex situ* propagation of ferns and lycophytes   265
10.6 Regional and ecosystem-level conservation   265
10.7 Conservation of fern and lycophyte taxa   268
10.8 Genetics in fern and lycophyte conservation strategies   269
10.9 Protection and restoration   270
10.10 Future directions in fern and lycophyte conservation   272
References   273

11 *Ex situ* conservation of ferns and lycophytes – approaches and techniques   284
VALERIE C. PENCE

11.1 Introduction   284
11.2 Methods for *ex situ* conservation   285
11.3 *In vitro* cultures and collections   290
11.4 *Ex situ* cryostorage of gametophytes   291
11.5 *Ex situ* cryostorage of sporophytes   292
11.6 *In vitro* collecting for *ex situ* conservation   293
11.7 Current status and future perspectives   294
References   296

Part IV   Systematics and evolutionary biology

12 Species and speciation   303
CHRISTOPHER H. HAUFLER

12.1 Introduction   303
12.2 Species concepts and definitions   304
12.3 Speciation   311
12.4 Summary and future prospects   322
References   324

13 Phylogeny and evolution of ferns: a paleontological perspective   332
GAR W. ROTHWELL AND RUTH A. STOCKEY

13.1 Introduction   332
13.2 Nature of the fossil record   335
13.3 Systematic relationships among ferns, fern-like plants, and other euphyllophytes   335
13.4 Moniliforms – the most ancient fern-like plants   337
13.5 Ophioglossid ferns   343
13.6 Marattioid ferns   345
13.7 Patterns of diversification among leptosporangiate ferns   346
13.8 Historical context, popular practices, and the upward outlook   358
References   360

14 Diversity, biogeography, and floristics 367
ROBBIN C. MORAN

14.1 Introduction 367
14.2 Historical review 367
14.3 Diversity 368
14.4 Long-distance dispersal 372
14.5 Vicariance 378
14.6 Floristics 379
14.7 Important questions 381
14.8 Future directions 382
References 383

15 Fern phylogeny 395
ERIC SCHUETTPELZ AND KATHLEEN M. PRYER

15.1 Introduction 395
15.2 Early vascular plant divergences 396
15.3 Early fern divergences 399
15.4 Early leptosproangiate divergences 403
15.5 Early polypod divergences 405
15.6 Divergences within eupolypods I 406
15.7 Divergences within eupolypods II 407
15.8 Future prospects 408
References 409

16 Fern classification 417
ALAN R. SMITH, KATHLEEN M. PRYER, ERIC SCHUETTPELZ, PETRA KORALL, HARALD SCHNEIDER, AND PAUL G. WOLF

16.1 Introduction and historical summary 417
16.2 Review of critical recent advances 420
16.3 Synthesis of current perspectives: the classification of ferns 421
16.4 Synthesis: lessons learned from morphology and molecular systematics, and unexpected surprises 445
16.5 Future goals and directions 448
References 448
Appendix A: Familial names applied to extant ferns 460
Appendix B: Index to genera 462

Index 468

Contributors

Naomi N. Arcand
Department of Geography, University of Colorado, Boulder, CO 80309, USA

Michael S. Barker
Department of Biology, Indiana University, Bloomington, IN 47405, USA

Cynthia Dassler
Museum of Biological Diversity, Ohio State University, Columbus, OH 43210, USA

Donald R. Farrar
Department of Ecology, Evolution, and Organismal Biology, Iowa State University, Ames, IA 50011, USA

Gerald J. Gastony
Department of Biology, Indiana University, Bloomington, IN 47405, USA

Jennifer M. O. Geiger
Department of Natural Sciences, Carroll College, Helena, MT 59625, USA

Christopher H. Haufler
Department of Ecology and Evolutionary Biology, University of Kansas, Lawrence, KS 66045, USA

Ryoko Imaichi
Department of Chemical and Biological Sciences, Faculty of Science, Japan Women's University, 8-1, Mejirodai 2-chome, Tokyo 112-8681, Japan

Petra Korall
Department of Phanerogamic Botany, Swedish Museum of Natural History, SE-104 05 Stockholm, Sweden

Klaus Mehltreter
Departamento de Sistemática Vegetal, Instituto de Ecología, A. C., Xalapa, Veracruz 91000, México

Robbin C. Moran
The New York Botanical Garden, Bronx, NY 10458, USA

Takuya Nakazato
Department of Biology, Indiana University, Bloomington, IN 47405, USA

Valerie C. Pence
Plant Conservation Division, Center for Research of Endangered Wildlife, Cincinnati Zoo and Botanic Garden, Cincinnati, OH 45220, USA

Kathleen M. Pryer
Department of Biology, Duke University, Durham, NC 27708, USA

Tom A. Ranker
Botany Department, University of Hawaii at Manoa, Honolulu, HI 96822, USA

Loren H. Rieseberg
Department of Botany, University of British Columbia, Vancouver V6T 1Z4, Canada and Department of Biology, Indiana University, Bloomington, IN 47405, USA

Jessie M. Roper
Department of Biology, Utah State University, Logan, UT 84322, USA

Gar W. Rothwell
Department of Environmental and Plant Biology, University of Ohio, Athens, OH 45701, USA

Harald Schneider
Natural History Museum, Cromwell Road, London SW7 5BD, UK

Jakob J. Schneller
Institut für Systematische Botanik, Universität Zürich, CH-8008 Zürich, Switzerland

Eric Schuettpelz
Department of Biology, Duke University, Durham, NC 27708, USA

Elizabeth Sheffield
School of Biological Sciences, University of Manchester, Manchester M13 9PT, UK

Chanda Skelton
Department of Ecology, Evolution, and Organismal Biology, Iowa State University, Ames, IA 50011, USA

Alan R. Smith
University Herbarium, University of California, Berkeley, CA 94720, USA

Ruth A. Stockey
Department of Biological Sciences, University of Alberta, Edmonton, AB T6G 2E9, Canada

Masamitsu Wada
Department of Biology, Tokyo Metropolitan University, Minami Osawa 1-1, Hachioji-shi, Tokyo 192-0397, Japan

James E. Watkins, Jr.
16 Divinity Avenue, Harvard University, Cambridge, MA 02138, USA

Paul G. Wolf
Department of Biology, Utah State University, Logan, UT 84322, USA

Preface

Over the past century, books on basic research into ferns and lycophytes have largely focused on particular topics, floras, or methods of study. Setting the stage for understanding fern structure and evolution was a three-volume masterpiece by Frederick O. Bower, published between 1923 and 1928 by Cambridge University Press, and titled simply, *The Ferns*. In 1950, Cambridge also published Irene Manton's magnum opus, *Problems of Cytology and Evolution in the Pteridophyta*, establishing a new era of exploring the genetics and evolution of ferns and lycophytes. Books concentrating on laboratory studies have included Adrian Dyer's multi-authored *The Experimental Biology of Ferns* and Valayamghat Raghavan's *Developmental Biology of Fern Gametophytes*. Others, such as the detailed and well illustrated *Ferns and Allied Plants* published in 1982 by Rolla and Alice Tryon, were more systematically focused. Several books have captured the exchange of information at international conferences such as *The Phylogeny and Classification of Ferns* edited by A. C. Jermy, J. A. Crabbe, and B. A. Thomas in 1973, the *Biology of Pteridophytes* edited by A. Dyer and C. Page in 1985, a 1989 volume *Systematic Pteridology* edited by K. H. Shing and K. U. Kramer and based on a Beijing conference, and *Pteridology in Perspective* edited by J. M. Camus, M. Gibby, and R. J. Johns in 1996. These and others have synthesized ideas on particular areas of basic research, and helped to maintain excitement and communication about fern and lycophyte biology.

Distinctive from these books was one edited 70 years ago by Frans Verdoorn entitled *Manual of Pteridology*. This volume brought together authorities including Ingrid Andersson-Kottö, Lenette Atkinson, Carl Christensen, W. Döpp, Eric Holttum, and W. Zimmerman, who provided state-of-the-art summaries of topics ranging from chemistry, morphology, and anatomy, to cytology, genetics, and tropisms, and even synthesizing topics such as ecology, geography, paleobotany, classification, and phylogeny. In the tradition of Verdoorn's broad perspective, we have assembled contemporary scholars to present a broad perspective on the *Biology and Evolution of Ferns and Lycophytes*. Instead of using the classic name "pteridophyte," we chose this title because the preponderance of current evidence indicates that "pteridophytes" do not exclusively share a common ancestor. We wanted to incorporate all of the seed-free vascular plants, and we wanted to represent them using labels that captured their diversity without applying overly specialized names. Thus, we consider "ferns" to be equivalent to the "monilophytes" of other authors, and lycophytes to include the extant members of the Lycopodiaceae, Selaginellaceae, and Isoëtaceae.

We have organized our book into major topics that build from the individual and its biochemistry and structure, to genetics and populations, to interactions among individuals and the conservation of species, and conclude with perspectives on evolutionary history and classification. Beginning with a view from the laboratory, M. Wada summarizes decades of studies that have employed fern gametophytes to elucidate basic aspects of plant responses to light, as well as insights on cell structure and function. E. Sheffield shows how ferns can illuminate critical components of phase changes during plant life histories, and reviews the importance of studying species with independent gametophytes and sporophytes. R. Imaichi provides a tour through the anatomy and structure of fern and lycophyte sporophytes, concentrating on the self-perpetuating meristem and its role in establishing plant organization. Moving on to a consideration of genetics and reproduction, T. Ranker and J. Geiger review the dynamics of population biology and show how genetic variation is maintained and partitioned in nature. Integrating physiological responses and genetic consequences, J. Schneller reviews studies of antheridiogen, from both a laboratory and a field perspective. As analysis of entire genomes has become possible, studies of both chloroplast and nuclear DNA sequences have provided remarkable genetic and phylogenetic insights. P. Wolf and J. Roper summarize our current knowledge of the relatively compact chloroplast genome, and T. Nakazato and his colleagues provide new perspectives on the large and complex nuclear genomes of ferns. Shifting to ecology, K. Mehltreter reviews the adaptations of ferns to their environments, and D. Farrar and his co-authors develop provocative insights on the importance of studying gametophytes in their natural habitats. Given the critical topic of the ongoing mass extinction of species, this section concludes with a consideration of conservation, from both field and *ex situ* points of view. In the final section, C. Haufler considers the current status of defining species and discovering their origins, while G. Rothwell and R. Stockey demonstrate the value of fossil plants in considering the evolutionary history of ferns, synthesizing new data from recent discoveries with that from earlier material. R. Moran integrates ecology, population biology, and geology to show how dispersal and vicariance have contributed to current distributions of species. E. Schuettpelz and K. Pryer discuss recent advances in applying DNA sequences to constructing hypotheses of relationships among the diverse members of the ferns and lycophytes, and finally A. Smith and his colleagues use these phylogenetic trees to construct a revised and updated classification.

We recognize that a single volume cannot summarize all of the ongoing research relating to ferns and lycophytes, but by focusing attention on a diverse array of disciplines and approaches, the goal of capturing exciting, contemporary issues and casting a view to the future of each may spark innovative approaches and yield opportunities for new generations of researchers. With each of the chapters organized to review past work, explore current questions, and suggest productive directions for interested investigators, we hope to have developed a compendium that can serve as a benchmark, and one that can energize prospects for continued discoveries about these fascinating groups of organisms.

Acknowledgments

We thank Naomi Arcand for valuable editorial assistance.

Chapter 1 I would like to thank Dr. Chris Haufler for his effort of critical reading and English editing of this article. I also acknowledge Dr. Noriyuki Suetsugu for his critical reading of the article, checking its references, and drawing Figure 1.19, and Dr. Sam-Geun Kong and Mr. Hidenori Tsuboi for drawing Figures 1.6, 1.8, and 1.14 and taking the photographs of Figures 1.7, 1.16, and 1.18, respectively. The results discussed in this article were mainly supported by Mext and JSPS.

Chapter 3 I thank Professor Masahiro Kato of the Department of Botany of National Science Museum, Tokyo, for having participated in many joint projects with me, and for providing useful suggestions on my work. I am also indebted to Professor Judy Jernstedt of the University of California, Davis, for giving me a chance to work in her laboratory, where I obtained some of the data cited here.

Chapter 4 We are grateful for funding provided by the National Geographic Society, the National Science Foundation (DEB-9096282, DEB-9726607, DEB-9807053, DEB-9807054, DEB-0104962, DEB-0344522, and DEB-0343664), the University of Colorado Museum of Natural History, the University of Colorado Graduate School, the Botanical Society of America, and the University of Hawaii.

Chapter 6 Thanks to National Science Foundation grant DEB-0228432 for funding. The manuscript benefited from comments by Carol Rowe, Aaron Duffy, Josh Der, and Mark Ellis.

Chapter 7 This material is based upon work supported by the National Science Foundation under grant no. 0128926.

Chapter 8 I thank José Luis González Gálvez, Adriana Hernández Rojas, Leticia Monge González, and Javier Tolome for their assistance during field work and the Instituto de Ecología, A.C. (902-11-796) and CONACYT-SEMARNAT (2002-C01-0194) for financial support.

Chapter 11 This work has been funded in part by grants from the Institute of Museum and Library Services and from the Fairchild Tropical Botanic Garden.

Chapter 15 Funding was provided by a NSF CAREER award to Kathleen M. Pryer (DEB-0347840) and a NSF DDIG award to Kathleen M. Pryer and Eric Schuettpelz (DEB-0408077).

Chapter 16 We thank James Reveal, who kindly provided comments and suggestions on nomenclatural aspects of our work, especially the information summarized in the appendices. Our work was supported in part by National Science Foundation grants DEB-9616260 to A. R. S.; DEB-9615533, DEB-0089909, and DEB-0347840 to K. M. P., DEB-9707087 to P. G. W., and DEB-0408077 to E. S.; a post-doctoral fellowship from the Swedish Research Council (2003−2724) to P. K.; and a German Science Foundation grant SCHN 758/2-1 to H. S.

PART I DEVELOPMENT AND MORPHOGENESIS

1
Photoresponses in fern gametophytes

MASAMITSU WADA

1.1 Introduction

Fern gametophytes are ideal model systems for study of the mechanisms of photomorphogenesis from the standpoint of physiology, photobiology, and cell biology (Wada, 2003, 2007; Kanegae and Wada, 2006). Positive aspects of the fern system include the following. (1) Spores can be preserved at room temperature and they germinate under appropriate conditions within about a week in many species, becoming gametophytes that grow rapidly, at least in their critical early stages. (2) Gametophytes are nutritionally autonomous, facilitating ease of cultivation. (3) Gametophytes are not enclosed by other tissue, so that observation, light irradiation, and experimental manipulation are readily performed. (4) Each developmental step can be controlled synchronously because gametophytes are highly sensitive to light. Each step in development is completely dependent on light; indeed, without light, development does not proceed.

Since the nineteenth century, especially in Germany, fern gametophytes have been used (see Dyer, 1979a) to study photo-physiological phenomena, such as light dependent spore germination (Mohr, 1956a), differentiation from one-dimensional protonemata to two-dimensional prothalli (Mohr, 1956b), and intracellular dichroic orientation of phytochrome (Etzold, 1965). Even though fern gametophytes are very good materials for the study of both photobiology and cell biology, only a few laboratories use them presently, probably for the following reasons. (1) Although mutants can be obtained easily by phenomenological screening (gametophytes are haplophase), making crosses for genetic studies is difficult and time consuming. (2) The biochemistry is also challenging because

Biology and Evolution of Ferns and Lycophytes, ed. Tom A. Ranker and Christopher H. Haufler. Published by Cambridge University Press. © Cambridge University Press 2008.

collecting enough gametophyte tissue for biochemical analyses is difficult. (3) Molecular biological techniques are not yet established (e.g., stable transformation is not available, although transient gene expression is possible). (4) Most ferns are not commercially valuable plants, although some species, such as *Osmunda japonica*, *Pteridium aquilinum*, and *Matteuccia struthiopteris*, are edible and obtainable commercially in eastern Asia, or are used as ornamental plants, or for cleaning soil polluted by heavy metals including arsenic (Ma *et al.*, 2001).

Nevertheless, fern gametophytes have structural and physiological characteristics that seed plants do not have, making them more tractable systems for studying many phenomena that are common to ferns and seed plants. For example, we have analyzed factors controlling the pre-prophase band (PPB) formation and its disruption (Murata and Wada, 1989b, 1991a, 1991b, 1992) (Figure 1.1). The PPB is recognized as a factor controlling the attachment site of newly synthesized cell plates to mother cell walls (Mineyuki, 1999). It appears before prophase of the nuclear division cycle at the future site of cell plate fusion to the mother cell wall, but disappears before cell plate formation. The kind of information remaining at the PPB region has long been a mystery, as have the factors that determine the future cell plate attachment site and disrupt the PPB. To study this issue physiologically, Murata and Wada (1989b, 1991b, 1992) used a long protonemal cell cultured under red light in which cell division occurred at 40–60 μm from the tip where the division site is pre-determined by the PPB. During the period when the PPB was polymerizing, protonemal cells with a premature PPB were centrifuged to reposition the nucleus. A new PPB formed at the new nuclear site, distant from the original position, and then cell division occurred, suggesting that the nucleus must be close to the PPB polymerization site. In these cells the first PPB at the apical part did not de-polymerize even after cell division occurred, but if a dividing nucleus was returned to the former PPB site, the PPB de-polymerized. This result indicates that PPB de-polymerization requires a nucleus and/or surrounding cytoplasm. Experiments such as these could not be done using seed plant cells because long cells like protonemal cells are not found in seed plants, except in some special cases such as cambium cells, where cell division occurs periclinally, making them inappropriate for the experiment. Experiments using long protonemal cells were also performed to study the recovery of a nucleus elongated by cell centrifugation (Wunsch and Wada, 1989; Wunsch *et al.*, 1989).

To analyze the physiological characteristics at each step of the developmental process or during transitions from one step to another of photobiological responses in fern gametophytes, various tools and special techniques have been developed. These include microbeam irradiators to stimulate only a small

Photoresponses in fern gametophytes 5

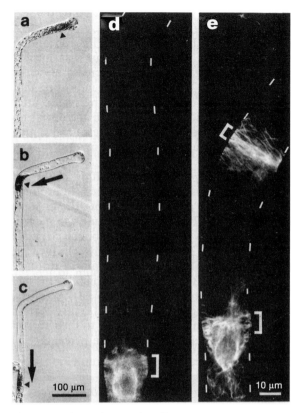

Figure 1.1 Photomicrographs of bent protonemata showing the effect of double centrifugation on pre-prophase band formation. (a)–(c) A bent protonema centrifuged parallel with the arrows to sediment cytoplasm, including a nucleus and chloroplasts. Note that a nucleus, indicated by small arrowheads, moved downward. (d), (e) Bent protonemata centrifuged 4 and 8.5 hours (d) and 4 and 12.5 hours (e) after the onset of blue light irradiation and fixed at 14.5 hours. The regions between the bend and the nucleus are shown. One pre-prophase band (marked with a bracket) was found in (d) and two bands were found in (e). The second centrifugation was applied before and during pre-prophase band formation, respectively. (After Murata and Wada, 1991b).

part of a cell and identify the photoreceptive site, i.e. the localization of photoreceptor molecules mediating a target phenomenon. The first machine was constructed in 1978 (Wada and Furuya, 1978) (Figure 1.2). Current microbeam projectors are now in their fourth or fifth generation, and are equipped with various accessories depending on their purpose (Iino et al., 1990; Yatsuhashi and Wada, 1990). Photoreceptive sites revealed by experiments using microbeam irradiators are summarized in Figure 1.3, and will be explained in the following sections.

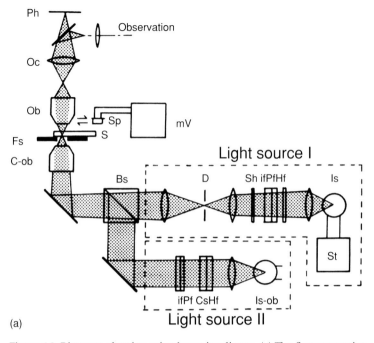

(a)

Figure 1.2 Diagrams showing microbeam irradiators. (a) The first generation irradiator. An ordinal light microscope was remodeled for microbeam irradiation by inserting a diaphragm, and another light source was added for observation. Ph, photographic camera; Oc, ocular lens; Ob, objective lens; S, specimen; Fs, focusing stage; C-ob, condenser objective lens; Sp, silicon photocell; mV, millivolt meter; Bs, beam splitter; D, diaphragm; Sh, shutter; if, interference filter; Pf, plastic filter; Hf, heat filter; Is, irradiation source; St, stabilizer; Cs, $CuSO_4$ solution; Is-ob, irradiation source for observation. (b) A third generation irradiator. Four different wavelength lights can be irradiated simultaneously at one point or two mixed lights can be irradiated side by side. CF, cut-off filter; Dp, depolarizer; Fs, field stop; HM, half-silvered mirror; IF, interference filter; IRV, infrared viewer; LS, light source; M, mirror; Obs, observation point; P, polarizer; PC, photographic camera; Sh, shutter; Sl, slit; St, sample stage; WF, water cell; WP, pump for circulation of cooled water to water cell. See Fig. 6 of Wada (2007) for a diagram of the second generation irraditor. ((a) After Wada and Furuya, 1978. (b) After Iino *et al.*, 1990.)

This chapter will focus on recent analyses performed mostly by my laboratory group using *Adiantum capillus-veneris*. I also include some results that have not been published but are based on a synthesis of nearly 40 years of my experience with fern gametophytes. Our knowledge, mostly obtained from *A. capillus-veneris*, assumes that this species follows a pattern of development that is typical of most ferns. However, because of the large diversity in species and gametophytes, numerical data such as the growth rate of protonemata mentioned here may

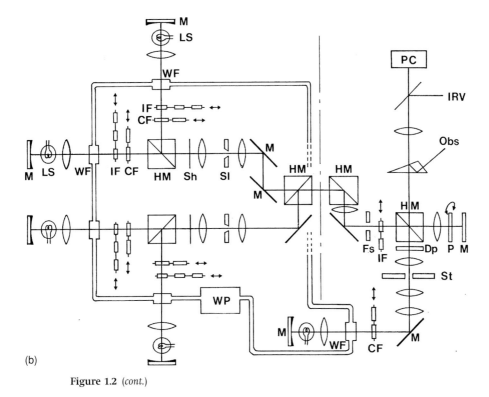

Figure 1.2 (cont.)

or may not be applicable to other fern species. For more information refer to books by Dyer (1979b) and Raghavan (1989) and the following reviews: Wada and Kadota (1989), Wada and Sugai (1994), Kanegae and Wada (2006), and Wada (2007).

1.2 Spore germination

There are two kinds of fern spores based on their color: one is green (chlorophyllous) and the other is brown (non-chlorophyllous). Green spores have chloroplasts even before water imbibition and, unless refrigerated, their germination ability (spore viability) does not persist long after harvest. See Raghavan (1989) for more information.

Most fern spore germination is light dependent. In a tetrahedral, non-chlorophyllous, dormant spore, the nucleus sits in one corner surrounded by three furrows. When spores are irradiated with red light after imbibition in the dark, they become round, and the nucleus, still in its corner position, divides, followed by cell division to produce large protonemal and small rhizoidal mother cells (Furuya et al., 1997). In *A. capillus-veneris*, *Pteris vittata*, and probably other

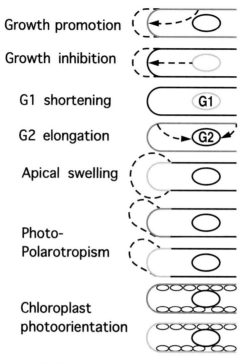

Figure 1.3 Photoreceptive sites for light-induced phenomena in an *Adiantum capillus-veneris* protonema. Light grey (blue) and dark grey (red) indicate photoreceptive sites of blue and red light photoreceptors, respectively. (After Wada, 2007.)

species, red-light induced germination is inhibited by far-red light in a red/far-red reversible manner, indicating the involvement of phytochrome (Sugai and Furuya, 1967; Furuya *et al.*, 1997). The red light effect is inhibited by blue light, on exposure before or after the red light treatment (Sugai and Furuya, 1967; Furuya *et al.*, 1997). The blue light inhibition effect, however, cannot be reversed instantaneously by subsequent exposure to a pulse of red light, suggesting the involvement of a blue light receptor, but not a phytochrome system. Inhibition can be prevented when the spores are kept in the dark for about a week (Sugai and Furuya, 1968; Furuya *et al.*, 1997). The time period required for prevention of blue light inhibition is very much reduced if the spores are irradiated with red light. The red light effect can be reversed by far-red light, indicating phytochrome dependence (Sugai and Furuya, 1968; Furuya *et al.*, 1997). The inhibitory effects of far-red and blue light could not be observed after the first mitosis in spores, suggesting that cell division is a crucial step for spore germination (Furuya *et al.*, 1997). Partial spore irradiation with red or blue microbeam lights showed that the blue light receptor is located in the nucleus, but the location of the red light photoreceptor could not be identified (Furuya *et al.*, 1997). The photoreceptors mediating spore germination (both phytochrome and blue light receptors) have not yet been identified, although several candidate genes have been cloned and sequenced. The details of fern photoreceptors will be described in Section 1.10.

1.3 Cell growth

In most homosporous ferns, after spore germination under red light, a filamentous protonemal cell grows at the apical dome towards a red light source without (or at least with a low frequency of) cell division. The cell is about 15–20 μm in diameter, although this varies with species and culture conditions. The nucleus is always located about 60 μm from the tip during cell growth in *A. capillus-veneris*, indicating that the nucleus migrates in the cell toward the tip, maintaining a constant distance (Figure 1.4) (Wada and O'Brien, 1975; Wada *et al.*, 1980). In a growing protonema, microtubule and microfilament strands connect the nucleus to the cortex of the apical and basal parts of the cell (Kadota and Wada, 1995), although how these cytoskeletal strands control nuclear migration is not yet known. The growth rate varies with species and also with environmental conditions in the same species. In the case of *A. capillus-veneris* under continuous red light (0.5 W m^{-2} s^{-1}) at 25 °C the protonemata grew at an average rate of about 200 μm/day (Wada, 1988a).

In some species (including those of *Anemia*, *Osmunda*, and *Lygodium*), even under red light conditions, gametophytes germinate as two-dimensional prothallia, and no protonemal stage is observed (Raghavan, 1989). In *Ceratopteris*, when spore germination was induced by white light irradiation for 1 day after imbibition and then the spores were kept in the dark, gametophytes germinated as a two-dimensional, strap-shaped prothallium in four cell-columns. A cell mass proliferated at the apical part of the gametophyte and each cell at the basal part of the cell mass grew in the dark parallel to the cell polarity (Murata *et al.*, 1997). In this species, cells can grow in the dark, similar to protonemal cells of *A. capillus-veneris* grown under red light, but the cells are not protonemata.

The cell diameter under red light is reasonably constant. How do cells know the diameter and how do they maintain it? At the basal part of the apical dome of protonemata, a circular array of microtubules and microfilaments is observed (Murata *et al.*, 1987; Kadota and Wada, 1992b). Because this will be discussed in detail in Section 1.6, it is sufficient to note that these cytoskeletal structures play a key role in maintaining a constant diameter, as has been well established in higher plant cells (Shibaoka, 1994).

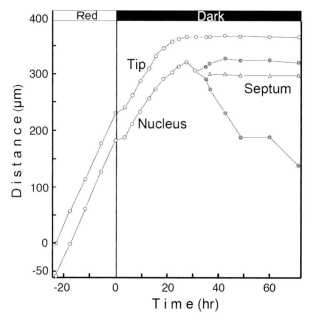

Figure 1.4 Time courses of interacellular nuclear movement and apical growth during the cell cycle in *Adiantum capillus-veneris* protonema. A red-light grown protonema was transferred into the dark to induce cell division and the positions of a nucleus and the tip of the protonema were traced under infrared light microscopy. (After Wada *et al.*, 1980.)

1.3.1 *Growth cessation*

When protonemal cells cultured under continuous red light are transferred to darkness, the growth rate is reduced and ultimately cell growth stops (Kadota and Furuya, 1977). Simultaneously, the nucleus moves a short distance towards the cell base and after a period of time cell division occurs (Figure 1.5) (Wada *et al.*, 1980). The timing of cell division is controlled by phytochrome (Wada and Furuya, 1972), as will be discussed in Section 1.5. The timing of growth retardation and the length of cell growth during the dark period before cell division occurs are also controlled reversibly by red/far-red light irradiation just before transferring to the dark (Kadota and Furuya, 1977). Protonemal cell growth may be controlled in conjunction with the timing of cell division by the same phytochrome system. It is curious, however, that the timing of cell division is delayed by far-red light whereas growth retardation is advanced and, consequently, growth is reduced. This result indicates that cell cycle advancement and cell growth retardation (i.e., cell growth) are not really parallel. Protonemal cell growth is also controlled by blue light; this will be discussed in Section 1.5.

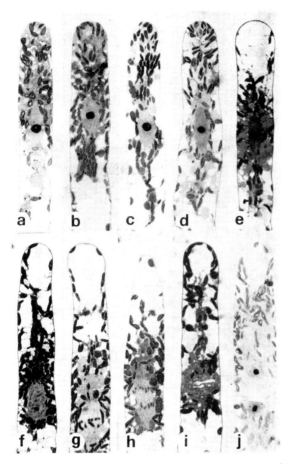

Figure 1.5 Light micrographs of a longitudinal section of *Adiantum capillus-veneris* protonemata during cell cycle in the dark. Nuclear migration towards the base of the cell and structural changes of organelle patterns are shown. (After Wada *et al.*, 1980.)

1.3.2 Resumption of cell growth

When red-light grown protonemata are transferred into the dark, cell division occurs in the apical region of the linear protonemata (Wada and Furuya, 1972). If protonemal cells are kept in the dark after cell division for several days, almost all cytoplasm in the apical cells, including chloroplasts and a nucleus, moves toward the cell plate at the basal end, so that the cells are occupied by a large vacuole and become transparent. These cells appear dormant and neither grow nor divide until light is provided. It is not known how long the cells can survive without light. When protonemata are irradiated with red light continuously, the nucleus moves toward the cell tip, the cytoplasm disperses over the entire cell, and then elongation resumes at the apex of the cell (Kadota and Furuya, 1981). When red-light grown protonemata kept in the dark for 3 days were used, 24 hours of irradiation was required for all protonemata to recover and grow normally (Kadota and Furuya, 1981). Resumption of cell growth is controlled in a phytochrome red/far-red light reversible manner. However, full reversibility by far-red light is lost when cells are irradiated by red light for longer than 4 s at 4.6 W m^{-2}, although it is not dependent on an escape reaction (Kadota and Furuya, 1981). This is an unusually rapid and sensitive response compared to other phytochrome-dependent phenomena in fern gametophytes, e.g., the timing of cell division, which is induced by transferring protonemata from red light to darkness and is still reversible by far-red light after irradiation by red light for 10 minutes (Wada and Furuya, 1972).

Red light controlled protonemal growth, cessation, and resumption of growth discussed in this section are mediated by phytochrome, but it is not yet known which of the three phytochrome genes cloned in *A. capillus-veneris* mediates these phenomena.

1.4 Phototropism and polarotropism

The direction of protonemal growth is controlled by light. Protonemal cells grow toward a red light source (phototropism) or perpendicular to a vibration plane of polarized red light, following a phenomenon known as "polarotropism" (Bünning and Etzold, 1958; Etzold, 1965). Phototropism and polarotropism are phenomenologically different responses because in phototropism a protonema grows toward a light source whereas in polarotropism a protonema grows perpendicular to the polarized incident light and to its vibration plane, regardless of the direction of incident light (Figure 1.6). Thus, the protonema grows toward the side that absorbs more light. Polarotropism occurs because of the orderly intracellular arrangement of phytochrome molecules attached to the plasma membrane. However the two light-induced tropisms may be considered equivalent if the direction of growth is determined by the highest concentration of the far-red light absorbing form of phytochrome (Pfr) in the protonemal apical dome (explained in the following section). Hence, here I treat these two responses as one physiological phenomenon controlled by the same phytochrome molecular species and the same mechanism. When polarotropism is induced in protonemata growing on the surface of an agar medium, growth is not only perpendicular to the vibration plane but also toward the light source of the polarized light. In order to avoid phototropism under polarized light, to irradiate cells evenly without reflection and refraction, and to eliminate the lens effect caused when cylindrical protonemal cells are elevated above the growing

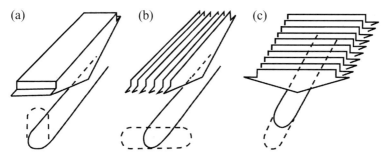

Figure 1.6 Schematic illustration of phototropism and polarotropism in fern protonemata growing horizontally at the tip. When a protonema (solid line) is irradiated with non-polarized red light from above, the protonema grows at its tip towards the light source (a, dotted line). If the red light is polarized, the protonema no longer grows towards the light source but grows horizontally, perpendicular to the vibration plane of the polarized red light (b and c, dotted lines). The vibration plane of the polarized light is parallel to the cell axis in (b) and perpendicular in (c).

surface, all experiments were performed using protonemata cultured on an agar medium covered with a cover slip, or under similar submerged conditions.

When the direction of incident red light is changed, a change in the direction of protonemal growth toward the new light source can be detected about 1 hour after the light treatment. However, intracellular events show that the cells respond instantaneously to the new light by modifying the cytoskeletal pattern and subsequently the pattern of microfibrils (Wada et al., 1990). Analytical studies of the response have focused on either polarotropism or microbeam-induced phototropism, because the responses induced by these methods can be controlled more accurately. The tropistic curvature of protonemata is very sharp when induced through whole cell irradiation by polarized light or by partial cell irradiation with a microbeam (Figure 1.7), but is more rounded when the whole cell is irradiated with ordinary light. It is likely that in the former instances light absorption by phytochrome molecules is restricted to a very small area compared to that in the latter.

Etzold (1965) proposed a hypothesis to explain the polarotropic response in *Dryopteris* protonemata. According to this hypothesis the red light absorbing form of phytochrome (Pr) is localized at the cell periphery (close to the plasma membrane) and has a transition moment parallel to the plasma membrane. Growth occurs in the portion of the cell where the highest concentration of phytochrome is transformed to the far-red light absorbing form (Pfr) by red light absorption (Figure 1.8). Based on this hypothesis, when polarized red light vibrating perpendicular to the cell axis is applied to the apical dome of protonemata, the protonemata grow straight as before at their tip, because the phytochromes

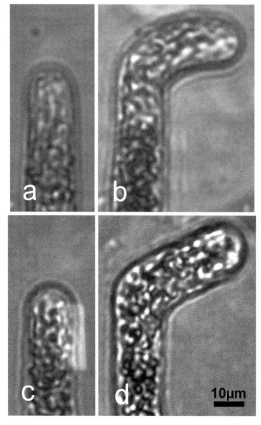

Figure 1.7 Light micrographs of red-light grown protonemata showing tropistic responses. (a), (b) Phototropism induced by whole cell irradiation with red light (0.1 W m^{-2}, continuously) from the right-hand side (a, before irradiation; b, 14 hours after changing the light direction). The tropistic response is rather gradual, not at a sharp angle. (c), (d) Phototropism induced by microbeam (10 × 20 μm, 5 W m^{-2}, 30 s) irradiation at the right side of the protonema (c, start of irradiation; d, 14 hours after the microbeam irradiation). The tropistic response is at a sharp angle, not gradual.

with transition moments parallel to the vibration plane of the polarized light are localized only at the tip of the apical dome. When the vibration plane is twisted (no longer perpendicular to the cell axis), a portion of the apical dome whose plasma membrane becomes parallel to the direction of the new vibration plane can absorb more polarized light, and can then become a new growing cell tip.

According to Etzold's hypothesis the transition moment of Pfr should be perpendicular to the plasma membrane (Etzold, 1965). The photo-conversion

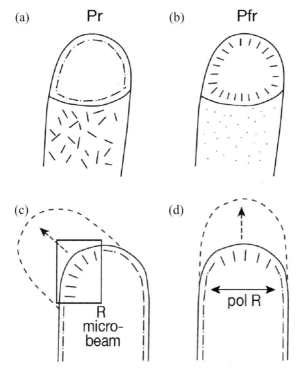

Figure 1.8 Schematic illustrations of the intracellular arrangement of the red light absorbing form (Pr) and far-red light absorbing form (Pfr) of Acneo1 and their rearrangement during tropistic responses. The transition moment of Pr (inactive form) is parallel to the plasma membrane (a) and that of Pfr (active form) is perpendicular (b). When half of the apical dome of the protonema is irradiated with a red microbeam (c), Acneo1 molecules change their transition moment from Pr to Pfr and cell growth changes direction towards the irradiated side (dotted line in c). When the protonema is irradiated with polarized red light vibrating perpendicular to the cell axis, Pr at the tip of the cell changes to Pfr and growth is straight (d).

of the transition moment between Pr parallel and Pfr perpendicular to the plasma membrane by red and far-red light irradiation respectively, was confirmed through chloroplast movement in *Mougeotia* by Haupt and his colleagues (Haupt et al., 1969). In *A. capillus-veneris* we determined the actual photoreceptive site in a protonema by microbeam irradiation with polarized red light at various portions of the cell and found that the cell margin, especially at the basal part of the apical dome, is the most effective site for polarized light absorption (Wada et al., 1981). The dichroic orientation of phytochrome was also confirmed in the polarotropism of *A. capillus-veneris* protonemata by very precise analyses (Kadota et al., 1982, 1985; Wada et al., 1983). Recently neochrome1 (neo1) (formerly called phytochrome 3 (phy3)) was discovered in *A. capillus-veneris*

(Nozue *et al.*, 1998) and was shown to be the photoreceptor of fern phototropism and polarotropism induced by red light (Kawai *et al.*, 2003, see more details in the following sections).

Phototropism and polarotropism can be induced by blue light also, although, because blue light inhibits protonemal cell growth, it is better to irradiate with red light simultaneously to stimulate cell growth (Kadota *et al.*, 1979, 1989). In these experiments, blue light was applied unilaterally or as polarized light, but red light was applied vertically in the former case and as non-polarized light in the latter case to avoid the directional influence of red light. The blue light receptor of this phenomenon is not yet known, although phototropins are plausible candidates.

1.4.1   Change of cell structure during tropistic responses

Circular arrays of microtubules and microfilaments, as well as the microfibril pattern at the subapical portion of protonemata, change during phototropism (Wada *et al.*, 1990; Kadota and Wada, 1992a, 1992b). When polarotropism was induced by polarized red light vibrating 45° to the cell axis, the cortical array of microtubules became oblique within 30 minutes after irradiation to the direction of bending, but if the vibration plane was 70°, the microtubule array disappeared. After 1 hour, the tropistic response could be observed using a microscope. By 2 hours after polarotropism induction, the microfibril rearrangement of the innermost layer of the cell wall became oblique to the former growing axis (Wada *et al.*, 1990). During phototropism, reorganization of the microfilament structure precedes that of the microtubule structure (Kadota and Wada, 1992a), suggesting that the microtubule array is influenced by the microfilament array. Interestingly, this hypothesis was confirmed by experiments using cytoskeletal inhibitors (Kadota and Wada, 1992b). Colchicine and amiprophosmethyl disrupted the microtubule array but not the microfilament array. In contrast, cytochalasin B disrupted both arrays, indicating that the microtubule array depends on the array of microfibrils. Taken together, phototropism and polarotropism must occur through sequential changes: the microfilament array controls the direction of the microtubule array, which controls the direction of microfibril arrangement, and finally microfibrils restrict the cell diameter and the direction of cell growth.

1.5   Cell division

When red-light grown protonemata are transferred into darkness or white light, cell division occurs synchronously at the apical portion of the protonemata (Wada and Furuya, 1970, 1972). Structural changes during the cell

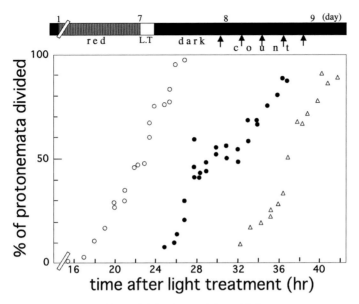

Figure 1.9 Timing of cell division induced in red-light grown protonemata of *Adiantum capillus-veneris* by irradiation with short pulses (10 min) of blue (open circles), red (filled circles) and far-red (triangles) light before transfer to darkness. Blue light shortens the cell cycle but far-red light lengthens it compared with the direct transfer from red light to the dark. L.T, light treatment. (After Wada, 2007, modified from Wada and Furuya, 1972.)

division were analyzed using thin sections (see the details in Wada and O'Brien, 1975; Wada et al., 1980). The timing of cell division is light dependent. Blue light induces cell division (Wada and Furuya, 1974) (Figure 1.9), whereas red light inhibits cell division (Wada and Furuya, 1970). When protonemata grow under red light, their cell cycle remains at an early G1 phase (Miyata et al., 1979). Altogether, the cell cycle of protonemata is controlled by light. Blue light shortens the period of the G1 phase but far-red light lengthens the G2 phase (Miyata et al., 1979). Within a restricted range, the stronger the intensity of blue light or the longer the light irradiation, the shorter the G1 phase. This light-dependent protonemal cell cycle may be a very useful system for studying the regulation mechanisms of the plant cell cycle precisely from a physiological and cell biological standpoint. In this system, cell division is controlled by light and occurs synchronously in almost all protonemata. The most difficult aspect of this model system is the collection of sufficient protonemal cells if biochemical studies are intended.

The protonemal cell cycle induced by darkness can be reversed by red light irradiation, if the cell cycle is still in the G1 phase (Wada et al., 1984). When

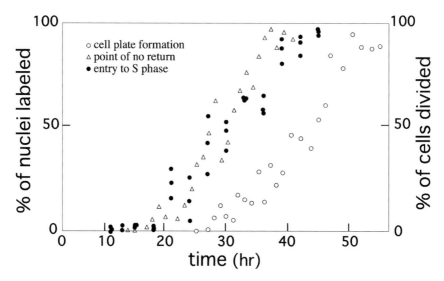

Figure 1.10 "The point of no return" in cell division in red-light grown *Adiantum capillus-veneris* protonemata. Under continuous red light, protonemata remain at the beginning of the G1 phase. When the red-light grown protonemata are transferred into the dark, cell division is induced. However, the cell cycle stops and then returns to the beginning of the G1 phase, if the protonemata are irradiated with continuous red light before entering the S phase. Triangles, open circles, and filled circles show the point of no return, timing of cell division, and entry to the S phase, respectively. (After Wada, 2007, modified from Wada *et al.*, 1984.)

red-light grown, single-celled protonemata are transferred into darkness, the cell cycle starts and cell growth slows. However, the protonemata begin to elongate again and the cell division trigger is cancelled if the protonemata are returned to red light after the transfer to darkness. But if the timing of the transfer to red light is delayed to beyond the entry to the S phase, cell division occurs and the cell cycle cannot return to the G1 phase even under continuous red light. The switch between "returnable" and "unreturnable" is called a "point of no return" (Wada *et al.*, 1984) (Figure 1.10). The point of no return in blue light induced cell division is different from that of cell division induced in the dark. In this case the point of no return by red light irradiation is in the middle of the G1 phase, much earlier than the entry to the S phase (Wada, unpublished data). The mechanisms and pathways of cell cycle progress may be different for blue light induced cell division and dark induced cell division.

Intracellular localization of the blue light receptor for the induction of cell division was studied by microbeam irradiation (Wada and Furuya, 1978). Various portions of the tip of red-light grown, single-celled protonemata were irradiated with a short pulse of a blue light microbeam through a slit (30 μm in width). The time at which cell division occurred after microbeam irradiation was calculated. The portion that induced the highest frequency of cell division was located about 60 μm from the tip, where a spindle-shaped nucleus is usually found (Wada and Furuya, 1978). To confirm that the photoreceptive site is on or in the nucleus or near the nuclear region, protonemal cells were centrifuged to move the nuclei basipetally. The region now containing a nucleus or the former region now lacking a nucleus was irradiated with blue light through a narrow slit (Kadota *et al.*, 1986). The results clearly showed that the new nuclear region, rather than the former region, was responsible for initiating cell division, indicating that photoreceptors were localized within the cytoplasm mass near the nucleus. We are not yet sure what receives the blue light, nor do we know whether the photoreceptors are within the nucleus or outside. One or more cryptochrome(s) of five (*Ac*CRY 1 to 5) already cloned in *A. capillus-veneris* (Kanegae and Wada, 1998; Imaizumi *et al.*, 2000) may be a photoreceptor candidate for this response. Depending on the distribution of GUS-*Ac*CRY nucleocytoplasma (Imaizumi *et al.*, 2000), *Ac*CRY3/4 are plausible candidates for this response (Kanegae and Wada, 2006).

1.6 Apical cell bulging

When red-light grown protonemata are irradiated with blue light continuously, the apical part of the protonemata starts to swell and becomes bulbous prior to cell division (Wada *et al.*, 1978). The cell plate occurs usually near the neck of the swollen part (Wada and Furuya, 1970; Wada and O'Brien, 1975), that is, at the junction between the bulbous and filamentous parts of the protonemata. Apical bulging has been recognized as the event that initiates two-dimensional growth (Mohr, 1956b). Using microbeam irradiation, the apical dome of protonemal cells was identified as the location of blue light reception for this phenomenon, but not in or around the nucleus. Polarized blue light irradiation shows a dichroic effect: polarized light vibrating parallel to the plasma membrane is more effective than that vibrating perpendicular to the plasma membrane. These data indicate that the blue light receptor should be localized on the plasma membrane and should be arranged dichroically parallel to the plasma membrane (Wada *et al.*, 1978). The evidence appears to indicate that both cell division and apical cell bulging can be induced by a single application of blue light, and that these responses are sequential phenomena along one signal transduction pathway, starting with apical bulging leading to cell division. However, by careful analyses of intracellular localization of photoreceptor(s) by microbeam experiments, it was discovered that the two phenomena are completely independent from each other (Wada and Furuya, 1978; Wada *et al.*, 1978).

The photoreceptor for cell division is localized in the nuclear region (Wada and Furuya, 1978) but that for apical bulging is on the plasma membrane (Wada et al., 1978). Presumably the photoreceptors for apical bulging are phototropins, although that remains to be determined through further studies.

Typical apical cell bulging is not observed when cell division is induced in the dark (Wada et al., 1980) or even under weak white light, although there is a tendency for the apical part of the protonemal cell to swell slightly when cell division occurs under any light conditions, in the dark or under red light.

Investigations of cytoskeletal changes during apical bulging revealed that the diameter of protonemal cells was regulated by actin filaments, microtubules, and microfibrils. Grown under red light, microtubules (Murata et al., 1987) and microfilaments (Kadota and Wada, 1989a) at the subapical part of protonemata showed a circular arrangement. When these red-light grown protonemata were transferred to blue light, the circular array of microtubules and microfilaments disappeared prior to apical bulging. Both structures disappeared at about the same time (Kadota and Wada, 1992a). Under red light, the arrangement of microfibrils in the innermost layer of the cell wall at the subapical part of the protonemata was perpendicular to the cell axis. Thus, the microfibrils were parallel to the circular arrays of microtubules and microfilaments. When transferred to blue light, however, the microfibrils changed to a random arrangement (Murata and Wada, 1989a) (Figure 1.11). The circular microtubule array disappeared before the microfibril pattern became random, prior to the detection of apical bulging, indicating that the cortical microtubule array regulates the microfibril arrangement, which, in turn, controls the cell diameter. Apical cell bulging is also induced by disruption of the microtubule array by colchicine and amiprophosmethyl (Murata and Wada, 1989c), confirming the regulation of the cell diameter by microtubules.

1.7 Chloroplast movement

Chloroplast photorelocation has been well known since the nineteenth century in groups ranging from algae to seed plants (Wada et al., 1993, 2003). Under weak light, chloroplasts move toward a light source or to a brighter part of a cell (accumulation response) for efficient light absorption. Under excess light, they move away from the light (avoidance response), preventing photo-damage of chloroplasts (Figure 1.12). When *Arabidopsis* mutants deficient in avoidance response were kept under strong light (1400 μmol m^{-2} s^{-1}) for more than 10 hours, the leaves became necrotic because the chloroplasts were destroyed and mesophyll cells became seriously damaged (Kasahara et al., 2002). Hence, the avoidance response is very important for plant survival. Augustynowicz and

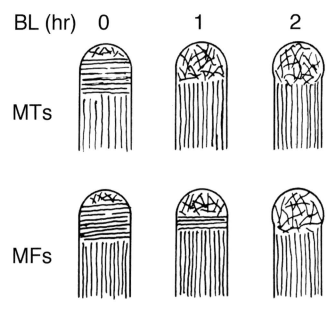

Figure 1.11 Diagrams showing changes in the arrangement of cortical microtubules (MTs) and microfibrils (MFs). The pattern of cortical microtubules at the apical part of the protonemata grown under red light is modified by blue light irradiation. The transverse arrangement of microtubules disappears and a random arrangement becomes dominant. The pattern of microfibril arrangement follows the change in microtubule pattern, suggesting that the microfibril pattern is controlled by microtubules. The numbers indicate hours after blue light irradiation. (After Murata and Wada, 1989a.)

Gabrys (1999) reported the ecological significance of chloroplast movement in fern sporophytes. They reported that *A. capillus-veneris* and *Pteris cretica* showed clear photorelocation movement under both strong and weak light. However, *Adiantum caudatum*, found in high light environments, does not show photorelocation movement and *Adiantum diaphanum*, living in shady environments, shows only weak photorelocation movement. Augustynowicz and Gabrys (1999) concluded that chloroplast photorelocation is only found in plants living in environments with fluctuating light intensities. Using a microscope, we detected chloroplast photorelocation movement only in very young leaves of *A. capillus-veneris* (Kawai *et al.*, 2003). I am not certain that the chloroplast avoidance response occurs under strong light in adult fern leaves, even if they live under fluctuating light conditions, but it is clear that almost all fern gametophytes so far tested show both the accumulation response and the avoidance response. In general, blue light stimulates both accumulation and avoidance responses in many fern species, except in the genus *Pteris* (Kadota *et al.*, 1989; Kagawa and Wada, 2002).

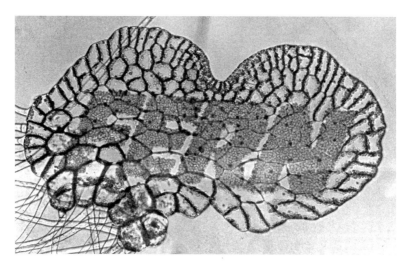

Figure 1.12 Chloroplast photorelocation movement in a prothallium of *Adiantum capillus-veneris*. The portions outside the letters "FERN" were irradiated with strong light and chloroplasts moved away from the area, showing an avoidance response. The portions within the letters were irradiated with weak light to induce an accumulation response.

In lower plants, such as *Mougeotia* (Haupt *et al.*, 1969), *Mesotaenium* (Kraml *et al.*, 1988), and *Physcomitrella* (Kadota *et al.*, 2000), red light is also effective. We focus here on chloroplast movement in fern gametophytes. For other plant species refer to Haupt, 1999; Wada *et al.*, 1993; or Wada *et al.*, 2003.

1.7.1   Cell sensitivity to light

Chloroplast movement in *A. capillus-veneris* has been studied using linear protonemal cells or two-dimensional prothalli for more than 20 years. Both accumulation and avoidance responses can be induced by irradiating whole gametophytes, but for analytical studies, partial cell irradiation using a microbeam irradiator or polarized light irradiation is more efficient and useful. The sensitivity of chloroplast movement to light varies among cells or even among parts of a single cell. The basal cell of a two-celled protonema is more sensitive than that in a single-celled and still growing protonema, although the reason for this is not known. The upper part of the basal cell is more sensitive than the lower part, probably because the upper part is physiologically younger than the lower part. We have observed a similar situation in two-dimensional gametophytes, i.e., a newly made region of a gametophyte is more sensitive than old cells located near the gametophyte base, although we do not have numerical data on the sensitivity. All the data presented below were obtained using highly sensitive cells or cell parts.

1.7.2 Induction of chloroplast movement

When part of a long protonemal cell (Yatsuhashi et al., 1985) or a two-dimensional gametophyte (Kagawa and Wada, 1994) was irradiated with a red or blue light microbeam (either slit, spot, or rectangular in shape, from a few micrometers to 10 or 15 μm in diameter) at a weak fluence rate (e.g., 1 W m^{-2}), chloroplasts outside the beam moved towards the light irradiated area. When the fluence rate of blue light was increased above 10 W m^{-2}, chloroplasts moved away from the beam. Red light does not induce an avoidance response within the range of fluence rate which is reasonable for physiological function (Yatsuhashi et al., 1985).

Polarized light is also very effective at inducing chloroplast photorelocation in fern protonemata (Figure 1.13) (Yatsuhashi et al., 1987a, 1987b). When the sides of protonemal cells sandwiched between an agar surface and a cover slip were irradiated with polarized red or blue light, chloroplasts moved depending on the vibration plane of the polarized light. If the vibration plane was perpendicular to the protonemal axis, chloroplasts moved towards the light source. But if the vibration plane was parallel to the cell axis, theoretically the chloroplasts should not have moved in any direction but should stay in their original position. To obtain these results, however, polarized light should be applied exactly from the side of the protonema as a parallel light beam hitting the cell perpendicularly. If the light source is a long fluorescent lamp, for example, protonemata can easily be irradiated obliquely or, in extreme cases, from their tip or base, so the chloroplasts move on both sides of the protonemata.

When polarized light was applied from the direction of the protonemal tip or base along the growing axis, i.e., parallel to the cell axis, chloroplasts moved to the cell wall parallel to the vibration plane of the polarized light (Yatsuhashi et al., 1987a, 1987b). The reason why polarized light induces such an effect on chloroplast movement is that the photoreceptors mediating chloroplast movement are localized on the plasma membrane or attached to the plasma membrane through other membrane proteins. In this case a transition moment of the photoreceptor is arranged parallel, perpendicular, or obliquely with the plasma membrane as in the case of polarotropism (Yatsuhashi et al., 1987a, 1987b). Because the photoreceptors mediating chloroplast movement and polarotropism were identified as those mentioned in Section 1.10, the intracellular arrangement of photoreceptors should be the same. The distinction between the two phenomena is the different sites where photoperception and the response occur in the protonemata. The apical part of protonemata is the site for polarotropism and the linear part is the site for chloroplast movement. This can be explained easily with the drawing shown in Figure 1.14. In the case of phytochrome it

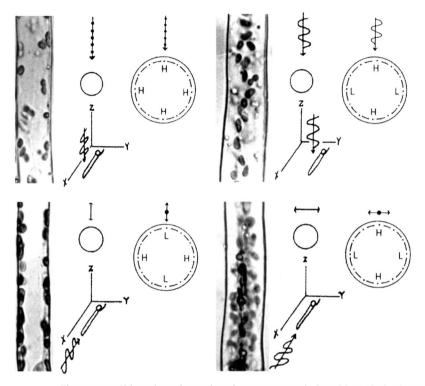

Figure 1.13 Chloroplast photorelocation movement induced by polarized red light with different vibration planes irradiated from different directions as shown in each scheme. The chloroplast distribution can be explained by tetrapolar distribution of high (H) and low (L) densities of accelerated photoreceptor molecules. The transition moments of photoreceptors shown in this diagram occur before light absorption. Circles are cross-sections of protonemata. Arrows and bars indicate the direction of polarized light and their vibration planes. The chloroplast distributions shown in the photographs were those observed from the z axis. (Modified from Wada and Kagawa, 2001.)

has been proposed that the red light absorbing form of phytochrome (Pr) has its transition moment parallel to the plasma membrane, but the far-red light absorbing form of phytochrome (Pfr) is perpendicular (Etzold, 1965). All dichroic effects observed in ferns, mosses, and some algae could be explained with these arrangements of Pr and Pfr. But the transition moments of Pr and Pfr are not necessarily at exactly 90°, i.e., perpendicular with or parallel to the cell walls. If the transition moment of Pr molecules in protonemata is close to parallel rather than perpendicular to the plasma membrane as shown in Figure 1.14, parallel-polarized light has a tendency to be absorbed more than perpendicular light, and vice versa.

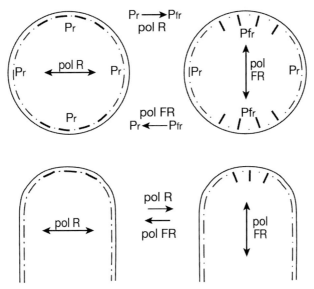

Figure 1.14 Schematic illustrations of Acneo1 arrangement in protonemata of *Adiantum capillus-veneris*. The transition moment of the red light absorbing form (Pr) is parallel to the plasma membrane and that of the far-red light absorbing form (Pfr) is perpendicular. Cross-sections (top) and longitudinal sections (bottom) of protonemata are shown. Double arrowhead lines indicate the vibration plane of polarized light. The transition moment of Acneo1 parallel to the vibration plane of polarized light can absorb the light efficiently and convert between Pr and Pfr repeatedly.

These polarized light effects cannot be seen clearly unless protonemal cells are submerged in water, because polarized light cannot penetrate into a cell evenly by refraction. The vibration plane may be randomized, and polarized light does not reach some parts of the other side of the cells because of the lens effect of the cylindrical protonemata when cells are held above the agar surface.

1.7.3 How chloroplasts sense different fluence rates

How do chloroplasts know the difference in brightness between two areas irradiated with different fluence rates? To analyze this question we irradiated two adjacent parts of a protonema 30 μm each in length with different fluence rates of red or blue light (Yatsuhashi *et al.*, 1987b). Sets of different light fluence microbeams were prepared and adjacent parts of protonemata were irradiated simultaneously and continuously for 2 hours and then observed. One group of light conditions was prepared so that the fluence ratio of the two microbeams was constant but the fluence levels of the two beams differed. A second set of conditions established different fluence ratios but held the fluence

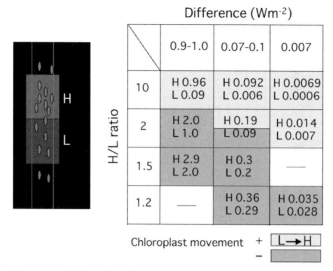

Figure 1.15 Chloroplasts detect the ratio of two fluence rates of light irradiated at two adjacent areas, rather than the absolute difference between the two, to determine the direction of movement. Two adjacent areas of a single protonema were irradiated with different sets of fluence rates of light (left panel), such that the ratios were the same, for example 10 or 2, but the differences were one tenth, and one hundredth. After 2 hours of continuous irradiation, chloroplast movement was detected to assess whether they moved in response to a higher fluence rate or not. The slightly shaded combinations show positive movement from weak light to strong light, but the strongly shaded combinations show no movement. Chloroplasts can detect very small ratios between two different fluence rates.

levels of the two beams constant (Figure 1.15). Each protonema was observed under a microbeam irradiator to determine in which microbeam irradiated areas chloroplasts moved. The results were very clear. Chloroplasts moved from weak light to strong light depending on their ratio but not on the difference in fluence rates. Chloroplasts can detect a small difference in ratio at a threshold of around 1.2–1.5 for blue light and 1.5–2.0 for red light (Yatsuhashi et al., 1987b).

1.7.4 Signals transferred from a photoreceptor to chloroplasts

Signal transduction pathways of photoreceptors are not well understood but it is well established that phytochrome and cryptochrome move into nuclei and control the expression of light-mediated genes (Sakamoto and Nagatani, 1996; Guo et al., 1999; Kleiner et al., 1999). However, the phototransduction pathway of chloroplast movement has not been clarified. To discover whether gene expression was involved in chloroplast movement, a long protonemal cell was

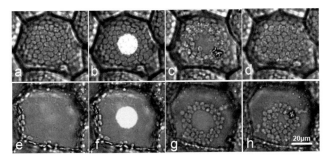

Figure 1.16 Chloroplast movement induced by microbeam irradiation. (a) Weak light adapted gametophytes of *Adiantum capillus-veneris*, (b) irradiated with a microbeam of strong light. (c) Chloroplasts move away from the beam (1.5 hour after the microbeam irradiation) and (d) return when the light is switched off. (e) A dark-adapted gametophyte kept in the dark for 3 days, (f) partly irradiated with a strong microbeam light. (g) Chloroplasts show an accumulation response towards the microbeam but remain outside the beam (photograph taken 1.5 hour after the start of irradiation) and (h) do not enter until the light is switched off, because of the high fluence rate of the beam light. The fluence rate of both microbeams was 15 W m^{-2}, the diameter of the microbeam was 20 µm.

separated into two parts and the portion of the cell containing the nucleus was removed. Even the enucleated cells showed chloroplast photorelocation under either red or blue light irradiation, indicating that signals from photoreceptors were transferred directly to chloroplasts (Wada, 1988b). Thus, chloroplast photorelocation is a very simple system for studying signal transduction pathways that may occur through the cell surface along the plasma membrane.

When the center of a prothallial cell cultured under weak white light was irradiated with a microbeam (e.g., 10 µm in width) of high fluence blue light (e.g., 10 W m^{-2}), chloroplasts in the beam migrated outside of the beam. However, as soon as they reached the edge of the beam they stopped moving and remained along the border of the beam spot (Kagawa and Wada, 1999). When a similar experiment was performed using dark-adapted prothallial cells, in which all chloroplasts had moved to the anticlinal walls, chloroplasts began to move toward the beam, displaying an accumulation response, but they stopped moving at the border of the beam and remained outside the beam (Kagawa and Wada, 1999). These two experiments indicate that the signal for an accumulation response can be transmitted over a long distance, at least to the edge of the cell, but the signal for an avoidance response can only be transmitted a short distance. As soon as the microbeam of the strong light was switched off, chloroplasts rushed into the former beam-irradiated area (Figure 1.16), indicating that the lifetime of the avoidance signal was very short but that of the accumulation response was long. Moreover, under strong light conditions, the accumulation response signal could be released simultaneously with that of the avoidance response. We do not know what the signals are for the two responses, nor do we know whether they are the same or different substances. There have been long debates on whether the signal for chloroplast movement is a calcium ion (see Wada *et al.*, 2003 for more details), but we do not have any positive evidence so far on the calcium ion theory.

1.7.5 Life of activated photoreceptors

The chloroplast accumulation response can be induced even by a short (a few seconds) pulse of blue or red light in a dark-adapted cell if the light intensity is high enough. The information on light perception persists in the light irradiated area for a period of time after the light pulse ends, and chloroplasts move towards the light irradiated area in the dark. This observation raises the question whether (1) the photo-activated photoreceptor continues to work as an active form after removing the light source and releasing the signals or (2) the light signal is transferred from the photoreceptors to the next components and the photoreceptors are already non-functional. The red/far-red photo-reversibility of phytochrome that mediates the red light induced chloroplast movement is an ideal system to answer this question.

When a short microbeam pulse of red light was aimed at the center of a dark-adapted gametophyte cell, chloroplasts along the anticlinal walls moved to the beam-irradiated area and remained there for a period of time and then returned to the anticlinal walls (Kagawa and Wada, 1994). Similar experiments were performed using protonemata (Yatsuhashi and Kobayashi, 1993; Kagawa *et al.*, 1994; Figure 1.17). When chloroplast movement was induced by a pulse of polarized light applied horizontally, chloroplasts moved toward the upper and lower sides of the protonemal cell and stayed there for a period of time and returned to their original position. If far-red light was applied after chloroplast movement was induced by red light, all chloroplasts that were still moving or had already reached the upper side of the cell immediately started to move back (Kagawa *et al.*, 1994). This indicates that without far-red light, phytochrome molecules activated by red light remained in the far-red light absorbing form (Pfr) and retained chloroplasts at the site of highest Pfr concentration. In addition, the signal transferred from the light irradiated area to the anticlinal walls is not the photoreceptor itself but represents other factors (or components) activated by Pfr. It is not possible to conduct similar experiments to confirm the results using blue light, because we do not have any means of converting activated blue light receptors to an inactive state to cancel the blue light effect. However, considering that the blue light receptor for chloroplast movement and

Photoresponses in fern gametophytes 29

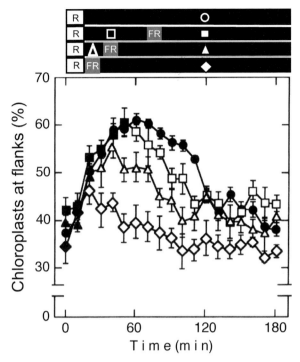

Figure 1.17 Effects of non-polarized far-red light, applied at various times after irradiation with polarized red light, on the orientation of chloroplast movement. Protonemata were irradiated with vertically polarized red light (3 W m^{-2}, 10 min) and then with non-polarized far-red light (5 W m^{-2}, 10 min) immediately after the red light, or after an intervening dark period of 10 or 40 min. Solid symbols represent the response after red light irradiation without far-red light irradiation. Open symbols represent the response after far-red light irradiation. The upper panel shows the experimental schedule for the light treatment. (Modified from Kagawa et al., 1994.)

the red light receptor described here as phytochrome belong to the same family of proteins (see Section 1.10), a similar situation should be working in both systems.

1.7.6 Speed of chloroplast movement

The velocities of chloroplast movement were measured for both avoidance and accumulation responses. The velocities of the accumulation response in two-dimensional gametophytes of *A. capillus-veneris* were always constant at about 0.3 μm min^{-1} regardless of whether they were induced by red or blue light with different fluence rates (Kagawa and Wada, 1996). However, in *Arabidopsis* leaves the velocities of the avoidance response are dependent on fluence rate: the higher the fluence rate of blue light, the faster the chloroplasts move (Kagawa and Wada, 2004). Moreover, the velocity seems to depend on the total amount of photoreceptor (phototropin2) (Kagawa and Wada, 2004). Whether ferns have a similar system is not known and similar experiments should be performed using fern gametophytes. However, this is not possible at the present time because crossing gametophytes is difficult and transformation has proven to be challenging in fern systems.

1.7.7 Mechanism of chloroplast movement

The machinery of chloroplast movement has not yet been clarified in any plant species, although an actin–myosin system has been proposed based on experimental results using inhibitors for actin (e.g., cytochalasin B) and myosin (e.g., N-ethylmaleimide) (see Wada *et al.*, 2003). In the moss *Physcomitrella patens* microtubules as well as actin filaments are involved in movement (Sato *et al.*, 2001), but ferns do not use microtubules (Kadota and Wada, 1992d). The circular structure of F-actin was observed along the edge of chloroplasts on the surface of plasma membranes in *A. capillus-veneris* protonemata (Kadota and Wada, 1989a, 1989b). The structure was found only when chloroplasts gathered at a light irradiated area and settled there. It could not be located on a chloroplast under darkness or on chloroplasts located far from a microbeam irradiated area (Kadota and Wada, 1992c), suggesting that the structure plays a role in keeping the chloroplast in the light irradiated area. Before chloroplasts retreat from the irradiated area, the structure disappears (Kadota and Wada, 1992c). We are not sure whether a similar structure of microfilaments is involved with chloroplast movement itself.

1.8 Nuclear movement

Nuclei also move towards light, although the speed of movement is very slow compared to that of chloroplasts. Under weak white light conditions, nuclei remain in the central part of the light-facing side of prothallial cells. Under darkness, nuclei move to the anticlinal wall just as chloroplasts do (Figure 1.18) (Kagawa and Wada, 1993). Polarized red and blue light is also effective, in the same way that the chloroplast accumulation response occurs, although the velocity is more than ten times slower than that of chloroplast movement (Kagawa and Wada, 1995). Very recently Tsuboi *et al.* (2007) found, using AcNEO1 and AcPHOT2 mutants, that neochrome1 is the photoreceptor for red light induction and those for blue light induced nuclear movement are phototropins.

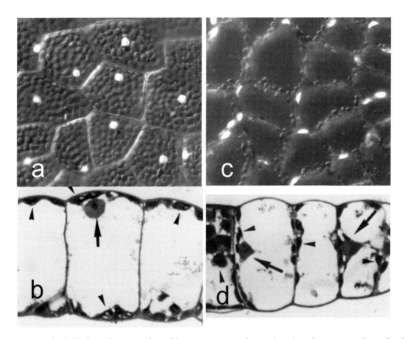

Figure 1.18 Light micrographs of intact gametophytes (a, c) and cross-sections (b, d) showing intracellular nuclear positioning under different light conditions. Nuclei positioned at the cell surface under weak white light (a, b), but at anticlinal walls under darkness (c, d, kept in the dark for 3 and 2 days, respectively) or under strong light conditions (data not shown).

1.9 Reproductive organs

Developmental and genetic studies of fern reproductive organs are interesting and important, not only for fern investigators but also for plant biologists in general. Recent experimental research has been well reviewed by Banks (1999). The isolation and identification of antheridiogens in different species have been of concern for a long time (Yamauchi et al., 1996). However, the effects of light on differentiation or development of reproductive organs have not been studied, except for a few reports (Gemmrich, 1986; Schraudolf, 1967). In the case of *Pteris vittata* antheridial formation is inhibited by red and blue light and promoted by far-red light. Red and far-red reversibility of this phenomenon indicates a phytochrome involvement (Gemmrich, 1986). The photoresponse necessitates Ca^{2+} and NO_3^- (Gemmrich, 1988), although the reason for requiring these inorganic ions for antheridiogenesis has not been studied.

1.10 Photoreceptors for photomorphogenesis

In *A. capillus-veneris*, four genes of the phytochrome protein family have been cloned, i.e., two conventional phytochrome sequences, phytochrome1

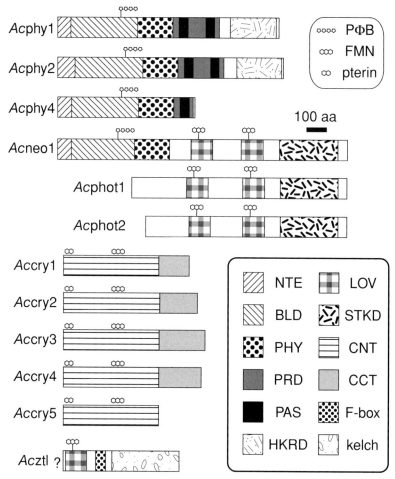

Figure 1.19 Photoreceptors in *Adiantum capillus-veneris*. Two conventional phytochromes and a short phytochrome, two phototropins and one chimera photoreceptor between phytochrome chromophore-binding domain and phototropin, five cryptochromes and a FKF/LKP/ZTL family protein were found. FMN, flavin mononucleotide; NTE, N-terminal extension; BLD, bilin lyase domain, PHY, PHY domain; PRD, PAS-related domain; PAS: PAS domain; HKRD, histidine kinase-related domain; LOV, light, oxygen and voltage domain; STKD, serine/threonine kinase domain; CNT, cryptochrome N-terminal domain; CCT, cryptochrome C-terminal domain.

(*AcPHY1*) and *AcPHY2*, one chimeric sequence consisting of a phytochrome chromophore binding part and whole phototropin genes (*AcPHY3*, renamed Acneochrome1 (*AcNEO1*)), and a short N-terminal sequence (*AcPHY4*). In the case of *AcPHY4*, it is not clear whether a full sequence of conventional phytochrome is present because a long retrotransposon sequence is inserted after the *AcPHY4* sequence and the downstream has not yet been analyzed (Nozue *et al.*, 1997). As blue light receptors, two phototropin genes (*AcPHOT1* and *AcPHOT2*) and five cryptochrome genes (*AcCRY1* to *AcCRY5*) were cloned and sequenced (Figure 1.19). In *Arabidopsis thaliana* another blue light receptor was found, ZTL/FKF/LKP/ADO (Kiyosue and Wada, 2000; Nelson *et al.*, 2000; Somers *et al.*, 2000; Jarillo *et al.*, 2001a; Schultz *et al.*, 2001; Imaizumi *et al.*, 2005). This protein is involved in flowering time as members of clock genes. In *A. capillus-veneris* EST we found a fragment of this family gene (Yamauchi *et al.*, 2005), and it may function as a clock related gene in ferns too. Although orthologs of *Arabidopsis* photoreceptor genes were found in ferns, the function of these photoreceptors is not necessarily the same as those in *Arabidopsis*. For example, phototropins mediate stomatal opening in *Arabidopsis* (Kinoshita *et al.*, 2001) but may not in ferns (Doi *et al.*, 2006). For more details about fern photoreceptors, see Wada (2003), Suetsugu and Wada (2005), and Kanegae and Wada (2006).

1.10.1 *Phytochromes*

Phytochrome function in seed plants, especially in *Arabidopsis* and in rice, has been studied intensively ever since phytochrome family proteins and genes were identified (Abe *et al.*, 1989; Sharrock and Quail, 1989), because (1) the whole genome sequence revealed that only phyA to E phytochrome family genes exist in *Arabidopsis*, and only phyA to C in rice, and (2) many phytochrome mutants are available for analyzing the function of each phytochrome molecule in these plants, even double, triple, quadruple mutants (Franklin *et al.*, 2003). In contrast, fern phytochrome studies have not advanced greatly, even though several phytochrome genes have been cloned and sequenced (Okamoto *et al.*, 1993; Kanegae and Wada, 2006). Before seed plant phytochrome genes were cloned and sequenced, the intracellular distribution of functional phytochrome was only known in *Mougeotia* and ferns from microbeam studies and/or dichroic effects found under polarized light irradiation (see review by Wada *et al.*, 1993). However, these dichroic effects in *A. capillus-veneris* were mediated by Acneo1, not by any conventional phytochrome (Kawai *et al.*, 2003), indicating that the function of conventional phytochrome is not yet clarified in ferns.

Adiantum capillus-veneris has two conventional phytochrome genes *AcPHY1* and *AcPHY2*, and a sequence corresponding to the N-terminus chromophore-binding domain called *AcPHY4* (Kanegae and Wada, 2006). cDNA sequences encoding the chromophore-binding domain of the N-terminus like *AcPHY4* is also found in several other fern species (Tsuboi and Suetsugu, unpublished data), suggesting that it may have a specific function. In addition to chloroplast movement and phototropism regulated by Acneo1, the red light controlled phenomena known so far in fern gametophytes are spore germination, apical cell growth, and the timing of cell division. Given that these phenomena are normal in *Acneo1* mutant lines and that the red light effect can be reversed by far-red light, these phenomena must be controlled by conventional phytochromes (Kadota and Wada, 1999). We do not have any mutant line defective for these phytochrome-controlled phenomena, because our strategy for screening these mutants is not realistic. A mutant deficient in spore germination or cell growth does not germinate or grow under red light. If mutant spores do not germinate, it is almost impossible to distinguish them from immature or dead spores to identify them as a mutant. But still, it must be possible because fern spores germinate if gibberellins or antheridiogen is applied (Sugai *et al.*, 1987).

1.10.2 Phototropins

Phototropin (phot1) was found for the first time as a blue light receptor with two LOV domains (chromophore binding domains of phototropin) mediating phototropism under low light conditions in *Arabidopsis* (Huala *et al.*, 1997). Thereafter another member of phototropin (phot2) was discovered (Jarillo *et al.*, 1998) and identified as a photoreceptor for chloroplast avoidance movement (Jarillo *et al.*, 2001b; Kagawa *et al.*, 2001). Phot1 and phot2 redundantly mediate the accumulation response (Sakia *et al.*, 2001) and phot2 alone is necessary for chloroplast accumulation on the cell bottom in the darkness (Suetsugu *et al.*, 2005b). Besides these phenomena, phototropins mediate stomata opening (Kinoshita *et al.*, 2001), leaf expansion (Sakai *et al.*, 2001; Sakamoto and Briggs, 2002) and growth inhibition of hypocotyls (Folta and Spalding, 2001) in *Arabidopsis*. In *A. capillus-veneris* it was also confirmed that phot2 (Acphot2) was a photoreceptor of chloroplast avoidance response like that in *Arabidopsis thaliana* (Kagawa *et al.*, 2004), but it has not yet been clarified whether Acphot1 and/or Acphot2 mediate other phenomena that are under phototropin control in *Arabidopsis thaliana*. As mentioned above, phototropins may not work as photoreceptors in the opening of fern stomata because fern stomata do not respond to blue light even though phototropins are expressed in fern guard cells (Doi *et al.*, 2006).

We screened two *Acphot2* mutants in *A. capillus-veneris* as lines deficient in chloroplast avoidance movement (Kagawa *et al.*, 2004). Sequence analysis revealed that one blue high light-dependent chloroplast movement mutant *BHC7* has two regions containing nucleotide deletions. *BHC8* has a deletion of 26 nucleotides and two extra nucleotides. In fern mutant lines backcrossing is very difficult, almost impossible at the moment (at least in the case of *A. capillus-veneris*). Hence, we confirmed that the *Acphot2* mutations were the real causes of the functional deficiency by recovery of the mutant lines under transient expression of wild type *PHOT2* cDNA (Kagawa *et al.*, 2004).

This rescue system by transient expression of *AcPHOT2* in these two mutant lines was applicable to study of the functional domains and/or important amino acids in phototropin function (Kagawa *et al.*, 2004). We made deletion constructs of the N-terminus or C-terminus part of *AcPHOT2* with the 35S promoter of cauliflower mosaic virus, bombarded them into *Acphot2* mutant lines, and observed whether chloroplast avoidance response could be rescued or not. The results were obtained only 2 days after bombardment; similar experiments using stable transformants of *Arabidopsis* may require more than 6 months. It was found that the LOV1 region is not necessary for the *Ac*phot2-dependent chloroplast avoidance response. C-terminal deletion series revealed that deletions of fewer than 20 residues from the C terminus did not affect function but deleting more than 40 residues eliminated function. When a cysteine residue in the LOV2 domain where the chromophore binds was changed to alanine, phototropin was no longer functional, but if the change was in LOV1 the phototropin was functional. Similar results were obtained (Christie *et al.*, 2002) for *Arabidopsis thaliana* phot1 function in phototropism.

1.10.3 Neochrome

The neochrome gene was cloned as a member of the phytochrome gene family (so that it was formally called phytochrome3), but the sequence of the C-terminus of this gene differed significantly from other conventional phytochrome genes, and appeared to be related to phototropin genes (Nozue *et al.*, 1998). It is curious that the *AcNEO1* gene has no intron either in phytochrome or phototropin related regions, although the *AcPHOT2* gene has 22 introns, as is the case for *Arabidopsis*. These results indicate that *AcNEO1* was not assembled by simple fusion of phytochrome and phototropin genes, but at least the phototropin region was reverse transcribed from spliced cDNA and integrated into the region near the first intron of a phytochrome gene. The genes for the conventional phytochrome of *A. capillus-veneris* do not have any intron in the chromophore binding domain (Okamoto *et al.*, 1993), so it cannot be determined whether the phytochrome chromophore binding domain of neo1 is also reverse transcribed.

Distribution of the neochrome sequence in ferns was studied by polymerase chain reaction (PCR) using several species from ancient groups (such as *Osmunda*) to recent groups (such as *Dryopteris*) based on the fern phylogenetic tree constructed by Hasebe *et al.* (1995) (Kawai *et al.*, 2003). Among recent ferns, the neochrome sequence could be detected easily but in ancient species we could

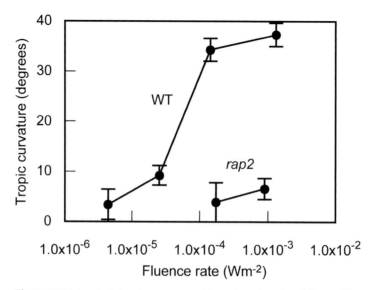

Figure 1.20 The phototropic response of Acneo1 mutant (rap2) leaves. Mutant leaves are less sensitive to dim white light than wild type (WT) leaves. Young etiolated leaves of Acneo1 and wild type plants were irradiated continuously for 4.5 hours from one side with white light of various fluence rates. Note, the existence of Acneo1 makes this fern very sensitive to light. (Modified from Kawai et al., 2003.)

not find any neochrome sequence even though phytochrome and/or phototropin sequences were found. These results suggest that neochrome arose recently in fern evolution, probably after tree ferns originated.

Neochrome function was studied using red-light aphototropic mutants (rap) screened by a deficiency for red-light induced phototropism in protonemata (Kadota and Wada, 1999). All rap mutants were defective in red-light induced chloroplast photorelocation movement as well as phototropism, although both phenomena could be induced by blue light. If the wild type NEO1 cDNA with a cauliflower mosaic virus 35S promoter was introduced into rap mutant cells by particle bombardment and expressed transiently, both red light mediated phenomena were rescued, indicating that neo1 is the photoreceptor for red-light induced chloroplast movement and possibly phototropism. It was very interesting that the rap mutants showed a deficiency for phototropism in sporophyte leaves under red light although blue-light induced phototropism was normal (Kawai et al., 2003). This result indicates that neo1 may be the photoreceptor of phototropism in both gametophytes and sporophytes, although the mechanism of tropistic response in apical cell growth of protonemata and that of bending in multicellular leaves must be quite different.

More interestingly, the sensitivity to white light in the phototropic response of rap mutants was greatly reduced in leaves compared to that of wild type plants

(Figure 1.20) (Kawai *et al.*, 2003). We studied fluence response relationships in the phototropism of *rap* mutants and wild type leaves that had been cultured under total darkness. Because *A. capillus-veneris* leaves show phototropism under blue as well as red light, we tested the phototropic response under white light. The sensitivities of mutant and wild type plants to white light differed more than 10 times, and nearly 100 times. The reason why the sensitivity to white light was so high in wild type leaves is unknown. We recently confirmed that a similar phenomenon occurs in *Arabidopsis* (Kanegae *et al.*, 2006). When *AcNEO1* cDNA was transformed into *phot1 phot2* double mutant plants of *Arabidopsis* that are deficient in phototropism, the transformants showed a positive phototropic response to a red light source as well as to blue light, although *Arabidopsis* wild type plants show phototropism only under blue light. When red and blue light was supplied simultaneously to these transformed plants, the degree of tropistic response was very high compared to when either blue or red light was supplied independently. We do not yet know how blue and red light work synergistically to make the sensitivity so high. What we do know is that neochrome functions through the serine threonine protein kinase domain in the phototropin region under red or blue light. Mutant lines deficient in the protein kinase domain do not show any phototropism under red and blue light (Kanegae *et al.*, 2006). I suggest that the phytochrome domain and the phototropin domain function synergistically because the kinase domain itself is active (or functional) without phytochrome and phototropin chromophore binding domains, and chromophore binding domains work as regulatory factors, one for inhibition under darkness and the other for promotion under light conditions. But this hypothesis needs to be confirmed by experimental data.

Ferns may have occupied a new niche under the canopy of gymnosperm forests in the Jurassic by using neochrome to become highly sensitive to white light (Kawai et al., 2003). Based on molecular data and fossil records Schneider *et al.* (2004) reported that modern fern genera diversified in the Cretaceous after angiosperms had become dominant in terrestrial ecosystems. It is reasonable to suggest that neochrome played a major role in modern fern evolution under forest canopies.

Surprisingly, *AcNEO1*-like sequences were also found in *Mougeotia* (*MsNEO1*, *MsNEO2*, Suetsugu *et al.*, 2005a), a green algae made famous by Professor Wolfgang Haupt who studied the mechanism of phytochrome-dependent chloroplast movement for many years (Wada *et al.*, 1993). *MsNEO1* and *MsNEO2* genes have many introns in both the phytochrome and phototropin regions, and many of them share intron sites with *Mougeotia* phytochromes and phototropins, suggesting that the ancestor of *MsNEO1* and *MsNEO2* was formed by simple fusion of the two photoreceptor genes, although it has not yet been determined which gene appeared first. More interestingly, *Msneo1* and *Msneo2* can replace *Acneo1* function. When either *Msneo1* or 2 was expressed transiently in *A. capillus-veneris neo1* mutants (*rap2* line, lacking red-light induced chloroplast movement) (Kawai *et al.*, 2003), accumulation movement in *rap2* was rescued (Suetsugu *et al.*, 2005a). Judging from intron sites and genome sequences, neochromes of both species must have arisen independently, indicating that these chimeric photoreceptors have a very similar structure and could have been very useful for seed-free plants living under forest canopies.

1.10.4 Cryptochromes

Adiantum capillus-veneris has five cryptochrome genes (Kanegae and Wada, 1998; Imaizumi *et al.*, 2000) whereas *Arabidopsis* has only two (Li and Yang, 2006). Cryptochromes consist of a photolyase-related sequence in the N-terminus with a C-terminus extension. Trp-277, which is important for photolyase function, is not conserved in all fern cryptochromes, indicating that fern cryptochromes may not have any photolyase function, as is the case for *Arabidopsis* cryptochromes (Li and Yang, 2006). Even when fern cryptochromes were (1) expressed in photolyase-deficient *E. coli*, (2) exposed to UV light, and (3) irradiated with photoreactivating light (blue light for 1 hour), the blue light did not change the survival rate of the *E. coli* compared to the control without blue light irradiation. This result indicates that fern cryptochromes do not function as photolyase in *E. coli* cells (Imaizumi *et al.*, 2000). The expression of each cryptochrome mRNA under a variety of light conditions and in different cell and gametophyte stages was studied (Imaizumi *et al.*, 2000). For example, *CRY3* mRNA expression level is higher in protonemata and sporophytes than in spores or prothallia, and *CRY5* mRNA is mainly expressed in sporophytes. *CRY4* is down regulated during spore germination by phytochrome. Intracellular localization was also examined precisely using a *GUS-CRY* construct introduced through particle bombardment. GUS-cry3 and 4 are clearly localized in the nucleus and GUS-cry4 is predominantly found in the nucleus. On the other hand, cry1, cry2, and cry5 are found in the cytoplasm (Imaizumi *et al.*, 2000).

Intracellular photoreceptive sites of blue light responses were studied physiologically using a microbeam irradiator. These techniques showed that the blue light receptors mediating inhibition of spore germination (Furuya *et al.*, 1997) and promoting cell division (Wada and Furuya, 1978) should be localized in or very close to the nuclei. Ferns have two phototropins and one neochrome as blue light receptors, but these photoreceptors are mainly localized on the plasma membrane, not in the nucleus (Kadota *et al.*, 1982; Sakamoto and Briggs, 2002; Kawai *et al.*, 2003). Taken together, the blue light perception of these phenomena might be mediated by cry3 and/or cry4. However, we do not have any conclusive evidence for this hypothesis at the moment. To address this open question we need to screen for mutants deficient in blue light inhibition of red-light induced spore germination. We have tried to find spores which germinate under continuous blue light after the induction of germination by a red light pulse, but so far we have not succeeded. From phylogenetic analysis of the five *CRY* family genes, *CRY 1* and *CRY 2*, and *CRY 3* and *CRY 4* consist of different subfamily groups, suggesting that *CRY 3* and *CRY 4* might function redundantly. That is a reasonable explanation for why there are no mutant spores defective in the blue light effect for germination.

1.11   Concluding remarks

Almost all the early developmental processes of fern gametophytes, such as cell growth, direction of cell growth, cell division and its timing, are controlled by light (branch formation (Wada *et al.*, 1998) and negative phototropism of rhizoids (Tsuboi *et al.*, 2006) have not been discussed in this chapter). Without light signals they are not able to progress to the next developmental stages, indicating that we can control early developmental stages synchronously with light. At early developmental stages, gametophytes consist of only two different cell types, a protonemal cell and a rhizoid, so they are very uniform. Since rhizoid cells do not differentiate further, gametophyte differentiation is restricted to protonemal cells. Because all the developmental processes mentioned above occur synchronously in protonemal cells, as far as we can see under a microscope, gametophytes are suitable for studying each of the processes of cell differentiation, at least from the stand points of cell biology, photobiology, and physiology. However, because molecular biological techniques are not yet established in fern gametophyte systems, few people are involved in fern studies. To advance fern studies, the top priorities are to (1) increase the number of people who study fern gametophytes, (2) establish molecular techniques in fern gametophytes, and (3) accumulate data such as EST for various stages of gametophyte development. A possible technique of gene introduction into spores through the spore coat during imbibition was reported using *Marsilea* (Klink and Wolniak, 2001) and *Ceratopteris* (Stout *et al.*, 2003), but so far we have not replicated the technique using *A. capillus-veneris*. On the other hand, gene silencing using DNA fragments is available in *A. capillus-veneris* (Kawai-Toyooka *et al.*, 2004) and *Ceratopteris* (Ratherford *et al.*, 2004), similar to an RNA interference (RNAi) technique in various organisms. The technique is very simple and easy. PCR-amplified double stranded DNA fragments of a target gene (either cDNA or genome DNA) can be introduced into gametophyte cells with a hygromycin phosphotransferase gene driven by a cauliflower mosaic virus 35S promoter by particle bombardment. The effectiveness is different depending on the genes transformed but it is a relatively useful technique (Kawai-Toyooka et al., 2004). Expressed sequence tag (EST) libraries in ferns are available for *A. capillus-veneris* (Yamauchi et al., 2005) and *Ceratopteris* (Salmi et al., 2005). These techniques and databases of fern genomes are not yet used frequently, but may contribute to fern studies in the near future.

References

Abe, H., Takio, K., Titani, K., and Furuya, M. (1989). Amino-terminal amino acid sequences of pea phytochrome II fragments obtained by limited proteolysis. *Plant Cell Physiology*, **30**, 1089–1097.

Augustynowicz, J. and Gabrys, H. (1999). Chloroplast movement in fern leaves: correlation of movement dynamics and environmental flexibility of the species. *Plant Cell and Environment*, **22**, 1239–1248.

Banks, J. A. (1999). Gametophyte development in ferns. *Annual Review of Plant Physiology and Plant Molecular Biology*, **50**, 163–186.

Bünning, E. and Etzold, H. (1958). Über die Wirkung von polarisiertem Licht auf keimende Sporen von Pilzen, Moosen und Farnen. *Berichte der Deutschen Botanischen Gesellschaft*, **71**, 304–306.

Christie, J. M., Swartz, T. E., Bogomolni, R. A., and Briggs, W. R. (2002). Phototropin LOV domains exhibit distinct roles in regulating photoreceptor function. *Plant Journal*, **32**, 205–219.

Doi, M., Wada, M., and Shimazaki, K. (2006). The fern *Adiantum capillus-veneris* lacks stomatal responses to blue light. *Plant and Cell Physiology*, **47**, 748–755.

Dyer, A. (1979a). The culture of fern gametophytes for experimental investigation. In *The Experimental Biology of Ferns*, ed. A. Dyer. London: Academic Press, pp. 254–305.

Dyer, A. (1979b). *The Experimental Biology of Ferns*. London: Academic Press.

Etzold, H. (1965). Der Polarotropismus und Phototropismus der Chloronemen von *Dryopteris filix-mas* (L.) Schott. *Planta*, **64**, 254–280.

Folta, K. and Spalding, E. (2001). Unexpected roles for cryptochrome 2 and phototropin revealed by high-resolution analysis of blue light-mediated hypocotyl growth inhibition. *Plant Journal*, **26**, 471–478.

Franklin, K. A., Davis, S. J., Stoddart, W. M., Vierstra, R. D., and Whitelam, G. C. (2003). Mutant analyses define multiple roles for phytochrome C in *Arabidopsis* photomorphogenesis. *Plant Cell*, **15**, 1981–1989.

Furuya, M., Kanno, M., Okamoto, H., Fukuda, S., and Wada, M. (1997). Control of mitosis by phytochrome and a blue-light receptor in *Adiantum* spores. *Plant Physiology*, **113**, 677–683.

Gemmrich, A. R. (1986). Antheridiogenesis in the fern *Pteris vittata*. I. Photocontrol of antheridium formation. *Plant Science*, **43**, 135–140.

Gemmrich, A. R. (1988). Ion requirement of red light and blue light mediated inhibition of antheridiogenesis in the fern *Pteris vittata*. *Plant Science*, **58**, 159–164.

Guo, H., Duong, H., Ma, N., and Lin, C. (1999). The *Arabidopsis* blue light receptor cryptochrome 2 is a nuclear protein regulated by a blue light-dependent post-transcriptional mechanism. *Plant Journal*, **19**, 279–287.

Hasebe, M., Wolf, P. G., Pryer, K. M., Ueda, K., Ito, M., Sano, S., Gastony, G. J., Yokoyama, J., Manhart, J. R., Murakami, N., Crane, E. H., Haufler, C. H., and Hauk, W. D. (1995). Fern phylogeny based on *rbc*L nucleotide sequences. *American Fern Journal*, **85**, 134–181.

Haupt, W. (1999). Chloroplast movement from phenomenology to molecular biology. *Progress in Botany*, **60**, 3–36.

Haupt, W., Mortel, G., and Winkelnkemper, I. (1969). Demonstration of different dichroic orientation of phytochrome Pr and Pfr. *Planta*, **88**, 183–186.

Huala, E., Oeller, P. W., Liscum, E., Han, I. S., Larsen, E., and Briggs, W. R. (1997). *Arabidopsis* NPH1: a protein kinase with a putative redox-sensing domain. *Science*, **278**, 2120–2123.

Iino, M., Shitanishi, K., Kadota, A., and Wada, M. (1990). Phytochrome-mediated phototropism in *Adiantum* protonemata. I. Phototropism as a function of the lateral Pfr gradient. *Photochemistry and Photobiology*, **51**, 469–476.

Imaizumi, T., Kanegae, T., and Wada, M. (2000). Cryptochrome nucleocytoplasmic distribution and gene expression are regulated by light quality in the fern *Adiantum capillus-veneris*. *Plant Cell*, **12**, 81–96.

Imaizumi, T., Schultz, T. F., Harmon, F. G., Ho, L. A., and Kay, S. A. (2005). FKF1 F-box protein mediated cyclic degradation of a repressor of CONSTANTS in *Arabidopsis*. *Science*, **309**, 293–297.

Jarillo, J. A., Ahmad, M., and Cashmore, A. R. (1998). NPL1 (accession No. AF053941): a second member of the NPH serine/threonine kinase family of *Arabidopsis* (PGR98–100). *Plant Physiology*, **117**, 719.

Jarillo, J. A., Capel, J., Tang, R. H., Yang, H. Q., Alonso, J. M., Ecker, J. R., and Cashmore, A. R. (2001a). An *Arabidopsis* circadian clock component interacts with both CRY1 and phyB. *Nature*, **410**, 487–490.

Jarillo, J. A. Gabrys, H., Capel, J., Alonso, J. M., Ecker, J. R., and Cashmore, A. R. (2001b). Phototropin-related NPL1 controls chroloplast relocation induced by blue light. *Nature*, **410**, 952–954.

Kadota, A. and Furuya, M. (1977). Apical growth of protonemata in *Adiantum capillus-veneris*. I. Red far-red reversible effect on growth cessation in the dark. *Development, Growth, and Differentiation*, **19**, 357–365.

Kadota, A. and Furuya, M. (1981). Apical growth of protonemata in *Adiantum capillus-veneris*. IV. Phytochrome-mediated induction in non-growing cells. *Plant and Cell Physiology*, **22**, 629–638.

Kadota, A. and Wada, M. (1989a). Photoinduction of circular F-actin on chloroplast in a fern protonemal cell. *Protoplasma*, **151**, 171–174.

Kadota, A. and Wada, M. (1989b). Circular arrangement of cortical F-actin around the subapical region of a tip-growing fern protonemal cell. *Plant and Cell Physiology*, **30**, 1183–1186.

Kadota, A. and Wada, M. (1992a). Reorganization of the cortical cytoskeleton in tip-growing fern protonemal cells during phytochrome-mediated phototropism and blue light-induced apical swelling. *Protoplasma*, **166**, 35–41.

Kadota, A. and Wada, M. (1992b). The circular arrangement of cortical microtubules around the subapex of tip-growing fern protonemata is sensitive to cytochalasin B. *Plant and Cell Physiology*, **33**, 99–102.

Kadota, A. and Wada, M. (1992c). Photoinduction of formation of circular structures by microfilaments on chloroplasts during intracellular orientation in protonemal cells of the fern *Adiantum capillus-veneris*. *Protoplasma*, **167**, 97–107.

Kadota, A. and Wada, M. (1992d). Photoorientation of chloroplasts in protonemal cells of the fern *Adiantum* as analyzed by use of a video-tracking system. *Botanical Magazine (Tokyo)*, **105**, 265–279.

Kadota, A. and Wada, M. (1995). Cytoskeletal aspect of nuclear migration during tip growth in the fern *Adiantum* protonemal cell. *Protoplasma*, **188**, 170–179.

Kadota, A. and Wada, M. (1999). Red light-aphototropic (rap) mutants lack red light-induced chloroplast relocation movement in the fern *Adiantum capillus-veneris*. *Plant and Cell Physiology*, **40**, 238–247.

Kadota, A., Wada, M., and Furuya, M. (1979). Apical growth of protonemata in *Adiantum capillus-veneris*. III. Action spectra for the light effect on dark cessation of apical growth and the intracellular photoreceptive site. *Plant Science Letters*, **15**, 193–201.

Kadota, A., Wada, M., and Furuya, M. (1982). Phytochrome-mediated phototropism and different dichroic orientation of Pr and Pfr in protonemata of the fern *Adiantum capillus-veneris* L. *Photochemistry and Photobiology*, **35**, 533–536.

Kadota, A., Wada, M., and Furuya, M. (1985). Phytochrome-mediated phototropism of *Adiantum capillus-veneris* L. protonemata as analyzed by microbeam irradiation with polarized light. *Planta*, **165**, 30–36.

Kadota, A., Fushimi, Y., and Wada, M. (1986). Intracellular photoreceptive site for blue light-induced cell division in protonemata of the fern *Adiantum* – further analysis by polarized light irradiation and cell centrifugation. *Plant and Cell Physiology*, **27**, 989–995.

Kadota, A., Kohyama, I., and Wada, M. (1989). Polarotropism and photomovement of chloroplasts in the fern *Pteris* and *Adiantum* protonemata: evidence for the possible lack of dichroic phytochrome in *Pteris*. *Plant and Cell Physiology*, **30**, 523–531.

Kadota, A., Sato, K., and Wada, M. (2000). Intracellular chloroplast photorelocation in the moss *Physcomitrella patens* is mediated by phytochrome as well as by a blue-light receptor. *Planta*, **210**, 932–937.

Kagawa, T. and Wada, M. (1993). Light-dependent nuclear positioning in prothallial cells of *Adiantum capillus-veneris*. *Protoplasma*, **177**, 82–85.

Kagawa, T. and Wada, M. (1994). Brief irradiation with red or blue light induces orientational movement of chloroplasts in dark-adapted prothallial cells of the fern *Adiantum*. *Journal of Plant Research*, **107**, 389–398.

Kagawa, T. and Wada, M. (1995). Polarized light induces nuclear migration in prothallial cells of *Adiantum capillus-veneris* L. *Planta*, **196**, 775–780.

Kagawa, T. and Wada, M. (1996). Phytochrome- and blue light-absorbing pigment-mediated directional movement of chloroplasts in dark-adapted prothallial cells of fern *Adiantum* as analyzed by microbeam irradiation. *Planta*, **198**, 488–493.

Kagawa, T. and Wada, M. (1999). Chloroplast-avoidance response induced by high-fluence blue light in prothallial cells of the fern *Adiantum capillus-veneris* as analyzed by microbeam irradiation. *Plant Physiology*, **119**, 917–923.

Kagawa, T. and Wada, M. (2002). Blue light-induced chloroplast relocation. *Plant and Cell Physiology*, **43**, 367–371.

Kagawa, T. and Wada, M. (2004). Velocity of chloroplast avoidance movement is fluence rate dependent. *Photochemical and Photobiological Sciences*, **3**, 592–595.

Kagawa, T., Kadota, A., and Wada, M. (1994). Phytochrome-mediated photoorientation of chloroplasts in protonemal cells of the fern *Adiantum* can be induced by brief irradiation with red light. *Plant and Cell Physiology*, **35**, 371–377.

Kagawa, T., Sakai, T., Suetsugu, N., Oikawa, K., Ishiguro, S., Kato, T., Tabata, S., Okada, K., and Wada, M. (2001). *Arabidopsis* NPL1: a phototropin homolog controlling the chloroplast high-light avoidance response. *Science*, **291**, 2138–2141.

Kagawa, T., Kasahara, M., Abe, T., Yoshida, S., and Wada, M. (2004). Function analysis of Acphot2 using mutants deficient in blue light-induced chloroplast avoidance movement of the fern *Adiantum capillus-veneris* L. *Plant and Cell Physiology*, **45**, 416–426.

Kanegae, T. and Wada, M. (1998). Isolation and characterization of homologues of plant blue-light photoreceptor (cryptochrome) genes from the fern *Adiantum capillus-veneris*. *Molecular and General Genetics*, **259**, 345–353.

Kanegae, T. and Wada, M. (2006). Photomorphogenesis of ferns. In *Photomorphogenesis in Plants and Bacteria*, ed. E. Schäfer and F. Nagy, 3rd edn. Dordrecht: Springer, pp. 515–536.

Kanegae, T., Hayashida, E., Kuramoto, C., and Wada, M. (2006). A single chromoprotein with triple chromophores acts as both a phytochrome and a phototropin. *Proceedings of the National Academy of Sciences of the United States of America*, **103**, 17997–18001.

Kasahara, M., Kagawa, T., Oikawa, Suetsugu, N., Miyao, M., and Wada, M. (2002). Chloroplast avoidance movement reduces photodamage in plants. *Nature*, **420**, 829–832.

Kawai, H., Kanegae, T., Christensen, S., Kiyosue, T., Sato, Y., Imaizumi, T., Kadota, A., and Wada, M. (2003). Responses of ferns to red light are mediated by an unconventional photoreceptor. *Nature*, **421**, 287–290.

Kawai-Toyooka, H., Kuramoto, C., Orui, K., Motoyama, K., Kikuchi, K., Kanegae, T., and Wada, M. (2004). DNA interference: a simple and efficient gene-silencing system for high-throughput functional analysis in the fern *Adiantum*. *Plant and Cell Physiology*, **45**, 1648–1657.

Kinoshita, T., Doi, M., Suetsugu, N., Kagawa, T., Wada, M., and Shimazaki, K. (2001). phot1 and phot2 mediate blue light regulation of stomatal opening. *Nature*, **414**, 656–660.

Kiyosue, T. and Wada, M. (2000). LKP1 (LOV kelch protein 1): a factor involved in the regulation of flowering time in *Arabidopsis*. *Plant Journal*, **23**, 807–815.

Kleiner, O., Kircher, S., Harter, K., and Batschauer, A. (1999). Nuclear localization of the *Arabidopsis* blue light receptor cryptochrome 2. *Plant Journal*, **19**, 289–296.

Klink, V. P. and Walniak, S. M. (2001). Centrin is necessary for the formation of the motile apparatus in spermatids of *Marsilea*. *Molecular Biology of the Cell*, **12**, 761–776.

Kraml, M., Buttner, G., Haupt, W., and Herrman, H. (1988). Chloroplast orientation in *Mesotaenium*: the phytochrome effect is strongly potentiated by interaction with blue light. *Protoplasma*, **S1**, 172–179.

Li, Q. H. and Yang, H. Q. (2006). Cryptochrome signaling in plants. *Photochemistry and Photobiology*, **83**, 94–101.

Ma, L. Q., Komar, K. M., Tu, C., Zhang, W., Cai, Y., and Kennelley, E. D. (2001). A fern that hyperaccumulates arsenic. *Nature*, **409**, 579.

Mineyuki, Y. (1999). The preprophase band of microtubules: its function as a cytokinetic apparatus in higher plant. *International Review of Cytology*, **187**, 1–49.

Miyata, M., Wada, M., and Furuya, M. (1979). Effects of phytochrome and blue-near ultraviolet light-absorbing pigment on duration of component phases of the cell cycle in *Adiantum* gametophytes. *Development, Growth and Differentiation*, **21**, 577–584.

Mohr, H. (1956a). Die Beeinflussung der Keimung von Farnsporen durch Licht und andere Factoren. *Planta*, **46**, 534–551.

Mohr, H. (1956b). Die Abhängigkeit des Protonemawachstums und der Protonemapolarität bei Farnen vom Licht. *Planta*, **47**, 127–158.

Murata, T. and Wada, M. (1989a). Organization of cortical microtubules and microfibril deposition in response to blue-light induced apical swelling in a tip-growing *Adiantum* protonemal cell. *Planta*, **178**, 334–341.

Murata, T. and Wada, M. (1989b). Re-organization of microtubules during preprophase band development in *Adiantum* protonemata. *Protoplasma*, **151**, 73–80.

Murata, T. and Wada, M. (1989c). Effects of colchicines and amiprophos-methyl on microfibril arrangement and cell shape in *Adiantum* protonemal cells. *Protoplasma*, **151**, 81–87.

Murata, T. and Wada, M. (1991a). Re-formation of the preprophase band after cold-induced depolymerization of microtubules in *Adiantum* protonemata. *Plant and Cell Physiology*, **32**, 1145–1151.

Murata, T. and Wada, M. (1991b). Effects of centrifugation on preprophase-band formation in *Adiantum* protonemata. *Planta*, **183**, 391–398.

Murata, T. and Wada, M. (1992). Cell cycle specific disruption of the preprophase band of microtubules in fern protonemata: effects of displacement of the endoplasm by centrifugation. *Journal of Cell Science*, **101**, 93–98.

Murata, T., Kadota, A., Hogetsu, T., and Wada, M. (1987). Circular arrangement of cortical microtubules around the subapical part of a tip-growing fern protonema. *Protoplasma*, **141**, 135–138.

Murata, T., Kadota, A., and Wada, M. (1997). Effects of blue light on cell elongation and microtubule orientation in dark-grown gametophytes of *Ceratopteris richardii*. *Plant and Cell Physiology*, **38**, 201–209.

Nelson, D. C., Lasswell, J., Rogg, L. E., Cohen, M. A., and Bartel, B. (2000). FKF1, a clock-controlled gene that regulates the transition to flowering in Arabidopsis. *Cell*, **101**, 331–340.

Nozue, K., Kanegae, T., and Wada, M. (1997). A full length Ty3/Gypsy-type retrotransposon in the fern *Adiantum*. *Journal of Plant Research*, **110**, 495–499.

Nozue, K., Kanegae, T., Imaizumi, T., Fukuda, S., Okamoto, H., Yeh, K. C., Lagarias, J. C., and Wada, M. (1998). A phytochrome from the fern *Adiantum* with features of the putative photoreceptor NPH1. *Proceedings of the National Academy of Sciences of the United States of America*, **95**, 15826–15830.

Okamoto, H., Hirano, Y., Abe, H., Tomizawa, K., Furuya, M., and Wada, M. (1993). The deduced amino acid sequence of *Adiantum* phytochrome reveals consensus motifs with phytochrome B from seed plants. *Plant and Cell Physiology*, **34**, 1329–1334.

Raghavan, V. (1989). *Developmental Biology of Ferns*. New York: Cambridge University Press.

Ratherford, G., Tanurdzic, M., Hasebe, M., and Banks, J. A. (2004). A systemic gene silencing method suitable for high throughput, reverse genetic analyses of gene function in fern gametophytes. *BMC Plant Biology*, **4**, 6.

Sakai, T., Kagawa, T., Kasahara, M., Swartz, T. E., Christie, J. M., Briggs, W. R., Wada, M., and Okada, K. (2001). *Arabidopsis* nph1 and npl1: blue light receptors that mediate both phototropism and chloroplast relocation. *Proceedings of the National Academy of Sciences of the United States of America*, **98**, 6969–6974.

Sakamoto, K. and Briggs, W. R. (2002). Cellular and subcellular localization of phototropin 1. *Plant Cell*, **14**, 1723–1735.

Sakamoto, K. and Nagatani, A. (1996). Nuclear localization activity of phytochrome B. *Plant Journal*, **10**, 859–868.

Salmi, M. L., Bushart, T. J., Stout, S. C., and Roux, S. J. (2005). Profile and analysis of gene expression changes during early development in germinating spores of *Ceratopteris richardii*. *Plant Physiology*, **138**, 1734–1745.

Sato, Y., Wada, M., and Kadota, A. (2001). Choice of tracks, microtubules and/or actin filaments for chloroplast photo-movement is differentially controlled by phytochrome and a blue light receptor. *Journal of Cell Science*, **114**, 269–279.

Schneider, H., Schuettpelz, E., Pryer, K. M., Cranfill, R., Magallón, S., and Lupia, R. (2004). Ferns diversified in the shadow of angiosperms. *Nature*, **428**, 553–557.

Schraudolf, H. (1967). Die Steuerung der Antheridiogenbildung in *Polypodium crassifolium* L. (*Pessopteris crassifolia* Underw. and Maxon) durch Licht. *Planta*, **76**, 37–46.

Schultz, T. M., Kiyosue, T., Yanovsky, M., Wada, M., and Kay, S. A. (2001). A role of LKP2 in the circadian clock of *Arabidopsis*. *Plant Cell*, **13**, 2659–2670.

Sharrock, R. A. and Quail, P. H. (1989). Novel phytochrome sequences in *Arabidopsis thaliana*: structure, evolution, and differential expression of a plant regulatory photoreceptor family. *Genes and Development*, **3**, 1745–1757.

Shibaoka, H. (1994). Plant hormone-induced changes in the orientation of cortical microtubules: alterations in the cross-linking between microtubules and the plasma membrane. *Annual Review of Plant Physiology and Plant Molecular Biology*, **45**, 527–544.

Somers, D. E., Schultz, T. F., Milnamow, M., and Kay, S. (2000). ZEITLUPE encodes a novel clock-associated PAS protein from *Arabidopsis*. *Cell*, **101**, 319–329.

Stout, S. C., Clark, G. B., Archer-Evans, S., and Roux, S. J. (2003). Rapid and efficient suppression of gene expression in a single-cell model system, *Ceratopteris richardii*. *Plant Physiology*, **131**, 1165–1168.

Suetsugu, N. and Wada, M. (2005). Photoreceptor gene families in lower plants. In *Handbook of Photosensory Receptors*, ed. W. R. Briggs and J. L. Spudich. Weinheim: Wiley-VCH Verlag, pp. 349–369.

Suetsugu, N., Mittmann, F., Wagner, G., Hughes, J., and Wada, M. (2005a). A chimeric photoreceptor gene, NEOCHROME, has arisen twice during plant evolution. *Proceedings of the National Academy of Sciences of the United States of America*, **102**, 13705–13709.

Suetsugu, N., Kagawa, T., and Wada, M. (2005b). An auxilin-like J-domain protein, JAC1, regulates phototropin-mediated chloroplast movement in *Arabidopsis thaliana*. *Plant Physiology*, **139**, 151–162.

Sugai, M. and Furuya, M. (1967). Photomorphogenesis in *Pteris vittata*. I. Phytochrome-mediated spore germination and blue light interaction. *Plant and Cell Physiology*, **8**, 737–748.

Sugai, M. and Furuya, M. (1968). Photomorphogenesis in *Pteris vittata*. II. Recovery from blue-light-induced inhibition of spore germination. *Plant and Cell Physiology*, **9**, 671–680.

Sugai, M., Nakamura, K., Yamane, H., Sato, Y., and Takahashi, N. (1987). Effects of gibberellins and their methyl esters on dark germination and antheridium formation in *Lygodium japonicum* and *Anemia phyllitidis*. *Plant and Cell Physiology*, **28**, 199–202.

Tsuboi, H., Suetsugu, N., and Wada, M. (2006). Negative phototropic response of rhizoid cells in the fern *Adiantum capillus-veneris*. *Journal of Plant Research*, **119**, 505–512.

Tsuboi, H., Suetsugu, N., Kawai-Toyooka, H., and Wada, M. (2007). Phototropins and neochrome1 mediate nuclear movement in the fern *Adiantum capillus-veneris*. *Plant and Cell Physiology*, **48**, 892–896.

Wada, M. (1988a). Chloroplast proliferation based on their distribution and arrangement in *Adiantum* protonemata growing under red light. *Botanical Magazine (Tokyo)*, **11**, 555–561.

Wada, M. (1988b). Chloroplast photoorientation in enucleated fern protonemata. *Plant and Cell Physiology*, **29**, 1227–1232.

Wada, M. (2003). Blue light receptors in fern and moss. In *Comprehensive Series in Photoscience*, Vol. 3, *Photoreceptors and Light Signaling*, ed. A. Batchauer. Cambridge: Elsevier, pp. 329–342.

Wada, M. (2007). The fern as a model system to study photomorphogenesis. *Journal of Plant Research*, **120**, 3–16.

Wada, M. and Furuya, M. (1970). Photocontrol of the orientation of cell division in *Adiantum*. I. Effects of the dark and red periods in the apical cell of gametophytes. *Development, Growth and Differentiation*, **12**, 109–118.

Wada, M. and Furuya, M. (1972). Phytochrome action on the timing of cell division in *Adiantum* gametophytes. *Plant Physiology*, **49**, 110–113.

Wada, M. and Furuya, M. (1974). Action spectrum for the timing of photo-induced cell division in *Adiantum* gametophytes. *Physiologia Plantarum*, **32**, 377–381.

Wada, M. and Furuya, M. (1978). Effects of narrow-beam irradiations with blue and far-red light on the timing of cell division in *Adiantum* gametophytes. *Planta*, **138**, 85–90.

Wada, M. and Kadota, A. (1989). Photomorphogenesis in lower plants. *Annual Review of Plant Physiology and Plant Molecular Biology*, **40**, 169–191.

Wada, M. and Kagawa, T. (2001). Light-controlled chloroplast movement. In *Photomovement, Comprehensive Series in Photosciences*, Vol. 1, ed. D.-P. Häder and M. Lebert. Amsterdam: Elsevier, pp. 897–924.

Wada, M. and O'Brien, T. P. (1975). Observations on the structure of the protonema of *Adiantum capillus-veneris* L. undergoing cell division following white-light irradiation. *Planta*, **126**, 213–227.

Wada, M. and Sugai, M. (1994). Photobiology of ferns. In *Photomorphogenesis in Plants*, ed. R. E. Kendrick and G. H. M. Kronenberg, 2nd edn. Dordrecht: Kluwer, pp. 783–802.

Wada, M., Kadota, A., and Furuya, M. (1978). Apical growth of protonemata in *Adiantum capillus-veneris*. II. Action spectra for the induction of apical swelling and the intracellular photoreceptive site. *Botanical Magazine (Tokyo)*, **91**, 113–120.

Wada, M., Mineyuki, Y., Kadota, A., and Furuya, M. (1980). The changes of nuclear position and distribution of circumferentially aligned cortical microtubules during the progression of cell cycle in *Adiantum* protonemata. *Botanical Magazine (Tokyo)*, **93**, 237–245.

Wada, M., Kadota, A., and Furuya, M. (1981). Intracellular photoreceptive site for polarotropism in protonema of the fern *Adiantum capillus-veneris* L. *Plant and Cell Physiology*, **22**, 1481–1488.

Wada, M., Kadota, A., and Furuya, M. (1983). Intracellular localization and dichroic orientation of phytochrome in plasma membrane and/or ectoplasm of a centrifuged protonema of fern *Adiantum capillus-veneris*. *Plant and Cell Physiology*, **24**, 1441–1447.

Wada, M., Hayami, J., and Kadota, A. (1984). Returning dark-induced cell cycle to the beginning of G1phase by red light irradiation in fern *Adiantum* protonemata. *Plant and Cell Physiology*, **25**, 1053–1058.

Wada, M., Murata, T., and Shibata, M. (1990). Changes in microtubule and microfibril arrangement during polarotropism in *Adiantum* protonemata. *Botanical Magazine (Tokyo)*, **103**, 391–401.

Wada, M., Grolig, F., and Haupt, W. (1993). Light-oriented chloroplast positioning. Contribution to progress in photobiology. *Journal of Photochemistry and Photobiology, B, Biology*, **17**, 3–25.

Wada, M., Nozue, K., and Kadota, A. (1998). Cytoskeletal pattern changes during branch formation in a centrifuged *Adiantum* protonema. *Journal of Plant Research*, **111**, 53–58.

Wada, M., Kagawa, T., and Sato, Y. (2003). Chloroplast movement. *Annual Review of Plant Biology*, **54**, 455–468.

Wunsch, C. and Wada, M. (1989). Nuclear recovery from centrifugation-caused elongation: involvement of the microfilament system in the nuclear plasticity. *Journal of Plant Research*, **111**, 389–398.

Wunsch, C., Kurachi, M., Kikumoto, M., Tashiro, H., and Wada, M. (1989). Detection of intranuclear forces by the use of laser optics during the recovery process of elongated interphase nuclei in centrifuged protonemal cells of *Adiantum capillus-veneris*. *Journal of Plant Research*, **111**, 399–405.

Yamauchi, T., Oyama, N., Yamane, H., Murofushi, N., Schraudolf, H., Pour, M., Furber, M., and Mander, L. N. (1996). Identification of antheridiogens in *Lygodium circinnatum* and *Lygodium flexuosum*. *Plant Physiology*, **111**, 741–745.

Yamauchi, D., Sutoh, K., Kanegae, H., Horiguchi, T., Matsuoka, K., Fukuda, H., and Wada, M. (2005). Analysis of expressed sequence tags in prothallia of *Adiantum capillus-veneris*. *Journal of Plant Research*, **118**, 223–227.

Yatsuhashi, H. and Kobayashi, H. (1993). Dual involvement of phytochrome in light-oriented chloroplast movement in *Dryopteris sparsa*. *Journal of Photochemistry and Photobiology, B, Biology*, **19**, 25–31.

Yatsuhashi, H. and Wada, M. (1990). High-fluence rate responses in the light-oriented chloroplast movement in *Adiantum* protonemata. *Plant Science*, **68**, 87–94.

Yatsuhashi, H., Kadota, A., and Wada, M. (1985). Blue- and red-light action in photoorientation of chloroplasts in *Adiantum* protonemata. *Planta*, **165**, 43–50.

Yatsuhashi, H., Hashimoto, T., and Wada, M. (1987a). Dichroic orientation of photoreceptors for chloroplast movement in *Adiantum* protonemata. Non-helical orientation. *Plant Science*, **51**, 165–170.

Yatsuhashi, H., Wada, M., and Hashimoto, T. (1987b). Dichroic orientation of phytochrome and blue-light photoreceptor in *Adiantum* protonemata as determined by chloroplast movement. *Acta Physiologia Plantarum*, **9**, 163–173.

2

Alternation of generations

ELIZABETH SHEFFIELD

2.1 Introduction

What is meant by the term "alternation of generations"? There is no consensus on this, but a plethora of definitions and interpretations. For example: "The alternation of a sexual phase and an asexual phase in the life cycle of an organism. The two phases, or generations, are often morphologically, and sometimes chromosomally, distinct." This is the current Encyclopedia Britannica version, one of the broadest, and one of the most defensible. One alternative is: "The succession of multicellular haploid and diploid phases in some sexually reproducing organisms . . ." (Purves *et al.*, 2004). The latter is typical of the definitions found in biological textbooks, and as we shall see, restricts the process too much to be useful to fern biologists. The essential feature of the process upon which most authors agree is the presence of distinct multicellular forms. This distinguishes a set of organisms from those with only a single multicellular phase (such as humans, which reproduce, at least at present, via single-celled gametes that, on fusion, generate a multicellular phase morphologically comparable with the parent form that generated the gametes). Organisms with a single multicellular phase include those like ourselves, where the conspicuous phase is diploid ("diplonts"), and those in which the haploid phase is the only one with more than single cells ("haplonts").

The possession of two different free-living forms allows each to exploit different environments. The tiny spores of the ferns allow genes to travel far beyond the immediate vicinity of the parent. As Farrar *et al.* (Chapter 9) remark, this allows one generation of such organisms to fulfill an exploratory role. It may

Biology and Evolution of Ferns and Lycophytes, ed. Tom A. Ranker and Christopher H. Haufler. Published by Cambridge University Press. © Cambridge University Press 2008.

50   Elizabeth Sheffield

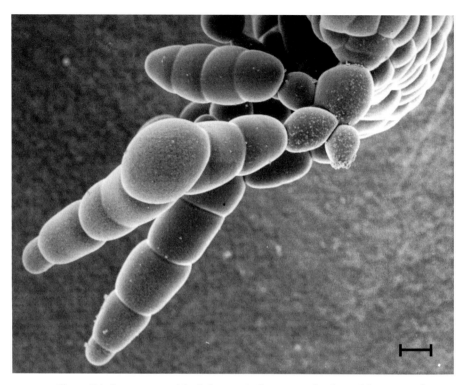

Figure 2.1 Gemmae – multicellular vegetative propagules formed by gametophytes of some ferns (in this case *Vittaria*). (LTSEM, scale bar 40 μm.)

turn out that the new environment proves unsuitable for the alternate generation, but some ferns have developed mechanisms to cope with this. For example, many taxa have tough, long-lived gametophytes capable of persisting indefinitely through vegetative growth and proliferation (see Figure 2.1 and Chapter 9).

Although some fern gametophytes may have a greater degree of stress tolerance than the sporophytes of the same species (see Chapter 9 for a review), and most organisms with an alternation produce morphologically distinct forms, this is not always the case. Alternation is not attended by differences in appearance in many algae – described as being isomorphic. It is not immediately obvious why isomorphic life cycles have persisted, as there would seem to be no advantage conferred by alternating between identical life forms. When careful field studies have been undertaken, however, it has become clear that there are significant ecological differences between the two phases of isomorphic species (e.g., Dyck and DeWreede, 1995). Even in the absence of clear-cut differences in ecological requirements of the multicellular stages of isomorphic taxa, there are invariably substantial differences in the cytological and physiological status of their unicellular propagules. Diploid cells have double the DNA and are usually larger than their haploid counterparts, and Cavalier Smith (1978) drew attention to the likely physiological and ecological consequences of size differences for single cells (such as gametes and spores), another factor favoring the maintenance of biphasic life cycles.

Evolution has favored many algae, bryophytes, and fungi with biphasic life cycles characterized by substantial development of both the haploid and diploid phases that generate very different independent organisms (see Bell, 1994), as in the ferns and lycophytes. Such organisms are referred to as heteromorphic. Whether the final form of each generation is similar or entirely different, it seems clear that the biphasic life cycles have persisted since the earliest days of such organisms because they offer opportunities to exploit an environment more efficiently together than either phase could do alone (Hughes and Otto, 1999). They may also reduce the cost of sex. Richerd *et al.* (1993, 1994) pointed out that if the duration of each phase is equal, biphasic organisms will require sexual union only half as often as haplonts or diplonts. They showed that this intrinsic advantage favors biphasic life cycles whenever the cost of sex is high. The cost is certainly high in organisms unable to ensure that male gametes can be delivered reliably and accurately to female gametes, such as ferns (see Chapters 5 and 9). Mable and Otto (1998) pointed out that this cost can also be reduced through the evolution of asexual reproduction and, as we know (e.g., Moran, 2004) and will see later in this chapter, many ferns also have effective and sophisticated processes to exploit this route.

In summary, the "alternation of generations" refers to a reproductive cycle of certain vascular plants, fungi, and protists in which each phase consists of one of two separate, free-living organisms: a gametophyte, which is often but by no means always genetically haploid, and a sporophyte, which is often, but not always, genetically diploid. The gametophyte generation produces gametes by mitosis. Two gametes, originating from different organisms of the same species or from the same organism, combine to produce a zygote, which is one way to produce a sporophyte generation (see later). This sporophyte produces spores by meiosis, which germinate and develop into gametophytes of the next generation. There are many variations on this theme, but first we need to examine how the literature serves these fundamental steps. In most published work, this takes the form of a diagram and text outlining the "life cycle."

2.2 "The" fern life cycle

Having written a basic textbook myself, I am only too well aware of the tensions that arise between the need for simplicity and a desire to provide a complete and accurate account. Illustrations of the life cycle of ferns remain dogged with elements that can mislead, however, such as sperm apparently swimming from an antheridium directly into an archegonium from the same gametophyte (e.g., Figure 26.12 in Solomon *et al.*, 2005, Figure 29.13 in Raven *et al.*, 2005, Figure 29.14 in Freeman, 2005, Figure 28.20 in Savada *et al.*, 2008, and Figure 29.11 in Campbell and Reece, 2004). Inclusions in the text do clarify matters in some, for example the last book helpfully includes "Although this illustration shows an egg and sperm from the same gametophyte, a variety of mechanisms promote cross-fertilization between gametophytes." However, the visual presentations in textbooks may leave a long-lasting impression. This is especially worrisome when some illustrators juxtapose gametangia apparently from the same gametophyte in such proximity that the sperm literally have nowhere else to go but into the waiting archegonium (e.g., Figure 29.20 in Purves *et al.*, 2004), so it is good that the figure in the newest edition of the book (Savada *et al.*, 2008) has been re-drawn. However, it is clear that the encouragement given to textbook illustrators in my last review (Sheffield, 1994) to show separate parent gametophytes was largely ignored. I remain convinced that newcomers to the delights of fern life cycles would be better prepared for the genetic consequences of fertilization if life cycles were shown with gametes issuing from more transparently separate parents, as is most certainly the norm in nature (e.g., Ranker *et al.*, 1996; see Chapter 4). Perhaps this would avoid authors trying to figure out fern distributions assuming that they have "small self-fertilizing spores" (e.g., Wild and Gagnon, 2005).

We once needed to turn to specialist literature (e.g., Dyce, 1993) to find helpful examples such as that shown in Figure 2.2. Happily, versions of this have now been adopted by other authors and are starting to appear in the primary literature (e.g., Marrs and Watt, 2006). Figure 2.2 summarizes events in the life cycle of bracken, a fern that does generate butterfly-shaped gametophytes ("prothalli") (see Chapter 9) and routinely forms sporophytes from two separate gametophytes (Wolf *et al.*, 1988; see Figures 2.3 and 2.4). This summary outlines events in *Pteridium*, and should not be assumed to be accurate for other homosporous ferns, or indeed any heterosporous ferns (which seldom receive more than a passing mention in basic textbooks).

2.3 Historical summary

By 1851 Hofmeister had figured out the steps outlined in the fern life cycle just described (see Kaplan and Cooke, 1996, for review). As Mayr (1982) pointed out, it was Hofmeister's insight that all plant life cycles shared common elements that prepared botanists to accept Darwin's principle of common descent. Hofmeister's contribution is seldom recognized, but Farrar *et al.*

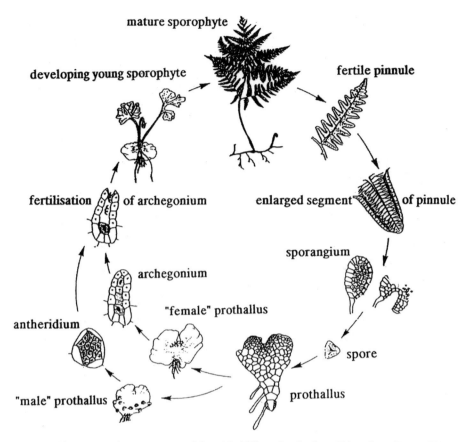

Figure 2.2 Diagram summarizing "the" life cycle of a fern. Taken from Dyce, 1993. (Adapted from *Biological Sciences Review*, **1** (1988), 36–39, drawn by F. J. Rumsey.)

(Chapter 9) review some of the excellent work that followed, which documented the (at-the-time surprising) role of the gametophyte and the significance of the cytological and morphological changes that attend each stage of the cycle. During the ensuing period, studies of living species and of fossils gradually came together. While paleobotany hardly figured in the punctuated equilibrium debate of the 1970s and 1980s, it has more recently been recognized that fossils, especially those preserved in exquisite detail in beds such as the Rhynie chert, provide insight that allow interpretation of the evolutionary pressures that led to the splendid array of current lycophyte and fern life cycles. *Cooksonia* and related early tracheophytes with sporophyte-dominated, heteromorphic life cycles are now thought to have produced all modern day homosporous tracheophytes (Gerrienne *et al.*, 2006), whereas at least ten distinct homosporous lineages are thought to have generated heterosporous plants (DiMichele and Bateman, 1996).

54 Elizabeth Sheffield

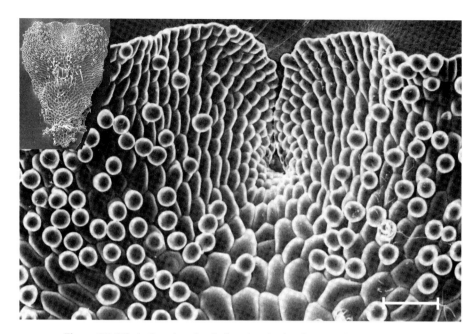

Figure 2.3 Whole (inset) and apical region (main picture) of a young gametophyte ("prothallus") of bracken (*Pteridium*) bearing numerous antheridia. (LTSEM, scale bar 100 μm.)

Efforts to understand the life cycles of the first lycophytes and ferns were at first frustrated by an apparent lack of fossils representing the gamete-producing plants. Thanks to discoveries in the early 1980s, early misinterpretation and assumptions about the ca. 400 million year old plants from which current day lycophytes and ferns arose have gradually been expunged. It is now relatively well accepted that there was an early divergence in vascular plants. The (extinct) rhyniophytes, thought to have represented the earliest vascular land plants, are believed to have been characterized by a more or less isomorphic alternation of generations. Their gametophytes were multilayered, with vascular tissue and a cutinized epidermis, bearing several-centimeter-high gametangiophores (see Taylor *et al.*, 2005, for review). The Eutracheophyta (including all extant vascular plants) probably all had reduced and thalloid gametophytes, having already started the trend towards the heteromorphic, sporophyte-dominant alternation of generations described above (see Gerrienne *et al.*, 2006, for review). It is significant that rhyniophyte gametophytes were unisexual – producing either antheridia or archegonia. Sexual dimorphism in these gametophytes therefore mirrors the situation in present-day lycophytes and ferns, and is inconsistent with earlier models of the bisexual haploid phase of early land plants (rather

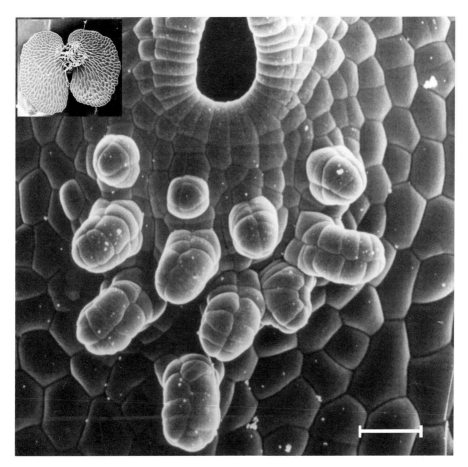

Figure 2.4 Whole (inset) and apical region (main picture) of a more mature gametophyte than that shown in Figure 2.3, also bracken (*Pteridium*), bearing developing and mature archegonia. (LTSEM, scale bar 100 μm.)

like established diagrams of the fern life cycle) (see Taylor *et al.*, 2005, for review).

2.4 Variations on a theme

The regular alternations described so far were historically treated as the "normal" fern and lycophyte life cycle, with departures that avoided one or more of the steps shown in the life cycle regarded as being aberrant or abnormal. The lycophytes and ferns have been pivotal in our understanding that there is more than one way to produce a new generation, and that the changes in ploidy levels that commonly attend the transition from one phase to another need not occur.

2.4.1 Apogamy

Sporophytes arise in most plants from zygotes formed by the union of two gametes. In approximately 10% of ferns this is never the case (Walker, 1984). Although gametophytes, and often functional antheridia, are produced, the sporophytes in these plants arise most frequently from somatic cells located behind the apical meristem. This is the tissue that traditionally generates archegonia in ferns, and in some of these taxa (e.g., *Pteris cretica*, Laird and Sheffield, 1986) archegonia still form, but do not function. The names given to the process are many and varied, and include agamospory (Walker, 1985), apomixis (e.g., Lovis, 1977), and apogamy. Detailed discussion of the terms would not be helpful to our theme and, as with the definitions of alternation of generations, their use depends heavily upon which definition you choose. Agamospory and apomixis are the terms favored by workers who are concerned with the process that generates the spores of such ferns (e.g., see Gastony and Haufler, 1976; Gastony and Windham, 1989; and references therein). If one accepts the simplest definition of apogamy – the production of a sporophyte without sexual fusion (Sheffield and Bell, 1987), it makes it easier to start to understand the significance of this method of alternation.

There are essentially two types of apogamy. One is that routinely shown in the life cycle of the ferns referred to above, the other is an induced process – sometimes called facultative or induced apogamy (see Raghavan, 1989). The latter is of interest both as a tool for plant breeders and propagators (e.g., Martin *et al.*, 2006) and as confirmation that the environment can have a powerful influence on the alternation of generations (this will be a recurrent theme throughout the chapter). It does, however, generate plants that have double the number of chromosomes usual for the species, and while it may be of horticultural significance, it is doubtful that it plays a role in natural populations.

2.4.2 Facultative apogamy

This relates to lycophyte and fern taxa that usually reproduce sexually, but which can be persuaded to generate sporophytes without sexual fusion. There is no single recipe for this, indeed the triggers may be completely dissimilar even within the ferns. *Lycopodium* (Freeberg, 1957) and many ferns (Farlow, 1874) can be induced to form sporophytes without fertilization if they are deprived of the water so essential to the operation of the process. High levels of illumination can promote it (e.g., Lang, 1898), and some ferns have been reported to become apogamous when exposed to high concentrations of sugars (Whittier and Steeves, 1960; but see also Menendez *et al.*, 2006). In many others apogamy is favored when nutrients and sucrose concentrations are reduced (e.g., heterosporous ferns, Mahlberg and Baldwin, 1975; *Osmunda regalis* and *Pteris ensiformis*, Fernandez *et al.*, 1999; *Pityrogramma calomelanos*, Martin *et al.*, 2006). Sugars do not promote apogamy in *Equisetum*, but the plant growth regulators kinetin and benzyladenine do have an effect (Ooya, 1974; Kuriyama *et al.*, 1990). Kinetin and benzyladenine promote apogamy in *Pityrogramma calomelanos* (Martin *et al.*, 2006) and ethylene and other plant growth regulators promote apogamy in some ferns (e.g., Whittier, 1966; Elmore and Whittier, 1975). Exogenous and endogenous gibberellins may have a positive or negative role, depending on the species or genus (Jimenez *et al.*, 2001).

Just how these conditions and additives work to promote apogamy is presently unknown, but it is exciting that we are starting to obtain molecular and biochemical tools that should allow us to develop this understanding. For example, several genes have been cloned that regulate hormone responses in the "C" fern (*Ceratopteris richardii*) (see Banks, 1999, for a review). This semi-aquatic fern has emerged as a productive model system for studying developmental processes (Hickock *et al.*, 1987; Nakazato *et al.*, 2006) and it is exciting that apogamy has recently been induced in this fern (Cordle *et al.*, 2007). As the latter authors suggest, this provides a tractable experimental system for understanding the gene network that controls the switching from one generation to another. We are also at last able to examine plant growth regulator (PGR) effects by looking at the endogenous levels of these molecules. Until recently the only clues we could get about the role of PGRs was through application of exogenous substances (either the PGRs themselves, or inhibitors of them). Now we can measure endogenous hormone levels and the techniques used to study apogamy in *Dryopteris affinis* ssp. *affinis* could provide useful information about the process of facultative apogamy (Menendez *et al.*, 2006).

2.4.3 Obligate apogamy

Dryopteris affinis ssp. *affinis* needs no prompting to generate sporophytes apogamously, but we know that their induction is stimulated by auxins and gibberellins (Menendez *et al.*, 2006). In common with many other apogamous species, this taxon is widely distributed and very successful. *Bommeria pedata*, for example, has a considerably more extensive distribution than most of its sexually reproducing congeners (Gastony and Haufler, 1976). It is usually held that the reason for this success is that the apogamous process avoids the need for water in excess of the hydration needs of each plant. Apogamous taxa are certainly successful in drier habitats than their sexually reproducing relatives, but this may well reflect the rapid growth and maturation rate of their gametophytes in at least equal measure to their lack of requirement of water for

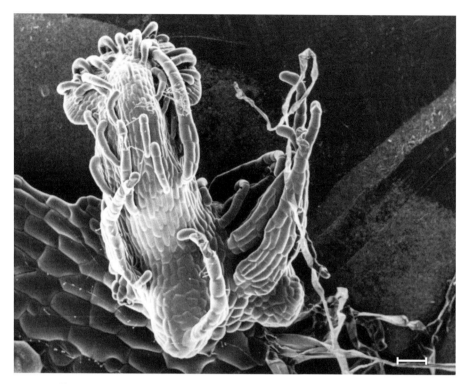

Figure 2.5 Apogamous sporophyte of *Pteris cretica*. The plant has formed directly from vegetative cells of the gametophyte, not from a fertilized egg, and so has the same chromosome number as the cells from which it arose. (LTSEM, scale bar 100 μm.)

fertilization. Non-apogamous taxa with fast-growing/maturing gametophytes are also very successful (e.g., *Pteridium*, bracken fern).

The life cycle of obligate apogamous ferns is characterized by two consecutive events. First, during the formation of spores (sporogenesis), there is an avoidance of the reduction in chromosome number that normally attends meiosis (see Raghavan, 1989, for a comprehensive review of this process). This means that the spores (and hence the subsequent gametophytes) have the same chromosome number as the parent plant. Next, as in the facultative process described above, a sporophyte forms without the union of two gametes, directly from the gametophytic tissue (Figure 2.5). This generates a sporophyte with the same chromosome number as the gametophyte (and of course, the sporophyte that generated the spore). This explains the earlier remark that alternation of generations does not necessarily mean changes in chromosome number, but there are at least two pathways of sporogenesis in obligate apogamous ferns and the reader is referred to Walker (1985) and Gastony and Windham (1989) for details. This production of "clones" rather than genetically variable offspring might seem pointless in plants so adept at asexual reproduction, but it is critical to remember that spores have enormously greater dispersal potential than the more cumbersome bulbils, plantlets, offshoots, and stolons that clone non-apogamous taxa. The important message in the current context is that there is nothing abnormal or aberrant in apogamous ferns, and that alternation of generations for obligate apogamous species allows them to disseminate successful genotypes at enormously greater distances than would be possible using strictly vegetative growth.

It is exciting that two obligate apogamous laboratory strains of the model fern Ceratopteris have now been described (Cordle et al., 2007), as this adds to our suite of methods with which to explore the mechanisms of this process. It is clear that a differentiating cell in a gametophyte can follow one of multiple pathways: vegetative gametophyte, gamete, or vegetative sporophyte (apogamy), depending on the developmental cues to which it responds. Hopefully we now have techniques that will allow us to discover exactly how those cues are received and interpreted, and perhaps also to manipulate responses in other ferns and lycophytes.

2.4.4 Apospory

Apogamy in gametophytes meets its functional counterpart in apospory – the formation of gametophytes from sporophytic tissue in the absence of meiosis or spore formation (Figure 2.6). This usually takes the form of outgrowths from sporophytic organs (e.g., leaves or scales) that therefore generate gametophytes with the same ploidy level as the plant from which they arose. The absence of any observation of compensating mechanisms similar to those seen in the obligate apogamous ferns means that although this process may be of utility to the fern horticulturist or breeder (Sheffield, 1992) it is probably of little significance in nature. That does not diminish the significance of the process with regard to alternation of generations, as it provides more evidence that ploidy level is unconnected with the activation of genes concerned with a particular morphology.

Experimental systems characterizing the requirements for the process are more similar to each other than those associated with apogamy, and low "nutrient" (especially sugars) level is a ubiquitous element (e.g., Materi and Cumming, 1991), as is interruption of communication between cells and/or damage (see Raghavan, 1989, for a review). The age of the tissue that generates the gametophytes also has a strong influence, with juvenile sporophytic tissue apparently far more readily influenced to generate gametophytes than older material (see Sheffield and Bell, 1987, for review, and also Ambrozic-Dolinsek et al., 2002). The latter report concerns Platycerium bifurcatum, which has emerged as a good model

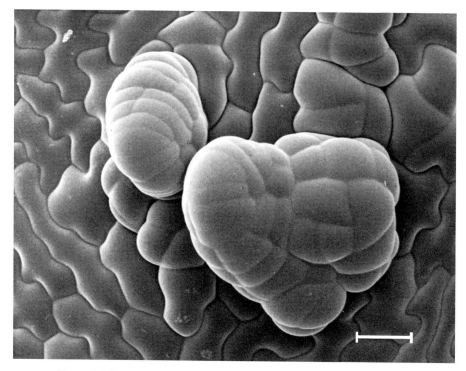

Figure 2.6 Two aposporous outgrowths from the surface of a juvenile leaf of a bracken (*Pteridium*) plant. The leaf had been detached from the parent plant 5 days previously, and the outgrowths would have gone on to form gametophytes with the same chromosome number as the gametophyte cells from which they arose. (LTSEM, scale bar 20 μm.)

system in which to examine the phenomenon of apospory. The elegant culture system of Teng and Teng (1997), for example, allows experimenters to manipulate the life cycle stage of regenerants. Single cells and aggregates of up to 100 leaf cells in suspension culture generate aposporous gametophytes, larger aggregates form sporophyte tissue directly. This system is considerably more amenable to biochemical and molecular study than the whole leaf or rhizome based systems of earlier authors (e.g., Mehra and Sulklyan, 1969; Hirsch, 1975).

2.4.5 "Conventional" alternation of generations

The phenomena of apogamy and apospory, although easily induced and in some taxa the norm, have little impact on alternation of generations in the vast majority of lycophytes and ferns. The life cycle shown in Figure 2.2 shows the structures more conventionally concerned with alternation of generations in bracken fern, and which we can use as the basis of a consideration of the process in non-apogamous lycophytes and ferns. The reason for choosing bracken is that we know more about this fern than any other.

There are two main reasons for the plethora of papers on this fern. One is that it contains toxins and carcinogens (e.g., Simán et al., 1999, 2000; Schmidt et al., 2005), which adds to the problems caused by the other reason, which is that it is a vigorous and invasive weed (see Smith and Taylor, 2000, for review). The "it" to which I refer is the genus, within which there are now four species generally recognized (see Marrs and Watt, 2006, for review, and Thomson et al., in press) but taxonomic distinctions are not linked to any substantial differences in life cycle processes in this genus. Small differences, however, between taxa may have far-reaching consequences.

One example of this concerns the production of spores – the event that could be considered as the start of the alternation process from sporophyte to gametophyte. There has been keen interest in this process and much progress made over the thirteen years since my last review (Sheffield, 1994). One explanation for this is that it has become clear that some fern spores, including both northern and southern hemisphere bracken, contain compounds that could be injurious to human and animal health (Simán and Sheffield, 2006) and the other is that there are important genetic, taxonomic and ecological consequences of spore formation and dispersal (see Chapter 4). Spores are one of the vehicles that affect gene flow – the movement of genes from one location to another. By measuring genes or their products in established plants (see Chapter 4) we can infer events that produced them. Although outcrossing is very clearly the norm for bracken (not intragametophytic selfing) (Wolf et al., 1988, 1990; Korpelainen, 1995), allozyme studies indicate very disparate levels of gene flow between Laurasian bracken populations. Levels of genetic exchange between populations of the bracken taxon in Britain are sufficiently high that "aquilinum" should be regarded as a single panmictic entity (Wolf et al., 1991; Thomson et al., in press). However, in contrast, allozymes of geographically close Scandinavian bracken populations (presumably at least predominantly "pinetorum") indicate low levels of gene flow. This is thought to result from the very limited spore production in the stands sampled (Korpelainen, 1995; Thomson et al., in press).

2.5 Sporogenesis

What do we know about the factors that control spore production? In her seminal paper, Conway (1957) suggested that spore production in bracken is "undoubtedly" influenced by the age of the plant, the developmental stage of the frond, seasonal weather conditions, and environmental factors. The latter certainly include shade, and differences in spore production between shaded and unshaded fronds were noticed over a century ago (see Daniels, 1986). Bracken fronds partly covered by a canopy have a correspondingly lower spore output (Schwabe, 1951) and the same is true for some other ferns (e.g., *Polystichum acrostichoides* and *Cyathea caracasana*, Greer and McCarthy, 2000; Arens and Baracaldo, 2000). Steeves and Wetmore (1953) studied (green) spore production in *Osmunda cinnamomea* plants in heavy woodland and open areas and found the same for this species, and the only reports of increased photosynthetically available radiation (PAR) inhibiting spore production in ferns is that relating to an in vitro study on *O. cinnamomea*, where high PAR inhibited sporophyll differentiation (Harvey and Caponetti, 1972). It therefore seems likely that the PAR levels used in the latter study were higher than plants are likely to experience in the natural environment, as all subsequent studies indicate promotion of fertility by increased light.

Genotype can have a strong influence on sporogenesis, so it is critical that observations and experiments are conducted on genetically identical or matched samples. One such experiment used clones – two pieces taken from rhizomes of *Polypodium vulgare* and then cultivated in natural and glasshouse conditions of controlled temperature and light (Simán and Sheffield, 2002). The population of clones raised indoors, with higher temperatures and PAR than that experienced by the outdoor clones, behaved very differently. The plants grown outside produced the usual singe pulse of fertile fronds each season; those grown inside generated fertile fronds continuously. Intriguingly, these plants also produced new fronds in distinct pulses approximately every 3 months, rather than incrementally, as if some element of the normal year-long reproductive cycle had been accelerated (Figure 2.7). Although the mechanism at work was not determined, this study showed clearly that the normal year-long process of fern spore production and release can be interrupted by experimental manipulation of environmental factors.

So light and temperature have a powerful influence on sporogenesis, but the studies reported above cannot separate out the effects of higher PAR and temperature, as both vary considerably between heavy woodland and open areas, outside and inside a glasshouse. A preliminary study involving clones of bracken cultivated in growth chambers reported that each factor independently promotes fertility (Wynn *et al.*, 2000). Rhizomes were excavated in the middle of winter and cultivated in pots subjected to two different temperature regimes, and two levels of PAR. They generated fronds that bore the first stages of spore-forming tissue in less than 9 weeks – again showing that the usual cycle of spore production can be interrupted, as bracken in the UK normally generates spores

Alternation of generations 63

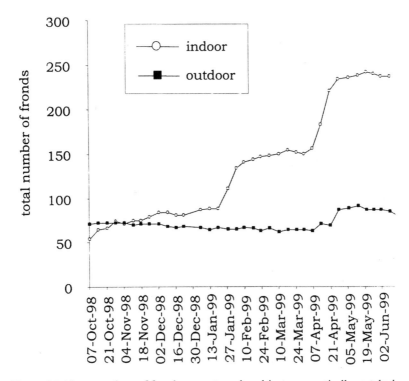

Figure 2.7 Mean numbers of fronds per pot produced in two genetically matched cultivated populations of the fern *Polypodium vulgare* plotted against time. Open circles represent the indoor population, grown in higher temperatures and light than the outdoor population (closed circles), and show three distinct waves of frond recruitment. (From Simán, 2000.)

in late summer. Significantly more fertile material was generated by plants in the higher temperatures and light, and there appeared to be some differences between genotypes. The possibility, however, that development of fertile material reflects the nutritional status of the rhizomes, and therefore the environmental conditions at the sites from which they were excavated, could not be ruled out.

Later experiments with the same rhizomes and genotypes clarified this (Wynn, 2002). Rhizomes originated from two contrasting environments in the UK. One was 550 kilometers north of the other, and experiences a considerably harsher climate. Bracken in the more northerly location had never been seen to generate spores, whereas plants from the southerly collection site routinely did so. Plants of two distinct genotypes used in the experiments conducted a year earlier had been maintained in cultivation in potting compost and their responses to environmental variables was measured in comparison with plants raised from newly excavated rhizomes from the source populations. Higher fertility was again noted in plants grown in the highest light and temperatures, and there were clear-cut differences between genotypes that were unrelated to behavior observed in the field. A genotype never observed to spore in Scotland, at the more northerly site, showed significantly greater fertility than a plant that spored profusely in its native site in Manchester, when cultivated in identical conditions. Although the rate of development of sporing material was similar in both plants, this shows that the amount of sporogenous tissue is under genetic control.

These experiments also showed that transplantation/rhizome disturbance is not required to trigger formation of fertile material. Tyson *et al.* (1995) studied nutrient flow through bracken rhizomes that generated spore-bearing fronds in cultivation. Sheffield (1996) hypothesized that disturbance, stress, and/or interference with sink-source nutrient flow might therefore promote sporing. Wynn *et al.*'s experiments included rhizomes that were both newly excavated and undisturbed, and that were supplied with both high and low nutrients and water. Newly excavated rhizomes yielded significantly less, not more, reproductive tissue, and the different nutrient and water treatments had no discernable effect on sporangial development. It may be that the differences between the conditions used were insufficient to reveal an effect, as authors have reported significant effects of water availability on fertility in other ferns. The rheophyte *Thelypteris angustifolia*, for example, shows reduced fertile-frond production in response to dry conditions (Sharpe, 1996). As bracken tolerates flooded conditions poorly, and characteristically grows on low-nutrient soils (Pakeman *et al.*, 1996), the expectation would be that this fern would have a high tolerance to a range of nutrient and water availability, but that copious amounts of water might be too much of a good thing. Further experiments to establish this would be useful in view of current predictions on climate change.

Unlike the vast majority of lycophyte and fern taxa, most of the bracken plants currently growing in temperate regions produce spores rarely. Bracken growing in warmer locations, e.g., the Hawaiian Islands (Sheffield *et al.*, 1995) sporulates profusely – perhaps adding weight to the hypothesis that the genus originated in the tropics (Page, 1990). Given the intolerance of frost in this species and the increase in frost-free days and temperature predicted by most climate change models (e.g., Hulme and Jenkins, 1998) the interaction of PAR (which may fall with increased cloud cover) and water (likely increased rainfall) is likely to be an important influence on future levels of bracken sporogenesis (see Kendall *et al.*, 1995). It seems very likely that environmental conditions currently experienced by bracken plants in temperate regions are not conducive to the completion of this part of the life cycle before winter ensues, but that the plants are capable of becoming fully fertile if conditions improve. In this respect the story echoes that of the Killarney fern, *Trichomanes speciosum*, in which current genetic variability within temperate sites occupied by gametophyte populations is attributed to spores produced in more favorable environmental conditions (Rumsey et al., 1999). In Europe this species currently disperses almost exclusively via gemmae (see Figure 2.1; Sheffield, 1994 and Chapter 9 for a review) and alternation of generations is undoubtedly extremely rare at present.

2.5.1 Spore release and dispersal

Once spores have been formed, it stands to reason that they can only effect the next stage in the alternation process if they are released, dispersed, and alight somewhere conducive to germination and gametophyte growth. This presents a huge challenge for propagules that lack vectors which target such sites (cf. flowering plants with insect pollinators). There is scant evidence for productive interactions between animal vectors and fern spores, but very little study of this seems to have been made. Spore-feeding was one nutritional strategy for insects in early terrestrial ecosystems (Habgood et al., 2004), however, and there have been reports of insects feeding on various taxa of ferns (see Srivastava et al., 1997, for review). As the latter paper reports, some lepidopterans do feed on mature spores of bracken, despite their chemical arsenal (Alonso-Amelot et al., 2001), but the consequences for the spores have not been explored. Hemipterans dining on immature spores are clearly of no benefit to the ferns involved (e.g., Balick et al., 1978) and in general it is likely that herbivores are not as helpful to ferns as detritivores. Given that coprolites of early terrestrial detritivores resemble droppings of modern day insect taxa, it seems possible that insects do contribute to dispersal of fern spores, but we must await the necessary experiments. Only one case has been reported of birds possibly specializing on ferns as a major component of their diet. This was the case of the extinct Hawaiian, flightless bird *Thambetochen chauliodous* ("moa-nalo"), for which evidence was obtained from coprolites showing a high density of fern spores (James and Burney, 1997).

The vast majority of fern and lycophyte spores are undoubtedly dispersed by wind or water. The inadequacies of these agents are considerable. The endangered status of the aquatic quillwort *Isoetes sinensis* in China, for example, does not reflect low viable spore output, or poor germination, but poor dispersal (Wang et al., 2005). Dispersal in air, as used by most lycophytes and ferns, is generally rather more efficient.

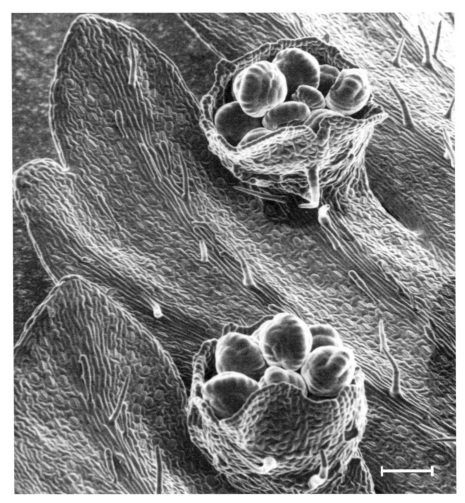

Figure 2.8 Underside of a fertile frond of *Dennstaedtia cicutaris* showing the cup-like indusia that first protect the developing sporangia, then allow them to emerge into the dry air such that the mechanism shown in Figure 2.9 is triggered. (Image courtesy of Professor E. G. Cutter.) (SEM, scale bar 1.0 mm.)

Most taxa have specialized structures that first protect the developing sporangia (Figure 2.8), then ensure an effective launch for spores via hygroscopic movements (Figures 2.9 and 2.10). The process has been well understood for over a century and has even inspired the production of biomimetic microactuators (Borno *et al.*, 2006). For a full and engaging account of "spore shooting" the reader seeking details is recommended to read Chapter 2 in Moran (2004). From the moment they are catapulted into the air, the spores are reliant on air or water currents for transportation.

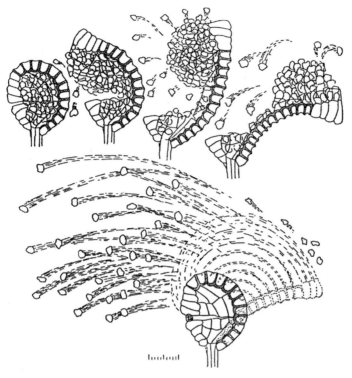

Figure 2.9 Diagram showing the catapult mechanism that launches spores from fertile fronds of ferns. (From Moran, 2004.)

We might expect ferns on forest floors to face the greatest challenge, but careful study of soil samples taken at distances from hay-scented ferns (*Dennstaedtia punctilobula*) reveal plentiful spores up to 50 meters from their point of origin (Penrod and McCormick, 1996). The highest spore rain fell within a few meters of the sporing fronds, as observed in research on ferns ranging from other woodland taxa to tree ferns (e.g., Conant, 1978; Peck *et al.*, 1990; Schneller, 1998; Simán, 2000). Tree-fern spores are those most likely to experience conditions conducive to establishment, and most form a long-lived spore bank (Dyer and Lindsay, 1992). With some notable exceptions, for example, bracken (Lindsay *et al.*, 1995), almost every lycophyte and fern studied seems to generate spores that survive well in soil, and almost every substratum sampled contains spores. This holds true for different vegetation (e.g., del Ramirez-Trejo *et al.*, 2004) and substratum types, including gravel (Ranker *et al.*, 1996) and tree bark (Ranal, 2004). The relatively recent recognition of the importance and potential value of spore banks is encouraging. Researchers are now focusing on conservation methods that take this into account (see Dyer and Lindsay, 1996), including

68 Elizabeth Sheffield

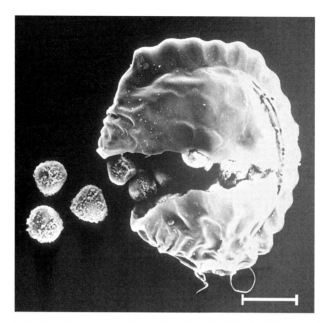

Figure 2.10 Ripe sporangium of bracken (*Pteridium*) which has split open, catapulted out spores and returned to its original position (see Figure 2.8). (LTSEM, scale bar 50 μm.)

restoration of pre-existing vegetation from the soil diaspore bank after weed (bracken!) control (e.g., Ghorbani *et al.*, 2003).

The lucky spore that arrives directly in, or finds its way out of, the spore bank and into the right conditions for germination leads us to events covered in later chapters in this book. Whether it encounters antheridiogen (see Chapter 5) or not, what type of gametophyte it is (see Chapter 13) and whether its gametes can find a mate (see Chapter 8) will all dictate whether it manages to finish the cycle and complete an alternation of generations.

References

Alonso-Amelot, M. E., Oliveros, A., Calcagno, M. P., and Arellano, E. (2001). Bracken adaptation mechanisms and xenobiotic chemistry. *Pure and Applied Chemistry*, 73, 549–553.

Ambrozic-Dolinsek, J., Camloh, M., and Bohanec, J. Z. (2002). Apospory in leaf culture of staghorn fern (*Platycerium bifurcatum*). *Plant Cell Reports*, 20, 791–796.

Arens, N. C. and Baracaldo, P. S. (2000). Variation in tree fern stipe length with canopy height: tracking preferred habitat through morphological change. *American Fern Journal*, 90, 1–15.

Balick, M. J., Furth, D. G., and Cooper-Driver, G. (1978). Biochemical and evolutionary aspects of arthropod predation in ferns. *Oecologia*, 35, 55–89.

Banks, J. A. (1999). Gametophyte development in ferns. *Annual Review of Plant Physiology*, **50**, 163–186.

Bell, G. (1994). The comparative biology of the alternation of generations. In *Lectures on Mathematics in the Life Sciences: Theories for the Evolution of Haploid–Diploid Life Cycles*, Vol. 25, ed. M. Kirkpatrick. Providence, RI: American Mathematical Society, pp. 1–26.

Borno, R. T., Steinmeyer, J. D., and Maharbiz, M. M. (2006). Transpiration actuation: the design, fabrication and characterization of biomimetic microactuators driven by the surface tension of water. *Journal of Micromechanics and Microengineering*, **16**, 2375–2383.

Campbell, N. A. and Reece, J. B. (2004). *Biology*, 7th edn. San Francisco, CA: Benjamin Cummings.

Cavalier Smith, T. (1978). Nuclear volume control by nucleoskeletal DNA, selection for cell volume and cell growth rate, and the solution of the DNA C-value paradox. *Journal of Cell Science*, **34**, 247–278.

Conant, D. S. (1978). A radioisotope technique to measure spore dispersal of the tree fern *Cyathea arborea* Sm. *Pollen et Spores*, **20**, 583–593.

Conway, E. (1957). Spore production in bracken (*Pteridium aquilinum* (L.) Kuhn). *Journal of Ecology*, **45**, 273–284.

Cordle, A. R., Irish, E. E., and Cheng, C. L. (2007). Apogamy induction in *Ceratopteris richardii*. *International Journal of Plant Science*, **168**, 361–369.

Daniels, R. E. (1986). Studies in the growth of *Pteridium aquilinum* (L.) Kuhn (bracken). 2. Effects of shading and nutrient application. *Weed Research*, **26**, 121–126.

del Ramirez-Trejo, M., Perez-Garcia, B., and Orozco-Segovia, A. (2004). Analysis of fern spore banks from the soil of three vegetation types in the central region of Mexico. *American Journal of Botany*, **91**, 682–688.

DiMichele, W. A. and Bateman, R. M. (1996). Plant paleoecology and evolutionary inference: two examples from the Paleozoic. *Review of Paleobotany and Palynology*, **90**, 223–247.

Dyce, J. W. (1993). *The Cultivation and Propagation of British Ferns*, Special Publication 3, 2nd edn. London: British Pteridological Society.

Dyck, L. J. and DeWreede, R. E. (1995). Patterns of seasonal demographic changes in the alternate isomorphic stages of *Mazzaella splendens* (Gigartinales, Rhodophyta). *Phycologia*, **34**, 390–395.

Dyer, A. F. and Lindsay, S. (1992). Soil spore banks of temperate ferns. *American Fern Journal*, **82**, 69–123.

Dyer, A. F. and Lindsay, S. (1996). Soil spore banks – a new resource for conservation. In *Pteridology in Perspective*, ed. J. M. Camus, M. Gibby, and R. J. Johns. Kew: Royal Botanic Gardens, pp. 153–160.

Elmore, H. W. and Whittier, D. P. (1975). The involvement of ethylene and sucrose in the inductive and developmental phases of apogamous bud formation in *Pteridium* gametophytes. *Canadian Journal of Botany*, **53**, 375–381.

Farlow, W. (1874). An asexual growth from the prothallus of *Pteris cretica*. *Quarterly Journal of the Microscopical Society*, **14**, 266–272.

Fernandez, H., Bertrand, A. M., and Sanchez-Tames, R. (1999). Biological and nutritional aspects involved in fern multiplication. *Plant Cell, Tissue and Organ Culture*, **56**, 211–214.

Freeberg, J. A. (1957). The apogamous development of sporelings of *Lycopodium cernuum* L., *L. complanatum* var. *flabelliforme* Fernald and *L. selago* L. in vitro. *Phytomorphology*, **7**, 217–229.

Freeman, S. (2005). *Biological Science*, 2nd stdt. edn. Upper Saddle River, NJ: Pearson Prentice Hall.

Gastony, G. J. and Haufler, C. H. (1976). Chromosome number and apomixis in the fern genus *Bommeria* (Gymnogrammaceae). *Biotropica*, **8**, 1–11.

Gastony, G. J. and Windham, M. D. (1989). Species concepts in pteridophytes: the treatment and definition of agamosporous species. *American Fern Journal*, **79**, 65–77.

Gerrienne, P., Dilcher, D. L., Bergamaschi, S., Milagres, I., Pereira, E., and Rodrigues, M. A. C. (2006). An exceptional specimen of the early land plant *Cooksonia paranensis* and a hypothesis on the life cycle of the earliest tracheophytes. *Review of Paleobotany and Palynology*, **142**, 123–130.

Ghorbani, J., Das, P. M., Das, A. B., Hughes, J. M., McAllister, H. A., Pallai, S. K., Pakeman, R. J., Marrs, R. H., and Le Duc, M. G. (2003). Effects of restoration treatments on the diaspore bank under dense *Pteridium* stands in the UK. *Applied Vegetation Science*, **6**, 189–198.

Greer, G. K. and McCarthy, B. C. (2000). Patterns of growth and reproduction in a natural population of the fern *Polystichum acrostichoides*. *American Fern Journal*, **90**, 60–76.

Habgood, K. S., Hass, H., Kerp, H. (2004). Evidence for an early terrestrial food web: coprolites from the Early Devonian Rhynie chert. *Transactions of the Royal Society of Edinburgh, Earth Sciences*, **94**, 371–389.

Harvey, W. H. and Caponetti, J. D. (1972). In vitro studies on the induction of sporogenous tissue on leaves of cinnamon fern. I. Environmental factors. *Canadian Journal of Botany*, **50**, 2673–2682.

Hickock, L. G., Warne, L. K., and Slocum, M. K. (1987). *Ceratopteris richardii*: applications for experimental plant biology. *American Journal of Botany*, **74**, 1304–1316.

Hirsch, A. M. (1975). The effect of sucrose on the differentiation of excised fern leaf tissue into either gametophytes or sporophytes. *Plant Physiology*, **56**, 390–393.

Hughes, J. S. and Otto, S. P. (1999). Ecology and the evolution of biphasic life cycles. *The American Naturalist*, **154**, 306–320.

Hulme, M. and Jenkins, G. J. (1998). *Climate Change Scenarios for the UK: Scientific Report*. UKCIP Technical Report No 1. Norwich: Climate Research Unit.

James, H. F. and Burney, D. A. (1997). The diet and ecology of Hawaii's extinct flightless waterfowl: evidence from coprolites. *Biological Journal of the Linnean Society*, **62**, 279–297.

Jimenez, V. M., Guevara, E., Herrera, J., and Bangerth, F. (2001). Endogenous hormone levels in habituated nucellar *Citrus* callus during the initial stages of regeneration. *Plant Cell Reports*, **20**, 92–100.

Kaplan, D. R. and Cooke, T. J. (1996). The genius of Wilhelm Hofmeister: the origin of causal-analytical research in plant development. *American Journal of Botany*, **83**, 1647–1660.

Kendall, A., Page, C. N., and Taylor, J. A. (1995). Linkages between bracken sporulation rates and weather and climate in Britain. In *Bracken: An Environmental Issue*, ed. R. T. Smith and J. A. Taylor. Aberystwyth: International Bracken Group, Special Publication No. 2, pp. 77–81.

Korpelainen, H. (1995). Mating system and distribution of enzyme genetic-variation in bracken (*Pteridium aquilinum*). *Canadian Journal of Botany*, **73**, 1611–1617.

Kuriyama, A., Sugawara, Y., Matsushima, H., and Takeuchi, M. (1990). Production of sporophytic structures from gametophytes by cytokinin in *Equisetum arvense*. *Naturwissenschaften*, **77**, 31–32.

Laird, S. and Sheffield, E. (1986). Antheridia and archegonia of the apogamous fern *Pteris cretica*. *Annals of Botany*, **57**, 139–143.

Lang, W. H. (1898). On apogamy and the development of sporangia upon fern prothalli. *Philosophical Transactions of the Royal Society, London*, **110**, 187–236.

Lindsay, S., Sheffield, E., and Dyer, A. F. (1995). Dark germination as a factor limiting the formation of soil spore banks by bracken. In *Bracken: An Environmental Issue*, ed. R. T. Smith and J. A. Taylor. Aberystwyth: International Bracken Group, Special Publication No. 2, pp. 47–51.

Lovis, J. D. (1977). Evolutionary patterns and processes in ferns. In *Advances in Botanical Research*, ed. R. D. Preston and H. W. Woolhouse. New York: Academic Press, pp. 229–415.

Mable, B. K. and Otto, S. P. (1998). The evolution of life cycles with haploid and diploid phases. *Bioessays*, **20**, 453–462.

Mahlberg, P. G. and Baldwin, M. (1975). Experimental studies on megaspore viability, parthenogenesis and sporophyte formation in *Marsilea, Pilularia* and *Regnellidium*. *Botanical Gazette*, **136**, 269–273.

Marrs, R. H. and Watt, A. S. (2006). Biological flora of the British isles No. 245 List Br. Vasc. Pl. (1958) no. 8, 1: *Pteridium aquilinum* (L.) Kuhn. *Journal of Ecology*, **94**, 1272–1321.

Martin, K. P., Sini, S., Zhang, C.-L., Slater, A., and Madhusoodanan, P. V. (2006). Efficient induction of apospory and apogamy in vitro in silver fern (*Pityrogramma calomelanos* L.). *Plant Cell Reports*, **25**, 1300–1307.

Materi, D. M. and Cumming, B. G. (1991). Effects of carbohydrate deprivation on rejuvenation, apospory and regeneration in ostrich fern (*Matteuccia struthiopteris*) sporophytes. *Canadian Journal of Botany*, **69**, 1241–1245.

Mayr, E. (1982). Adaptation and selection. *Biologisches Zentralblatt*, **101**, 161–174.

Mehra, P. N. and Sulklyan, D. S. (1969). In vitro studies on apogamy, apospory and controlled differentiation of rhizome segments of the fern *Ampelopteris prolifera* (Retz.) Copel. *Botanical Journal of the Linnean Society*, **62**, 431–443.

Menendez, V., Villacorta, N. F., and Revilla, M. A. (2006). Exogenous and endogenous growth regulators on apogamy in *Dryopteris affinis* (Lowe) Fraser-Jenkins ssp. *affinis*. *Plant Cell Reports*, **25**, 85–91.

Moran, R. C. (2004). *A Natural History of Ferns*. Portland, OR: Timber Press.

Nakazato, T., Jung, M. K., Housworth, E. A., Riesberg, L. H., and Gastony, G. J. (2006). Genetic map-based analysis of genome structure in the homosporous fern *Ceratopteris richardii*. *Genetics*, **173**, 1585–1597.

Ooya, N. (1974). Induction of apogamy in *Equisetum arvense*. *Botanical Magazine (Tokyo)*, **87**, 253–259.

Page, C. N. (1990). Taxonomic evaluation of the fern genus *Pteridium* and its active evolutionary state. In *Bracken Biology and Management*, ed. J. A. Thomson and R. T. Smith. Sydney: Australian Institute of Agricultural Science, pp. 23–34.

Pakeman, R. J., Marrs, R. H., Howard, D. C., Barr, C. J., and Fuller, R. M. (1996). The bracken problem in Great Britain; its present extent and future changes. *Applied Geography*, **16**, 65–86.

Peck, J. H., Peck, C. J., and Farrar, D. R. (1990). Comparative life history studies and the distribution of pteridophyte populations. *American Fern Journal*, **80**, 126–142.

Penrod, K. A. and McCormick, L. H. (1996). Abundance of viable hay-scented fern spores germinated from hardwood forest soils at various distances from a source. *American Fern Journal*, **86**, 69–79.

Purves, W. K., Sadava, D., Orians, G. H., and Heller, H. C. (2004). *Life: the Science of Biology*. New York: W. H. Freeman.

Raghavan, V. (1989). *Developmental Biology of Fern Gametophytes*. Cambridge: Cambridge University Press.

Ranal, M. A. (2004). Bark spore bank of ferns in a gallery forest of the ecological station of Pangua, Uberlandia-MG, Brazil. *American Fern Journal*, **94**, 57–69.

Ranker, T. A., Gemmill, C. E. C., Trapp, P. G., Hambleton, A., and Ha, K. (1996). Population genetics and reproductive biology of lava-flow colonising species of Hawaiian *Sadleria* (Blechnaceae). In *Pteridology in Perspective*, ed. J. M. Camus, M. Gibby, and R. J. Johns. Kew: Royal Botanic Gardens, pp. 581–598.

Raven, P. H., Evert, R. F., and Eichorn, S. E. (2005). *Biology of Plants*. New York: W. H. Freeman.

Richerd, S., Couvet, D., and Valero, M. (1993). Evolution of the alternation of haploid and diploid phases in life cycles. II. Maintenance of the haplo-diplontic cycle. *Journal of Evolutionary Biology*, **6**, 263–280.

Richerd, S., Perrot, D., Couvet, M., Valero, M., and Kondrashov, A. S. (1994). Deleterious mutations can account for the maintenance of the haplo-diploid cycle. In *Genetics and Evolution in Aquatic Organisms*, ed. A. R. Beaumont. New York: Chapman and Hall, pp. 263–280.

Rumsey, F. J., Vogel, J. C., Russell, S. J., Barrett, J. A., Gibby, M. (1999). Population structure and conservation biology of the endangered fern *Trichomanes speciosum* Willd. (Hymenophyllaceae) at its northern distributional limit. *Biological Journal of the Linnean Society*, **66**, 333–344.

Savada, D., Heller, C., Orians, G., Purves, W. K., and Hillis, D. M. (2008). *Life: The Science of Biology*, 8th edn. Gordonsville, VA: Sinauer.

Schmidt, B., Rasmussen, L. H., Svendsen, G. W., Ingerslev, F., and Hansen, H. C. B. (2005). Genotoxic activity and inhibition of soil respiration by ptaquiloside, a bracken fern carcinogen. *Environmental Toxicology and Chemistry*, **24**, 2751–2756.

Schneller, J. J. (1998). How much genetic variation in fern populations is stored in the spore banks? A study of *Athyrium filix-femina* (L.) Roth. *Journal of the Linnean Society*, **127**, 195–206.

Schwabe, W. W. (1951). Physiological studies in plant nutrition. XVI. The mineral nutrition of bracken. Part 1. Prothallial culture and the effects of phosphorus and potassium supply on leaf production in the sporophyte. *Annals of Botany*, **15**, 417–446.

Sharpe, J. M. (1996). Growth and demography of sporophytes of *Thelypteris angustifolia* in the Luquillo rainforest of Puerto Rico. In *Pteridology in Perspective*, ed. J. M. Camus, M. Gibby, and R. J. Johns. Kew: Royal Botanic Gardens, pp. 667–668.

Sheffield, E. (1992). Apogamy and apospory: their potential uses in breeding and propagation. In *Fern Horticulture, Past, Present and Future Perspectives*, ed. J. M. Ide, A. C. Jermy, and A. Paul. Andover: Intercept, pp. 189–193.

Sheffield, E. (1994). Alternation of generations in ferns: mechanisms and significance. *Biological Reviews*, **69**, 331–343.

Sheffield, E. (1996). From pteridophyte spore to sporophyte in the natural environment. In *Pteridology in Perspective*, ed. J. M. Camus, M. Gibby, and R. J. Johns. Kew: Royal Botanic Gardens, pp. 541–549.

Sheffield, E. and Bell, P. R. (1987). Current studies of the pteridophyte life cycle. *Botanical Reviews*, **53**, 442–490.

Sheffield, E., Wolf, P. G., and Ranker, T. A. (1995). Genetic analysis of bracken in the Hawaiian Islands. In *Bracken: An Environmental Issue*, ed. R. T. Smith and J. A. Taylor. Aberystwyth: International Bracken Group, Special Publication No. 2, pp. 29–32.

Simán, S. E. (2000). Fern spores and human health. Unpublished Ph.D. Thesis, University of Manchester.

Simán, S. E. and Sheffield, E. (2002). *Polypodium vulgare* plants sporulate continuously in a non-seasonal glasshouse environment. *American Fern Journal*, **92**, 30–38.

Simán, S. E. and Sheffield, E. (2006). Growth impairment of human cells by fern spore extracts. *Fern Gazette*, **17**, 287–291.

Simán, S. E., Povey, A., and Sheffield, E. (1999). Human health risks from fern spores? – a review. *Fern Gazette*, **15**, 275–287.

Simán, S. E., Povey, A. C., O'Connor, P. J., Ward, T. H., Margison, G. P., and Sheffield, E. (2000). Fern spore extracts can damage DNA. *British Journal of Cancer*, **83**, 69–73.

Solomon, E. P., Berg, L. R., and Martin, D. W. (2005). *Biology*. Belmont, CA: Brooks/Cole-Thomson Learning.

Smith, R. T. and Taylor, J. A. (2000). *Bracken: An Environmental Issue*. Leeds: University of Leeds.

Srivastava, D. S., Lawton, J. H., and Robinson, G. S. (1997). Spore-feeding: a new, regionally vacant niche for bracken herbivores. *Ecological Entomology*, **22**, 475–478.

Steeves, T. A. and Wetmore, R. H. (1953). Morphogenetic studies on *Osmunda cinnamomea* L. – some aspects of the general morphology. *Phytomorphology*, **3**, 339–354.

Taylor, T. N., Kerp, H., and Hass, H. (2005). Life history biology of early land plants: deciphering the gametophyte phase. *Proceedings of the National Academy of Sciences of the United States of America*, **102**, 5892–5897.

Teng, W. L. and Teng, M. C. (1997). In vitro regeneration patterns of *Platycerium bifurcatum* leaf cell suspension culture. *Plant Cell Reports*, **16**, 820–824.

Thomson, J. A., Mickel, J. A., and Mehltreter, K. (in press). Taxonomic status and relationships of bracken ferns (*Pteridium*: Dennstaedtiaceae) of Laurasian affinity in Central and North America. *Botanical Journal of the Linnean Society*.

Tyson, M. J., Sheffield, E., and Callaghan, T. V. (1995). An overview of ^{134}Cs and ^{85}Sr transport, allocation and concentration studies in artificially propagated bracken. In *Bracken: An Environmental Issue*, ed. R. T. Smith and J. A. Taylor. Aberystwyth: International Bracken Group, Special Publication No. 2, pp. 38–42.

Walker, T. G. (1984). Chromosomes and evolution in pteridophytes. In *Chromosomes in Evolution of Eukaryotic Groups*, Vol. 2, ed. A. K. Sharma and A. Sharma. Boca Raton, FL: CRC Press.

Walker, T. G. (1985). Some aspects of agamospory in ferns – the Braithwaite system. *Proceedings of the Royal Society of Edinburgh Section B, Biological Sciences*, **86**, 59–86.

Wang, J. Y., Gitura, R. W., and Wang, Q. F. (2005). Ecology and conservation of the endangered quillwort *Isoetes sinensis* in China. *Journal of Natural History*, **39**, 4069–4079.

Whittier, D. P. (1966). The influence of growth substances on the induction of apogamy in *Pteridium* gametophytes. *American Journal of Botany*, **53**, 882–886.

Whittier, D. P. and Steeves, T. A. (1960). The induction of apogamy in the bracken fern. *Canadian Journal of Botany*, **40**, 1525–1531.

Wild, M. and Gagnon, D. (2005). Does lack of available suitable habitat explain the patchy distributions of rare calcicole fern species? *Ecography*, **28**, 191–196.

Wolf, P. G., Haufler, C. H., and Sheffield, E. (1988). Electrophoretic variation and mating system of the clonal weed *Pteridium aquilinum* (L.) Kuhn (Bracken). *Evolution*, **42**, 1350–1355.

Wolf, P. G., Sheffield, E., and Haufler, C. H. (1990). Genetic attributes of bracken as revealed by enzyme electrophoresis. In *Bracken Biology and Management*, ed. J. A. Thomson and R. T. Smith. Hawthorn, Victoria: Australian Institute of Agriculture and Science, pp. 71–78.

Wolf, P. G., Sheffield, E., and Haufler, C. H. (1991). Estimates of gene flow, genetic substructure and population heterogeneity in bracken (*Pteridium aquilinum*). *Biological Journal of the Linnean Society*, **42**, 407–423.

Wynn, J. M. (2002). Factors contributing to the regeneration of bracken (*Pteridium aquilinum* (L.) Kuhn by spores. Unpublished Ph.D. Thesis, University of Manchester.

Wynn, J. M., Small, J. L., Pakeman, R. J., and Sheffield, E. (2000). An assessment of genetic and environmental effects on sporangial development in bracken [*Pteridium aquilinum* (L.) Kuhn] using a novel quantitative method. *Annals of Botany*, **85** (Suppl. 2), 113–115.

3

Meristem organization and organ diversity

RYOKO IMAICHI

3.1 Introduction

Vascular plants are classified into two groups, microphyllous lycophytes and megaphyllous euphyllophytes (ferns including whisk ferns and horsetails, and seed plants). This classification is based on comparative morphology (Kenrick and Crane, 1997), and it is consistent with recent molecular phylogenetic analyses (Qiu and Palmer, 1999; Pryer *et al.*, 2001; 2004, Qiu *et al.*, 2006; see Chapter 15). Based on this classification, it seems likely that the stem, the leaf, and the root evolved independently in both plant groups. In addition, the root-producing organs called the rhizophore and rhizomorph have evolved only in lycophytes (Kato and Imaichi, 1997). The evolutionary origins of these organs have been proposed mainly based on comparative morphology and anatomy of extant as well as fossil plants (Gifford and Foster, 1989; Stewart and Rothwell, 1993).

Each organ develops through a series of individual morphogenetic events (ontogeny), so attention should be and has been focused on the role of developmental changes during evolutionary diversification. If the ontogeny of a given organ is modified by addition or deletion of specific morphogenetic events to or from the original morphogenetic series, or is modified by alteration of the timing of morphogenetic events, such as retardation or acceleration (heterochrony), the final organ shape could change, leading to evolution of specialized and novel organs (e.g., Gould, 1977; Kluge, 1988; Imaichi and Kato, 1992). Therefore, comparison of morphogenetic events (development) among organs in the context of

Biology and Evolution of Ferns and Lycophytes, ed. Tom A. Ranker and Christopher H. Haufler. Published by Cambridge University Press. © Cambridge University Press 2008.

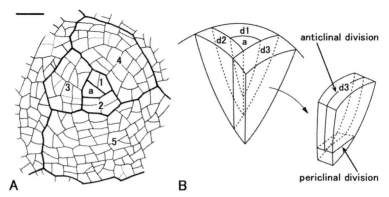

Figure 3.1 Diagrams of a shoot apical meristem. (A) Surface view of *Angiopteris lygodiifolia* SAM (modified from Imaichi, 1986). (B) Three-dimensional view to show apical segmentation. The apical cell (a) produces derivative cells (d1–d3) or merophytes (1–4, outlined by heavy lines) in a spiral. Scale bar 100 μm for (A).

phylogeny should offer clues for improving the evolutionary scenarios of certain organs.

The meristem, which is a self-perpetuating tissue, is the most striking feature of the plant body. The apical meristem is responsible for the indeterminate growth of stems and roots, and the initiation of leaves, and it is involved in shoot and root branching. The marginal meristem contributes to leaf lamina growth. The intercalary meristem is necessary for growth of stems and leaves of some monocots. Through the action of these meristems, all plant organs are formed. Therefore, consideration of meristem organization is a prerequisite for understanding organ evolution. Comparative development focusing on meristem behavior across ferns and lycophytes can help to clarify the evolution and demarcation of a variety of organs in early vascular plants.

3.2   Stem

3.2.1   *Organization of shoot apical meristems of ferns and lycophytes*

Shoot apical meristems (SAMs) of ferns, including whisk ferns and horsetails, are distinguished by a prominent apical cell that functions as the single initial cell for the entire shoot (Figure 3.1, Hébant-Mauri, 1975; Bierhorst, 1977; Imaichi, 1986). They contrast with seed plant SAMs, which have plural apical initial cells instead (Steeves and Sussex, 1989; Buvat, 1989; Lyndon, 1998). The fern apical cell is usually tetrahedral (the exception is *Pteridium aquilinum*, in which it is lenticular; Ogura, 1972), and cuts off derivatives regularly from three lateral faces following clockwise or counterclockwise helical sequences (Figure 3.1). The

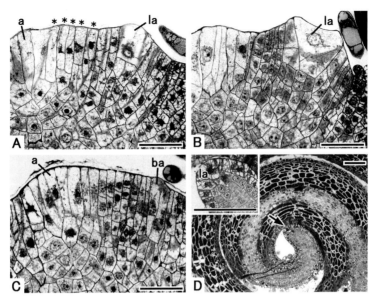

Figure 3.2 Leaf and bud development in *Dicranopteris dichotoma* (A–C) and *Oleandra pistillaris* (D). All longitudinal sections. (A), (B) SAMs with just initiated (A) and somewhat developed (B) leaf primordia. Asterisks in (A) indicate elongated surface cells called prismatic cells. (C) SAM with just initiated lateral shoot (bud). (D) Well-developed leaf with apical crosier (arrow). The inset shows an enlarged apical portion of the leaf. a, shoot apical cell; ba, bud apical cell; la, leaf apical cell. Scale bar 50 μm for (A)–(C), 100 μm for (D).

regularity of the segmentation pattern of the apical cell was also confirmed by detecting differences in cellulose alignment when observed with polarized light (Lintilhac and Green, 1976). Each immediate derivative proliferates by periclinal and anticlinal divisions to form cell packets called merophytes or segments (Figure 3.1). These periclinal divisions usually take place unequally, resulting in a SAM configuration with elongate cells called prismatic cells at its surface (Figure 3.2A).

Unlike single-type fern SAMs, microphyllous lycophytes show two contrasting types of SAM structure: Selaginellaceae SAMs with a single apical cell (Figure 3.3A), and Lycopodiaceae and Isoëtaceae SAMs with plural initial cells (Figures 3.3B, 3.7C). Although Selaginellaceae SAMs are generally considered to be like those of ferns, their apical cells actually differ from fern SAMs in the narrow shape with two to five cutting faces (Siegert, 1974; Hagemann, 1980; Dengler, 1983; Imaichi and Kato, 1989, 1991). The plural initial cells of the Lycopodiaceae and Isoëtaceae SAMs have been interpreted either as a special type of SAM that differs from those of seed plants (Popham, 1951, 1960; Klekowski, 1988),

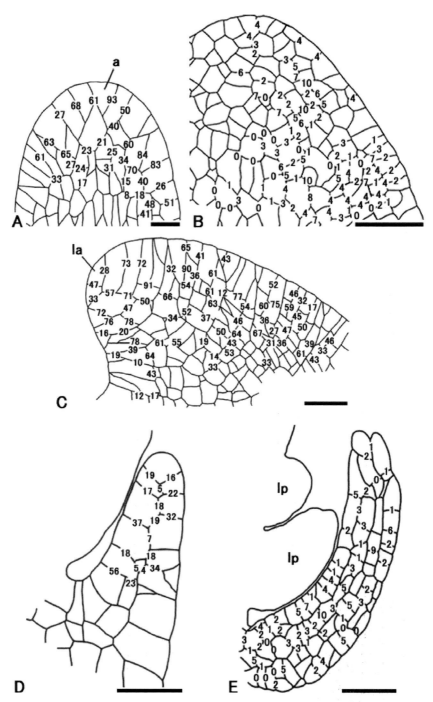

Figure 3.3 Line drawings of cell walls of SAMs (A, B) and leaf primordia (C–E). Numbers in the middle of walls indicate plasmodesmatal (PD) numbers per 1 μm^2 of the counted wall. (A) *Selaginella martensii* with single apical cell (a). (B) *Lycopodiella cernua* with plural apical initial cells. (C) *Hypolepis punctata* with a single leaf apical cell (la). (D) *Selaginella martensii*. (E) *Lycopodium clavatum*. a, shoot apical cell; la, leaf apical cell; lp, leaf primordium. Scale bar 20 µm for (A), 50 µm for (B) and (C), 10 µm for (D) and (E).

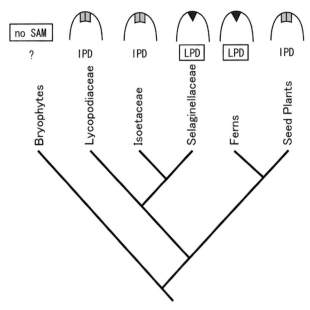

Figure 3.4 Phylogenetic tree of land plants with SAM structures and two character states of plasmodesmatal network in SAMs (modified from Imaichi and Hiratsuka, 2007). IPD, interface-specific plasmodesmatal network; LPD, lineage-specific plasmodesmatal network. Triangles and rectangles in the SAM diagrams indicate the apical cell and apical initial cells, respectively.

or as typical seed-plant SAMs (Newman, 1965; Freeberg and Wetmore, 1967; Stevenson, 1976b; Philipson, 1990). Whether the two types of lycophyte SAMs are morphologically comparable to those of ferns and seed plants should be investigated.

Cooke et al. (1996) reported that fern SAMs with single apical cells and seed-plant SAMs with plural apical initial cells have different plasmodesmatal (PD) networks – lineage specific (LPD) versus interface specific (IPD). The LPD is characterized by high PD densities (PD numbers per unit area in cell walls of the SAM), and the IPD is characterized by low PD densities. There is no transition type between LPD and IPD. Recently, Imaichi and Hiratsuka (2007) compared PD networks of SAMs for 17 families and 24 species, including microphyllous lycophytes, and concluded that there is a strong correlation between SAM structure (single versus plural apical initial cells) and PD network (LPD versus IPD). Surprisingly, among microphyllous lycophytes, Selaginellaceae SAMs with a single apical cell show LPD (fern type), whereas Isoëtaceae and Lycopodiaceae SAMs with plural initial cells show IPD (seed-plant type) (Figure 3.4, Imaichi and Hiratsuka, 2007). Lycophyte SAMs with plural initial cells are comparable to seed-plant SAMs in meristem configuration and PD network. Although vascular-plant

SAMs have been traditionally classified into three (Newman, 1965) or five (Popham, 1951, 1960) types, depending on the number and position of initial cells and the directions in which they divide, now they can be classified into two in terms of PD network, i.e., apical cell based and apical initial cell based SAMs.

3.2.2 Apical cell versus plural initial cells

What is the essential difference between SAMs with single and plural apical initial cells? The role of the apical cell as the single initial cell was once in doubt, based on experimental and quantitative analyses, i.e., low mitotic activity and high DNA content (endopolyploidy), but now we have reached the general explanation that the mitotic activity and DNA content of apical cells differ depending on species and shoot age (e.g., Gifford, 1983; Lyndon, 1998). In conclusion, the apical cell of fern SAMs functions as an initial or founder cell, although it divides more slowly than adjacent cells.

Some authors advise describing fern and lycophyte SAMs on a zonation basis, based on recognition of cytohistochemical zones similar to those of seed-plant SAMs, ignoring whether the SAMs have single or plural initial cells (Stevenson, 1976a, 1976b; Steeves and Sussex, 1989). In the zonal concept, the apical cell or apical initial cells are regarded as one or some of the distinctive cells that organize the initiating region or the promeristem (the apical initial cells and their most recent derivatives, *sensu* Sussex and Steeves, 1967) of the apical meristem (McAlpin and White, 1974). Using this concept, the apical cell or apical initial cells need not be distinguished from adjacent cells in the promeristem.

Recent molecular developmental genetic analyses identified two cell groups from the promeristem of the model eudicot, *Arabidopsis thaliana*: (1) a stem cell population (apical initial cells) expressing the *CLAVATA3* gene, and (2) the underlying organizing center (OC) expressing the *WUSCHEL (WUS)* gene (Bowman and Eshed, 2000). A balanced population of stem cells is maintained by an autoregulatory feedback loop between *WUS* and *CLV3* genes. Some questions remain to be answered. Do fern SAMs with single apical cells have a similar feedback loop mechanism? If so, is stem cell number the only difference between SAMs with an apical cell and apical initial cells? Do lycophyte SAMs with plural apical initial cells (seed-plant type) have the autoregulatory feedback loop found in *A. thaliana*? So far, there have been no studies on the regulation system of the apical cell or apical initial cells in either fern or lycophyte SAMs. Recent genomic analyses showed that *Selaginella* species have *CLV3* homologous genes (Floyd and Bowman, 2007), but it is still unknown whether these genes are involved in regulation of the stem cell population. Comparison of the molecular mechanism maintaining stem cell populations is required across vascular plants, especially ferns and lycophytes, to understand better the differences in SAMs with single or plural apical initial cells.

3.2.3 Evolution of organization of the shoot apical meristem

Based on expected mutation rates in initial cells, it has been assumed that SAMs with plural initial cells probably have an advantage over SAMs with a single apical cell, because the increased number of initial cells can result in increased somatic mutation buffering when mutation rates are high (Klekowski, 1988). Phylogenetically, it is also assumed that SAMs with the single apical cell are primitive and SAMs with plural initial cells evolved from them, because the bryophyte sister to vascular plants commonly bears a single apical cell, though in the gametophyte generation (Mishler and Churchill, 1984; Kato and Imaichi, 1997). Bryophyte sporophytes have an unbranched axial body with no SAMs, but they have either an apical cell or no apical cell when young, depending on the class (Schuster, 1984; Crum, 2001). Although it remains an open question whether the bryophyte apical cell is homologous to that of pteridophytes (Friedman et al., 2004), it is possible that the sporophytes of ancestors of lycophytes and euphyllophytes (ferns and seed plants, Pryer et al., 2004) recruited the bryophyte apical cell (Kato and Akiyama, 2005). If this is the case, SAMs with plural initial cells evolved in lycophytes and seed plants independently in different clades (Figure 3.4).

Fossil evidence is quite limited, but the apex of the Devonian fossil vascular plant, *Rhynia gwynn-vaughanii*, appears to have several rectangular merophytes in place of a single apical cell (Edwards, 1994). This is not congruent with the general account that the apical cell based SAM is more primitive. Taking into account PD network traits along with the vascular plant phylogeny, an alternative evolutionary scenario seems equally plausible for SAM evolution – the apical cell based SAMs with LPD were derived from IPD SAMs with plural initial cells by reduction in initial cell numbers associated with loss of secondary PD networks (Figure 3.4). Ferns with LPD SAMs have only primary PD networks, which are formed in cell plates during cytokinesis, while seed-plant IPD SAMs can form secondary PD networks in addition to primary PD networks, the former of which are inserted into pre-existing walls when the walls expand during development (van der Schoot and Rinne, 1999). If secondary PD networks were lost, the need for all adjacent cells to communicate via their PD networks would exert a strong selection pressure for maintaining single apical cells in plants that only have primary PD networks (Cooke et al., 1996). It is difficult to distinguish secondary from primary PD networks, but the elegant work by Gunning (1978) demonstrated the absence of secondary PD networks in the fern genus *Azolla*. We must clarify whether PD networks can develop secondarily in pre-existing

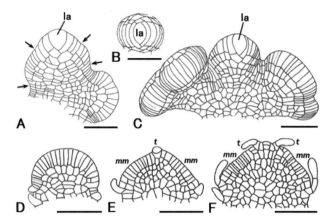

Figure 3.5 Drawings of leaf lamina development of *Lygodium japonicum* (modified from Mueller, 1982a). (A)–(C) Leaf apical cell (la) and pinna initiation (between arrows), and pinna formation by marginal meristem. (D)–(F) Lobe formation of the first leaf. la, leaf apical cell; t, trichome; mm, marginal meristem. Scale bar 100 μm.

walls in Lycopodiaceae and Isoëtaceae with IPD SAMs. Furthermore, comparative analyses of bryophyte PD networks in a phylogenetic framework are crucial to clarify the evolution of various SAM structures.

3.3 Leaf

Megaphylls and microphylls are generally thought to have evolved independently in association with the evolution of stems. Megaphylls are simple or compound with branched vascular bundles and have a small parenchymatous portion (leaf gap) at the divergence point of the leaf trace from the stem bundle. Microphylls are scaly or needle-like with unbranched vascular bundles and no leaf gap. These two leaf types differ markedly in development, suggesting different origins and evolutionary pathways (Gifford and Foster, 1989).

3.3.1 *Megaphyll development in ferns (euphyllous)*

At inception, the leaf primordium arises from the apical flank of the SAM as a small mound (leaf buttress) comprising a group of surface (prismatic) and subsurface cells. Soon after, a single surface cell becomes enlarged and undergoes oblique divisions to cut off a single leaf apical cell in the leaf buttress (Figure 3.2A, B; Imaichi, 1988). The leaf apical cell is commonly lenticular with two cutting faces (Figure 3.5B; Hébant-Mauri, 1975; Bierhorst, 1977; Imaichi, 1982; Lee, 1989), and rarely with three cutting faces in some genera (i.e., *Angiopteris, Osmunda, Botrychium*) (Guttenberg, 1966; Imaichi and Nishida,

1986). Leaf apices with the single leaf apical cell and its immediate derivatives are comparable to the SAM in organization, so they are sometimes designated leaf apical meristems (LAMs). The LAM also resembles the SAM in having high plasmodesmatal densities equivalent to the SAM (cf. Figure 3.3C in this chapter with Figure 3A in Imaichi and Hiratsuka, 2007). The long retention of the active LAM in the leaf primordium is responsible for the prolonged leaf apical growth, resulting in the coiled crosier (Figure 3.2D). Surprisingly, in climbing leaves of some ferns, e.g., *Lygodium japonicum*, the leaf apical cell is maintained permanently at the apex of indeterminate adult leaves several meters long (Figure 3.5A, C; Mueller, 1982b). Long retention of LAMs causes acropetal tissue differentiation in leaves, as shown in the stem terminated by the SAM.

During or after the period of LAM activity, the marginal meristem or the marginal blastozone (*sensu* Hagemann & Gleissberg, 1996) is formed next to the LAM (Figure 3.5; Mueller, 1982b; Hagemann, 1984). In many ferns, lamina growth is restricted to the marginal meristem with the youngest lamina portion in the margins. The marginal meristem is fractionated to form the pinnae and pinnules, and at the final stage, the marginal meristem continues to grow without fractionation (Hagemann, 1984). Although it is still unclear what underlying mechanisms lead to fractionation of the marginal meristem, when the lamina is bilobed as in *Lygodium japonica*, the marginal meristem is divided by the cessation of cell proliferation in its middle portion (Figure 3.5D, E; Mueller, 1982b).

3.3.2 Microphyll development in lycophytes

Like fern leaves, lycophyte microphylls arise from the apical flank of SAMs, but from a lower site than the leaf primordium in fern SAMs (Figures 3.6A, 3.7A; Freeberg and Wetmore, 1967; Dengler, 1983). The most remarkable feature of microphyll development is the "lack of both the LAM and the marginal meristem," resulting in basipetal tissue differentiation (Dengler, 1983). Replica SEM observations on a growing *Selaginella* leaf clearly show features common to microphylls (Figure 3.6): (1) the leaf initiation involves several dermal cells arranged in a horizontal line to form a plate-like protrusion, (2) there are neither apical meristems nor marginal meristems, and (3)ced lamina expansion is attributed to cell proliferation over the entire leaf primordium. Leaf development of Isoëtaceae and Lycopodiaceae has received little attention since the review by Guttenberg (1966). Preliminary examination shows that the leaf primordia of *Lycopodium* (Figure 3.7B, C) and *Isoëtes* species with needle-like leaves are also plate-like at inception (R. Imaichi, unpublished data). The leaf primordia of Selaginellaceae have lower PD densities than SAMs, without high PD densities in the apical portions (cf. Figure 3.3A, D). Lycopodiaceae leaf primordia have very low PD densities identical to SAMs (Figure 3.3B, E).

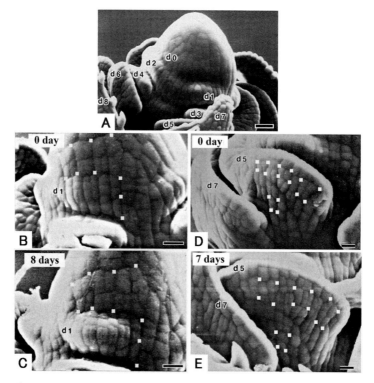

Figure 3.6 SEM images of microphyll development of *Selaginella martensii*. (A) Dorsal view of SAM with growing dorsal leaf primordia (d0–d8). Small white rectangles indicate given positions of growing leaves to help show cell proliferation. (B), (C) Development of first dorsal leaf (d1). (C) Image taken 8 days after (B). (D), (E) Development of fifth youngest leaf primordium (d5). (E) Image taken 7 days after (D). Scale bar 50 μm for (A), 20 μm for (B)–(E).

3.3.3 Origin and evolution of megaphylls and microphylls

The megaphyll is assumed to have evolved from three-dimensional branched axes (telomes) via three events: (1) overtopping to form main and lateral axes, (2) planation to form branched lateral axes into a single plane, and (3) webbing to form the mesophyll (Zimmerman, 1952, cited in Stewart and Rothwell, 1993). Fern SAMs develop LAMs from their apical flanks, both of which have equivalent single apical cells. In some fern genera with SAMs having plural apical initial cells when old, the LAMs have plural leaf apical initial cells instead of the single leaf apical cell (e.g., *Angiopteris*, Guttenberg, 1966; Ogura, 1972). The strong similarity in meristem configuration and branching (division) manner noted above between the LAM and SAM, as well as the longevity of the LAM in fern leaves, is consistent with the evolutionary hypothesis that the megaphyll and the stem originated similarly from telomic axes having apical

Meristem organization and organ diversity 85

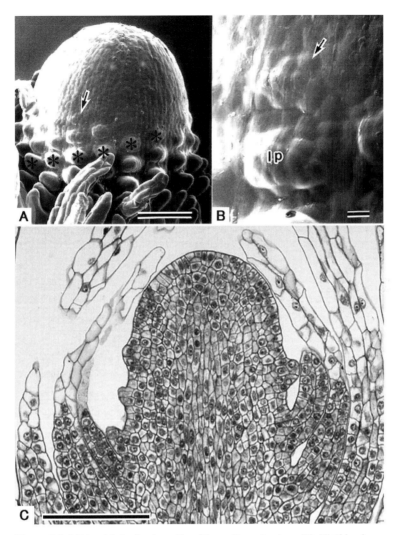

Figure 3.7 SAMs with leaf primordia of *Lycopodium clavatum*. (A), (B) Side view. (B) Enlarged image of a just initiating leaf primordium (arrow). Asterisks show parastichy. (C) Median longitudinal section. lp, leaf primordium. Scale bar 100 μm for (A) and (C), 10 μm for (B).

meristems (Zimmerman, 1959, cited in Stewart and Rothwell, 1993). However, considering the longevity of fern leaves, there is another interpretation: it is a rather specialized character of fern leaves, which was later acquired under strong selection to a vining growth habit (Kaplan and Groff, 1995).

The question about how "webbing" occurred remains to be solved. Hagemann (1984) stressed that the ability of marginal meristems to spread out by incorporating neighboring meristematic tissues, which is commonly found in angiosperm leaf development, might have played a key role in webbing. However, fern leaves have no such characters. Accordingly, Hagemann (1984) developed a strong argument against the telome theory of Zimmermann (1959, cited in Stewart and Rothwell, 1993), because there is no real basis for what he called "webbing" of telome-like entities (Hagemann and Gleissberg, 1996).

There have been three competing hypotheses to explain microphyll evolution: (1) reduction from telomes, (2) enation (lateral outgrowth) on telomes, and (3) sterilized sporangia. Each is critically dependent on putative microphyll homologies (Kenrick and Crane, 1997). The above developmental data do not support the telome reduction theory and seem to favor the enation theory over the sterilized sporangium theory. Developmental molecular genetic analyses are also consistent with the enation hypothesis (Floyd and Bowman, 2006). However, it is still difficult to deny the sterilized sporangium hypothesis owing to scanty molecular genetics and comparative developmental data for Lycopodiaceae sporangia.

Recent molecular genetic analyses suggest that KMOX–ARP interactions regulate the balance between indeterminate and determinate growth of leaves. In model angiosperm taxa such as *Arabidopsis*, *KNOX* genes are expressed in SAMs to maintain their indeterminacy, and the *ARP* genes in leaf initiation sites, P0, to maintain the *KNOX*-off state (e.g. Floyd and Bowman, 2007). Harrison *et al.* (2005) showed that KNOX and ARP proteins are overlapped in *Osmunda* megaphyll primordia, and claimed that co-localization of KNOX and ARP may reflect the delayed determinancy of fern leaves. On the other hand, *KNOX and ARP* genes are expressed in SAMs and microphyll primordia in *Selaginella*, respectively (Harrison *et al.*, 2005). It is interesting that the megaphyll and the microphyll, whose evolutionary origins are different from each other, similarly have KNOX-ARP interaction mechanisms.

3.4 Shoot branching, dichotomous versus monopodial

3.4.1 *Ferns*

Shoots of ferns usually branch dichotomously or laterally (monopodially); however, tree ferns, such as *Dicksonia* (Hébant-Mauri, 1975) and ferns with massive stems, such as *Angiopteris* (Bower, 1923), show no branching at all. In dichotomous branching, a shoot is bifurcated into two equal axes; in lateral branching, a shoot is divided into strong and weak axes, the latter of which sometimes remain dormant as lateral buds (Figure 3.8). From the perspective of gross morphology, there is a sharp distinction between dichotomous and lateral branching in relation to leaf insertions; dichotomous branching occurs in no spatial relation to the leaf inserted in the stem, whereas lateral branches

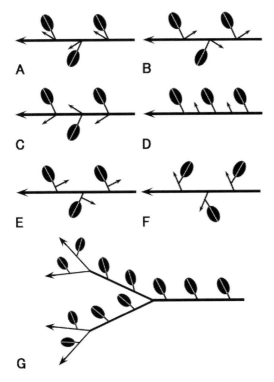

Figure 3.8 Diagrams of various shoot branching patterns of ferns. (A) Axillary branching. (B)–(D) Extra-axillary branching. (E) Epipetiolar branching. (F) Leaves arising from short shoots. (G) Dichotomous branching.

are in close spatial proximity to the leaf sites mentioned below. In other words, there is no correlation between the phyllotaxis and branch taxis in dichotomizing shoots, although several patterns of combination of both taxis types are found in shoots with lateral branchings.

Fern lateral buds show various insertion sites: (1) axillary, (2) extra-axillary, (3) interfoliar (alternate with leaves), and (4) epipetiolar. This is in marked contrast to angiosperm lateral buds, which arise exclusively from leaf axils (Esau, 1977). Axillary buds typical of seed-plant shoots are uncommon in ferns and are restricted to the Hymenophyllaceae (Figure 3.8A; Hébant-Mauri, 1984, 1990) and *Helminthostachys* (Ophioglossaceae) (Kato et al., 1988). Extra-axillary buds occur near each leaf base from their lateral or abaxial sites (Figure 3.8B, C; *Diplopterygium*, Soma, 1966; *Histiopteris*, Imaichi, 1980; *Davallia*, Croxdale, 1976; *Stromatopteris*, Hébant-Mauri and Veillon, 1989; *Lomagramma*, Hébant-Mauri and Gay, 1993). Interfoliar buds are independent of leaves and are arranged on the stem as alternating leaves in different orthostichies between leaves and buds (*Microgramma*, Hirsh and Kaplan, 1974) or in one orthostichy of leaves and buds

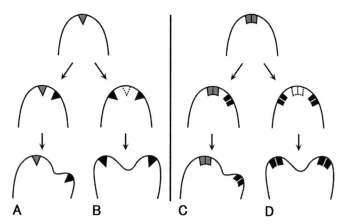

Figure 3.9 Diagrams showing two patterns of shoot branching in SAMs with a single apical cell (A, B) and with plural apical initial cells (C, D). (A), (C) Lateral branching. (B), (D) Dichotomous branching.

(Figure 3.8D; *Dicranopteris*, Soma, 1966). Epipetiolar buds arise from somewhat upper portions of the petiole, and are mostly found in dennstaedtioid ferns (Figure 3.8E; Imaichi, 1980; Troop and Mickel, 1968). In an extreme case, *Pteridium* (Dasanayake, 1960) and *Hypolepis* (Imaichi, 1982, 1983) appear to have leaves on the short (lateral) shoots, not on long (main) shoots (Figure 3.8F). It is noteworthy that each species has either dichotomous or lateral branching systems, but not both in an individual. In contrast, shoots of *Dicranopteris nitida* (Gleicheniaceae) show both types of branching (dichotomous and lateral) in one individual (Hagemann and Schulz, 1978).

Developmentally, all lateral buds arise, as do leaf primordia, from the apical flank of the SAM, involving a group of surface prismatic and subsurface cells (Figure 3.2C). The bud apical cell, although tetrahedral, is cut off by oblique divisions from one prismatic cell, although its formation is often delayed when the bud is dormant (Hébant-Mauri and Gay, 1993). Wardlaw (1946) claimed that the bud meristem was a consequence of extreme unequal division of the SAM and called it a detached meristem. Epipetiolar buds in dennstaedtioid ferns were once considered adventitious foliar buds (Troop and Mickel, 1968), but they are initiated at the base of the leaf primordium in the shoot tip (Imaichi, 1980, 1982, 1983). In horsetails, bud meristems arise from the extra-axillary position, i.e., alternate to leaf primordia, as lateral buds (Frankenhäuser, 1987). It is clear that all lateral buds in ferns, including horsetails, form in a similar manner irrespective of the various sites; the original apical cell is retained and a new apical cell of the lateral branch is formed at the apical flank (Figure 3.9A; Hébant-Mauri, 1993).

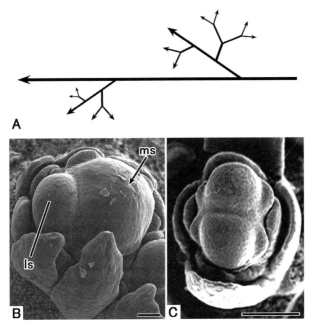

Figure 3.10 Shoot branching pattern of *Lycopodium complanatum*. (A) Diagram showing unequal and equal branching. (B), (C) SEM images of shoot tips. The shoot apex divides unequally (B) and equally (C). ls, lateral shoot apex; ms, main shoot apex. Scale bar 50 μm. (Courtesy of Ikumi Honma.)

Dichotomous branching occurs in quite a different manner from lateral branching in fern SAMs. It had been interpreted as forming either by (1) equal division of the shoot apical cell itself (Bierhorst, 1977), or (2) new formation of two apical cells after cessation of the original apical cell (Kato and Imaichi, 1997, and references therein). To date, careful examination suggests that the latter manner is most common (Figure 3.9B; Mueller, 1982a; Imaichi, 1984). In conclusion, the most striking morphological difference between dichotomous and lateral branchings is whether the apical cell disappears and is replaced by two new apical cells or whether the original apical cell is retained as is.

3.4.2 Lycophytes

Lycophyte shoots with microphylls branch either equally or unequally depending on plant group, like fern shoots. Interestingly, in some Lycopodiaceae species, equal and unequal branching occurs in one individual (Figure 3.10A; e.g., *Lycopodium complanatum*, *L. tristachyum*) (Guttenberg, 1966). This is a remarkable contrast to the fern shoots that show dichotomous or lateral branching depending on species. Since Troll's (1937) definition, equal and unequal branching modes in lycophytes have been traditionally called isotomous or anisotomous dichotomy, respectively. Isotomy occurs by equal division of the shoot apex, and anisotomy results from unequal growth of a once isotomously bifurcated shoot apex (Troll, 1941).

The developmental anatomy of shoot branching in lycophytes has been poorly examined, but available data are not in accordance with Troll's anisotomy; unequal branching is not produced by unequal growth of once equally divided SAMs, but by unequal division of SAMs. In unequal branching in Selaginellaceae shoots with the apical cell based SAM, lateral shoot primordia arise from the apical flank of the SAM retaining the original apical cell, and the lateral apical cell forms from one of the primordium surface cells (Siegert, 1974; Imaichi and Kato, 1989). This developmental manner is identical to lateral shoot formation in ferns (Figure 3.9A). In unequal branching of Lycopodiaceae SAMs with plural initial cells, the original (main) SAM appear to be retained and new SAMs for lateral shoots form from the apical flank (Figures 3.9C, 3.10B). In equal branching of Lycopodiaceae shoots, two new SAMs appear to be formed after the cessation of the original SAM in equal branching (Figures 3.9D, 3.10C; Guttenberg, 1966). Such unequal and equal bifurcation in lycophyte SAMs are similar morphologically to the lateral and dichotomous branching in fern SAMs, respectively, regardless of the different meristem organizations (Figure 3.9). In the Isoëtaceae, *Stylites* shows apparently equal branching, but *Isoëtes* does not branch (Bierhorst, 1977). There are no developmental data on *Stylites*. In conclusion, there appears to be no difference in meristem branching behavior between the megaphyllous fern and microphyllous lycophyte SAMs. Stems of both the plant groups may be comparable to each other.

3.5 Roots

The root is an important organ for anchoring the aerial shoot in the soil and absorbing inorganic nutrients from it. Comparative anatomy and fossil evidence combined with phylogenetic mapping suggest that the root evolved at least twice – once each within the lycophytes and euphyllophytes (ferns and seed plants) (Friedman *et al.*, 2004, and references therein). Roots of ferns and lycophytes have similar adventitious and endogenous origins in stems or special root-producing organs, with poorly developed embryonic roots. The most prominent trait characterizing fern and lycophyte roots is whether the root branches laterally (monopodially) or dichotomously (Figure 3.12 below; Gifford and Foster, 1989).

3.5.1 *Initiation and branching of fern roots*

The root apical meristem (RAM) of ferns originates in shoot tips near the procambium, grows through the stem cortex, at last exiting through the

Meristem organization and organ diversity 91

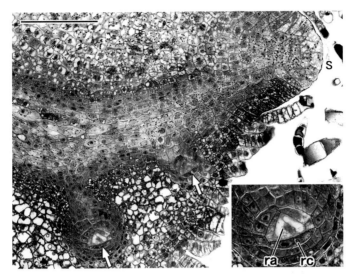

Figure 3.11 Longitudinal section of a shoot tip of *Hypolepis punctata*, showing root initiation. White arrows indicate endogenous root primordia. The inset shows an enlarged figure of the root apical meristem (RAM). s, SAM; ra, root apical cell; rc, root cap. Scale bar 200 μm.

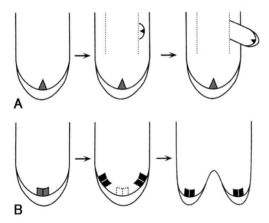

Figure 3.12 Diagrams showing two patterns of root initiation and branching. (A) Endogenous origin and lateral branching in megaphyllous euphyllophytes. (B) Exogenous and terminal branching in microphyll lycophytes. (Modified from Kato and Imaichi 1997.)

stem epidermis (Figure 3.11; Stevenson, 1976c). The main root produces lateral roots below the root tip, with new RAMs initiating from endodermal cells of the vascular bundles (Figure 3.12A; Ogura, 1972; Barlow, 2002). Fern RAMs have tetrahedral single apical cells that regularly cut off derivatives on four faces with the basal derivative (root cap cells) facing outwards. The RAMs of Marattiaceae and

Osmundaceae – in which the SAMs sometimes have no apparent apical cells – are distinctive in having no discernible apical cells (Guttenberg, 1966). Determining whether the root apical cell is mitotically active or inactive has been a controversial topic (Gifford, 1983). The central question is whether the root apical cell is equivalent to the mitotically inactive "quiescent center" (QC) found commonly in angiosperm roots (Clowes, 1961). The most prominent hypothesis at present is that fern RAMs have no QC, and the QC evolved later in the seed-plant clade (Barlow et al., 2004).

There have been few reports on PD networks only for fern RAMs. *Azolla* and *Dryopteris* have high PD densities (Cooke et al., 1996) comparable to those of fern SAMs, suggesting a strong correlation between RAM and SAM structures and PD networks in ferns.

3.5.2 Lycophyte root apical meristems and their branching

Lycophyte RAMs show a greater diversity of apical organization than fern RAMs. RAMs show layered (Isoëtaceae, Figure 3.13C) or non-layered (Selaginellaceae) structures (Guttenberg, 1966; Ogura, 1972; Yi and Kato, 2001). Lycopodiaceae RAMs were interpreted as having the layered structure (Ogura, 1972), but some actually show a non-layered structure with a mass of initial cells (Figure 3.13B). In Selaginellaceae, RAMs possess no apparent apical cell (Guttenberg, 1966), or have single apical cell (Imaichi and Kato, 1989).

All Selaginellaceae and Isoëtaceae roots branch dichotomously, whereas Lycopodiaceae roots superficially show both dichotomous and unequal branching (Figure 3.13D). RAM division in lycophytes has been poorly examined, with only one recent paper on Isoëtacean roots (Yi and Kato, 2001). RAMs are divided into two because of a small intervening group of non-meristematic cells derived from apical initial cells of the outer layer of the RAM by periclinal divisions, and two new RAMs are formed on either side of these non-meristematic cells (Figure 3.12C, D). These non-meristematic cells are similar to "pavement cells" that intervene to split two dividing apical meristems of the shoot (Barlow et al., 2004). In this sense, RAMs have a similar branching mode to SAMs, that is, the apex ceases meristematic activity and is replaced by two new apices.

In contrast, information about RAM meristem behavior in unequal branching is totally lacking. The question is whether (1) the original RAM is retained as is, and lateral small RAMs newly form in the flank (comparable to the unequal branching of the shoot apex), or (2) the meristem is replaced by newly formed RAMs of unequal size. Research to clarify the meristem behavior in RAM branching is urgently needed.

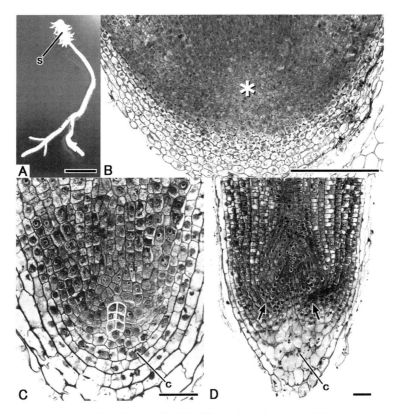

Figure 3.13 Root branching and RAM of *Lycopodium clavatum* (A, B) and dichotomously branching roots of *Isoëtes asiatica* (C, D, modified from Yi and Kato, 2001). (A) Equally and unequally branched roots arising from a stem segment. (B) Median longitudinal section of a RAM with a mass of lightly stained initial cells (asterisk). (C) Incipient root branching with two meristem groups separated by two files of three cells (demarcated by white lines). (D) Two newly formed RAMs (arrows) are still covered by the original root cap. c, root cap; s, stem segment. Scale bar 1 cm for (A), 200 μm for (B), 50 μm for (C) and (D).

3.5.3 Evolution of roots

Nothing is known about the evolutionary origin of roots. Recent molecular genetics shows that angiosperm RAMs and SAMs are controlled by similar mechanisms to maintain stem cell populations, suggesting that roots may be derived from a developmental program associated with the SAM (references cited in Friedman *et al.*, 2004), but genetic data are lacking for fern and lycophyte roots. Roots are similarly endogenous in ferns and lycophytes, regardless of whether they branch endogenously (ferns) or exogenously (lycophytes). Of particular importance is determining how the root shifted its initiation site from surface to interior tissue during evolution. Barlow *et al.* (2004) argued that

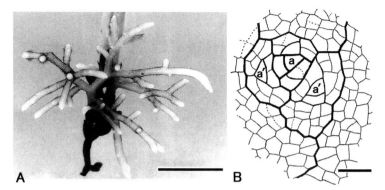

Figure 3.14 Irregularly branched rhizomes (A) and line drawing of a rhizome apex in transectional view (B) of *Psilotum nudum* in hydroponic culture. The original apical cell (a) produces merophytes outlined by heavy lines, and new apical cells (a′) are formed in merophytes. (Modified from Takiguchi *et al.*, 1997.) Scale bar 1 cm for (A), 50 μm for (B).

the evolution of the quiescent center could have resulted in a change of the branching pattern from dichotomous to a so-called herring-bone pattern with endogenous lateral roots; pavement cells intervene to separate two branches of lycophyte roots that may be comparable to the nascent quiescent center. If this is the case, lycophyte roots with dichotomous branching could help clarify the origin of endogenous roots.

3.6 Psilotalean rhizomes (subterranean axes)

The Psilotaceae (whisk ferns) are unique in having no roots throughout their life history. In *Psilotum*, enation-like leaves form in the aerial stems but not in the subterranean rhizomes. There had been controversy concerning whether such a simple body plan represents a primitive character. However, recent molecular phylogenetic analyses showed clearly that members of the Psilotaceae should be classified as ferns (Pryer *et al.*, 2001, 2004; see Chapter 15).

The most intriguing morphological trait is a tangled subterranean rhizome that branches frequently in equal or unequal fashion, with no regularity (Figure 3.14A). This contrasts with the regular dichotomously branching aerial stems. Such irregularity in the rhizome was once argued to be the result of injuries when growing in the soil, caused by the absence of any protecting tissue, such as a root cap (Bierhorst, 1954). However, Takiguchi *et al.* (1997) clarified that the rhizome shows a similar complex branching system when cultured hydroponically without obstacles, suggesting that irregular branching is an inherent feature of the Psilotaceae rhizome.

The Psilotaceae rhizome has an apical cell based meristem, but surprisingly several additional apical cells are found in merophytes of the original apical cell (Figure 3.14B). Because these additional apical cells soon produce their own derivatives, it is often difficult to trace which one is the original among several apical cells. Some apical cells develop arbitrarily as new apical meristems, but some others do not and become inactive. Furthermore there is no rule about which direction and how fast these new meristems grow as lateral branches, resulting in the complex rhizome branching system. It is worth noting that superficially dichotomous branching of Psilotaceae rhizomes is caused by the rapid growth of a lateral branch, not through true dichotomy in apical cell behavior. In contrast, the apical meristem of the aerial stems (SAMs) shows the typical organization of fern SAMs. When bifurcated, the apical cell disappears prior to the formation of two new apical cells. This is similar to dichotomous branching in other fern SAMs (Figure 3.9B). Interestingly, some branches of subterranean rhizomes transform directly into aerial stems by an unknown mechanism.

The complex SAM behavior in Psilotaceae rhizomes was interpreted as a crude character compared to other fern SAMs showing lateral or dichotomous branching, and was regarded as an independent unique organ (Takiguchi *et al.*, 1997). However, the aerial SAM with leaves exhibits regulated apical organization including dichotomous branching, like other fern SAMs. Combined with the fact that the site of the lateral branch is strictly regulated in the fern SAMs mentioned above, the lack of leaves in *Psilotum* rhizomes might facilitate irregular branching as shown in fossil Filicalean ferns (Holmes, 1989) owing to a loss of the constraint provided by the leaf–branch combination.

3.7 Root-producing organs, rhizophores and rhizomorphs

The rhizophore is a leafless, root-producing axial organ unique to the Selaginellaceae (Figure 3.15). It is a historically controversial structure, variously interpreted as a root-producing organ not equivalent to other organs, a transformed stem, or the proximal portion of a root (aerial root) that branches to subterranean roots (Kato and Imaichi, 1997, and references therein). The rhizophore concept has been revised by the work of Imaichi and Kato (1989, 1991) on a temperate *Selaginella* species of moderate size. The rhizophore arises exogenously at each branching point of the lateral shoots. Its apical meristem, which was once called the angle meristem (Jernstedt *et al.*, 1994), has a prominent apical cell with three cutting faces (Figure 3.15B). The apical cell soon disappears (Figure 3.15C) and two new root meristems arise in inner tissues of the rhizophore tip (Figure 3.15D). There is a gap in development between rhizophores

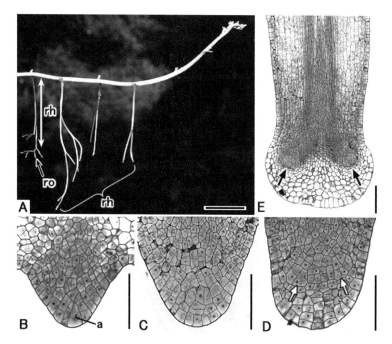

Figure 3.15 Rhizophores of *Selaginella caudata* (large tropical species, A) and *S. kraussiana* (moderately sized species, B–E, longisections). (A) Once to three-times dichotomously branched rhizophores (rh) arising from every branching site of lateral shoots. Roots (ro) arise from tips of rhizophores. (B)–(E) Endogenous root development in rhizophores. The apical cell in a just formed rhizophore (B) disappears soon (C), and two root meristems (arrows) are initiated in the inner tissue of the rhizophore tip (D, E). a, apical cell. Scale bar 5 cm for (A), 50 μm for (B)–(D), 100 μm for (E).

and endogenous roots. Tropical *Selaginella* species with large rhizophores usually branch dichotomously three to five times (Figure 3.15A). The branching manner is identical to that of stem dichotomy: prior to bifurcation the apical cell becomes indistinguishable, ceasing oblique divisions (Figure 3.9B). In conclusion, rhizophores are a root-producing, leafless, capless axial organ that is autonomously and dichotomously branched in large species and depauperately unbranched in small species, such as *S. uncinata* and *S. kraussiana*.

The Isoëtaceae rhizomorph is another root-producing organ. It shows anatomical features common to the Carboniferous Lepidodendrid rhizomorph (Karrfalt, 1984; Stewart and Rothwell, 1993). Roots are formed by the activity of the basal meristem. The basal meristem is an enigma, and has been interpreted as a primary meristem, or as a cambium, or a part of a cambium (Paolillo, 1982, and references therein). However, recent detailed examinations of the initiation and growth of root primordia from the basal meristem, and embryonic roots of

Isoëtes species show that the basal meristem has both organogenetic (primary) and thickening (secondary) meristem attributes (Yi and Kato, 2001). The basal meristem does not correspond to any known meristem in other vascular plants.

Rhizomorphs and rhizophores are sometimes compared, but the *Selaginella* rhizophore and *Isoëtes* rhizomorph show differences in: (1) initiation pattern (exogenous or endogenous origin), (2) growth pattern (definite versus indefinite growth), (3) meristem structure (dome shaped with apical cell versus linear basal meristem with a layer of thin initials), and (4) number of roots produced (two versus many per apex) (Yi and Kato, 2001). The evolutionary relationships of these two root-producing organs, which are unique to the heterosporous ligulate lycopods, need further developmental studies.

3.8 Summary and future goals

Comparative development focusing on meristem behavior has helped define and demarcate plant organs and clarify their origins. It is noteworthy that fern stems with megaphylls (comparable to an exogenous stem branch) have roots that branch endogenously, whereas lycophyte stems with microphylls (not comparable to an exogenous stem branch) have roots that branch exogenously, as the stem does. This may confirm that at least stems and leaves are of telomic origin in ferns, although the origin of fern roots remains an open question. In lycophytes, stems and roots may have been derived from telomic axes, but the leaves were not.

Although developmental studies are still lacking, especially for lycophytes, there seems to be a general rule for behavior of apical meristems of axial organs (stems, roots, *Psilotum* rhizomes, and *Selaginella* rhizophores) when they divide. In dichotomous branching, the original apical cell or apical initial cells disappear and are replaced by two new apical cells or two groups of apical initial cells (Figures 3.9B, D, 3.12B). In lateral branching, the original apical cell or apical initial cells are retained and a new apical cell or initial cells are formed exogenously or endogenously in lateral positions (Figures 3.9A, 3.12A). These equal and unequal divisions are also seen in the megaphyll marginal meristem. The equal division associated with cessation of a central part of the marginal meristem results in the bilobed lamina; the unequal division retaining the original marginal meristem results in lateral pinna formation. From these facts, it is plausible that telomic apical meristems acquired the capacity to divide equally and unequally, before stems, leaves, and roots evolved independently in lycophytes and euphyllophytes. The most mysterious evolutionary event is endogenous branching of the meristem, i.e., root meristem initiation in the stem and root.

Meristem branching is similar in the stem, leaf and root, irrespective of whether the meristem has a single apical cell or plural apical initial cells. This might reinforce the organismal theory over the cell theory (Kaplan and Hagemann, 1991); the organismal theory interprets the living protoplasmic mass as a whole, rather than considering its constituent cells as the basic unit. In the cell theory, cell lineages are more significant, whereas for the organismal theory relative special positions are more significant. However, meristem structures correlate well to PD networks, suggesting that meristems with the single or plural initial cells are under different regulatory systems. Furthermore, it remains to be determined whether apical cell based and plural initial cell based meristems differ only in stem cell numbers or whether there are other essential differences.

Extant comparative developmental data show that lycophytes have greater variation in organ diversity and meristem behavior than do ferns. Nevertheless, developmental information on stems, roots, and root-producing organs in lycophytes is very fragmentary, and there are still no evolutionary hypotheses for their origins. Our preliminary examination suggests that the increase or decrease in meristem size could play an important role in meristem division or organ development in lycophytes. To elucidate the meristem behavior, including a change in size, certain markers such as genes expressed in the meristem itself are necessary. Molecular developmental genetics is very helpful for revealing organ identity, and should be extended to studies of ferns and lycophytes. A combination of developmental morphogenesis and developmental molecular genetics in a phylogenetic framework should yield results.

References

Barlow, P. W. (2002). Cellular patterning in root meristems: its origins and significance. In *Plant Roots, The Hidden Half*, ed. Y. Waisel, A. Eshel, and U. Kafkafi, 3rd edn., New York: Marcel Dekker, pp. 49–82.

Barlow, P. W., Löck, H. B., and Löck, J. (2004). Pathways towards the evolution of a quiescent centre in roots. *Biologia, Bratislava*, **59** (suppl. 13), 21–32.

Bierhorst, D. W. (1954). The subterranean sporophytic axes of *Psilotum nudum*. *American Journal of Botany*, **41**, 732–739.

Bierhorst, D. W. (1977). On the stem apex, leaf initiation and early leaf ontogeny in filicalean ferns. *American Journal of Botany*, **64**, 125–152.

Bower, F. O. (1923). *The Ferns*, Vol. 1, *Analytical Examination of the Criteria of Comparison*. Cambridge: Cambridge University Press.

Bowman, J. L. and Eshed, Y. (2000). Formation and maintenance of the shoot apical meristem. *Trends in Plant Science*, **5**, 110–115.

Buvat, R. (1989). *Ontogeny, Cell Differentiation, and Structure of Vascular Plants*. Heidelberg: Springer-Verlag.

Clowes, F. A. L. (1961). *Apical Meristems*. Oxford: Blackwell Scientific.

Cooke, T. D., Tilney, M. S., and Tilney, L. G. (1996). Plasmodesmatal networks in apical meristems and mature structures: geometric evidence for both primary and secondary formation of plasmodesmata. In *Membranes: Specialized Functions in Plants*, ed. M. Smallwood, J. P. Knox, and D. J. Bowles, pp. 471–488. Oxford: Bios Scientific.

Croxdale, J. G. (1976). Origin and early morphogenesis of lateral buds in the fern *Davallia*. *American Journal of Botany*, 63, 226–238.

Crum, H. A. (2001). *Structural Diversity of Bryophytes*. Ann Arbor, MI: The University of Michigan.

Dasanayake, M. D. (1960). Aspects of morphogenesis in a dorsiventral fern, *Pteridium aquilinum* (L.) Kuhn. *Annals of Botany*, 24, 317–329.

Dengler, N. G. (1983). The developmental basis of anisophylly in *Selaginella martensii*. I. Initiation and morphology of growth. *American Journal of Botany*, 70, 181–192.

Edwards, D. (1994). Towards an understanding of pattern and process in the growth of early vascular plants. In *Shape and Form in Plants and Fungi*, ed. D. S. Ingram and A. Hudson. London: Academic Press, pp. 39–59.

Esau, K. (1977). *Anatomy of Seed Plants*. 2nd edn. New York: Wiley.

Floyd, S. K. and Bowman, J. L. (2006). Distinct developmental mechanisms reflect the independent origins of leaves in vascular plants. *Current Biology*, 16, 1911–1917.

Floyd, S. K. and Bowman, J. L. (2007). The ancestral developmental tool kit of land plants. *International Journal of Plant Sciences*, 168, 1–35.

Frankenhäuser H. von. (1987). Morphogenetische und histogenetische Studien am Vegetationskegel der Equiseten. I. Achsenmeristem und Verzweigung. *Beiträge zur Biologie der Pflanzen*, 62, 369–404.

Freeberg, J. A. and Wetmore, R. H. (1967). The lycopsida – a study in development. *Phytomorphology*, 17, 78–91.

Friedman, W. E., Moore, R. C., and Purugganan, M. D. (2004). The evolution of plant development. *American Journal of Botany*, 91, 1726–1741.

Gifford, E. M., Jr. (1983). Concept of apical cells in bryophytes and pteridophytes. *Annual Review of Plant Physiology*, 34, 419–440.

Gifford, E. M. and Foster, A. S. (1989). *Morphology and Evolution of Vascular Plants*, 3rd edn. New York: Freeman.

Gould, S. J. (1977). *Ontogeny and Phylogeny*. Cambridge, MA: Belknap Press of Harvard University Press.

Gunning, B. E. S. (1978). Age-related and origin-related control of the numbers of plasmodesmata in cell walls of developing *Azolla* roots. *Planta*, 143, 181–190.

Guttenberg, H. von. (1966). *Histogenese der pteridophyten*. Berlin-Nikolassee: Gebröder Borntraeger.

Hagemann, W. (1980). Über den Verzweigungsvorgang bei *Psilotum* und *Selaginella* mit Anmerkungen zum Begriff der Dichotomie. *Plant Systematics and Evolution*, 133, 181–197.

Hagemann, W. (1984). Morphological aspects of leaf development in ferns and angiosperms. In *Contemporary Problems in Plant Anatomy*, ed. R. A. White and W. C. Dickison. Orlando, FL: Academic Press, pp. 301–349.

Hagemann, W. and Gleissberg, S. (1996). Organogenetic capacity of leaves: the significance of marginal blastozones in angiosperms. *Plant Systematics and Evolution*, **199**, 121–152.

Hagemann, W. and Schulz, U. (1978). Wedelanlegung und Rhizomverzweigung bei einigen Gleicheniaceen. *Botanische Jahrbücher Systematic*, **99**, 380–399.

Harrison, C. J., Corley, S. B., Moylan, E. C., Alexander, D. L., Scotland, R. W., and Langdale, J. A. (2005). Independent recruitment of a conserved developmental mechanism during leaf evolution. *Nature*, **434**, 509–514.

Hébant-Mauri, R. (1975). Apical segmentation and leaf initiation in the tree fern *Dicksonia squarrosa*. *Canadian Journal of Botany*, **53**, 764–772.

Hébant-Mauri, R. (1984). Branching patterns in *Trichomanes* and *Cardiomanes* (hymenophyllaceous ferns). *Canadian Journal of Botany*, **62**, 1336–1343.

Hébant-Mauri, R. (1990). The branching of *Trichomanes proliferum* (Hymenophyllaceae). *Canadian Journal of Botany*, **68**, 1091–1097.

Hébant-Mauri, R. (1993). Cauline meristems in leptosporangiate ferns: structure, lateral appendages, and branching. *Canadian Journal of Botany*, **71**, 1612–1624.

Hébant-Mauri, R. and Gay, H. (1993). Morphogenesis and its relation to architecture in the dimorphic clonal fern *Lomagramma guianensis* (Aublet) Ching (Dryopteridaceae). *Botanical Journal of the Linnean Society*, **112**, 257–276.

Hébant-Mauri, R. and Veillon, J. M. (1989). Branching and leaf initiation in the erect aerial system of *Stromatopteris moniliformis* (Gleicheniaceae). *Canadian Journal of Botany*, **67**, 407–414.

Hirsch, A. M. and Kaplan, D. R. (1974). Organography, branching, and the problem of leaf versus bud differentiation in the vining epiphytic fern genus *Microgramma*. *American Journal of Botany*, **61**, 217–229.

Holmes, J. (1989). Anomalous branching patterns in some fossil Filicales: implications in the evolution of the megaphyll and the lateral branch, habit and growth pattern. *Plant Systematics and Evolution*, **165**, 137–158.

Imaichi, R. (1980). Developmental studies on the leaf and the extra-axillary bud of *Histiopteris incisa*. *Botanical Magazine (Tokyo)*, **93**, 25–38.

Imaichi, R. (1982). Developmental study on *Hypolepis punctata* (Thunb.) Mett. I. Initiation of the first and the second petiolar buds in relation to early leaf ontogeny. *Botanical Magazine (Tokyo)*, **95**, 435–453.

Imaichi, R. (1983). Developmental study on *Hypolepis punctata* (Thunb.) Mett. II. Initiation of the third petiolar bud. *Botanical Magazine (Tokyo)*, **96**, 159–170.

Imaichi, R. (1984). Developmental anatomy of the shoot apex of leptosporangiate ferns. I. Leaf ontogeny and shoot branching of *Dennstaedtia scabra*. *Journal of Japanese Botany*, **59**, 367–380.

Imaichi, R. (1986). Surface-viewed shoot apex of *Angiopteris lygodiifolia* Ros. (Marattiaceae). *Botanical Magazine (Tokyo)*, **99**, 309–317.

Imaichi, R. (1988). Developmental anatomy of the shoot apex of leptosporangiate ferns. II. Leaf ontogeny of *Adiantum capillus-veneris* (Adiantaceae). *Canadian Journal of Botany*, **66**, 1729–1733.

Imaichi, R. and Hiratsuka, R. (2007). Evolution of shoot apical meristem structures in vascular plants with respect to plasmodesmatal network. *American Journal of Botany*, **94**, 1911–1921.

Imaichi, R. and Kato, M. (1989). Developmental anatomy of the shoot apical cell, rhizophore and root of *Selaginella uncinata*. *Botanical Magazine (Tokyo)*, **102**, 369–380.

Imaichi, R. and Kato, M. (1991). Developmental study of branched rhizophores in three *Selaginella* species. *American Journal of Botany*, **78**, 1694–1703.

Imaichi, R. and Kato, M. (1992). Comparative leaf development of *Osmunda lancea* and *O. japonica* (Osmundaceae): heterochronic origin of rheophytic stenophylly. *Botanical Magazine (Tokyo)*, **105**, 199–213.

Imaichi R. and Nishida M. (1986). Developmental anatomy of three-dimensional leaf of *Botrychium ternatum* (Thunb.) Sw. *Botanical Magazine (Tokyo)*, **99**, 85–106.

Jernstedt, J. A., Cutter, E. G., and Lu P. (1994). Independence of organogenesis and cell pattern in developing angle shoots of *Selaginella martensii*. *Annals of Botany*, **74**, 343–355.

Kaplan, D. R. and Groff, P. A. (1995). Developmental themes in vascular plants: functional and evolutionary significance. *Monographs in Systematic Botany from the Missouri Botanical Garden*, **53**, 111–145.

Kaplan, D. R. and Hagemann, W. (1991). The relationship of cell and organism in vascular plants. *BioScience*, **41**, 693–703.

Karrfalt, E. E. (1984). The origin and early development of the root-producing meristem of *Isoetes andicola* L. D. Gomez. *Botanical Gazette*, **145**, 372–377.

Kato, M. and Akiyama H. (2005). Interpolation hypothesis for origin of the vegetative sporophyte of land plants. *Taxon*, **54**, 443–450.

Kato, M. and Imaichi, R. (1997). Morphological diversity and evolution of vegetative organs in pteridophytes. In *Evolution and Diversification of Land Plants*, ed. K. Iwatsuki and P. H. Raven. Tokyo: Springer-Verlag, pp. 27–43.

Kato, M., Takahashi, A., and Imaichi, R. (1988). Anatomy of the axillary bud of *Helminthostachys zeylanica* (Ophioglossaceae) and its systematic implications. *Botanical Gazette*, **149**, 57–63.

Kenrick, P. and Crane, P. R. (1997). *The Origin and Early Diversification of Land Plants. A Cladistic Study*. Washington, DC: Smithsonian Institution Press.

Klekowski, E. J., Jr. (1988). *Mutation, Developmental Selection, and Plant Evolution*. New York: Columbia University Press.

Kluge, A. G. (1988). The characterization of ontogeny. In *Ontogeny and Systematics*, ed. C. J. Humphries. New York: Columbia University Press, pp. 57–81.

Lee Y.-H. (1989). Development of mantle leaves in *Platycerium bifurcatum* (Polypodiaceae). *Plant Systematics and Evolution*, **165**, 199–209.

Lintilhac, P. M. and Green, P. B. (1976). Patterns of microfibrillar order in a dormant fern apex. *American Journal of Botany*, **63**, 726–728.

Lyndon, R. F. (1998). *The Shoot Apical Meristem*. Cambridge: Cambridge University Press.

McAlpin, B. W. and White, R. A. (1974). Shoot organization in the Filicales: the promeristem. *American Journal of Botany*, **61**, 562–579.

Mishler, B. D. and Churchill, S. P. (1984). A cladistic approach to the phylogeny of the "bryophytes". *Brittonia*, **36**, 406–424.

Mueller, R. J. (1982a). Shoot morphology of the climbing fern *Lygodium* (Schizaeaceae): general organography, leaf initiation, and branching. *Botanical Gazette*, **143**, 319–330.

Mueller, R. J. (1982b). Shoot ontogeny and the comparative development of the heteroblastic leaf series in *Lygodium japonicum* (Thunb.). SW. *Botanical Gazette*, **143**, 424–438.

Newman, I. V. (1965). Pattern in the meristems of vascular plants. III. Pursuing the patterns in the apical meristem where no cell is a permanent cell. *Journal of the Linnean Society of London (Botany)*, **59**, 185–214.

Ogura, Y. (1972). *Comparative Anatomy of Vegetative Organs of the Pteridophytes*. Berlin: Gebröder Borntraeger.

Paolillo, D. J. (1982). Meristems and evolution: developmental correspondence among the rhizomorphs of the lycopsids. *American Journal of Botany*, **69**, 1032–1042.

Philipson, W. R. (1990). The significance of apical meristems in the phylogeny of land plants. *Plant Systematics and Evolution*, **173**, 17–38.

Popham, R. A. (1951). Principal types of vegetative shoot apex organization in vascular plants. *The Ohio Journal of Science*, **51**, 249–270.

Popham, R. A. (1960). Variability among vegetative shoot apices. *Bulletin of the Torrey Botanical Club*, **87**, 139–150.

Pryer, K. M., Schneider, H., Smith, A. R., Cranfill, R., Wolf, P. G., Hunt, J. S., and Sipes, S. D. (2001). Horsetails and ferns are a monophyletic group and the closest living relatives to seed plants. *Nature*, **409**, 618–622.

Pryer, K. M., Schuettpelz, E., Wolf, P. G., Schneider, H., Smith, A. R., and Cranfill, R. (2004). Phylogeny and evolution of ferns (Monilophytes) with a focus on the early leptosporangiate divergences. *American Journal of Botany*, **91**, 1582–1598.

Qiu, Y.-L. and Palmer, J. D. (1999). Phylogeny of early land plants: insights from genes and genomes. *Trends in Plant Science*, **4**, 26–30.

Qiu, Y.-L, Li, L., Wang, B., Chen, Z., Knoop, V., Groth-Malonek, M., Dombrovska, O., Lee, J., Kent, L., Rest, J., Estabrook, G. F., Hendry, T. A., Taylor, D. W., Testa, C. M., Ambros, M., Crandall-Stotler, B., Duff, R. J., Stech, M., Frey, W., Quandt, D., and Davisk, C. C. (2006). The deepest divergences in land plants inferred from phylogenomic evidence. *Proceedings of the National Academy of Sciences of the United States of America*, **103**, 15511–15516.

Schuster, R. M. (1984). Comparative anatomy and morphology of the Hepaticae. In *New Manual of Bryology*, ed. R. M. Schuster. Nichinan: The Hattori Botanical Laboratory, pp. 760–891.

Siegert, A. (1974). Die Verzweigung der Selaginellen unter Beröcksichtigung der Keimungsgeschichte. *Beiträge zur Biologie der Pflanzen*, **50**, 21–112.

Soma, K. (1966). On the shoot apices of *Dicranopteris dichotoma* and *Diplopterygium glaucum*. *Botanical Magazine (Tokyo)*, **79**, 457–466.

Steeves, T. A. and Sussex, I. M. (1989). *Patterns in Plant Development*, 2nd edn. Cambridge: Cambridge University Press.

Stevenson, D. W. (1976a). The cytohistological and cytohistochemical zonation of the shoot apex of *Botrychium multifidum*. *American Journal of Botany*, **63**, 852–856.

Stevenson, D. W. (1976b). Observations on phyllotaxis, stelar morphology, the shoot apex, and gemmae of *Lycopodium lucidulum* Michaux (Lycopodiaceae). *Botanical Journal of the Linnean Society*, **72**, 81–100.

Stevenson, D. W. (1976c). Shoot apex organization and origin of the rhizome-borne roots and their associated gaps in *Dennstaedtia cicutaria*. *American Journal of Botany*, **63**, 673–678.

Stewart, W. N. and Rothwell, G. W. (1993). *Paleobotany and the Evolution of Plants*, 2nd edn. Cambridge: Cambridge University Press.

Sussex, I. M. and Steeves, T. A. (1967). Apical initials and the concept of promeristem. *Phytomorphology*, **17**, 387–391.

Takiguchi, Y., Imaichi, R., and Kato, M. (1997). Cell division patterns in the apices of subterranean axis and aerial shoot of *Psilotum nudum* (Psilotaceae): morphological and phylogenetic implications for the subterranean axis. *American Journal of Botany*, **84**, 588–596.

Troll, W. (1937). *Vergleichende Morphologie der höheren Pflanzen*, band 1, teil 1. Berlin: Gebröder Borntraeger. (Reprinted 1967, Königstein.)

Troop, J. E. and Mickel, J. T. (1968). Petiolar shoots in the dennstaedtioid and related ferns. *American Fern Journal*, **58**, 64–70.

van der Schoot, C. and Rinne, P. (1999). The symplasmic organization of the shoot apical meristem. In *Plasmodesmata – Structure, Function, Role in Cell Communication*, ed., A. J. E. van Bel and W. J. P. van Kesteren. Heidelberg: Springer-Verlag, pp. 225–242.

Wardlaw, C. W. (1946). Experimental and analytical studies of Pteridophytes. VIII. Further observation on bud development in *Matteuccia Struthiopteris, Onoclea sensibilis*, and species of *Dryopteris*. *Annals of Botany*, **9**, 117–132.

Yi, S.-Y. and Kato, M. (2001). Basal meristem and root development in *Isoetes asiatica* and *Isoetes japonica*. *International Journal of Plant Sciences*, **162**, 1225–1235.

PART II GENETICS AND REPRODUCTION

4

Population genetics

TOM A. RANKER AND JENNIFER M. O. GEIGER

4.1 Introduction

William Henry Lang (1923) and Irma Andersson (later Andersson-Köttö; e.g., Andersson, 1923, 1927; Andersson-Köttö, 1929, 1930, 1931) were pioneers in the field of fern genetics. Lang (1923) was the first to demonstrate simple Mendelian inheritance in a fern with his experimental study of the inheritance of entire versus incised leaf margins in *Scolopendrium vulgare*. Andersson studied inheritance in ferns and was the first to introduce the use of an agar-based growth medium for the experimental study of fern gametophytes (Andersson, 1923). These pioneers paved the way for future explorations of fern and lycophyte population genetics.

In considering how ferns and lycophytes develop and maintain genetic variation, contemporary investigators have used an array of techniques to explore several primary, intertwining topics such as the population genetic implications of reproductive biology (including genetic load), genetic diversity and structure of populations, gene flow and divergence, and the genetics of dispersal and colonization. The goal of this chapter is to review the fern and lycophyte population genetic literature across these broad categories, to provide a synthesis of current knowledge, and to suggest possible future directions of study. We will focus primarily on homosporous taxa because little population genetic research has been conducted on heterosporous taxa.

Biology and Evolution of Ferns and Lycophytes, ed. Tom A. Ranker and Christopher H. Haufler. Published by Cambridge University Press. © Cambridge University Press 2008.

4.2 Population genetics and reproductive biology

An understanding of the reproductive biology of individuals and populations is fundamental for discussing the genetics of populations. Several valuable reviews and summaries have been published covering various aspects of population genetics and reproductive biology, including Klekowski (1969, 1971, 1972a, 1979), Haufler (1987, 1992, 2002), Hedrick (1987), Soltis et al. (1988a), Holsinger (1990, 1991), Werth and Cousens (1990), and Soltis and Soltis (1990a). Most of the research conducted prior to the early 1980s focused on the development of sexuality in gametophytes in culture (e.g., Stokey and Atkinson, 1958; Atkinson and Stokey, 1964; see also the extensive bibliography of gametophyte studies of Pérez-García and Riba, 1998), with only a few studies of natural gametophyte populations (e.g., Schneller, 1979). Observations of gametophyte sexuality have been used to infer the mating systems likely to be operating in nature and their probable population genetic consequences. Beginning in the mid-1980s, there was an increasing research emphasis on the use of biochemical and molecular genetic techniques to assess population genetic variation directly and then infer the mating systems that may have produced such levels and patterns of variation (e.g., see Ranker et al., 2000 and references cited therein).

Homosporous ferns and lycophytes have three possible modes of sexual reproduction (following Klekowski, 1969).

(1) Intragametophytic selfing – the union of sperm and egg from the same bisexual gametophyte. This form of selfing would result in a completely homozygous sporophyte and is not available to heterosporous taxa, including all seed plants.
(2) Intergametophytic selfing – the union of sperm and egg from different gametophytes arising from the same parental sporophyte. This form of selfing is analogous to selfing in seed plants.
(3) Intergametophytic crossing – the union of sperm and egg from different gametophytes each arising from a different sporophyte. This is analogous to outcrossing in seed plants.

The expected population genetic consequences of a predominance of each of these sexual reproductive modes, in the order listed above, are from lesser to greater levels of heterozygosity. Klekowski and Baker (1966) proposed that, "It may be a valid generalization that many, if not most, homosporous fern taxa have the capacity to produce complete homozygotes by self-fertilization and frequently do so in nature" (p. 153) (see also Chapters 2 and 7). This proposal was based partly on studies of laboratory cultured gametophytes of fewer than 12 species of homosporous ferns, which demonstrated that isolated, bisexual gametophytes were as capable of producing viable sporophytes via intragametophytic selfing as were gametophytes in cultures of two or more individuals that were reproducing via intergametophytic mating. If intragametophytic selfing predominates in natural populations of homosporous taxa, most fern and lycophyte populations and species would harbor little, if any, genetic diversity. Klekowski and Baker (1966) also noted an apparent correlation between homospory and high chromosome numbers versus that between heterospory and low chromosome numbers. The putative polyploidy of homosporous taxa would allow for the maintenance of high levels of genetic diversity stored across duplicated chromosome sets in the face of continual selfing in bisexual gametophytes (see Chapter 7). Klekowski (1972a) proposed that novel genotypes could be formed by the occasional pairing of homoeologous chromosomes during meiosis (i.e., allosyndetic pairing). (NB: Homoeologous chromosomes are partially homologous chromosomes such as those sets in an allopolyploid species that were inherited from its two parental species and that generally encode for the same genes.)

For intragametophytic selfing to predominate in populations and for species to be genetically viable over the long term, we might expect such populations to express several gametophytic and genetic characteristics. First, most gametophytes should be bisexual. Second, intragametophytic selfing events would result in the production of completely homozygous sporophytes. Under such extreme, continual selfing, deleterious or lethal alleles expressed in either the gametophyte or sporophyte generation should be purged from surviving populations and, therefore, levels of genetic load would be low (Muller, 1950; Dobzhansky, 1970; Wallace, 1970; Klekowski, 1988). Third, if the high chromosome numbers of homosporous taxa were indicative of polyploidy, such that genetic variation was stored across homoeologous sets of chromosomes, occasional homoeologous chromosome pairing would be necessary to form novel genotypes, thus maintaining the evolutionary genetic flexibility of populations in the face of environmental change (Klekowski, 1972a). Fourth, if intragametophytic selfing predominated in populations, and if species with the lowest chromosome numbers in their genus were actually genetically diploid, we would expect to find low to non-existent levels of genetic diversity within populations. These statements can be considered hypotheses for testing, and indeed they have driven significant investigations into the reproductive biology and population genetics of ferns and lycophytes over the past several decades.

Are most homosporous fern gametophytes bisexual? No. Numerous laboratory studies and several studies of natural populations of homosporous fern gametophytes have found that most sexual gametophytes are either antheridiate or archegoniate at any one time and usually less than about 10% are bisexual (see Chapters 2 and 9). For example, Ranker et al. (1996) examined 1343 and 916 sexual gametophytes of *Sadleria cyatheoides* and *S. pallida*, respectively, over 15–17 weeks of growth on mineral-enriched agar medium. Only 5.6% and 8.8%, respectively, were bisexual. Similarly, in a sample of *Sadleria* gametophytes taken from natural populations, which was probably a mix of the two species, only 8.5% were bisexual (Ranker and Houston, 2002). Greer (1993) sowed spores of *Aspidotis densa* onto native soil at combinations of two soil textures (fine versus coarse) and two spore densities (low versus high) and assessed the sexuality of subsequent gametophyte populations. The percentage of bisexual gametophytes across the different conditions ranged from 0% to 1.5%. Most other studies of experimental and natural gametophyte populations have found generally similar results (see Chapter 9 and references therein). There have been some exceptions, however. For example, Quintanilla et al. (2005) observed laboratory reared gametophytes of *Culcita macrocarpa* and *Woodwardia radicans* over the course of 51 weeks and found that the frequency of bisexual gametophytes varied over time from 0% to ca. 40% in *C. macrocarpa* and from 0% to nearly 80% in *W. radicans*.

Do most homosporous fern populations harbor high levels of genetic load? Studies have generally found that diploids tend to carry higher levels of genetic load than polyploids and that non-colonizing diploid species tend to carry higher levels of genetic load than colonizing diploid species.

Klekowski (e.g., 1970a, 1972b, 1979, and references therein) devised techniques for assessing the degree to which individual sporophytes carry deleterious, recessive, alleles (gametophytic or sporophytic). As stated by Klekowski (1979): "The fundamental question asked in these studies is whether a given sporophyte genotype is heterozygous for deleterious genetic combinations which, when present in the haploid gametophyte or diploid homozygous sporophyte generations, decrease the viability of that generation" (p. 144). The most commonly employed strategy is to facilitate intragametophytic selfing by growing gametophytes in isolation and assessing the extent to which bisexual isolates can produce new sporophytes. The number of non-sporophyte producing bisexual gametophytes as a percentage of the total number of isolated bisexual gametophytes (all grown from spores from a single sporophyte) can be used as an approximation of genetic load (e.g., see Peck et al., 1990). A few studies have gone a step further and assessed the viability of sporophytes produced to screen for sublethal, deleterious alleles (e.g., Klekowski, 1970a). To assess simply the ability of gametophytes to form sporophytes in culture and/or to explore further the nature of apparently deleterious alleles (i.e., gametophytic versus sporophytic, dominant versus recessive), isolate experiments are often coupled with experiments with pairs of gametophytes, with each gametophyte originating from a different sporophyte (see Klekowski, 1979, for a detailed explanation of the logic of these studies). For example, in studies of the diploid *Osmunda regalis*, Klekowski (1970a, 1973) found that most or all isolated, bisexual gametophytes were unable to form new sporophytes, whereas pairs of unrelated gametophytes usually could.

The general conclusion from studies of genetic load in homosporous ferns is that the levels observed are highly variable, taxon-specific, related to ploidy, and appear to correlate with ecology and recent population history. For example, Peck *et al.* (1990) (and see Chapter 13) conducted studies of genetic load of 11 diploid species, across eight genera, of homosporous ferns in Iowa, USA. Estimates of genetic load, as measured by the percentage of bisexual gametophytes unable to produce sporophytes, were 0, 0, 8, 12, 62, 62, 67, 93, 96, 96, and 98. Chiou *et al.* (2002) found a similar range of genetic load estimates in three species of epiphytic, diploid Polypodiaceae (10–100%). Chiou *et al.* (1998) found that in two species of *Elaphoglossum*, no isolated, bisexual gametophytes were able to produce new sporophytes whereas about 25% to 50% of paired, unrelated gametophytes were able to produce new sporophytes. Thus, levels of genetic load carried by diploid, sexual species are highly variable.

Masuyama (1979) compared the genetic load of diploid versus tetraploid plants of *Phegopteris decursive-pinnata* and found that all isolated, bisexual gametophytes grown from the diploid spores obtained from tetraploids produced sporophytes via intragametophytic selfing. By contrast, only 33% to 65% of the gametophytes grown from haploid spores obtained from diploids could produce new sporophytes. Comparable results were obtained in studies of diploid versus tetraploid plants of *Lepisorus thunbergianus* (Masuyama *et al.*, 1987) and *Pteris dispar* (Masuyama and Watano, 1990). Similarly, Chiou *et al.* (2002) found that 76% of the gametophytic progeny of a polyploid individual of *Campyloneurum angustifolium* were capable of undergoing successful intragametophytic selfing, whereas none of the progeny from a conspecific diploid could do so. Several studies of the tetraploid *Ceratopteris thalictroides* have found evidence for low levels of genetic load, with most isolated gametophytes able to form viable sporophytes via intragametophytic selfing (Klekowski, 1970b; Lloyd and Warne, 1978; Watano and Masuyama, 1991). Studies of other sexual tetraploids have also demonstrated the ability of isolated gametophytes to produce new sporophytes via intragametophytic selfing, including *Hemionitis pinnatifida* (Ranker, 1987), *Cosentinia vellea*, *Cheilanthes tinaei*, and *C. acrostica* (Pangua and Vega, 1996), *Asplenium trichomanes* ssp. *quadrivalens* (Suter *et al.*, 2000), *Asplenium septentrionale* (Aragón and Pangua, 2003), and *Asplenium csikii* (Vogel *et al.*, 1999). The general conclusion from these and other studies is that polyploid species either generally carry low levels of genetic load or that the genetic load is masked from expression because of the expression of dominant genes in duplicated sets of chromosomes (see Crow and Kimura, 1965; Otto and Marks, 1996).

Level of genetic load also appears to be related to the colonizing ability of individuals, with successful colonizers generally harboring little or no genetic load and non-colonizers harboring significant levels. See more details in Section 4.5.

Is there evidence of homologous chromosome pairing in sexually reproducing, diploid homosporous ferns? The hypothesis that at least occasional homologous chromosome pairing occurs to "release" genetic variation stored across duplicated chromosome sets (Klekowski, 1972a, 1979) was predicated on the hypothesis that all homosporous ferns are polyploids. The advent of isozyme electrophoresis and its application to ferns and lycophytes marked a major milestone in our understanding of many aspects of the nature of fern and lycophyte genetics. One of the most important findings was that species with the base chromosome number (x) for their group (i.e., genus and/or family) produced isozymic patterns that were consistent with them being genetic diploids (Gastony and Gottlieb, 1985; also see Haufler, 1987, 2002, and Haufler and Soltis, 1986 for thorough reviews and discussions). Species with greater than $2x$ sets of chromosomes typically exhibit isozymic expression that is consistent with genetic polyploidy. Among sexually reproducing ferns, evidence of homologous chromosomal pairing and recombination has only been discovered in a genetic polyploid, *Ceratopteris thalictroides* (Hickok, 1978a, 1978b). Otherwise, convincing evidence of homoeologous pairing has only been demonstrated for the apogamous triploid *Dryopteris nipponensis* (Ishikawa et al., 2003).

Do populations of homosporous ferns exhibit low levels of genetic diversity? The most commonly used technique for assessing population genetic diversity in ferns and lycophytes has been enzyme (also called isozyme or allozyme) electrophoresis. Commonly estimated measures of population genetic diversity are the mean number of alleles per locus (A), the percentage of loci examined that are polymorphic (P), and the expected heterozygosity assuming Hardy–Weinberg equilibrium (H_e). Table 4.1 shows data for 49 taxa of sexually reproducing, diploid, homosporous ferns for which values of these three parameters were available in the references cited or for which we could calculate values from the data given. Across all 49 taxa, mean A was 1.6 (range 1.0–2.8), mean P was 38.4 (range 0–80), and mean H_e was 0.182 (range 0.000–0.345). These mean values are similar to, and possibly higher than, those reported for populations of 468 species of seed plants (Table 4.1) reported by Hamrick and Godt (1990). Homosporous ferns as a group, therefore, do not harbor lower levels of genetic diversity than has been found in seed plants.

Table 4.1 *Population-level variation in sexual, diploid ferns and seed plants*

Values were obtained either directly from the literature cited or were calculated from the data provided therein. Blanks indicate data not available. A, mean number of alleles per locus; P, percentage of loci polymorphic, no criterion; H_e, mean expected heterozygosity under Hardy–Weinberg conditions.

Species	A	P	H_e	Reference
Adenophorus periens	2.8	80.0	0.213	Ranker, 1994
Adenophorus tamariscinus	2.2	55.3	0.146	Ranker, 1992b
Adenophorus tripinnatifidus	1.4	30.5	0.081	Ranker, 1992b
Asplenium montanum	1.3	26.7	0.338	Werth et al., 1985
Asplenium platyneuron	1.5	26.7	0.136	Werth et al., 1985
Asplenium rhizophyllum	1.1	13.3	–	Werth et al., 1985
Athyrium filix-femina var. *asplenioides*	2.0	32.8	0.115	Sciarretta et al., 2005
Blechnum spicant	1.4	23.6	0.024	Soltis and Soltis, 1988a
Bommeria elegans	1.4	33.3	0.295	Ranker, 1987
Bommeria hispida	2.6	61.5	0.206	Haufler and Soltis, 1984; Haufler, 1985; Ranker, 1987
Bommeria subpaleacea	1.5	38.5	0.162	Haufler, 1985
Bommeria ehrenbergiana	1.4	21.3	0.128	Haufler, 1985
Botrychium crenulatum	1.0	0.00	0.000	Hauk and Haufler, 1999
Botrychium lanceolatum	1.2	21.4	0.086	Hauk and Haufler, 1999
Botrychium lunaria	1.1	11.9	0.070	Hauk and Haufler, 1999
Botrychium multifidum var. *robustum*	2.0	55.6	0.193	Watano and Sahashi, 1992
Botrychium nipponicum	1.4	33.3	0.116	Watano and Sahashi, 1992
Botrychium pumicola	1.0	0.00	0.000	Hauk and Haufler, 1999
Botrychium simplex	1.0	3.30	0.017	Hauk and Haufler, 1999
Botrychium ternatum	2.1	55.6	0.216	Watano and Sahashi, 1992
Botrychium triangularifolium	1.0	0.0	0.000	Watano and Sahashi, 1992
Botrychium virginianum	1.3	15.3	0.225	Soltis and Soltis, 1986
Cheilanthes subcordata	1.9	65.0	0.345	Ranker, 1987
Dryopteris expansa	1.1	9.60	0.032	Soltis and Soltis, 1987a
Elaphoglossum bifurcatum	1.1	8.0	0.020	Eastwood et al., 2004
Elaphoglossum nervosum	1.0	1.0	0.002	Eastwood et al., 2004
Grammitis hookeri	1.5	26.4	0.066	Ranker, 1992b
Grammitis tenella	1.6	39.3	0.085	Ranker, 1992b
Gymnocarpium dryopteris ssp. *disjunctum*	1.8	58.6	0.186	Kirkpatrick et al., 1990
Hemionitis palmata	1.2	19.0	0.049	Ranker, 1992a

(cont.)

Table 4.1 (cont.)

Species	A	P	H_e	Reference
Osmunda cinnamomea	1.4	25.0	0.096	Li and Haufler, 1994
Osmunda claytoniana	1.3	15.4	0.062	Li and Haufler, 1994
Osmunda regalis	1.5	27.7	0.121	Li and Haufler, 1994
Pellaea andromedifolia	1.7	–	0.221	Gastony and Gottlieb, 1982
Pleopeltis astrolepis	2.2	68.2	0.153	Hooper and Haufler, 1997
Pleopeltis complanata	2.3	58.2	0.162	Hooper and Haufler, 1997
Pleopeltis crassinervata	2.7	69.1	0.252	Hooper and Haufler, 1997
Pleopeltis polylepis var. erythrolepis	2.0	54.5	0.179	Hooper and Haufler, 1997
Pleopeltis polylepis var. polylepis	2.4	65.1	0.201	Hooper and Haufler, 1997
Polypodium pellucidum	1.9	47.1	0.184	Li and Haufler, 1999
Polystichum acrostichoides	1.6	44.4	0.078	Soltis et al., 1990
Polystichum dudleyi	1.1	8.3	0.015	Soltis et al., 1990
Polystichum imbricans	1.8	50.0	0.147	Soltis and Soltis, 1990b
Polystichum lemmonii	1.3	25.0	0.045	Soltis et al., 1990
Polystichum lonchitis	1.1	6.7	0.005	Soltis et al., 1990
Polystichum munitum	1.6	39.4	0.111	Soltis et al., 1990
Polystichum otomasui	1.9	61.9	0.177	Maki and Asada, 1998
Pteridium aquilinum	1.5	34.9	0.098	Wolf et al., 1988, 1990
Sadleria cyatheoides	1.6	42.9	0.090	Ranker et al., 1996
Sadleria pallida	1.6	35.5	0.076	Ranker et al., 1996
Sphenomeris chinensis	1.4	26.9	0.039	Ranker et al., 2000
Grand mean	1.6	38.4	0.182	
SE	0.07	3.06	0.043	
Mean without Botrychium	1.8	39.2	0.135	
SE	0.07	3.22	0.014	
Mean of Botrychium only	1.3	19.6	0.287	
SE	0.12	6.54	0.199	
Mean of seed plants	1.5	34.2	0.113	Hamrick and Godt, 1990

The level of genetic diversity in plant populations is generally positively correlated with population size (e.g., Hamrick and Godt, 1990; Leimu et al., 2006); however, this is not always the case. Populations can become small and/or species rare for different historical reasons (Rabinowitz, 1981; Gitzendanner and Soltis, 2000), leaving populations with varying levels of genetic diversity in the long- or short-term. The study of Eastwood et al. (2004) is an example documenting a positive relationship between genetic diversity and population size. They found low levels of genetic diversity in several species of Elaphoglossum, which are restricted to the small island of St. Helena, comparable to what is often found in island species of seed plants (Table 4.1). By contrast, Ranker (1994) conducted an allozyme survey of one population of the rare Hawaiian endemic epiphyte, *Adenophorus periens*, that occurred in a forest covering a lava flow only 300–400 years old. *Adenophorus periens* had been collected in historical times on all of the main Hawaiian Islands, but in the early 1990s was only known from two populations on the island of Hawaii, with a few scattered individuals known from Kauai and Molokai (Ranker, 1994 and personal observations). The levels of allozymic diversity observed are among the highest ever recorded for any fern ($A = 2.8$, $P = 80.0$, and $H_e = 0.213$; Table 4.1). Apparently, the reduction in population size of this species has occurred too recently to impact levels of genetic diversity. A similar example of a recently restricted species, albeit heterosporous, still harboring high levels of genetic variation is the rare *Isoëtes sinensis* from China (Kang et al., 2005).

A long evolutionary history coupled with a formerly large population size can allow population lineages to maintain high levels of genetic diversity and to exhibit high levels of divergence across intraspecific lineages. Su et al. (2004) employed cpDNA sequence data from *atpß-rbcL* intergenic spacers to explore population genetic structure and phylogeographic patterns among modern relictual populations of *Alsophila spinulosa*. This species was distributed worldwide during the Jurassic Period (180 million years ago), became much more restricted during the Quaternary Period, and is now extremely rare in China due to the continued loss and fragmentation of habitat because of human destruction. Nevertheless, extant populations possess high levels of haplotype and nucleotide diversity and populations from different regions of China are extremely divergent from each other ($F_{ST} = 0.95$). Su et al. (2004) suggest that the high levels of genetic diversity within geographic regions may be caused by the accumulation of new mutations over the long evolutionary history of the species. They further suggest that the divergence of populations across different regions has been facilitated because different lineages accumulate different mutations, which has been maintained by low levels of inter-regional gene flow owing to a primarily inbreeding mating system.

Schneller and Holderegger (1996) discuss how a variety of historical and life-history phenomena could have impacted levels and patterns of genetic diversity in small populations of several species of *Asplenium* and *Polypodium vulgare* in the lowlands of Switzerland. For example, Holderegger and Schneller (1994) studied three small populations (9, 15, and 30 individuals each) of the autotetraploid *Asplenium septentrionale* in the vicinity of Lake Zürich. The populations were isolated from each other by 3–7 km and were ca. 40 km from the nearest larger populations in the Swiss Alps. Two populations were genetically uniform, one exhibited some genetic diversity, but each possessed at least one unique isozyme phenotype. The patterns of variation within and across populations suggested that there was zero gene flow among them and that each population primarily reproduced via intragametophytic selfing. The authors suggested that each population might have been the result of independent long-distance dispersal events from larger populations in the Alps.

What are the predominant mating systems operating in natural populations of ferns and lycophytes? Prior to the application of enzyme electrophoretic studies for estimating allelic frequencies and heterozygosities of natural populations of ferns and lycophytes, inferences of mating systems were made by studying gametophyte sexuality in the laboratory (see Lloyd, 1974) and, rarely, in nature (see Cousens et al., 1985, and discussion above on genetic load). As summarized by Haufler (1987) and Soltis and Soltis (1987b), such gametophytic studies are useful for showing the potential of what could be occurring in nature, but do not necessarily reflect what is really happening (e.g., see Chapter 13; Ranker and Houston, 2002). Allozymes have proven to be a powerful tool for inferring mating systems because they represent nuclear-encoded enzymes that are biparentally inherited and co-dominantly expressed. They can be used, therefore, to estimate the observed (H_o) and the expected (H_e) heterozygosities and, thus, the fixation index (Wright, 1943):

$$F = 1 - [H_o/H_e].$$

If F is primarily determined by mating behavior, it can be equated with an inbreeding coefficient (Wright, 1969). If H_o is not significantly different than H_e, then F is not significantly greater than zero and the population is assumed to be in Hardy–Weinberg equilibrium and, thus, randomly mating. If H_o is significantly greater than H_e, then F will be significantly less than zero (i.e., there is an excess of heterozygotes) and the population is assumed to be primarily outcrossing. If H_o is significantly less than H_e, then F will be significantly greater than zero (i.e., there is a deficiency of heterozygotes) and the population is assumed to be primarily inbreeding. Complete intragametophytic selfing would result in a value of F of 1.0.

Most sexually reproducing, diploid populations of homosporous ferns and lycophytes that have been surveyed allozymically exhibit F values that are either not significantly different than zero or are significantly less than zero, and, therefore, are either randomly mating or primarily outcrossing (Table 4.2). Mean F across the 24 species reported in Table 4.2 is 0.106 (range −0.077 to 0.962). Across populations within species, mating systems of sexual diploids are usually dominated by random mating or outcrossing, with a minority of taxa being primarily inbreeding or mixed-mating. See reviews and summaries of Haufler (1987, 1992, 2002) and Soltis and Soltis (1987b) for detailed examples and citations.

Table 4.2 *Mating systems, fixation index (F), differentiation (F_{ST}), genetic identity (I), and gene flow (Nm) estimates of populations of diploid, homosporous ferns available directly from the cited literature or calculated from data provided; blanks indicate that estimates or data were not available for a particular parameter*

Species	Mating system	mean F	mean F_{ST}[1]	range F_{ST}[1]	mean I	range I	mean Nm	range Nm[1]	Reference
Adenophorus tamariscinus	outcrossing	0.022	0.24	0.00–0.06	—	—	30.96	3.8–155.7	Ranker, 1992b
Adenophorus tripinnatifidus	outcrossing	−0.013	0.12	0.07–0.16	—	—	1.95	1.3–3.1	Ranker, 1992b
Athyrium filix-femina var. *asplenioides*	outcrossing	0.035	0.07	0.01–0.11	0.992	0.97–1.00	3.3	2.0–24.8	Sciarretta et al., 2005
Blechnum spicant	mixed	0.132	0.05	0.00–0.18	0.996	—	2.95	1.2–15.4	Soltis and Soltis, 1988a
Bommeria elegans	outcrossing	—	—	—	0.790	—	—	—	Ranker, 1987
Bommeria hispida	outcrossing	—	—	—	0.879	0.74–0.99	—	—	Haufler, 1985
Botrychium crenulatum	inbreeding	—	—	—	1.000	—	—	—	Hauk and Haufler, 1999
Botrychium dissectum	inbreeding	0.951	0.09	—	—	—	—	—	McCauley et al., 1985
Botrychium lanceolatum	inbreeding	—	—	—	0.953	0.85–1.00	—	—	Hauk and Haufler, 1999
Botrychium lunaria	inbreeding	—	—	—	0.958	0.88–1.00	—	—	Hauk and Haufler, 1999
Botrychium simplex	inbreeding	—	—	—	0.848	0.67–1.00	—	—	Hauk and Haufler, 1999
Botrychium virginianum	inbreeding	0.962	0.09	—	—	—	0.41	—	Soltis and Soltis, 1986, Soltis et al., 1988a
Cheilanthes gracillima	outcrossing	−0.184	0.264	—	—	—	0.846	0.05–2.13	Soltis et al., 1989
Cheilanthes subcordata	outcrossing	—	—	—	0.900	—	—	—	Ranker, 1987
Cystopteris bulbifera	outcrossing	—	—	—	0.826	—	—	—	Haufler et al., 1990
Cystopteris protrusa	outcrossing	—	—	—	0.834	—	—	—	Haufler et al., 1990

(cont.)

Table 4.2 (cont.)

Species	Mating system	mean F	mean F_{ST}[1]	range F_{ST}[1]	mean I	range I	mean Nm	range Nm[1]	Reference
Dryopteris expansa	mixed	0.335	0.21	–	–	–	0.83	–	Soltis and Soltis, 1987a
Grammitis hookeri	outcrossing	0.105	0.16	0.01–0.30	–	–	10.22	0.6–55.0	Ranker, 1992b
Grammitis tenella	outcrossing	0.031	0.70	0.00–0.24	–	–	7.23	0.8–22.2[2]	Ranker, 1992b
Gymnocarpium dryopteris ssp. *disjunctum*	outcrossing	−0.077	0.11	–	0.97	0.92–0.99	4.09	–	Kirkpatrick et al., 1990
Hemionitis palmata	mixed	0.256	0.70	0.00–0.94	0.87	0.71–1.00	0.52[2]	0.02–4.9[2]	Ranker, 1992a
Pellaea andromedifolia	outcrossing	–	–	–	0.94	–	–	–	Gastony and Gottlieb, 1985
Pleopeltis astrolepis	outcrossing	0.058	0.02	–	1.00	–	11.6	–	Hooper and Haufler, 1997
Pleopeltis complanata	outcrossing	0.117	0.04	–	1.00	–	6.2	–	Hooper and Haufler, 1997
Pleopeltis crassinervata	outcrossing	0.046	0.04	–	0.99	–	6.9	–	Hooper and Haufler, 1997
Pleopeltis polylepis var. *erythrolepis*	outcrossing	−0.009	0.09	–	0.98	–	3.6	–	Hooper and Haufler, 1997
Pleopeltis polylepis var. *polylepis*	outcrossing	0.013	0.07	–	0.98	–	2.4	–	Hooper and Haufler, 1997
Polystichum acrostichoides	outcrossing	0.036	–	–	0.998	0.996–0.999	12.69	–	Soltis and Soltis, 1990a; Soltis et al., 1990
Polystichum dudleyi	outcrossing	−0.075	–	–	0.969	0.912–1.000	10.78	–	Soltis and Soltis, 1990a; Soltis et al., 1990
Polystichum imbricans	outcrossing	0.033	–	–	0.974	0.948–0.993	2.20	–	Soltis and Soltis, 1987b; Soltis et al., 1990
Polystichum lemmonii	outcrossing	−0.033	–	–	0.989	0.979–0.999	0.43	–	Soltis and Soltis, 1990a; Soltis et al., 1990

Species	Breeding system							Reference
Polystichum lonchitis	outcrossing	−0.036	—	0.966	0.915–1.000	0.05	—	Soltis and Soltis, 1990a; Soltis et al., 1990
Polystichum munitum	outcrossing	0.052	0.024	0.957	0.908–0.994	24.00	—	Soltis and Soltis, 1987c; Soltis et al., 1990
Polystichum otomasui	outcrossing	0.049	0.108	0.971	0.929–0.989	2.07	—	Maki and Asada, 1998
Pteridium aquilinum	outcrossing							Wolf et al., 1991
British populations		0.123	0.110	0.994	0.987–1.000	36.51	—	
European populations		0.101	0.122	0.993	0.983–1.000	2.47	—	
Total		0.116	0.398	0.924	0.669–1.000	0.093	—	
Grand mean		0.106	0.184	0.952		6.13		

[1] Some studies employed G_{ST}. Ranges of F_{ST} and Nm are shown only for studies that calculated population pairwise values.

[2] For these species one or more pairwise values of F_{ST} were 0, thus calculations of Nm were not possible; the reported mean and maximum extent of the range of Nm do not include information from those pairs of populations.

A mixed-mating system is one wherein neither outcrossing nor inbreeding predominates (e.g., see Lande and Schemske, 1985). Although rare compared to the relative frequencies of either predominantly outcrossing or predominantly inbreeding populations, Holsinger (1991) showed that mixed-mating could be evolutionarily stable when self-fertilization evolves under certain levels of population density. Documented examples of sexual, diploid ferns with mixed-mating systems include *Dryopteris expansa* (Soltis and Soltis, 1987a), *Blechnum spicant* (Soltis and Soltis, 1988a), *Hemionitis palmata* (Ranker, 1992a), *Sadleria cyatheoides* and *S. pallida* (Ranker et al., 1996), *Sphenomeris chinensis* (syn. *Odontosoria chinensis*; Ranker et al., 2000), *Sticherus flabellatus* (Keiper and McConchie, 2000), and *Adiantum capillus-veneris* (Pryor et al., 2001). Ecologically, these mixed-mating species seem to share the attribute of having at least some populations or individuals that colonize disturbed places and others that grow in seemingly more stable microhabitats (see Ranker et al., 2000). Keiper and McConchie (2000) provided evidence from AFLP data (see Vos et al., 1995) that colonizing populations of *Sticherus flabellatus* are primarily inbreeding but also that larger, established populations occasionally exhibit outcrossing. Pryor et al. (2001) were the first to employ microsatellites in studying fern population genetics and they discovered evidence of a mixed-mating system in *Adiantum capillus-veneris* in Great Britain and Ireland.

Holsinger (1987) developed a statistical technique to estimate intragametophytic selfing (IGS) rates in homosporous plants based on estimates of genotype frequencies in populations. A number of studies have employed Holsinger's method, including species that essentially span the phylogenetic diversity of ferns and lycophytes (see Chapter 15): *Blechnum spicant* (Soltis and Soltis, 1988a); *Huperzia miyoshiana*, *Lycopodium annotinum*, and *L. clavatum* (Soltis and Soltis, 1988b); *Blechnum spicant*, *Botrychium virginianum*, *Dryopteris expansa*, *Polystichum imbricans*, and *P. munitum* (Soltis et al., 1988a); *Equisetum arvense* (Soltis et al., 1988b); *Pteridium aquilinum* (Wolf et al., 1988); *Cheilanthes gracillima* (Soltis et al., 1989); *Gymnocarpium dryopteris* ssp. *disjunctum* (Kirkpatrick et al., 1990); *Hemionitis palmata* (Ranker, 1992a); *Botrychium* (*Sceptridium*) *multifidum* var. *robustum* and *B.* (*S.*) *ternatum* (Watano and Sahashi, 1992); five species of *Pleopeltis* (Hooper and Haufler, 1997); *Polystichum otomasui* (Maki and Asada, 1998); and *Sphenomeris chinensis* (Ranker et al., 2000). Soltis and Soltis (1992) presented a summary of IGS estimates of 20 species, including some of those listed in other citations here. Overwhelmingly, these studies have supported the conclusions based on analyses of *F*-values that most populations of most species are primarily outcrossing and exhibit zero intragametophytic selfing. In the review of IGS rates of 20 species, Soltis and Soltis (1992) found a highly significant correlation between IGS rates and *F*-values, suggesting that intragametophytic selfing is the primary contributor to the fixation index in the taxa studied, rather than intergametophytic selfing.

There are several interesting exceptions to the generality of high outcrossing rates in ferns and these appear to relate to the colonization ability of a species (see below for discussion) or having subterranean gametophytes. All species of Ophioglossaceae are homosporous and have subterranean gametophytes. St. John (1949) and Tryon and Tryon (1982) suggested that the subterranean habit of Ophioglossaceae gametophytes might inhibit outcrossing, because of their potential isolation from other gametophytes. McCauley et al. (1985) employed enzyme electrophoresis to estimate selfing rates in three populations of *Botrychium dissectum*. Their estimate of the inbreeding coefficient, F_{IS} (equivalent to the weighted mean of the fixation index (F) across populations), was 0.951; that is, 95% selfing and only 5% outcrossing. Similar high rates of inbreeding, intragametophytic selfing, and allelic fixation were estimated for *Botrychium virginianum* (Soltis and Soltis, 1986; Soltis et al., 1988a). Watano and Sahashi (1992) reported inbreeding in *Botrychium (Sceptridium) multifidum* var. *robustum*, *B. (S.) nipponicum*, *B. (S.) ternatum*, and *B. (S.) triangularifolium*. Similarly, Hauk and Haufler (1999) provided isozymic evidence that populations of *Botrychium lanceolatum* and *B. simplex* are primarily inbreeding. Thus, there is ample support for the hypothesis of St. John (1949) that having subterranean gametophytes facilitates intragametophytic selfing.

In one of the few population genetic studies of a heterosporous fern, Vitalis et al. (2002) provided evidence from microsatellites that the water fern *Marsilea strigosa* reproduces almost entirely via intergametophytic selfing.

4.3   Genetic structure of populations

Population genetic studies of ferns and lycophytes report that genetic variation is structured within and among populations in the same way that it is in other groups of organisms. The apparent primary determinants of genetic structure are various life-history and ecological characteristics such as mating system, population size, dispersal and colonization ability, habitat diversity, and recent demographic history.

Population divergence in ferns and lycophytes has been measured by Wright's standardized variance in allele frequencies, F_{ST} (Wright, 1965, 1978; Nei, 1977), and/or Nei's unbiased genetic identity, I (Nei, 1978). Because F_{ST} is usually calculated as a weighted average for all alleles at a locus, it is equivalent to G_{ST}, the gene diversity among populations (Nei, 1973, 1977; Wright, 1978; Swofford and Selander, 1989) and, thus, studies that employ either of these measures can be compared.

Population-level measures of F_{ST} and I for a variety of fern taxa are reported in Table 4.2. Mean population pairwise values were either taken directly from the literature cited or were calculated from the data provided. The grand mean of F_{ST} values for 19 taxa was 0.184 (range 0.07 to 0.70), which is intermediate to the mean G_{ST} values reported for populations of seed plants with wind dispersal (0.14) and those reported for seed plant taxa with seed dispersal other than wind (range 0.22 to 0.28). Thus, in spite of generally high levels of interpopulational genetic identities (grand mean $I = 0.952$), species of homosporous ferns can exhibit levels of population genetic structure comparable to that of many species of seed plants.

4.4   Gene flow and divergence

The exchange of genes between populations (so-called gene flow) is a powerful force in evolution. If levels of gene flow are sufficiently low between two populations, they can be expected to diverge genetically as a result of genetic drift, even in the absence of natural selection (Wright, 1931). Specifically, if a proportion m of a population of effective size N is replaced each generation by migrants from another population, the two populations may diverge genetically if $Nm < 1.0$. Values of $Nm > 1.0$ (or even $2Nm > 1.0$) will tend to maintain genetic homogeneity between populations at selectively neutral loci. Specific predictions can be made about the probability of divergence of populations over time, given various levels of gene flow, when simplifying assumptions are made in theoretical models. Such models assume random mating and no mutation or natural selection (Wright, 1931, 1943, 1951; Slatkin, 1985a, 1985b). Slatkin and Barton (1989), however, demonstrated that even these assumptions could be relaxed. (Hey (2006) provides an insightful review of the literature, suggesting that speciation may occur in the face of gene flow in concert with the action of natural selection.)

Two methods have been commonly used in studies of ferns and lycophytes to estimate values of Nm between populations. One employs the approximate relationship between F_{ST} and Nm:

$$F_{ST} \approx 1/[4Nm + 1]$$

(Wright, 1931, 1943, 1951; Dobzhansky and Wright, 1941). Slatkin (1985b) devised a method to estimate Nm from the distribution and frequency of rare alleles. Both methods have been employed in the study of ferns, but Slatkin and Barton (1989) suggested that the F_{ST} method might actually be preferred over the private-allele method when based on enzyme electrophoretic data.

Table 4.2 summarizes estimates of interpopulational Nm for 24 species of ferns, most of which were estimated using the F_{ST} method and all of which used isozymes as genetic markers. The grand mean is 6.13, with a range across taxa of 0.05 to 155.7. Perhaps most notable is that most estimates are well above 1.0, which is consistent with the idea that ferns readily disperse via wind-blown spores.

Aside from dispersal and colonization ability, one of the primary determinants of effective gene flow between populations is mating system. Somewhat surprisingly, across the studies listed in Table 4.2, there is not a significant association between estimates of Nm and F (Spearman's rank correlation, $r_s = 0.0044$, $P = 0.49$), suggesting that other factors besides mating behavior may have a significant impact on gene flow. The most obvious cases where there does seem to be a causal effect between mating and interpopulational gene flow are found in the inbreeding species *Botrychium virginianum* and in species that exhibit mixed mating behavior across different populations. In *B. virginianum*, mean $F = 0.962$, with some variable loci exhibiting F values of 1.000 (Soltis and Soltis, 1986). The predominance of intragametophytic selfing in this species with subterranean gametophytes could account for the extremely low estimate of gene flow between populations ($Nm = 0.41$; Soltis *et al.*, 1988a). In mixed-mating *Hemionitis palmata*, estimates of Nm between pairs of populations ranged from effectively zero (0.02) to well over 1.0 (4.9), presumably because outcrossing populations incorporate genes from new migrants better than inbreeding populations (Ranker, 1992a).

Although some cases of repeated long-range dispersal and interpopulational gene flow in ferns have been hypothesized or documented (see below; Ranker *et al.*, 1994), there is some evidence that gene flow is negatively proportional to distance between populations. Certainly this is what one would predict from the probable leptokurtic dispersal of spores away from the parental plant (e.g., Peck *et al.*, 1990). For example, Ranker *et al.* (2000) examined isozyme variation in populations of *Sphenomeris chinensis* on all of the main high islands of the Hawaiian Islands. The mean Nm value between pairs of populations within islands was 13.1 (range 1.9–22.8), whereas the mean across islands of 6.9 (range 2.3–20.9) was significantly less (Mann–Whitney U-test, $P = 0.000$). Similar patterns of gene flow were observed within and across islands for four species of Hawaiian endemic grammitid ferns (Ranker, 1992b).

4.5 Population genetics of dispersal and colonization

Understanding the population genetics of dispersal and colonization has important implications for nearly every field of natural history, including biogeography, ecology, phylogeny, speciation theory, and epidemiology. Because of their presumably highly dispersible, wind-blown spores, ferns and lycophytes should generally have the capacity for long-distance dispersal. Not surprisingly, the extent to which species are effective dispersers depends on a wide range of life-history, adaptive, and genetic characteristics and their interplay with the biotic and physical environments. Clearly, some ferns and lycophytes are capable of dispersing long distances, as evidenced by the fact that they are common elements of the floras of isolated, oceanic islands and that they are often among the first colonizers of open and newly available habitat, especially in the tropics. For example, Schneider et al. (2005) explored the molecular phylogenetic relationships of the Hawaiian endemic clade *Diellia* (Aspleniaceae) to other members of the family. They estimated that the divergence time of the *Diellia* lineage from its closest relative coincided with the estimate of a renewal of Hawaiian terrestrial life at ca. 23 Myr ago (Clague, 1996; Price and Clague, 2002), following a 10 Myr lull in the production of new islands, such that essentially all pre-existing terrestrial life on older islands would have gone extinct due to island subsidence. Thus, this lineage of ferns was among the first to colonize the newly produced, mid-oceanic, isolated islands.

What are the genetic attributes of colonizing species of ferns and lycophytes and what are the genetic consequences of colonization? Several studies have shown that isolated, peripheral populations of some fern and lycophyte species, as well as populations of species that habitually colonize new habitats, are capable of intragametophytic selfing and, thus, harbor low levels of genetic load. Such taxa are consistent with "Baker's Law," which loosely states that self-compatible species should be better colonizers than self-incompatible species (Baker, 1955, 1967; Stebbins, 1957). Among diploids these include *Blechnum spicant* (Cousens, 1979), *Asplenium platyneuron* (Crist and Farrar, 1983), *Pteris multifida* (Watano, 1988), *Lygodium microphyllum* and *L. japonicum* (Lott et al., 2003), and *Dryopteris carthusiana* (Flinn, 2006). Not all colonizing species, however, have this ability. For example, Ranker et al. (1996) discovered that only 0.5% to 5.0% of isolated gametophytes of the lava-flow colonizing species *Sadleria cyatheoides* were able to produce new sporophytes. They provided evidence from isozymes suggesting that, although this species may have some degree of inbreeding, it is primarily outcrossing. Because polyploids reproduce more successfully via intragametophytic selfing than most diploids, polyploid ferns may be more effective dispersers and colonizers (see Masuyama, 1979; Watano, 1988; Masuyama and Watano, 1990).

4.6 Summary and future prospects

In summary, population genetic studies across a wide range of taxa have revealed several common attributes shared by most species. Most populations and species of homosporous ferns and lycophytes are genetically diploid and are primarily outcrossing. Inbreeding is relatively rare and is primarily restricted to taxa with subterranean gametophytes, but interesting exceptions have been discovered among taxa with epigeal gametophytes. Inbreeding is also common in polyploids. In spite of apparently high levels of interpopulational gene flow, populations can exhibit significant genetic structure that may relate to ecological diversity, isolation-by-distance, history of colonization, mating system, population size, and other factors.

Although the field of population genetics within the context of the study of ferns and lycophytes has made significant progress over the last several decades, this area of inquiry is still in its infancy. Compared to the number of known species of ferns and lycophytes, and especially in light of their importance in tropical ecosystems, relatively few species have been studied. Thus, there is still a pressing need for population genetic studies to provide more substantial evidence concerning patterns and processes of evolution in these important groups of land plants.

Areas of population genetic research that have been little explored, if at all, in ferns and lycophytes include studies of the genetics of reproductive fitness with respect to ecological diversity, the relationship between ploidy, fitness, and genetics, and outbreeding depression (but see Schneller, 1996). Also, although numerous studies have applied DNA-based techniques to molecular systematics, few have applied DNA markers to the study of population level questions. Genomic methods have yet to be applied at the population level to any taxa of ferns or lycophytes. The burgeoning field of ecological genomics holds great potential for understanding the interplay between genetic diversity within and among populations and such important evolutionary processes as adaptation and speciation. Knowledge of population genetic variation and the processes that generate and maintain genetic diversity are critical preludes for understanding the origin of species.

References

Andersson, I. (1923). The genetics of variegation in a fern. *Journal of Genetics*, **13**, 1–11.

Andersson, I. (1927). Note on some characters in ferns subject to Mendelian inheritance. *Hereditas*, **9**, 157–168.

Andersson-Köttö, I. (1929). A genetical investigation in *Scolopendrium vulgare*. *Hereditas*, **12**, 109–178.

Andersson-Köttö, I. (1930). Variegation in three species of ferns. *Zeitschrift fur induktive Abstammungs-und Vererbungs Lehre*, **56**, 115–201.

Andersson-Köttö, I. (1931). The genetics of ferns. *Bibliographica Genetica*, **8**, 269–294.

Aragón, C. E. and Pangua, E. (2003). Gender determination and mating system in the autotetraploid fern *Asplenium septentrionale* (L.) Hoffm. *Botanica Helvetica*, **113**, 181–193.

Atkinson, L. R. and Stokey, A. G. (1964). Comparative morphology of the gametophyte of homosporous ferns. *Phytomorphology*, **14**, 51–70.

Baker, H. G. (1955). Self-compatibility and establishment after 'long-distance' dispersal. *Evolution*, **3**, 347–349.

Baker, H. G. (1967). Support for Baker's Law – as a rule. *Evolution*, **4**, 853–856.

Chiou, W.-L., Farrar, D. R., and Ranker, T. A. (1998). Gametophyte morphology and reproductive biology in *Elaphoglossum* Schott. *Canadian Journal of Botany*, **76**, 1967–1977.

Chiou, W.-L., Farrar, D. R., and Ranker, T. A. (2002). The mating systems of some Polypodiaceae species. *American Fern Journal*, **92**, 65–79.

Clague, D. A. (1996). The growth and subsidence of the Hawaiian-Emperor volcanic chain. In *The Origin and Evolution of Pacific Island Biotas, New Guinea to Eastern Polynesia: Patterns and Processes*, ed. A. Keast and S. E. Miller. Amsterdam: SPB Academic Publishing, pp. 35–50.

Cousens, M. I. (1979). Gametophyte ontogeny, sex expression, and genetic load as measures of population divergence in *Blechnum spicant*. *American Journal of Botany*, **66**, 116–132.

Cousens, M. I., Lacey, D. G., and Kelly, E. M. (1985). Life history studies of ferns: a consideration of perspective. *Proceedings of the Royal Society of Edinburgh*, **86B**, 371–380.

Crist, K. C. and Farrar, D. R. (1983). Genetic load and long distance dispersal in *Asplenium platyneuron*. *Canadian Journal of Botany*, **61**, 1809–1814.

Crow J. and Kimura M. (1965). Evolution in sexual and asexual populations. *The American Naturalist*, **99**, 439–450.

Dobzhansky, T. (1970). *Genetics of the Evolutionary Process*. New York: Columbia University Press.

Dobzhansky, T. and Wright, S. (1941). Genetics of natural populations. V. Relations between mutation rate and accumulation of lethals in populations of *Drosophila pseudoobscura*. *Genetics*, **26**, 23–51.

Eastwood, A., Cronk, Q. C. B., Vogel, J. C., Hemp, A., and Gibby, M. (2004). Comparison of molecular and morphological data on St Helena: *Elaphoglossum*. *Plant Systematics and Evolution*, **245**, 93–106.

Flinn, K. M. (2006). Reproductive biology of three fern species may contribute to differential colonization success in post-agricultural forests. *American Journal of Botany*, **93**, 1289–1294.

Gastony, G. J. and Gottlieb, L. D. (1982). Evidence for genetic heterozygosity in a homosporous fern. *American Journal of Botany*, **69**, 634–637.

Gastony, G. J. and Gottlieb, L. D. (1985). Genetic variation in the homosporous fern *Pellaea andromedifolia*. *American Journal of Botany*, **72**, 257–267.

Gitzendanner, M. A. and Soltis, P. S. (2000). Patterns of genetic variation in rare and widespread plant congeners. *American Journal of Botany*, **87**, 783–792.

Greer, G. K. (1993). The influence of soil topography and spore-rain density on gender expression in gametophyte populations of the homosporous fern *Aspidotis densa*. *American Fern Journal*, **83**, 54–59.

Hamrick, J. L. and Godt, M. J. W. (1990). Allozyme diversity in plant species. In *Plant Population Genetics, Breeding, and Genetic Resources*, ed. A. H. D. Brown, M. T. Clegg, A. L. Kahler, and B. S. Weir. Sunderland, MA: Sinauer, pp. 43–63.

Haufler, C. H. (1985). Enzyme variability and modes of evolution in the fern genus *Bommeria*. *Systematic Botany*, **10**, 92–104.

Haufler, C. H. (1987). Electrophoresis is modifying our concepts of evolution in homosporous Pteridophytes. *American Journal of Botany*, **74**, 953–966.

Haufler, C. H. (1992). An introduction to fern genetics and breeding systems. In *Fern Horticulture: Past, Present and Future Perspectives*, ed. J. M. Ide, C. Jermy, and A. M. Paul. Andover: Intercept, pp. 145–155.

Haufler, C. H. (2002). Homospory 2002: an odyssey of progress in pteridophyte genetics and evolutionary biology. *Bioscience*, **52**, 1081–1093.

Haufler, C. H. and Soltis, D. E. (1984). Obligate outcrossing in a homosporous fern: field confirmation of a laboratory prediction. *American Journal of Botany*, **71**, 878–881.

Haufler, C. H. and Soltis, D. E. (1986). Genetic evidence suggests that homosporous ferns with high chromosome numbers are diploid. *Proceedings of the National Academy of Sciences of the United States of America*, **83**, 4389–4393.

Haufler, C. H., Windham, M. D., and Ranker, T. A. (1990). Biosystematic analysis of the *Cystopteris tennesseensis* (Dryopteridaceae) complex. *Annals of the Missouri Botanical Garden*, **77**, 314–329.

Hauk, W. D. and Haufler, C. H. (1999). Isozyme variability among cryptic species of *Botrychium* subgenus *Botrychium* (Ophioglossaceae). *American Journal of Botany*, **86**, 614–633.

Hedrick, P. W. (1987). Population genetics of intragametophytic selfing. *Evolution*, **41**, 137–144.

Hey, J. (2006). Recent advances in assessing gene flow between diverging populations and species. *Current Opinion in Genetics and Development*, **16**, 592–596.

Hickok, L. G. (1978a). Homoeologous chromosome pairing: frequency differences in inbred and intraspecific hybrid polyploid ferns. *Science*, **202**, 982–984.

Hickok, L. G. (1978b). Homoeologous chromosome pairing and restricted segregation in the fern *Ceratopteris*. *American Journal of Botany*, **65**, 516–521.

Holderegger, R. and Schneller, J. J. (1994). Are small isolated populations of *Asplenium septentrionale* variable? *Biological Journal of the Linnean Society*, **51**, 377–385.

Holsinger, K. E. (1987). Gametophytic self-fertilization in homosporous plants – development, evaluation, and application of a statistical-method for evaluating its importance. *American Journal of Botany*, **74**, 1173–1183.

Holsinger, K. E. (1990). The population genetics of mating system evolution in homosporous plants. *American Fern Journal*, **80**, 153–160.

Holsinger, K. E. (1991). Mass-action models of plant mating systems: the evolutionary stability of mixed mating systems. *The American Naturalist*, **138**, 606–622.

Hooper, E. A. and Haufler, C. H. (1997). Genetic diversity and breeding system in a group of neotropical epiphytic ferns (*Pleopeltis*; Polypodiaceae). *American Journal of Botany*, **84**, 1664–1674.

Ishikawa, H., Motomi, I., Watano, Y., and Kurita, S. (2003). Electrophoretic evidence for homoeologous chromosome pairing in the apogamous fern species *Dryopteris nipponensis* (Dryopteridaceae). *Journal of Plant Research*, **116**, 165–167.

Kang, M., Ye, Q., and Huang, H. (2005). Genetic consequence of restricted habitat and population decline in endangered *Isoetes sinensis* (Isoetaceae). *Annals of Botany*, **96**, 1265–1274.

Keiper, F. J. and McConchie, R. (2000). An analysis of genetic variation in natural populations of *Sticherus flabellatus* [R. Br. (St John)] using amplified fragment length polymorphism (AFLP) markers. *Molecular Ecology*, **9**, 571–581.

Kirkpatrick, R. E. B., Soltis, P. S., and Soltis, D. E. (1990). Mating system and distribution of genetic variation in *Gymnocarpium dryopteris* ssp. *disjunctum*. *American Journal of Botany*, **77**, 1101–1110.

Klekowski, E. J., Jr. (1969). Reproductive biology of the Pteridophyta. II. Theoretical considerations. *Botanical Journal of the Linnean Society*, **62**, 347–359.

Klekowski, E. J., Jr. (1970a). Populational and genetic studies of a homosporous fern – *Osmunda regalis*. *American Journal of Botany*, **57**, 1122–1138.

Klekowski, E. J., Jr. (1970b). Reproductive biology of the Pteridophyta. IV. An experimental study of mating systems in *Ceratopteris thalictroides* (L.) Brongn. *Journal of the Linnean Society, Botany*, **63**, 153–169.

Klekowski, E. J., Jr. (1971). Ferns and genetics. *Bioscience*, **21**, 317–322.

Klekowski, E. J., Jr. (1972a). Genetical features of ferns as contrasted to seed plants. *Annals of the Missouri Botanical Garden*, **59**, 138–151.

Klekowski, E. J., Jr. (1972b). Evidence against genetic self-incompatibility in the homosporous fern *Pteridium aquilinum*. *Evolution*, **26**, 66–73.

Klekowski, E. J. Jr. (1973). Genetic load in *Osmunda regalis* populations. *American Journal of Botany*, **60**, 146–154.

Klekowski, E. J., Jr. (1979). The genetics and reproductive biology of ferns. In *The Experimental Biology of Ferns*, ed. A. F. Dyer. London: Academic Press, pp. 133–170.

Klekowski, E. J., Jr. (1988). *Mutation, Developmental Selection, and Plant Evolution*. New York: Columbia University Press.

Klekowski, E. J., Jr. and Baker, H. G. (1966). Evolutionary significance of polyploidy in the Pteridophyta. *Science*, **153**, 305–307.

Lande, R. and Schemske, D. W. (1985). The evolution of self-fertilization and inbreeding depression in plants. I. Genetic models. *Evolution*, **39**, 24–40.

Lang, W. H. (1923). On the genetic analysis of a heterozygotic plant of *Scolopendrium vulgare*. *Journal of Genetics*, **13**, 167–175.

Leimu, R., Mutikainen, P., Koricheva, J., and Fischer, M. (2006). How general are positive relationships between plant population size, fitness and genetic variation? *Journal of Ecology*, **94**, 942–952.

Li, J. and Haufler, C. H. (1994). Phylogeny, biogeography, and population biology of *Osmunda* species: insights from isozymes. *American Fern Journal*, **85**, 105–114.

Li, J. W. and Haufler, C. H. (1999). Genetic variation, breeding systems, and patterns of diversification in Hawaiian *Polypodium* (Polypodiaceae). *Systematic Botany*, **24**, 339–355.

Lloyd, R. M. (1974). Mating systems and genetic load in pioneer and non-pioneer Hawaiian Pteridophyta. *Botanical Journal of the Linnean Society*, **69**, 23–35.

Lloyd, R. M. and Warne, T. R. (1978). The absence of genetic load in a morphologically variable sexual species, *Ceratopteris thalictroides* (Parkeriaceae). *Systematic Botany*, **3**, 20–36.

Lott, M. S., Volin, J. C., Pemberton, R. W., and Austin, D. F. (2003). The reproductive biology of the invasive ferns *Lygodium microphyllum* and *L. japonicum* (Schizaeaceae): implications for invasive potential. *American Journal of Botany*, **90**, 1144–1152.

Maki, M. and Asada, Y.-J. (1998). High genetic variability revealed by allozymic loci in the narrow endemic fern *Polystichum otomasui* (Dryopteridaceae). *Heredity*, **80**, 604–610.

Masuyama, S. (1979). Reproductive biology of the fern *Phegopteris decursive-pinnata*. I. The dissimilar mating systems of diploids and tetraploids. *Botanical Magazine (Tokyo)*, **92**, 275–289.

Masuyama, S. and Watano, Y. (1990). Trends for inbreeding in polyploid pteridophytes. *Plant Species Biology*, **5**, 13–17.

Masuyama, S., Mitui, K., and Nakato, N. (1987). Studies on intraspecific polyploids of the fern *Lepisorus thunbergianus*. (3) Mating system and the ploidy. *Journal of Japanese Botany*, **62**, 321–331.

McCauley, D. E., Whittier, D. P., and Reilly, L. M. (1985). Inbreeding and the rate of self-fertilization in a grape fern, *Botrychium dissectum*. *American Journal of Botany*, **72**, 1978–1981.

Muller, H. J. (1950). Our load of mutations. *The American Journal of Human Genetics*, **2**, 111–176.

Nei, M. (1973). Analysis of gene diversity in subdivided populations. *Proceedings of the National Academy of Sciences of the United States of America*, **70**, 3321–3323.

Nei, M. (1977). F-statistics and analysis of gene diversity in subdivided populations. *Annals of Human Genetics*, **41**, 225–233.

Nei, M. (1978). Estimation of average heterozygosity and genetic distance from a small number of individuals. *Genetics*, **89**, 583–590.

Otto, S. P. and Marks, J. C. (1996). Mating systems and the evolutionary transition between haploidy and diploidy. *Biological Journal of the Linnean Society*, **57**, 197–218.

Pangua, E. and Vega, B. (1996). Comparative study of gametophyte development in *Cosentinia* and *Anogramma* (Hemionitidaceae) and *Cheilanthes* (Sinopteridaceae). In *Pteridology in Perspective*, ed. J. M. Camus, M. Gibby, and R. J. John. Kew: Royal Botanic Gardens, pp. 497–508.

Peck, J. H., Peck, C. J., and Farrar, D. R. (1990). Comparative life history studies and the distribution of pteridophyte populations. *American Fern Journal*, **80**, 126–142.

Pérez-García, B. and Riba, R. (1998). Bibliografía sobre Gametofitos de Helechos y Plantas Afines. *Monographs in Systematic Botany from the Missouri Botanical Garden*, Vol. 70, ed. V. C. Hollowell. St. Louis, MO: Missouri Botanical Garden Press.

Price, J. P. and Clague, D. A. (2002). How old is the Hawaiian biota? Geology and phylogeny suggest recent divergence. *Proceedings of the Royal Society of London Series B, Biological Sciences*, **269**, 2429–2435.

Pryor, K. V., Young, J. E., Rumsey, F. J., Edwards, K. J., Bruford, M. W., and Rogers, H. J. (2001). Diversity, genetic structure and evidence of outcrossing in British populations of the rock fern *Adiantum capillus-veneris* using microsatellites. *Molecular Ecology*, **10**, 1881–1894.

Quintanilla, L. G., Pangua, E., Amigo, J., and Pajarón. S. (2005). Comparative study of the sympatric ferns *Culcita macrocarpa* and *Woodwardia radicans*: sexual phenotype. *Flora*, **200**, 187–194.

Rabinowitz, D. (1981). Seven forms of rarity. In *The Biological Aspects of Rare Plant Conservation*, ed. H. Synge. Chichester: Wiley, pp. 205–217.

Ranker, T. A. (1987). Experimental systematics and population biology of the fern genera *Hemionitis* and *Gymnopteris* with reference to *Bommeria*. Unpublished Ph.D. Thesis, University of Kansas, Lawrence, KS.

Ranker, T. A. (1992a). Genetic diversity, mating systems, and interpopulation gene flow in neotropical *Hemionitis palmata* L. (Adiantaceae). *Heredity*, **69**, 175–183.

Ranker, T. A. (1992b). Genetic diversity of endemic Hawaiian epiphytic ferns: implications for conservation. *Selbyana*, **13**, 131–137.

Ranker, T. A. (1994). Evolution of high genetic variability in the rare Hawaiian fern *Adenophorus periens* and implications for conservation management. *Biological Conservation*, **70**, 19–24.

Ranker, T. A. and Houston, H. A. (2002). Is gametophyte sexuality in the lab a good predictor of sexuality in nature? *Sadleria* as a case study. *American Fern Journal*, **92**, 112–118.

Ranker, T. A., Floyd, S. K., and Trapp, P. G. (1994). Multiple colonizations of *Asplenium adiantum-nigrum* onto the Hawaiian Archipelago. *Evolution*, **48**, 1364–1370.

Ranker, T. A., Gemmill, C. E. C., Trapp, P. G., Hambleton, A., and Ha, K. (1996). Population genetics and reproductive biology of lava-flow colonising species of Hawaiian *Sadleria* (Blechnaceae). In *Pteridology in Perspective*, ed. J. M. Camus, M. Gibby, and R. J. John. Kew: Royal Botanic Gardens, pp. 581–598.

Ranker, T. A., Gemmill, C. E. C., and Trapp, P. G. (2000). Microevolutionary patterns and processes of the native Hawaiian colonizing fern *Odontosoria chinensis* (Lindsaeaceae). *Evolution*, **54**, 828–839.

St. John, E. P. (1949). The evolution of the Ophioglossaceae of the eastern United States. *Quarterly Journal of the Florida Academy of Sciences*, **12**, 207–219.

Schneider, H., Ranker, T. A., Russell, S. J., Cranfill, R., Geiger, J. M. O., Aguraiuja, R., Wood, K. R., Grundmann, M., Kloberdanz, K., and Vogel, J. C. (2005). Origin and diversification of the Hawaiian fern genus *Diellia* Brack. (Aspleniaceae, Polypodiidae). *Proceedings of the Royal Society of London Series B, Biological Sciences*, **272**, 455–460.

Schneller, J. J. (1979). Biosystematic investigations on the Lady Fern (*Athyrium filix-femina*). *Plant Systematics and Evolution*, **132**, 255–277.

Schneller, J. J. (1996). Outbreeding depression in the fern *Asplenium ruta-muraria* L: evidence from enzyme electrophoresis, meiotic irregularities and reduced spore viability. *Biological Journal of the Linnean Society*, **59**, 281–295.

Schneller, J. J. and Holderegger, R. (1996). Genetic variation in small, isolated fern populations. *Journal of Vegetation Science*, **7**, 113–120.

Sciarretta, K. L., Potter Arbuckle, E., Haufler, C. H., and Werth, C. R. (2005). Patterns of genetic variation in southern Appalachian populations of *Athyrium filix-femina* var. *asplenioides* (Dryopteridaceae). *International Journal of Plant Science*, **166**, 761–780.

Slatkin, M. (1985a). Gene flow in natural populations. *Annual Review of Ecology and Systematics*, **16**, 393–430.

Slatkin, M. (1985b). Rare alleles as indicators of gene flow. *Evolution*, **39**, 53–65.

Slatkin, M. and Barton, N. H. (1989). A comparison of three indirect methods for estimating average levels of gene flow. *Evolution*, **43**, 1349–1368.

Soltis, D. E., and Soltis, P. S. (1986). Electrophoretic evidence for inbreeding in the fern *Botrychium virginianum* (Ophioglossaceae). *American Journal of Botany*, **73**, 588–592.

Soltis, D. E., and Soltis, P. S. (1987a). Breeding system of the fern *Dryopteris expansa*: evidence for mixed-mating. *American Journal of Botany*, **74**, 504–509.

Soltis, D. E., and Soltis, P. S. (1987b). Polyploidy and breeding systems in homosporous Pteridophyta: a reevaluation. *The American Naturalist*, **130**, 219–232.

Soltis, P. S., and Soltis, D. E. (1987c). Population structure and estimates of gene flow in the homosporous fern *Polystichum munitum*. *Evolution*, **41**, 620–629.

Soltis, P. S. and Soltis, D. E. (1988a). Genetic variation and population structure in the fern *Blechnum spicant* (Blechnaceae) from western North America. *American Journal of Botany*, **75**, 37–44.

Soltis, P. S. and Soltis, D. E. (1988b). Estimated rates of intragametophytic selfing in lycopods. *American Journal of Botany*, **75**, 248–256.

Soltis, P. S. and Soltis, D. E. (1990a). Genetic variation within and among populations of ferns. *American Fern Journal*, **80**, 161–172.

Soltis, P. S. and Soltis, D. E. (1990b). Evolution of inbreeding and outcrossing in ferns and fern-allies. *Plant Species Biology*, **5**, 1–11.

Soltis, D. E., and Soltis, P. S. (1992). The distribution of selfing rates in homosporous ferns. *American Journal of Botany*, **79**, 97–100.

Soltis, P. S., Soltis D. E., and Holsinger, K. E. (1988a). Estimates of intragametophytic selfing and interpopulational gene flow in homosporous ferns. *American Journal of Botany*, **75**, 1765–1770.

Soltis, P. S., Soltis, D. E., and Noyes, R. D. (1988b). An electrophoretic investigation of intragametophytic selfing in *Equisetum arvense*. *American Journal of Botany*, **75**, 231–237.

Soltis, P. S., Soltis, D. E., and Ness, B. D. (1989). Population genetic-structure in *Cheilanthes gracillima*. *American Journal of Botany*, **76**, 1114–1118.

Soltis, P. S., Soltis, D. E., and P. G. Wolf. (1990). Allozymic divergence and species relationships in North American *Polystichum* (Dryopteridaceae). *Systematic Botany*, **15**, 205–215.

Stebbins, G. L. (1957). Self fertilization and population variability in the higher plants. *The American Naturalist*, **91**, 337–354.

Stokey, A. G. and Atkinson, L. R. (1958). The gametophyte of the Grammitidaceae. *Phytomorphology*, **8**, 391–403.

Su, Y., Wang, T., Zheng, B., Jiang, Y., Chen, G., and Gu, H. (2004). Population genetic structure and phylogeographical pattern of a relict tree fern, *Alsophila spinulosa* (Cyatheaceae), inferred from cpDNA *atpB-rbcL* intergenic spacers. *Theoretical and Applied Genetics*, **109**, 1459–1467.

Suter, M., Schneller, J. J., and Vogel, J. C. (2000). Investigations into the genetic variation, population structure, and breeding systems of the fern *Asplenium trichomanes* subsp. *quadrivalens*. *International Journal of Plant Science*, **161**, 233–244.

Swofford, D. L. and Selander, R. B. (1989). *BIOSYS-1. A computer program for the analysis of allelic variation in population genetics and biochemical systematics, Release 1.7*. Urbana, IL: Illinois Natural History Survey.

Tryon, R. M. and Tryon, A. F. (1982). *Ferns and Allied Plants*. New York: Springer-Verlag.

Vitalis, R., Riba, M., Colas, B., Grillas, P., and Olivieri, I. (2002). Multilocus genetic structure at contrasted spatial scales of the endangered water fern *Marsilea strigosa* Willd. (Marsileaceae, Pteridophyta). *American Journal of Botany*, **89**, 1142–1155.

Vogel, J. C., Rumsey, F. J., Russell, S. J., Cox, C. J., Holmes, J. S., Bujnoch, W., Starks, C., Barrett, J. A., and Gibby, M. (1999). Genetic structure, reproductive biology and ecology of isolated populations of *Asplenium csikii* (Aspleniaceae, Pteridophyta). *Heredity*, **83**, 604–612.

Vos, P., Hogers, R., Bleeker, M., Reijans, M., Vandelee, T., Hornes, M., Frijters, A., Pot, J., Peleman, J., Kuiper, M., and Zabeau, M. (1995). AFLP: a new technique for DNA fingerprinting. *Nucleic Acids Research*, **23**, 4407–4414.

Wallace, B. (1970). *Genetic Load – Its Biological and Conceptual Aspects*. Englewood Cliffs, NJ: Prentice-Hall.

Watano, Y. (1988). High levels of genetic divergence among populations in a weedy fern, *Pteris multifida* Poir. *Plant Species Biology*, **3**, 109–115.

Watano, Y., and Masuyama, S. (1991). Inbreeding in natural populations of the annual polyploid fern *Ceratopteris thalictroides* (Parkeriaceae). *Systematic Botany*, **16**, 705–714.

Watano, Y., and Sahashi, N. (1992). Predominant inbreeding and its genetic consequences in a homosporous fern genus, *Sceptridium* (Ophioglossaceae). *Systematic Botany*, **17**, 486–502.

Werth, C. R. and Cousens, M. I. (1990). Summary: the contributions of population studies on ferns. *American Fern Journal*, **80**, 183–190.

Werth, C. R., Guttman, S. I., and Eshbaugh, W. H. (1985). Electrophoretic evidence of reticulate evolution in the Appalachian *Asplenium* complex. *Systematic Botany*, **10**, 184–192.

Wolf, P. G., Haufler, C. H., and Sheffield, E. (1988). Electrophoretic variation and mating system of the clonal weed *Pteridium aquilinum* (L.) Kuhn (Bracken). *Evolution*, **42**, 1350–1355.

Wolf, P. G., Sheffield, E., and Haufler, C. H. (1990). Genetic attributes of bracken as revealed by enzyme electrophoresis. In *Bracken Biology and Management*, ed. J. A. Thomson and R. T. Smith. Hawthorn, Victoria: Australian Institute of Agriculture and Science, Occasional Paper Number 40, pp. 71–78.

Wolf, P. G., Sheffield, E., and Haufler, C. H. (1991). Estimates of gene flow, genetic substructure and population heterogeneity in bracken (*Pteridium aquilinum*). *Biological Journal of the Linnean Society*, **42**, 407–423.

Wright, S. (1931). Evolution in Mendelian populations. *Genetics*, **16**, 97–159.

Wright, S. (1943). Isolation by distance. *Genetics*, **28**, 114–138.

Wright, S. (1951). The genetical structure of populations. *Annals of Eugenics*, **15**, 323–354.

Wright, S. (1965). The interpretation of population structure by F-statistics with special regard to systems of mating. *Evolution*, **19**, 395–420.

Wright, S. (1969). *Evolution and the Genetics of Populations*, Vol. 2, *The Theory of Gene Frequencies*. Chicago, IL: University of Chicago Press.

Wright, S. (1978). *Evolution and the genetics of populations*, Vol. 4, *Variability within and among natural populations*. Chicago, IL: University of Chicago Press.

5

Antheridiogens

JAKOB J. SCHNELLER

5.1 Introduction

In homosporous ferns individual gametophytes are generally able to form both antheridia and archegonia. No genetic regulation that determines the sex of the haploid generation has been demonstrated. Growth, temperature, light conditions, environmental characteristics, soil conditions, and, in many cases, antheridia-inducing substances can influence the development of antheridia and archegonia (Voeller, 1964; Miller, 1968; Voeller and Weinberg, 1969). We can therefore describe homosporous ferns as having labile sex expression (Korpelainen, 1998).

The antheridia-inducing substances are called antheridiogens, and are products (hormone-like substances) of the metabolism of prothalli. The term antheridiogen characterizes the function but not the chemical composition. In the literature there is a variety of different terms for antheridiogen, for instance, A-substance (Döpp, 1950, 1959, 1962), antheridogen (Pringle, 1961), pheromone (e.g., Näf et al., 1975; Scott and Hickok, 1987), and hormone (e.g., Näf, 1962; Näf et al., 1975; Raghavan, 1989). Schraudolf (1985) distinguished between the pheromonal (effective on neighboring individuals) and the hormonal (effective within an individual plant) phase of antheridiogens. Here, we will use antheridiogen, the term that is favored in the literature.

5.2 History of discovery

Döpp (1950) was the first to discover a naturally produced substance that induces antheridia formation in young prothalli. He showed that substrate

Biology and Evolution of Ferns and Lycophytes, ed. Tom A. Ranker and Christopher H. Haufler. Published by Cambridge University Press. © Cambridge University Press 2008.

from maturing prothallial cultures of bracken (*Pteridium aquilinum*) induced antheridia formation in young prothalli of its own species and those of *Dryopteris filix-mas*. The same was true also when using aquatic extractions. Döpp (1950) interpreted his observation by proposing the presence of a substance he called A-substance, which was active at very low concentrations and was water soluble. It was highly chemically and biologically stable. This pioneering work was supplemented by two later publications (Döpp, 1959, 1962). In 1959 Döpp tested *Cryptogramma crispa, Matteuccia struthiopteris, Gymnocarpium robertianum, Pellaea viridis, Notholaena sinuata, N. distans,* and *N. vellea*. Prothalli from all of these species reacted to the "A-substance" (antheridiogen) of *Pteridium aquilinum* by forming antheridia. Döpp (1959) observed that if a prothallus of *P. aquilinum* becomes older and develops a multilayered central part (i.e., becomes meristic), it produces antheridiogen but no longer reacts to it. However, in regeneration experiments, excising parts of female bracken gametophytes or applying incisions, Döpp (1959) realized that only the multilayered, meristic part did not react to the antheridiogen whereas severed parts became male. In 1956, Näf detected another antheridiogen that showed activity in the genus *Anemia* but not in the species reacting to the antheridiogen of *Pteridium* (Table 5.1). Schedlbauer and Klekowski (1972) found a third type of antheridiogen in *Ceratopteris*. The antheridiogen active in *Vittaria* has not yet been associated with one of the known main types of antheridiogen (Emigh and Farrar, 1977), nor has an additional, possibly different, type of antheridiogen active in *Asplenium* (Schneller and Hess, 1995).

5.3 General effect of antheridiogen

Since Döpp's initial discovery, the phenomenon of sex determination by antheridiogens has been studied by many different authors and under different experimental conditions (Pour *et al.*, 1998; see also reviews by Näf, 1979, and Yamane, 1998). Based on these investigations, the following general effects of antheridiogen can be described. Prothalli of multispore cultures will remain sterile until some have started to develop a meristic stage (more than one cell layer in the central part of the prothallus). They then start to produce antheridiogen, which diffuses into the environment. Prothalli that have not yet reached the meristic stage will react to the antheridiogen by producing antheridia and by having a slower rate of growth. Meristic prothalli in many fern species, however, are no longer sensitive to antheridiogen; this is true, for example, in *Pteridium aquilinum, Athyrium filix-femina, Sadleria cyatheoides, Bommeria* spp., and many other species (Table 5.1). In other taxa such as *Asplenium ruta-muraria, A. trichomanes,*

Table 5.1 *The occurrence of antheridiogens in different fern genera; species are only mentioned when differences of response were observed*

Genus or species	Activity		Dark germination		Reference
	A_{pt}	Own antheridiogen	A_{pt}	Own antheridiogen	
A_{pt}					
Adiantum	+				Voeller, 1964
Aglaomorpha	+				Näf, 1969
Anogramma	−				Baroutsis, 1976
Alsophila	−				Näf, 1960
Aspidotis densa		+			Greer, 1991
Asplenium (+ Ceterach)	−	+			Döpp, 1959; Schneller and Hess, 1995
Athyrium	+	+		+	Näf, 1956; Schneller, 1979
Blechnum brasiliense	+				Voeller, 1964
Blechnum gibbum		+			Näf, 1956
Blechnum occidentale	−				Voeller, 1964
Bommeria	+	+	+	+	Haufler and Welling, 1994
Campyloneuron	+	+	+	+	Chiou and Farrar, 1997
Cibotium	−				Voeller, 1964
Cryptogramma	+				Döpp, 1959; Pajaron et al., 1999
Culcita	−				Quintanilla et al., 2005
Cyathea podophylla	−				Chiou et al., 2000
Cyathea		+			Khare et al., 2006
Cyclosorus	−				Voeller, 1964
Cyrtomium	−				Voeller, 1964
Cystopteris	+	−			Haufler and Ranker, 1985
Davallia	−				Voeller, 1964
Dennstaedtia bipinnata	−				Voeller, 1964
Dennstaedtia punctilobula	+				Näf, 1959
Doodia	+				Voeller, 1964
Drynaria	−				Voeller, 1964
Dryopteris filix-mas	+	+	+	+	Döpp, 1950; Schneller, 1988
Dryopteris dilatata	−				Voeller, 1964
Elaphoglossum	−	−			Chiou et al., 1998
Gonophlebium	−				Voeller, 1964
Gymnocarpium	+				Döpp, 1959
Hemionitis	+				Voeller, 1964
Lepisorus	+	+	+	+	Chiou and Farrar, 1997
Matteuccia	+				Näf 1956; Döpp, 1959
Micrograma	+	+	+	+	Chiou and Farrar, 1997
Microlepia					Fellenberg-Kressel, 1969
Microsorium	−				Voeller, 1964

(cont.)

Table 5.1 (cont.)

Genus or species	Activity		Dark germination		Reference
	A_{pt}	Own antheridiogen	A_{pt}	Own antheridiogen	
Nephrolepis	+				Näf, 1960
Notholaena	+				Döpp, 1959
Onoclea	+	+			Näf, 1956
Osmunda	+				DeVol et al., 2005
Pellaea	+				Döpp, 1959
Phanerophlebia	−				Yatskievych, 1993
Phlebodium	+	−	−	−	Chiou and Farrar, 1997
Phymatosorus	+	+	+	+	Chiou et Farrar, 1997
Pityrogramma	−				Voeller, 1964
Pleopeltis		+			Hooper and Haufler, 1997
Polypodium	−	−			Näf, 1956
Polypodium crassifolium		−			Yatskievych, 1993
Polypodium subauriculatum		−			Yatskievych, 1993
Polypodium pellucidum	+	+	+		Chiou and Farrar, 1997
Polystichum	+				Näf, 1956
Pteridium	+	+			Döpp, 1950
Pteris cretica	−				Voeller, 1964
Pteris tremula	−				Voeller, 1964
Pteris longifolia	+				Voeller, 1964
Pteris vittata	+				Gemmrich, 1986
Sadleria		+			Holbrook-Walker and Lloyd, 1973
Sphaeropteris lepifera					Chiou, 1999
Tectaria macrodonta	−				Voeller, 1964
Tectaria heracleifolia	−				Voeller, 1964
Tectaria incisa	+				Näf et al., 1975
Thelypteris hexagonoptera	+	+			Näf, 1959
Thelypteris ovata		+			Nester-Hudson et al., 1997
Vittaria	+				Emigh and Farrar, 1977
Woodsia	+				Näf, 1959
Woodwardia	+				Näf, 1956
A_{an}					
Anemia	−	+	+		Näf, 1959; Näf, 1966
Lygodium	−	+	+		Näf, 1959; Sugai et al., 1987
Mohria	−	−			Näf, 1960
A_{ce}					
Ceratopteris	+	+			Schedlbauer and Klekowski, 1972

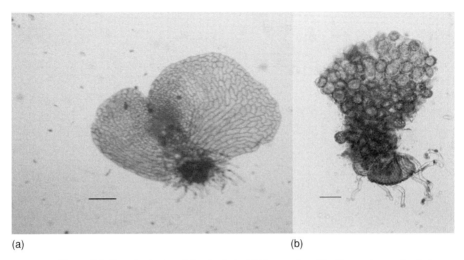

Figure 5.1 Gametophytes of *Ceratopteris*. (a) Hermaphrodite bisexual gametophyte grown when no antheridiogen is present. Stage of producing the pheromone (scale bar 100 μm). (b) Male gametophyte differentiated under the influence of antheridiogen (scale bar 20 μm).

Ceratopteris spp., *Dryopteris filix-mas, D. dilatata*, or *Anemia* spp. the ameristic parts of meristic gametophytes do not lose their sensitivity to antheridiogen and, therefore, the prothalli become hermaphroditic (Figure 5.1). In most hermaphrodites, the areas bearing antheridia and archegonia are separated by a sterile zone (Klekowski, 1969).

5.4 Occurrence of different antheridiogens and their chemical structure

Because the chemical structure of only one group of antheridiogens has been fully characterized, it still remains unknown how many different forms occur. Schedlbauer (1974) distinguished three main types based on their activity. The first type, usually called A_{pt}, is the antheridiogen produced by *Pteridium* and many other ferns (Table 5.1). Members of the Schizaeaceae produce the second type of antheridiogen, A_{an} ($\approx A_{ly}$). Näf (1956) first described A_{an} for *Anemia phyllitidis*. He showed that it belongs to a chemically different class from A_{pt} because it is not active in the species that react to A_{pt}. A_{an} also does not elicit a response in species of *Ceratopteris*, which produces the third type of antheridiogen (in some publications called A_{ce}) (Schedlbauer and Klekowski, 1972).

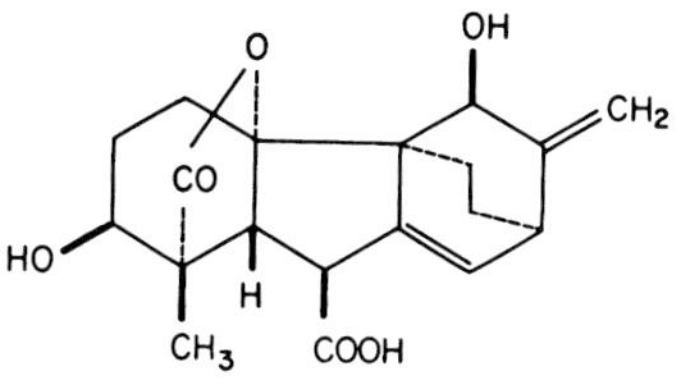

Figure 5.2 Structure of the antheridiogen of *Anemia phyllitidis* which is now called antheridic acid. (After Corey et al., 1986.)

Pringle et al. (1960) isolated the active form of A_{pt}, and examined its chemical properties (Pringle, 1961). It can be dissolved in water or in acetyl acetate and is destroyed by oxidation. A_{pt} behaves as a weak acid. However, to date it has still not been possible to isolate the hormone entirely or describe its chemical structure. Nester-Hudson et al. (1997) indicated that there are at least two principles involved in A_{pt} activity. As with all the antheridia-inducing substances, it is active at very low concentrations.

The antheridiogen of the Schizaeaceae can also be dissolved in water or acetyl acetate. In contrast to A_{pt}, antheridial formation is not inhibited by indole acetic acid (IAA). Experiments by Schraudolf (1962) revealed that in *Anemia* the influence of antheridiogen can be replaced by the gibberellic acid GA_3. He showed that very low concentrations of gibberellic acid (between 5×10^{-5} and 5×10^{-9} g ml^{-1}) were sufficient to induce antheridial formation in young prothalli. It was later shown that among seven structurally different gibberellic acids, GA_7 was the most active (effective at concentrations down to 5×10^{-9}) but also that other forms of GA could induce antheridia formation at low concentrations (Schraudolf, 1964, 1966). The structure of A_{an} was published by Nakanishi et al. (1971) and Corey et al. (1986) (Figure 5.2). Study of the activity of gibberellic acids and natural antheridiogens in the Schizaeaceae has revealed the existence of different, structurally similar, but not identical, antheridiogens (gibberellins), which occur in the family Schizaeaceae (summarized by Yamane, 1998). One of the important active forms is antheridic acid. Analogous differences are likely to occur also among the species that produce and respond to A_{pt}-type antheridiogens (Näf et al., 1962). The antheridiogens that initiate antheridia in *Pteridium aquilinum* and many other fern species (A_{pt}) are characterized by chemical variation among species, as demonstrated by Gemmrich (1986), who observed that antheridiogens produced by *Pteris vittata* and *Pteridium aquilinum* induce different responses in *Onoclea sensibilis*. Chemical differences have also been observed by different students of Hudson (1999). More recent studies concerning the relationship between antheridiogens and gibberellins are reviewed by Mander (2003).

The chemical structure of the antheridiogen produced by *Ceratopteris* species (A_{ce}) is unknown. However, this antheridiogen seems to be synthesized via a pathway that may include steps in common with gibberellin biosynthesis (Warne and Hickok, 1989). Wynne *et al.* (1998) revealed that this antheridiogen is related to gibberellic acid but is structurally and functionally different from those found in *Anemia*. Both A_{an} and A_{ce} can be inhibited by abscisic acid (Hickok, 1983; Schraudolf, 1987).

Schneller and Hess (1995) showed that young prothalli of *Asplenium ruta-muraria* became male when exposed to substrate from an old, mixed-sex culture or to conspecific older hermaphroditic prothalli. This is the first indication of an antheridiogen in the Aspleniaceae. The differences and/or similarities to the other antheridiogens have not yet been fully analyzed. Döpp (1959, 1962) showed that *Asplenium trichomanes* was not influenced by antheridiogen A_{pt}. Young prothalli of *A. ruta-muraria* do not react to gibberellic acid (GA_3) (Schneller and Hess, 1995) or to A_{pt} (Schneller, unpublished results).

Investigations by Scott and Hickok (1987) revealed differences in response to antheridiogen within species. Scott and Hickok (1987) showed that different strains of *Ceratopteris richardii* exhibited different sensitivities to the antheridiogen A_{ce}. Ranker (1987) described similar effects for *Hemionitis palmata*. Stevens and Werth (1999) observed dose-mediated responses to antheridiogen in *Onoclea sensibilis*. Different sensitivities may even be found between individual gametophytes within populations as shown for *Gymnocarpium dryopteris* ssp. *disjunctum* (Kirkpatrick and Soltis, 1992) and for *Sadleria cyatheoides* (Ranker *et al.*, 1996).

Antheridiogens can be inhibited by different substances depending on their chemical structure. A_{an} and A_{ce} become ineffective when treated with inhibitors of gibberellic acid (e.g., Warne and Hickok, 1989; Banks, 1999). Döpp (1962) showed that indole acetic acid (IAA) inhibited the action of A_{pt}. IAA also inhibits some gibberellic acids.

5.5 Experiments under laboratory conditions

Most studies of antheridiogens have been performed in laboratories. The majority of ferns can easily be cultured using artificial substrates. To produce aseptic cultures spores are surface sterilized using disinfectant, in most cases sodium hypochloride (4–6%; Dyer, 1979). Agar (usually 1%), supplemented with nutrient solutions has been the most commonly used artificial substrate (e.g., Döpp, 1964; Schraudolf, 1964; Voeller, 1964; Klekowski, 1969; Dyer, 1979). In mixed cultures, prothalli that become male due to antheridiogen induction remain male up to a size greater than that of prothalli that become female when reaching the meristic stage before antheridiogen is produced. These males may also become heart shaped, similar to female gametophytes (Näf et al., 1975).

One of the most frequently used plants to test the occurrence of antheridiogen has been *Onoclea sensibilis*. Näf (1965) and Näf et al. (1975) reported that gametophytes of this species failed to become male spontaneously in agar cultures. However, Rubin and Paolillo (1983) and Rubin et al. (1985) showed that when growing on soil without exogenously applied antheridiogen, gametophytes become male or female or hermaphroditic. Under such conditions the species seems to develop its own active antheridiogen. On agar, however, the antheridiogen will become active only when the substrate of mature prothalli is heat treated (Näf, 1965) or the culture is maintained for several weeks (Rubin et al., 1985).

There are some arguments that aseptic agar media with inorganic components may differentially influence the growth and development of the sex organs compared to natural conditions. In *Onoclea sensibilis* the development of antheridia and archegonia not only depends on the presence or absence of antheridiogen but also on the type of growth medium. Rubin et al. (1985) found that agar cultures promoted femaleness, whereas ash and soil cultures promoted maleness.

The gametangia and mainly the archegonia are negatively phototropic and are thus directed towards the substrate on agar. Agar media are partly transparent so the lower side of the prothallus may also be influenced by this somewhat unnatural condition (see Chapter 9). Some authors, therefore, have used sterilized or partly sterilized soil (i.e., pouring boiling water onto the soil) in attempts to produce substrates that may be closer to natural conditions (Schneller, 1979; Rubin and Paolillo, 1984). More recently Greer and McCarthy (1997) used sieved natural soil that had been steam sterilized. However, completely sterilized, autoclaved soil appeared toxic to fern spores (Rubin and Paolillo, 1984). A potential problem with using unsterilized or partly sterile soil is that the antheridia-inducing factor could be metabolic products from microorganisms and their interactions with fern prothalli. However, several studies have shown that the behavior of prothalli in nature is generally comparable to that found in completely sterile agar cultures (e.g., Von Aderkas, 1983; Ranker and Houston, 2002), although caution should be used when interpreting laboratory-based studies of gametophytes.

The concentration of antheridiogen that will induce the formation of antheridia varies among species. *Onoclea sensibilis* is the most sensitive species, and it reacts at hormone titrates of culture filtrates of approximately 1:250 000 to 1:300 000 (Voeller, 1964). *Pteridium aquilinum* requires a concentration that is about three times higher, and other species start to react at antheridiogen concentrations that are about 10 000 times higher (Näf, 1979). Comparable conditions are also found in the Schizaeaceae where differences in the reaction to varying concentrations of antheridiogen (and GA) can be seen, for instance, between *Anemia* and *Lygodium* (Näf, 1960) and also across species within the genera (Yamane, 1998).

Experiments with *Blechnum spicant* by Mendenez et al. (2006a, 2006b) revealed that antheridiogen-induced regulation of sex may be much more complicated than previously thought and many aspects are still not known. If grown from spores, prothalli start to produce antheridiogen just before becoming female. When regenerated from homogenized mature gametophytes, male gametophytes also produce antheridiogen. This has been interpreted as being dependent on the physiological conditions. Further investigation will be necessary to reveal the possible role in antheridiogen synthesis of the condition and the age of the gametophyte population.

Greer (1991, 1993), Greer and McCarthy (1997), and Korpelainen (1994) showed that not only the amount but also the origin of the spores and the spore rain density affected sex determination and growth. Efforts were taken to discover the range of activity of antheridiogens. Special culture conditions showed that the antheridiogen of one meristic *Pteridium aquilinum* gametophyte could be detected at about 25 cm from the source of production (Voeller and Weinberg, 1969) where it was still capable of inducing antheridia.

An interesting and additional role of antheridiogen was found in the genus *Vittaria*. Emigh and Farrar (1977) discovered that gemmae from *Vittaria* gametophytes were sensitive to a pheromone that is structurally and functionally different from A_{pt}. Although its chemical composition is unknown, it can be replaced by gibberellic acid and thus resembles the antheridiogen of the family Schizaeaceae (Emigh and Farrar, 1977).

In apogamous ferns that have been investigated, some species produce antheridiogen, some do not. Some gametophytes react whereas others lack any reaction to the antheridiogen. My experiments with the apogamous *Dryopteris affinis* group showed that older prothalli of the diploid and triploid genotypes induced antheridial formation (Schneller, unpublished results). In the genus *Bommeria* the triploid apogamous species *B. pedata* produces and reacts to antheridiogen (Haufler and Gastony, 1978). Yatskievych (1993) found that the apogamous *Cyrtomium falcatum* did not produce its own antheridiogen nor did it respond to A_{pt}, whereas the also apogamous species *C. fortunei* and *C. macrophyllum* produce antheridiogen (to which *Onoclea sensibilis* gametophytes respond) but do not react to it.

The genus *Ceratopteris* has become a model fern system for developmental, physiological, genetic, and molecular investigations (Chasan, 1992; Eberle et al.,

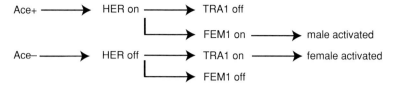

Figure 5.3 Genetic regulation of sex expression in *Ceratopteris* when antheridiogen is present or absent.

1995; Hickok and Warne, 1998). Several studies have evaluated the genetic regulation of sex expression in the gametophytes of this genus (Banks et al., 1993; Banks, 1994, 1997; Eberle and Banks, 1996). The epistatic pathway contains two master regulatory genes that regulate the sexual phenotype of the gametophyte (Figure 5.3). One includes the transformer gene (TRA) which, when active, simultaneously promotes femaleness (development of multilayered central part and archegonia) and represses maleness. The other includes the feminization (FEM1) gene which, when active, promotes maleness and represses femaleness (Banks, 1994, 1997). The factor that determines which of these two master sex-regulator genes is expressed first, is the presence or absence of A_{ce}. If A_{ce} is present it will activate the HER genes, which then repress TRA1. When A_{ce} is absent HER will be inactive and therefore cannot repress the TRA1 gene. The expression of TRA1 leads to the repression of the FEM1 gene.

Nothing is known presently about the genetic regulation of the sexuality of ferns reacting to A_{pt} and A_{an}, but new molecular methods have yielded remarkable progress (Banks, 1999). These studies promise to provide a much deeper understanding of the genetic regulation and the genetic and developmental processes of sex determination.

5.6 Dark germination: a further influence of antheridiogen

The influence of GA_3 in inducing dark germination of *Anemia* spores was first observed by Schraudolf (1964). The hormone replaced the light requirement in spore germination. The protonema formed pale, long, and thin cell rows a few cells long, which never became two dimensional but normally ended with an antheridium, although occasionally additional antheridia were formed on the side of the cell row. Different gibberellins such as GA_3, GA_4, and GA_9, together with the native antheridiogen A_{an}, also induced dark germination in *Anemia* and in *Lygodium* (Näf, 1966; Sugai et al., 1987). Weinberg and Voeller (1969) demonstrated the activity of the native antheridiogen A_{an} on different *Anemia* and *Lygodium* species and *Mohria caffrorum*. Schneller (1979, 1988) showed that the antheridiogen A_{pt} or closely related forms (produced by *Athyrium filix-femina* and

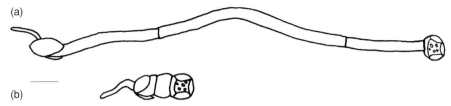

Figure 5.4 Dark germinated prothalli of *Athyrium filix-femina*, (a) responding to the species' own antheridiogen, and (b) responding to antheridiogen of *Dryopteris filix-mas*. A significant morphological difference can be seen (scale bar 40 μm).

Dryopteris filix-mas) induced dark germination in *A. filix-femina*, *D. filix-mas*, and *D. affinis*. His results revealed different reactions to congeneric antheridiogen compared to that from a different genus. When antheridiogen from *Dryopteris filix-mas* was used to induce dark germination in *Athyrium filix-femina*, the cells of dark germinated spores were short and only a few in number but one antheridium (sometimes more than one antheridum) occurred (Figure 5.4). When using the antheridiogen of *D. filix-mas* to induce dark germination of conspecific spores (and of *D. affinis*), the spores developed protonema with only short cells, which were not as long as those produced by dark germinated spores of *Athyrium filix-femina* under its own antheridiogen (Schneller, 1988). This result could be due to differences either in the concentration of the same antheridiogen or in the chemical structure of two somewhat different antheridiogens that are responsible for the different reactions. However, similar experiments have to be done with different species and genera to find out whether further interspecific differences to antheridiogen concentrations or chemical differences occur, especially in sympatric species. In *Bommeria*, Haufler and Welling (1994) observed germination and the differentiation of antheridia in the dark due to the influence of its own antheridiogen.

In the apogamous triploid *D. affinis*, older prothalli only weakly induced dark germination of conspecific spores. Therefore, the synthesis of antheridiogen probably results in a much smaller concentration compared to sexual *D. filix-mas*. Substrate or older prothalli of *D. filix-mas* and *A. filix-femina*, however, induced considerable dark germination and antheridia formation in *D. affinis* (Schneller, 1981).

Some species (e.g., *Pteridium aquilinum*, *Polystichum munitum*, *Polypodium feei*, and *P. crassifolium*) possess the ability for dark germination without the influence of pheromones like GA or antheridiogen (Schraudolf, 1967; Weinberg and Voeller, 1969; Näf et al., 1975). We do not know whether special environmental conditions may induce dark germination without the influence of antheridiogen in different species that have an antheridiogen system.

Figure 5.5 Sex and size of gametpophytes in nature. D, gametophytes of *Dryopteris* sp. Those without a letter belong to *Athyrium-filix-femina* (scale bar 45 μm).

Several experiments have shown that light and antheridiogen seem to have a somewhat antagonistic effect. Prothalli grown under low light conditions are more susceptible to the pheromone and form comparatively more antheridia then those grown under "normal" light conditions (Näf et al., 1975).

5.7 Antheridiogen in nature

Nearly all studies of antheridiogen have been carried out in the laboratory either on artificial substrate (agar) or on ash or pre-treated soil. Only a few observations and experiments have been carried out in nature. Tryon and Vitale (1977) provided the first evidence of the effects of antheridiogen in field studies. Their observations on the pattern of development and the sexuality in natural populations of *Asplenium pimpinellifolium* and *Lygodium heterodoxum* prothalli could be well explained by assuming the natural presence of antheridiogen. When studying different natural gametophyte populations of *Athyrium-filix femina*, *Blechnum* species, and *Dryopteris* species, different stages of sexuality could be seen (Cousens, 1979, 1981; Schneller, 1979, 1988; Hamilton and Lloyd, 1991) (Figure 5.5). Natural populations of developing gametophytes of *Athyrium filix-femina* contained many asexual young prothalli, some males, and only a few young females (Schneller, 1979). Older populations showed a much higher percentage of females and males. The well-developed populations were normally characterized by many smaller males (i.e., some with only 10–20 cells), most with numerous antheridia (Figures 5.6 and 5.7). These results showed a clear correspondence to the results obtained in culture (Schneller, 1979), and other studies comparing laboratory and field conditions also demonstrated a positive correlation between the two (Haufler and Soltis, 1984; Greer and McCarthy,

146   Jakob J. Schneller

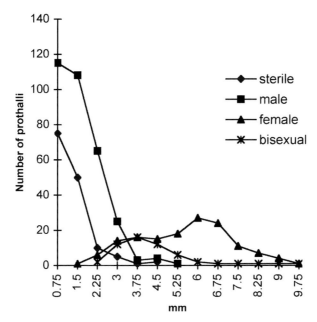

Figure 5.6 Diameter in millimeters of 638 gametophytes of *Athyrium filix-femina* collected in nature. Size differences of females and males are clearly shown.

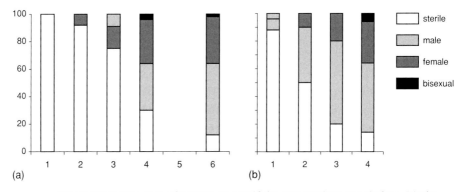

Figure 5.7 Comparison of natural and artificial gametophyte populations. (a) The development of sex in the laboratory. The abscissa shows the number of days after sowing. (b) The sex of different gametophyte populations collected in nature. They represent different stages of development which can be well correlated to those seen in culture.

1997; Ranker and Houston, 2002). The distance antheridiogen is dispersed from a source in nature is not known but experiments using *Polystichum acrostichoides* with natural steam-sterilized soil showed that it can be found up to 7.5 cm from a source (Greer and McCarthy, 1997), which is only about one third of the distance Voeller and Weinberg (1969) found in the laboratory.

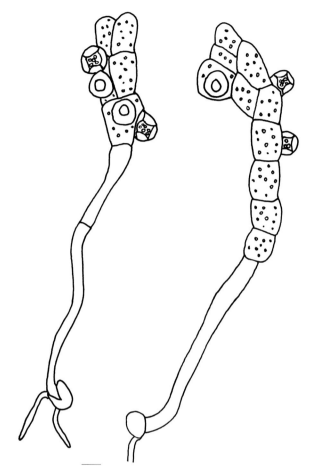

Figure 5.8 Male gametophytes collected in nature. The basal part has long thin and colorless cells, the upper part contains chlorophyll (scale bar 40 μm).

When collecting gametophyte populations from nature and looking carefully at the different individuals, the basal (earliest) cells of some prothalli were very long and thin (Figure 5.8) and nearly lacked chlorophyll (Schneller, 1988). They were interpreted as having germinated in the dark under the influence of antheridiogen. The first few thread-like cells grow in the dark towards the soil surface. Then, when the top of the cell row reaches the surface it begins to form normal sized green cells as found in prothalli growing in the light. My observations led to the assumption that the dark germinated protonemata show a positive phototactic reaction and will form regular gametophytes upon reaching the surface, if they are not too deeply buried in the soil. On untreated soil collected in a natural habitat of ferns (near Zürich) and brought into an adequate box in the laboratory, spores of *Dryopteris dilatata, Dryopteris filix-mas,*

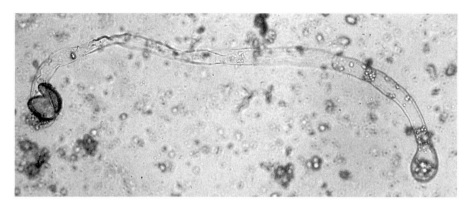

Figure 5.9 A dark germinated, under the influence of antheridiogen, spore of *Athyrium filix-femina* 1 month after the spore was buried in natural soil with gametophytes on the surface grown in the wild.

and *Athyrium filix-femina* started to germinate and many gametophytes developed. When the fastest growing gametophytes became female (or hermaphrodite) a few hundred fresh spores of *Athyrium filix-femina* were packed together with soil into a packet made of small fine-pored nylon netting. The packet was then buried ca. 2 cm below the soil surface with the gametophyte populations on the surface. After 4 weeks the packet was excavated and its content was investigated under a microscope. Many spores had germinated in the dark and usually formed one antheridium with viable spermatozoids (Figure 5.9).

When colonizing a new site in nature, the first prothallus/prothalli will, upon reaching the meristic stage, start to produce antheridiogen that will induce antheridia in the smaller (younger) prothalli, and also induce dark germination. However, this is only possible when two or more spores are present at the new site. Ferns like *Athyrium filix-femina* first develop into larger females and smaller males. When losing the meristic prothalli or when sporophytes are formed the antheridiogen source is exhausted, and older male prothalli will develop into functional females or in other species into hermaphrodites. In *A. filix-femina* when the archegonia are ripe, the antheridia of the same individual will be empty. Thus hermaphroditic gametophytes will be functionally female. Other species such as *Asplenium ruta-muraria* and *Dryopteris filix-mas* develop hermaphroditic gametophytes in mature cultures. When the antheridiogen source is removed the remaining male prothalli develop into hermaphrodites. If only one spore of an outbreeding species such as *A. filix-femina* reaches a new site, it will be unlikely to establish a new population because of genetic factors (i.e., inbreeding depression) (Schneller, 1979). To establish a new successful population, such species need at least two spores, which very likely will have originated from different sporophytes.

Chiou and Farrar (1997) have demonstrated that epiphytic fern species of the family Polypodiaceae possess an antheridiogen system, which suggests that dark germination may play a role in competition with bryophytes and higher plants.

5.8   Biological and evolutionary implications of the antheridiogen system

Self-fertilization is possible in functional, bisexual fern gametophytes. In such gametophytes, intragametophytic selfing would lead to completely homozygous sporophytes, thus ultimately reducing population genetic variability and making genetic recombination unlikely. Thus, the most obvious evolutionary implication of antheridiogen systems is that they promote outcrossing and reduce inbreeding.

Many investigations have shown that polyploids are more likely to self-fertilize than diploids (Klekowski, 1979; Masuyama and Watano, 1990; Watano and Masuyama, 1991; Schneller and Holderegger, 1996a; Vogel et al., 1999a; Soltis and Soltis, 2000; Chiou et al., 2002; Chiou, 2003). Polyploids may be better buffered against inbreeding and environmental changes due to heterosis and the presence of homoeologous genes. Antheridiogen activity is observed, however, in many tetraploid species (e.g., *Dryopteris filix-mas*, *Cystopteris tennesseensis*, and many others; Soltis and Soltis 1992), which are able to self fertilize successfully (Haufler and Ranker, 1985; Schneller et al., 1990). The antheridiogen systems in polyploids may simply represent the retention of ancestral traits inherited from their diploid progenitors.

Intragametophytic selfing may be advantageous for pioneer species especially in the early stages of colonization (Lloyd 1974; Crist and Farrar, 1982; Soltis and Soltis, 1990; Lott et al., 2003; Flinn, 2006). Selfing may also be an advantage when the availability of safe sites is low such that the chance of more than one spore arriving at a site is highly unlikely (Crist and Farrar, 1982). This is the case for rock inhabiting species that live in narrow crevices, for example some species of *Asplenium* and many members of the Pteridaceae. *Asplenium ruta-muraria*, *A. trichomanes*, and *A. septentrionale* are well-investigated examples of this strategy (Schneller and Holderegger, 1996a; Vogel et al., 1999b; Suter et al., 2000). Young populations of *A. ruta-muraria*, for instance, are composed of genetically identical, completely homozygous individuals, which arose from one (founder) prothallus. When genetically different spores arrive at a site, antheridiogen would promote outcrossing (Schneller and Holderegger, 1996b). When intragametophytic selfing in bisexual prothalli is predominant, one could argue that antheridiogen is not necessary. But Schneller and Hess (1995) observed the presence of antheridiogen in *A. ruta-muraria*, which predominantly propagates by intragametophytic selfing and may even develop mechanisms of outbreeding depression (Schneller, 1996). Schneller (1996) found that within gametophyte populations many males developed. Schneller and Hess (1995) suggested that the presence of an antheridiogen system in *A. ruta-muraria* could be a matter of optimal resource allocation, rather than the promotion of outcrossing per se, allowing female gametophytes to dedicate resources to egg and sporophyte formation while forcing neighboring gametophytes to spend resources on sperm cell production (see also Willson, 1981).

Antheridiogen not only influences gametophytes on the surface of a substrate, but it also mobilizes buried spores. Schraudolf (1985) argued that the hemispheric field of active antheridiogen has a diameter of about 10 cm in *Anemia phyllitidis*. How the different prothalli react may be related to competition between sexes. Males remain smaller and therefore do not compete much for space and light with female gametophytes.

Vittaria gemmae are special features of prothalli that allow vegetative dispersal of the haploid generation. Interestingly they are sensitive to antheridiogen. This may be a means to promote outbreeding whenever gametophytes formed by the gemmae are growing closely together (Dassler and Farrar, 2001). Emigh and Farrar (1977) suggested that the capability of *Vittaria* gemmae to form antheridia under the influence of the pheromone is related to sexual reproduction in addition to their role in vegetative reproduction.

5.9   Future goals

Although we have a good understanding of the role of antheridiogens, many questions remain unanswered. For example, are antheridiogen systems as common in tropical ferns (e.g., Korpelainen, 1994; Ranker *et al.*, 1996) as they appear to be in temperate taxa? How widespread are antheridiogens in members of the Aspleniaceae? How big is the active radius of the effect of antheridiogens under natural conditions?

A phenomenon that has not been well researched is the interspecific and/or intergeneric response to the pheromone. Is there some sort of antagonistic and competitive behavior? Can we see special mechanisms in sympatric species that may use the pheromone as a weapon in competition? The short-celled, dark germinated protonema of *Athyrium filix-femina* under the influence of the antheridiogen of *Dryopteris filix-mas* (Schneller, 1988) could be interpreted as a means to reduce the success of *Athyrium filix-femina* because in many cases the dark germinated prothalli do not reach the surface and therefore are lost.

Fast growing prothalli may also gain an advantage by causing more slowly growing, neighboring prothalli to become male, thus reducing competition from other females (Willson, 1981). The influence of antheridiogen seems to be more important in outbreeding species, because species that are able to self fertilize would not necessarily benefit from an antheridiogen response.

How many similarities or differences in sex determination based on antheridiogens will we find between different species? Can we see differences within species when growing under different environments? The experiments of Greer and McCarthy (1999) revealed that under more severe conditions, three of four species compensate for this disadvantage by exhibiting greater reproductive effort. What exactly is the influence of different nutritional conditions on the production of antheridiogen and the regulation of sex? How stable is antheridiogen under natural conditions? The seasonal timing of reproduction in temperate climates, similar to mosses (Greer, 1993; Hock et al., 2004), may have an important influence on the consequences of sex determination and predicting the survival of progeny (Khare, 2006). Kazmierczak (2003) showed that antheridiogen can be used to reveal developmental processes such as antheridial ontogeny. Future molecular investigations are promising in developing methods to learn more about the genetic regulation of sex determination and gain deeper insights into the functions of antheridiogen, for instance its influence on the phytochrome system (Banks, 1999).

References

Banks, J. A. (1994). Sex-determining genes in the homosporous fern *Ceratopteris*. *Development*, **120**, 1949–1958.

Banks, J. A. (1997). Sex determination in the fern *Ceratopteris*. *Trends in Plant Science*, **2**, 175–180.

Banks, J. A. (1999). Gametophyte development in ferns. *Annual Review of Plant Physiology and Plant Molecular Biology*, **50**, 163–186.

Banks, J. A., Hickok, L., and Webb, M. A. (1993). The programming of sexual phenotype in the homosporous fern *Ceratopteris richardii*. *International Journal of Plant Science*, **154**, 522–534.

Baroutsis, J. G. (1976). Cytology, morphology and developmental biology of the fern genus *Anogramma*. Unpublished Ph.D. Thesis, Indiana University, Bloomington, IN.

Chasan, R. (1992). *Ceratopteris*: a model plant for the 90s. *Plant Cell*, **4**, 113–115.

Chiou, W.-L. (1999). Gametophyte morphology and antheridiogen of *Sphaeropteris lepifera* (J. Sm.) Tryon and *Alsophila spinulosa* (Hook) Tryon (Cyatheaceae). XVI International Botanical Congress, Abstract Number 2530.

Chiou, W.-L. (2003). The reproductive biology of gametes of homosporous ferns. *Taiwanese Journal of Forest Science*, **18**, 67–73.

Chiou, W.-L. and Farrar, D. R. (1997). Antheridiogen production and response in Polypodiaceae species. *American Journal of Botany*, **84**, 633–640.

Chiou, W.-L., Farrar, D. R., and Ranker, T. (1998). Gametophyte morphology and reproductive biology in *Elaphoglossum*. *Canadian Journal of Botany*, **77**, 1967–1977.

Chiou, W.-L., Lee, P.-H., and Ying, S.-S. (2000). Reproductive biology of gametophytes of *Cyathea podophylla* (Hook.) Copel. *Taiwanese Journal of Forest Science*, **15**, 1–12.

Chiou, W.-L., Farrar, D. R., and Ranker, T. A. (2002). The mating systems of some Polypodiaceae species. *American Fern Journal*, **92**, 65–79.

Corey, E. J., Myers, A. G., Takahashi, N., Yamane, H., and Schraudolf, H. (1986). Constitution of antheridium-inducing factor of *Anemia phyllitidis*. *Tetrahedron Letters*, **27**, 5083–5084.

Cousens, M. I. (1979). Gametophyte ontogeny, sex expression, and genetic load as measures of population divergence in *Blechnum spicant*. *American Journal of Botany*, **66**, 116–132.

Cousens, M. I. (1981). *Blechnum spicant*: habitat and vigour of optimal, marginal, and disjunct populations, and field observations of gametophytes. *Botanical Gazette*, **142**, 251–258.

Crist, K. C. and Farrar, D. F. (1982). Genetic load and long distance dispersal in *Asplenium platyneuron*. *Canadian Journal of Botany*, **61**, 1809–1814.

Dassler, C. and Farrar, D. F. (2001). Significance of gametophyte from long-distance colonization by tropical, epiphytic ferns. *Brittonia*, **53**, 352–369.

DeVol, J., Rebert, A., Greer, G., and Dietrich, M. (2005). Effects of exogenous cytokinin application on morphological development of gametophytes of the fern *Osmunda regalis*. *Botanical Society of America, Pteridological Section, Botany 2005*, Abstract no. 192.

Döpp, W. (1950). Eine die Antheridienbildung bei Farnen fördernde Substanz in den Prothallien von *Pteridium aquilinum* (L.) Kuhn. *Berichte der Deutschen Botanischen Gesellschaft*, **63**, 139–147.

Döpp, W. (1959). Über eine hemmende und eine fördernde Substanz bei der Antheridienbildung in den Prothallien von *Pteridium aquilinum*. *Berichte der Deutschen Botanischen Gesellschaft*, **72**, 11–24.

Döpp, W. (1962). Weitere Untersuchungen über die Physiologie der Antheridienbildung bei *Pteridium aquilinum*. *Planta*, **58**, 483–508.

Dyer, A. F. (1979). The culture of fern gametophytes for experimental investigation. In *The Experimental Biology of Ferns*, ed. A. F. Dyer. New York: Academic Press, pp. 253–305.

Eberle, J. R. and Banks, J. A. (1996). Genetic interactions among sex-determining genes in the fern *Ceratopteris richardii*. *Genetics*, **142**, 973–985.

Eberle J., Nemachek, J., Wen, C.-K., Hasebe, M., and Banks, J. A. (1995). *Ceratopteris*: a model system for studying sex-determining mechanisms in plants. *International Journal of Plant Sciences*, **156**, 359–366.

Emigh V. D. and Farrar, D. R. (1977). Gemmae: a role in sexual reproduction in the fern genus *Vittaria*. *Science*, **198**, 297–298.

Fellenberg-Kressel, M. (1969). Untersuchungen über die Archegonien- und Antheridienbildung bei *Microlepia speluncae* (L.) Moore in Abhängigkeit von äusseren und inneren Faktoren. *Flora*, **160**, 14–39.

Flinn, K. M. (2006). Reproductive biology of three fern species may contribute to differential colonization success in post-agricultural forests. *American Journal of Botany*, **93**, 1289–1294.

Gemmrich, A. R. (1986). Antheridiogensynthesis in the fern *Pteris vittata* II. Hormonal control of antheridia formation. *Journal of Plant Physiology*, **125**, 155–166.

Greer, G. K. (1991). Sex expression and inferences on the mating systems of *Aspidotis*. Unpublished M.S. Thesis, Humboldt State University, Arcata, CA.

Greer, G. K. (1993). The influence of soil topography and spore-rain density on gender expression in gametophyte populations of the homosporous fern *Aspidotis densa* (Brack. in Wilkes) Lellinger. *American Fern Journal*, **84**, 54–59.

Greer, G. K. and McCarthy, B. C. (1997). The antheridiogen neighborhood of *Polystichum acrostichoides* (Dryopteridaceae) on a native substrate. *International Journal of Plant Science*, **158**, 764–768.

Greer, G. K. and McCarthy, B. C. (1999). Gametophytic plasticity among four species of ferns with contrasting ecological distributions. *International Journal of Plant Science*, **160**, 879–886.

Hamilton, R. G. and Lloyd, R. M. (1991). Antheridiogen in the wild. The development of fern gametophyte communities. *Functional Ecology*, **6**, 804–809.

Haufler, C. H. and Gastony, G. J. (1978). Antheridiogen and the breeding system in the fern genus *Bommeria*. *Canadian Journal of Botany*, **36**, 1594–1601.

Haufler, C. H. and Ranker, T. A. (1985). Differential antheridiogen responses and evolutionary mechanisms in *Cystopteris*. *American Journal of Botany*, **72**, 659–665.

Haufler, C. H. and Soltis, D. E. (1984). Obligate outcrossing in a homosporous fern: field observation of a laboratory prediction. *American Journal of Botany*, **71**, 878–881.

Haufler, C. H. and Welling, C. B. (1994). Antheridiogen, dark germination, and outcrossing mechanisms in *Bommeria* (Adiantaceae). *American Journal of Botany*, **81**, 616–671.

Hickok, L. G. (1983). Abscisic acid blocks antheridiogen – induced antheridium formation in gametophytes of *Ceratopteris*. *Canadian Journal of Botany*, **61**, 888–892.

Hickok L. G. and Warne, T. R. (1998). *C-Fern Manual*. University of Tennessee Research Foundation. Burlington, NC: Carolina Biological Supply Company.

Hock, Z., Szövényi, P., and Tóth, Z. (2004). Seasonal variation in the bryophyte diaspore bank of open grasslands on dolomite rock. *Journal of Bryology*, **26**, 285–292.

Holbrook-Walker, S. G., and Lloyd R. M. (1973). Reproductive biology and gametophyte morphology of the Hawaiian fern genus *Sadleria* (Blechnaceae) relative to habitat diversity and propensity for colonization. *Botanical Journal of the Linnean Society*, **67**, 157–174.

Hooper, E. A. and Haufler, C. H. 1997. Genetic diversity and breeding system in a group of neotropical epiphytic ferns (*Pleopeltis*; Polypodiaceae). *American Journal of Botany*, **84**, 1664–1674.

Hudson, J. E. N. (ed.) (1999). Texas Academy of Sciences Annual Meeting. www.shsu.edu/bio_jxn.

Kazmierczak, A. (2003). Induction of cell division and cell expansion at the beginning of gibberellin A_3-induced precocious antheridia formation in *Anemia phyllitidis* gametophytes. *Plant Science*, **165**, 933–939.

Khare, P. B., Behera, S. K., Srivastava, R., and Shukla, S. P. (2006). Studies on reproductive biology of a threatened tree fern, *Cyathea spinulosa* Wall. ex Hool. *Current Science*, **89**, 173–176.

Kirkpatrick, R. E. B. and Soltis, P. S. (1992). Antheridiogen production and response in *Gymnocarpium dryopteris* ssp. *disjunctum*. *Plant Species Biology*, **7**, 1–9.

Klekowski, E. J., Jr. (1969). Reproductive biology of the Pteridophyta II. Theoretical considerations. *Botanical Journal of the Linnean Society*, **62**, 347–359.

Klekowski, E. J., Jr. (1979). The genetics and reproductive biology in ferns. In *The Experimental Biology of Ferns*, ed. A. F. Dyer. New York: Academic Press, pp. 133–170.

Korpelainen, H. (1994). Growth, sex determination and reproduction of *Dryopteris filix-mas* (L.) Schott gametophytes under varying nutritional conditions. *Botanical Journal of the Linnean Society*, **114**, 357–366.

Korpelainen, H. (1998). Labile sex expression in plants. *Biological Reviews*, **73**, 157–180.

Lloyd, R. M. (1974). Mating system and genetic load in pioneer and non-pioneer Hawaiian Pteridophyta. *Botanical Journal of the Linnean Society*, **69**, 23–35.

Lott, M. S., Volin, J. C., Pemberton, R. W., and Austin, D. F. (2003). The reproductive biology of the invasive ferns *Lygodium microphyllum* and *L. japonicum* (Schizaeaceae): implications for invasive potential. *American Journal of Botany*, **90**, 1144–1152.

Mander, L. N. (2003). Twenty years of gibberellin research. *Natural Products Report*, **20**, 49–69.

Masuyama S, and Watano Y. (1990). Trend for inbreeding in polyploid pteridophytes. *Plant Species Biology*, **5**, 13–17.

Mendenez, V., Revilla, M. A., and Fernandez, H. (2006a). Growth and gender in the gametophyte of *Blechnum spicant*. *Plant Cell Tissue Organ Culture*, **86**, 47–53.

Mendenez, V., Revilla, M. A., Bernard, P., Gotor, V., and Fernandez, H. (2006b). Gibberellins and antheridiogen on sex in *Blechnum spicant*. *Plant Cell Report*, **25**, 1104–1110.

Miller, J. H. (1968). Fern gametophytes as experimental material. *Botanical Revue*, **34**, 361–440.

Näf, U. (1956). The demonstration of a factor concerned with the initiation of antheridia in polypodiaceous ferns. *Growth*, **20**, 91–105.

Näf, U. (1959). Control of antheridium formation in the fern species *Anemia phyllitidis*. *Nature*, **184**, 798–800.
Näf, U. (1960). On the control of antheridium formation in the fern species *Lygodium japonicum*. *Proceedings of the Society of Experimental Biology and Medicine*, **105**, 82–86.
Näf, U. (1962). Developmental physiology of lower archegoniates. *Annual Review of Plant Physiology*, **13**, 507–532.
Näf, U. (1965). On antheridial metabolism in the fern species *Onoclea sensibilis* L. *Plant Physiology*, **40**, 888–890
Näf, U. (1966). On dark germination and antheridium formation in *Anemia phyllitidis*. *Physiologia Plantarum*, **19**, 1079–1088.
Näf, U. (1969). On the control of antheridium formation in ferns. In *Current Topics in Plant Science*, ed. J. E. Gunckel. New York: Academic Press, pp. 1357–1360.
Näf, U. (1979). Antheridiogens and antheridial development. In *The Experimental Biology of Ferns*, ed. A. F. Dyer. New York: Academic Press, pp. 436–470.
Näf, U., Sullivan, J., and Cummins, M. (1962). New antheridiogen from the fern *Onoclea sensibilis*. *Science*, **163**, 1357–1358.
Näf, U., Nakanishi, K., and Endo, M. (1975). On the physiology and chemistry of fern antheridiogens. *Botanical. Review*, **41**, 315–359.
Nakanishi, K., Endo, M., and Näf, U. (1971). Structure of antheridium-inducing factor of the fern *Anemia phyllitidis*. *Journal of the American Chemical Society*, **93**, 5579–5581.
Nester-Hudson, J. E., Ladas, C., and McClurd, A. (1997). Gametophyte development and antheridiogen activity in *Thelypteris ovata* var. *lindheimeri*. *American Fern Journal*, **87**, 131–142.
Pajaron, S., Pangua E., and Garcia-Alvarez, L. (1999). Sexual expression and genetic diversity in populations of *Cryptogramma crispa* (Pteridaceae). *American Journal of Botany*, **86**, 964–973.
Pour, M., King, G. R., Monck, N. J. T., Morris, J. C., Zhang, H., and Mander, L. N. (1998). Synthetic and structural studies on novel gibberellins. *Pure and Applied Chemistry*, **70**, 351–354.
Pringle, R. B. (1961). Chemical nature of antheridiogen-A, a specific inducer of male sex organ in certain fern species. *Science*, **133**, 284.
Pringle, R. B., Näf, U., and Braun, A. C. (1960). Purification of a specific inducer of the male sex organ in certain fern species. *Nature*, **186**, 1066–1067.
Quintanilla, L., Pangua, E., Amigo, J., and Pajarom, S. (2005). Comparative study of the sympatric ferns *Culcita macrocarpa* and *Woodwardia radicans*: sexual phenotype. *Flora*, **200**, 187–194.
Raghavan, V. (1989). *Developmental Biology of Fern Gametophytes*. Cambridge: Cambridge University Press.
Ranker, T. A. (1987). Experimental systematics and population biology of the fern genus *Hemionitis* and *Gymnopteris* with reference to *Bommeria*. Unpublished Ph.D. Thesis, University of Kansas, Lawrence, KS.
Ranker, T. A. and Houston, H. A. (2002). Is gametophyte sexuality in the laboratory a good predictor of sexuality in nature? *American Fern Journal*, **92**, 112–118.

Ranker, T. A., Gemmill, C. E. C., Trapp, P. G., Hambleton, A., and Ha, K. (1996). Population genetics and reproductive biology of lava-flow colonising species of Hawaiian *Sadleria* (Blechnaceae). In *Pteridology in Perspective*, ed. J. M. Camus, M. Gibby, and R. J. John. Kew: Royal Botanic Gardens, pp. 591–598.

Rubin, G. and Paolillo, D. J. (1983). Sexual development of *Onoclea sensibilis* on agar and soil media without the addition of antheridiogen. *American Journal of Botany*, **70**, 811–815.

Rubin, G. and Paolillo, D. J. (1984). Obtaining sterilized soil for the growth of *Onoclea* gametophytes. *New Phytologist*, **97**, 621–628.

Rubin, G., Robson, D. S., and Paolillo, D. J. (1985). Effects of population density on sex expression in *Onoclea sensibilis* L. on agar and ashed soil. *Annals of Botany*, **55**, 201–215.

Schedlbauer, M. D. (1974). Biological specifity of the antheridiogen from *Ceratopteris thalictroides* (L.) Brogn.). *Planta*, **116**, 39–43.

Schedlbauer, M. D. and Klekowski, E. J. (1972). Antheridiogen activity in the fern *Ceratopteris thalictroides* (L.) Brogn. *Botanical Journal of the Linnean Society*, **65**, 399–413.

Schneller, J. J. (1979). Biosystematic investigation on the lady fern (*Athyrium filix-femina*). *Plant Systematics and Evolution*, **132**, 255–277.

Schneller, J. J. (1981). Bemerkungen zur Biologie der Wurmfarngruppe. *Farnblätter*, **7**, 9–17.

Schneller, J. J. (1988). Spore bank, dark germination and gender determination in *Athyrium* and *Dryopteris*. Results and implications for population biology of Pteridophyta. *Botanica Helvetica*, **98**, 77–86.

Schneller, J. J. (1996). Outbreeding depression in the fern *Asplenium ruta-muraria* L.: evidence from enzyme electrophoresis, meiotic irregularities and reduced spore viability. *Biological Journal of the Linnean Society*, **59**, 281–295.

Schneller, J. J. and Hess, A. (1995). Antheridiogen system in the fern *Asplenium ruta-muraria* (Aspleniaceae; Pteridophyta). *Fern Gazette*, **15**, 64–70.

Schneller, J. J. and Holderegger, R. (1996a). Genetic variation in small, isolated fern populations. *Journal of Vegetation Science*, **7**, 113–120.

Schneller, J. J. and Holderegger, R. (1996b). Colonisation events and genetic variability within populations of *Asplenium ruta-muraria* L. In *Pteridology in Perspective*, ed. J. M. Camus, M. Gibby, and R. J. Johns. Kew: Royal Botanic Gardens, pp. 571–580.

Schneller, J. J., Haufler, C. H., and Ranker, T. A. (1990). Antheridiogen and natural gametophyte populations. *American Fern Journal*, **80**, 143–152.

Schraudolf, H. (1962). Die Wirkung von Phytohormonen auf Keimung und Entwicklung von Farnprothallien. I. Auslösung der Antheridienbildung und Dunkelkeimung bei Schizaeaceen durch Gibberellinsäure. *Biologisches Zentralblatt*, **6**, 731–740

Schraudolf, H. (1964). Relative activity of gibberellins in the antheridium induction in *Anemia phyllititdis*. *Nature*, **201**, 98–99.

Schraudolf, H. (1966). Die Wirkung von Phytohormonen auf Keimung und Entwicklung von Farnprothallien. IV. Die Wirkung von unterschiedlichen Gibberellinsäuren und von Allo-Gibberellinsäure auf die Auslösung der Anteridienbildung einiger Polypodiaceen. *Plant Cell Physiology*, **7**, 277–289.

Schraudolf, H. (1967). Die Steuerung ders Antheridienbildung in *Polypodium crassifolium* L. durch Licht. *Planta*, **76**, 37–46.

Schraudolf, H. (1985). Action and phylogeny of antheridiogens, *Proceedings of the Royal Society of Edinburgh*, **86B**, 75–80.

Schraudolf, H. (1987). Antagonistic effects of abscisic acid and ABA analogous on hormone induces antheridium formation. *Journal of Plant Physiology*, **131**, 433–439.

Scott, R. J. and Hickok, L. G. (1987). Genetic analysis of antheridiogen sensitivity in *Ceratopteris richardii*. *American Journal of Botany*, **74**, 1872–1877.

Soltis, D. E. and Soltis, P. E. (1992). The distribution of selfing rates in homosporous ferns. *American Journal of Botany*, **76**, 97–100.

Soltis, P. S. and Soltis, D. E. (1990). Evolution of inbreeding and outcrossing in ferns and fern-allies. *Plant Species Biology*, **5**, 1–11.

Soltis, P. S. and Soltis, D. E. (2000). The role of genetic and genomic attributes in the success of polyploids. *Proceedings of the National Academy of Sciences of the United States of America*, **97**, 7051–7057.

Stevens, R. D. and Werth, C. W. (1999). Interpopulational comparison of dose-mediated antheridiogen response in *Onoclea sensibilis*. *American Fern Journal*, **89**, 221–231.

Sugai, M., Nakamura, K., Yamane, H., Sato, Y., and Takahashi, N. (1987). Effects of gibberellins and their methyl esters on dark germination and antheridium formation in *Lygodium japonicum* and *Anemia phyllitidis*. *Plant Cell Physiology*, **28**, 199–202.

Suter, M., Schneller, J. J., and Vogel, J. C. (2000). Investigations into the genetic variation, population structure and breeding systems of the fern *Asplenium trichomanes* subsp. *quadrivalens*. *International Journal of Plant Science*, **161**, 233–244.

Tryon, R. M and Vitale, G. (1977). Evidence for antheridiogen production and its mediation of a mating system in natural populations of fern gametophytes. *Botanical Journal of the Linnean Society*, **74**, 243–249.

Voeller, B. R. (1964). Antheridiogens in ferns. In *Regulateurs Naturels de la Croissance Vegetale*. Gif-sur-Yvette: Editions du CNRS, pp. 665–684.

Voeller, B. R. and Weinberg, E. S. (1969). Evolutionary and physiological aspects of antheridium induction in ferns. In *Current Topics in Plant Science*, ed. J. E. Gunckel. New York: Academic Press, pp. 77–93.

Vogel, J. C., Rumsey, F. J., Schneller, J. J., Barett, J. A., and Gibby, M. (1999a). Where are the glacial refugia in Europe? Evidence from pteridophytes. *Biological Journal of the Linnean Society*, **66**, 23–37.

Vogel, J. C., Rumsey, F. J., Russel, S. J., Cox, S. J., Holmes, J. S., Bujnoch, W., Stark, C. Battet, J. A., and Gibby, M. (1999b). Genetic structure, reproductive biology and ecology of isolated populations of *Asplenium csikii* (Aspleniaceae, Pteridophyta). *Heredity*, **83**, 604–612.

Von Aderkas, P. (1983). Studies of gametophytes of *Matteuccia struthiopteris*, ostrich fern in nature and culture. *Canadian Journal of Botany*, **61**, 3267–3270.

Warne, T. R. and Hickok L. G. (1989). Evidence of a gibberellin biosynthetic origin of *Ceratopteris* antheridiogen. *Plant Physiology*, **89**, 535–538.

Watano, Y. and Masuyama, S. (1991). Inbreeding in natural populations of the annual polyploid fern *Ceratopteris thalictroides* (Parkeriaceae). *Systematic Botany*, **16**, 705–714.

Weinberg, E. S. and Voeller, B. R. (1969). External factors inducing germination on fern spores. *American Fern Journal*, **59**, 153–167.

Willson, M. F. (1981). Sex expression in fern gametophytes: Some evolutionary possibilities. *Journal of Theoretical Biology*, **93**, 403–409.

Wynne, G. M., Mander, L. M., Goto, M., Yamane, H., and Omori, T. (1998). Biosynthetic origin of the antheridiogen, gibberellin A73 methylester, in ferns of the *Lygodium* genus. *Tetrahedron Letters*, **39**, 3877–3880.

Yamane, H. (1998). Fern antheridiogens. *International Revue of Cytology*, **184**, 1–32.

Yatskievych G. (1993). Antheridiogen response in *Phanerophlebia* and related genera. *American Fern Journal*, **83**, 30–36.

6

Structure and evolution of fern plastid genomes

PAUL G. WOLF AND JESSIE M. ROPER

6.1 Introduction

The concept of the genome, as the haploid complement of genes of an organism, is far from recent. The term genome is usually attributed to Hans Winkler in 1920 (Ledergerg and McCray, 2001). However, fine scale maps and understanding of the function of genes in the context of the genome did not begin until the 1970s after DNA sequencing techniques were developed. The term genome (and its corresponding genomics) can mean different things to different people (Ledergerg and McCray, 2001) but here we will focus on structural and evolutionary aspects of genomes in ferns. Although genomics is generally reserved for the main (nuclear) component of an organism, that topic is covered in Chapter 7. Instead we narrow the focus here to the chloroplast (i.e., plastid) genome. This small, well-defined genome is found in all green plants. Among land plants the plastid genome is highly conserved in structure and gene content (Palmer, 1985b). Compared to most nuclear genomes studied, plastid genomes contain a high proportion of DNA that codes for proteins and for RNA (ribosomal and transfer). Much of the non-coding regions (between protein-encoding genes) is transcribed and may well have important regulatory functions.

Because the plastid genome contains a high density of genes of well-studied processes, the genome is an excellent model for investigations into the relationship between genome structure and function. This field represents an ideal starting point leading into the much more complex field of the study of nuclear genomes. In this chapter, we start with a brief overview of plastid genomes, including their structure, function, and evolution. This is followed by a summary

Biology and Evolution of Ferns and Lycophytes, ed. Tom A. Ranker and Christopher H. Haufler. Published by Cambridge University Press. © Cambridge University Press 2008.

of the early work on plastid genome structure in ferns. We examine some recent studies that report complete plastid genome sequences, and the significance of cDNA sequences from the same genomes. We then present new and preliminary data on the evolution of a genome structure unique to some ferns. We finish with some ideas on prospects for future work.

The structure and evolution of plastid genomes is reviewed extensively elsewhere (Palmer, 1985a, 1987, 1991); here we begin with a brief summary. The plastid genome is usually contained within the plastids, most commonly in chloroplasts but also in amyloplasts, chromoplasts, and leucoplasts. The number of plastids per cell varies considerably across tissues and taxa, and the number of plastid genomes per cell varies during development. But even the most conservative estimates indicate that there are many thousands of copies of the plastid genome per cell (Palmer, 1987). Thus, the copy number of plastid DNA is several orders of magnitude higher than that of the nuclear genome, a feature that favors study of plastid genomes.

Most plastids are involved in photosynthesis. However, most proteins active in the photosynthesizing chloroplast (about 2000 of them) are nuclear encoded. Only a handful of the genes are retained in the plastid genome itself. These include those for proteins associated with photosystem I and II, chlorophyll biosynthesis, and the large subunit of ribulose-1,5-bisphosphate carboxylase/oxygenase (RUBISCO). Plastid genes also encode for proteins (or their subunits) associated with respiration, including those for NADH oxidoreductases and ATP synthase. Most plastid genomes also contain some of the genes required for transcription and translation, including genes for transfer RNAs, ribosomal RNAs, RNA polymerase subunits, and a translation factor. There are also several open reading frames that appear to be transcribed but the function of the products is currently unknown. The origins of plastid genomes from prokaryotes and subsequent transfer of genes to the nucleus is reviewed elsewhere (Martin and Herrmann, 1998; Martin and Miller, 1998; Martin *et al.*, 1992, 1998; Stoebe *et al.*, 1999).

The first restriction site maps of plastid genomes were published in the early 1980s (Palmer, 1985a), followed shortly by the first complete plastid genome sequences for the liverwort *Marchantia* (Ohyama *et al.*, 1986) and the angiosperm *Nicotiana* (tobacco) (Shinozaki *et al.*, 1986). Most plastid genome maps of land plants are circular and about 120–160 kb, including a region that is an inverted repeat (Kolodner and Tewari, 1979). The inverted repeat (IR) is typically 10–20 kb and includes the ribosomal RNA genes, as well as a few tRNA genes. The two copies of the IR separate a small single copy (SSC) region from a large single copy (LSC) region. The single copy regions can have two alternative orientations relative to each other (and the IR) and it appears that in most individual plants, equimolar amounts of the two orientations are found (Palmer, 1983). This phenomenon, known as flip-flop recombination, has been documented in the fern genus *Osmunda* (Stein et al., 1986). Although the map of most plastid genomes is circular, the condition of most genomes in most cells is probably linear (Bendich, 2004) and this may vary with developmental stage and taxon.

6.2 The golden age of fern chloroplast genomics

Early evolutionary studies of plastid genomes examined the basic morphology of the molecule. A combination of restriction enzyme digestion and Southern blot hybridization was used to generate physical maps and then to map gene locations. These elegant experiments provided a base knowledge of plastid genome structure and gene organization (Palmer and Stein, 1986; Stein et al., 1986, Hasebe and Iwatsuki, 1992; Olmstead and Palmer, 1994). However, genome mapping via these methods is laborious. The DNA of interest is cut with at least two restriction enzymes individually and in tandem. Agarose gel images of the resulting DNA fragments produce a banding pattern, which is then used to estimate the sizes of restriction fragments. The more restriction enzymes used, the finer the detail of the map produced. The DNA fragments are then transferred to a membrane filter where they are fixed into position. Labeled probes are washed over the membrane allowing the probes to bind to the fragmented DNA where there is homology between fragment and probe. Initially, restriction fragments themselves can be used to probe digests from different enzymes, and the double digests, to generate the map. Later, probes of previously characterized genes (cloned from tobacco, for example) are used to locate the genes on a newly characterized map.

Early studies of *Marchantia, Nicotiana, Pisum, Ginkgo, Osmunda,* and *Spinacia* found consistency in genome size and gene content among widely divergent species (Palmer, 1991). Small differences in genome size were attributed to variation in the size of the IR. Expansion of the IR changes single-copy genes to duplicate-copy which adds to genome length without changing the overall genome complexity (Palmer and Thompson, 1981; Palmer and Stein, 1986). Rearrangement events within the plastid genome are thought to be rare, making them phylogenetically useful (Olmstead and Palmer, 1994). One well-known example is a 30 kb inversion detected in the LSC of the bryophytes and lycophytes, relative to other land plants (Raubeson and Jansen, 1992), indicating that lycophytes are a sister to all other extant vascular plants.

Another important example of gene order variation was found within the inverted repeat of ferns. Cross hybridization studies of the fern *Adiantum* revealed a gene order within the inverted repeat that is highly rearranged compared to other vascular plants (Hasebe and Iwatsuki, 1990b, 1992). Hasebe and Iwatsuki (1990b, 1992) suggest that this gene order requires an expansion of the inverted repeat accompanied by at least two overlapping inversion events. The ferns *Polystichum* and *Cyathea* also have the rearranged inverted repeat gene order found in *Adiantum* (Stein et al., 1992). Raubeson and Stein (1995) attempted to characterize the individual rearrangements responsible for the gene order in *Adiantum* and other ferns. At the time of the study there was no robust phylogeny to guide taxon selection. Although more major clades were included than in previous studies, groups such as the filmy ferns were excluded. Two major changes in gene order were noted by Raubeson and Stein (1995). The first was that *Gleichenia* appeared to have a gene order different than that of *Osmunda* and *Adiantum*. The inclusion of the *ndhB* gene within the inverted repeat without a corresponding duplication of *rps7* and *rps12* suggests an inversion of these genes relative to the ancestral gene order and then an expansion of the IR. The result is a possible intermediate gene order in *Gleichenia*. All other taxa sampled had either the *Osmunda* or *Adiantum* gene order. The second finding was partial duplication of *chlL* in several taxa, including *Adiantum*. However, now that a robust phylogenetic hypothesis is available for ferns (Pryer et al., 2004; see Chapter 15), it appears that the distribution of the *chlL* duplication is not consistent with a single evolutionary event. Nevertheless, Raubeson and Stein (1995) moved us closer to a better understanding of fern plastid genome evolution.

6.3 The age of complete plastid genome sequences

Restriction site mapping can provide useful phylogenetic markers and reveal much of the gross morphology of plastid genomes. However, additional details and confirmation can be obtained with complete chloroplast genome sequences. With current technology, this is now easier than restriction site mapping, although obtaining purified chloroplast DNA is the biggest hurdle in the process. The first complete plastid genome sequence for a fern was that of *Adiantum capillus-veneris* (Wolf et al., 2003), using the clones that had earlier been developed for probing and mapping the genome (Hasebe and Iwatsuki 1990a, 1990b, 1992). Thus for this sequence, the genome was already isolated and purified. Sequencing involved end-sequencing each clone then primer walking. Subsequent genomes have required different isolation and purification techniques. Chloroplast DNA of the lycopod *Huperzia lucidula* was isolated using fluorescence activated cell sorting with subsequent rolling circle amplification to increase target DNA (Wolf et al., 2005). Alternative isolation techniques include traditional isolation in sucrose gradients (Palmer, 1986; Jansen et al., 2005) and creation of partial fosmid libraries (McNeal et al., 2006). Once a sample containing a high proportion of plastid DNA has been obtained, the next step is sequencing via shotgun cloning. Some contamination is not a problem, but as contamination increases, so does the number of wasted sequencing reactions. Contaminated sequences are usually discarded at the assembly stage. The trickiest part is to identify the boundaries of the IR, SSC and LSC, which often show up as misassemblies that can then be compiled manually. The final stage is to determine the possible function of each region of the sequence. This entails following the conventions in previously annotated genomes. With the first few plastid genome sequences published, there was some inconsistency in the naming of putative genes. However, this confusion has been eliminated with a very helpful online annotation program called DOGMA (Wyman et al., 2004) that makes the entire step much simpler and conveniently automated.

In addition to *Adiantum* and *Huperzia*, the plastid genomes of several other seed-free vascular plants have been sequenced, including *Angiopteris* (Roper et al., 2007), *Equisetum*, *Selaginella*, and *Isoëtes* (K. Karol et al., unpublished data). Some of these new sequences have provided useful structural data, but most of the new findings are at the sequence level. The first fern plastid genome from *Adiantum* (Wolf et al., 2003) showed some unusual patterns not previously seen in vascular plants. One such feature was some missing tRNA genes, which had only been observed in plastid genomes of non-photosynthetic plants (Wolfe et al., 1992). Another was a tRNA gene not previously observed in plants: a gene for the selenocysteine tRNA (trnSeC). One hypothesis to explain its presence was that the tRNA was post-transcriptionally modified to translate a different amino acid. This was consistent with the finding that several protein-coding genes contained internal stop codons, which also required modification to become functional. One such modification process is RNA editing, which has been reported from most plastid genomes. However, it appeared that the level of such modification in *Adiantum* was much higher than in other vascular plants. Subsequent sequencing of cDNAs from all plastid-encoding genes (Wolf et al., 2004) found 350 edited sites, correcting all internal stop codons from the genomic sequence, moving several start and stop positions, and altering a tRNA anticodon (thereby restoring one of the missing tRNAs). Unfortunately, no modification was detected in the trnSeC gene, so this remains a mystery. One problem with the cDNAs obtained by Wolf et al. (2004) was that they may have been produced from transcripts that were not fully mature and edited. Thus, the 350 sites that were detected comprised a minimum estimate, yet still more than ten times that detected for any other vascular plant. The highest level currently reported for a land plant is that of the hornwort *Anthoceros*, with 942 sites detected (Kugita et al., 2003).

An additional use of complete plastid genome sequences is for comparative sequence analysis. This can be used in the context of broad scale phylogenetic analyses (Nishiyama et al., 2004; Wolf et al., 2005) and also for designing universally useful primers (Small et al., 2005). In the next section, we show how new universal fern primers can be applied to PCR-based tests of comparative genome structure.

6.4 PCR mapping of fern plastid genomes

6.4.1 Background

As discussed earlier, the IR gene orders of *Adiantum*, *Polystichum*, and *Cyathea* are highly rearranged in comparison to the IR gene order of *Osmunda* (Hasebe and Iwatsuki, 1992; Stein et al., 1992; Raubeson and Stein, 1995). It should be emphasized that by 1995 it was realized that the *Adiantum* gene order was present in the largest clade (in terms of species numbers) of ferns, so it was not an isolated phenomenon. A model of the possible inversion events leading to the *Adiantum* gene order was proposed, indicating a minimum of two inversions (Hasebe and Iwatsuki, 1992). Figure 6.1 is an updated version of the Hasebe and Iwatsuki (1992) model, incorporating new gene order data from the complete genomes of *Adiantum* and *Angiopteris*. The gene order of *Angiopteris* appears to be the same as that in *Osmunda* (Roper et al., 2007). From the added detail provided by complete genome sequences, we can see what possible gene orders to expect after only one of the two inversions has occurred; a structure we refer to here as an "intermediate" gene order. Raubeson and Stein (1995) suggested that the gleichenioid ferns may contain such an intermediate gene order. However, some uncertainty still existed about the exact gene order within the taxa studied. With the complete sequence of *Adiantum* and *Angiopteris* now available, it is feasible to attempt mapping the IR of all major fern families using PCR-based approaches. Also, a robust phylogenetic framework is now available to guide taxon selection, a simplified version of which is presented in Figure 6.2 (and see Chapter 15).

There are several reasons to seek an understanding of the series of hypothesized inversions in fern plastid genomes. Although DNA sequence data have proven extremely valuable for phylogenetic studies within many groups of plants, phylogenetic signal is often lost for inferring deep divergences. There are several reasons for this outcome, one of which is that sequence data are simple and the probability of parallel and convergent substitutions among the four bases increases as one goes back in time. Conversely, the non-clock-like nature of complex genomic rearrangements suggests that such changes are unlikely to contain much homoplasy, and should therefore be of value for inferring deep branches (Helfenbein and Boore, 2004). However, this assumption has not been extensively tested. There is no reason to assume that genome rearrangements are

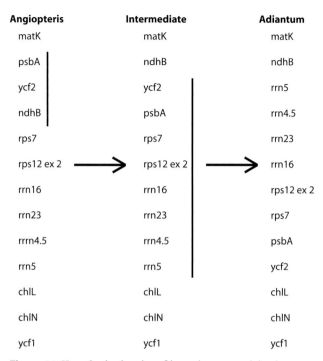

Figure 6.1 Hypothesized series of inversions to explain the reorganized IR region in polypod ferns. This model was first proposed by Hasebe and Iwatsuki (1992) and is modified here with the gene order data provided by the complete plastid genome sequences of *Adiantum* and *Angiopteris*. Vertical bars indicate the location of the inversion required to produce the next gene order.

free of homoplasy. Physical hotspots are known to occur on the plastid genome and temporal destabilization might also be a possibility. Have the putative inversions in fern plastid genomes occurred on different branches of the fern phylogeny, rendering them phylogenetically uninformative? If these inversion events occurred on the same branch, it would indicate temporal destabilization of the plastid genome, thus providing information about the nature of plastid genome evolution. The basis for asking this question is that we have what we believe to be a robust phylogenetic framework (Figure 6.2) and we can attempt to map inferred rearrangements of the plastid genome onto this framework. We first present some background and describe the technique, then present some preliminary findings.

The limit of information available from physical mapping using hybridization probes may have been reached due to constraints of the method itself. Probing relies on the presence of homologous regions. False positives occur when non-target regions are homologous enough to allow probe hybridization. False

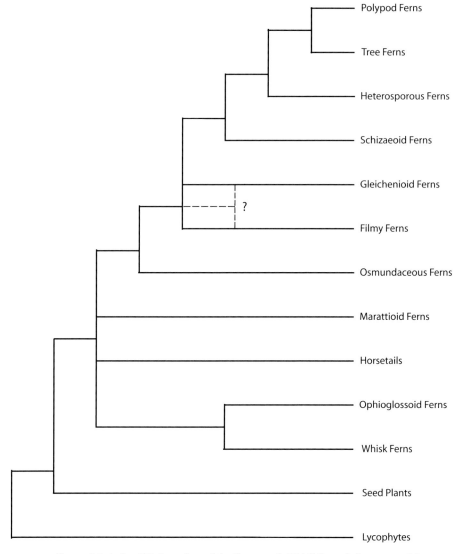

Figure 6.2 A simplified version of the Pryer et al. (2004) fern phylogeny used here as a framework.

negatives occur when the region of interest has a sequence that is highly diverged from that of the probe. In both cases, the results can be misleading. Basic knowledge of fern plastid gene order has been obtained with physical mapping, but current studies require information at a higher resolution.

Two alternatives to physical mapping exist: (1) complete genome sequencing and (2) using PCR to map regions of interest. Complete genome sequencing is currently in use and is very effective. However, like physical mapping, sequencing

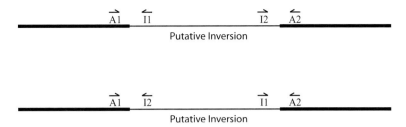

Figure 6.3 Location of PCR primers used to detect a putative inversion present in some taxa.

complete genomes can be overly time-consuming. PCR mapping, on the other hand, uses PCR to determine gene order within a targeted region of the genome relatively rapidly. First, universal primers are designed to amplify regions of interest. The PCR product is then sequenced to verify gene content and order. Because the procedure is relatively easy, many more taxa can be characterized than is feasible for complete genome sequencing. The PCR reactions may be more readily optimized than hybridization reactions. The technique is also less susceptible to variations in homology since false positives can be easily detected by direct sequencing of the PCR product. Universal PCR primers require the conservation of nucleotide order in small regions (25 bp is more than adequate). Furthermore, PCR mapping requires far less DNA than either filter hybridization or complete genome sequencing.

6.4.2 Methods

The general approach to PCR mapping is simple. Two sets of primers are needed to detect differences in gene order. The first primer set consists of the "anchored" primers, which are designed within genes close to, but not moved by, a putative inversion. The second primer set consists of the "inversion" primers, designed within genes moved by the putative inversion, and as close to the inversion boundaries as possible. Figure 6.3 illustrates these primer locations. To test for gene order, each anchored primer is combined with each inversion primer. The combination of working and failing PCR reactions indicate which gene order is present. For example, in Figure 6.3, if the gene order is A1, I1, I2, A2, then the primer combinations A1 + I1 and A2 + I2 would be positive, whereas the combinations A1 + I2 and A2 + I1 would be negative. This approach has been used successfully to examine the distribution of a large inversion in moss plastid genomes (Sugiura et al., 2003).

To apply PCR mapping to the IR of fern plastid genomes we designed two sets of primers as above. The anchored primers were located in *rpl2, rpl32, chlL*, and *chlB* (Figure 6.4). These genes are outside the IR and also outside the putative region that has been reorganized. Inversion primers were designed within *rrn5*, *rps7*, *rrn16*, and *psbA* (Figure 6.4). These primers can be used to determine whether a taxon has the gene order of either *Adiantum*, or *Angiopteris*, or a structure resulting from only one of the two inferred inversions. To compensate for changes in gene orientation, both forward and reverse inversion primers were designed.

We attempted to sample a representative from each major lineage of the leptosporangiate ferns. We included a member of the filmy ferns (*Trichomanes*), not included by Raubeson and Stein (1995). We determined gene order in the IR region for *Angiopteris*, *Osmunda*, *Trichomanes*, *Gleichenia*, *Lygodium*, *Marsilea*, *Dicksonia*, *Pteridium*, and *Adiantum*.

6.4.3 Results

Here we present a progress report. Details of primer sequences, PCR protocols, DNA sequences of genome segments, and complete maps will be published elsewhere. As in the earlier studies, we found that *Osmunda* and *Trichomanes* both have the *Angiopteris* gene order. *Marsilea*, *Dicksonia*, and *Pteridium* have the *Adiantum* gene order. *Lygodium* appears to have only one of the two inferred inversions, thus it has an "intermediate" gene order (Figure 6.1). The gene order of *Gleichenia* appears to be distinct, but has yet to be fully characterized. It may have only one of the two major inversions but with additional rearrangements, perhaps unique to the gleichenioid lineage. These results are similar to those of Raubeson and Stein (1995), except that the latter inferred the *Adiantum* structure for *Lygodium*, whereas we find the intermediate structure. We found that both *Lygodium* and *Gleichenia* share the gene order *chlL*, *trnN*, *trnR*, *rrn5* and *rpl32*, *trnN*, *trnR*, *rrn5*, as one moves from each end of the SSC into the IR, i.e., the *Angiopteris* structure. This indicates that the second inversion has not occurred. The problem then is determining the gene order at the LSC/IR boundary. This region is clearly a hotspot of structural evolution since the IR is subject to "ebb and flow" at its boundary (Goulding *et al.*, 1996). We are currently experimenting with additional primer combinations to test for distinct gene orders.

Figure 6.5 illustrates the evolutionary order of the taxa being studied as proposed by Pryer *et al.* (2004). Marking the changes in gene order described above on the phylogeny, we find three possible points for rearrangements. *Angiopteris*, *Osmunda*, and *Trichomanes* all share the ancestral gene order, similar to that of other vascular plants. Inversion 1 occurred at one of two possible positions: either on the branch leading to the common ancestor of the schizaeoid clade and its sister group, or on the branch leading to the common ancestor of the

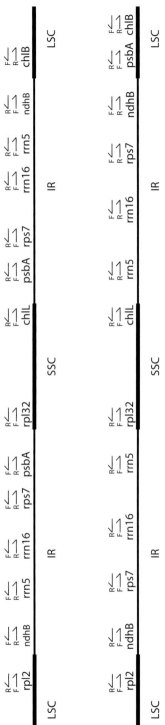

Figure 6.4 Primer locations for the current study of plastid genome organization in ferns: (a) *Adiantum* and (b) *Angiopteris*.

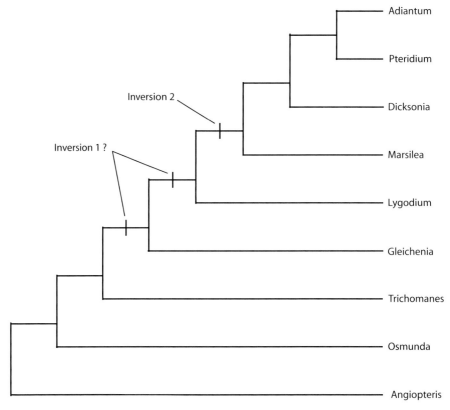

Figure 6.5 Current understanding of plastid genome inversions (see Figure 6.1) marked on a simplified phylogenetic framework.

gleichenioid clade and its sister. Inversion 2 occurred on the branch leading to the common ancestor of the heterosporous fern clade and its sister group. Note that either parsimonious acquisition of inversion 1 requires that the gleichenioid ferns are not sister to the filmy ferns, as is possible with the unresolved tree in Figure 6.2. That sister relationship was recovered in only some of the analyses by Pryer *et al.* (2004); see Chapter 15.

The gene order data gathered so far suggest that major rearrangement events do occur in intervals, rather than during temporal hot spots. This being the case, rearrangement events, when present, yield another potential phylogenetic marker to be considered. The one caveat we should add is that our inferences are based on the sampling of only one taxon per major lineage. Although this should be sufficient to infer the sequence of events involved in the major, already characterized, inversions, we could be missing other structural changes that are significant, and even phylogenetically informative within specific lineages.

6.5 Conclusions and prospects

Although our overall understanding of fern plastid genome structure and evolution is increasing, we still lack knowledge of the variation within many groups of ferns, where DNA sequence data from a sample of genes is the preferred approach for phylogenetic studies. In general, the plastid genome structure remains evolutionarily conservative (or perhaps there is convergence to a stable structure), yet in certain clades the structure can become destabilized. Thus, within some angiosperm families, the plastid genome structure provides a wealth of phylogenetic data (Cosner et al., 2004; Kim et al., 2005). In ferns, only a few groups have been examined extensively at the structural level, one excellent example being the tree ferns (Conant et al., 1994). With the shift to sequenced based approaches, it is likely that some potentially very useful data may be missed. In addition to the traditional studies of plastid genome structure and evolution, it is likely that future studies will examine other aspects of plastid "genomics," especially investigations into plastid proteins (Peltier et al., 2000; van Wijk, 2000; Leister, 2003) as well as regulation of plastid-encoded genes (Wu et al., 1993; Eberhard et al., 2002; Robbens et al., 2005). With the variety and combination of genomic tools currently available, it is likely that the next decade will open up exciting new avenues of investigation.

References

Bendich, A. J. (2004). Circular chloroplast chromosomes: the grand illusion. *The Plant Cell*, **16**, 1661–1666.

Conant, D. S., Stein, D. B., Valinski, A. E. C., Sudarsanam, P., and Ahearn, M. E. (1994). Phylogenetic implications of chloroplast DNA variation in the Cyatheaceae.1. *Systematic Botany*, **19**, 60–72.

Cosner, M. A., Raubeson, L. A., and Jansen, R. K. (2004). Chloroplast DNA rearrangements in Campanulaceae: phylogenetic utility of highly rearranged genomes. *BMC Evolutionary Biology*, **4**, 1–17.

Eberhard, S., Drapier, D., and Wollman, F. A. (2002). Searching limiting steps in the expression of chloroplast-encoded proteins: relations between gene copy number, transcription, transcript abundance and translation rate in the chloroplast of *Chlamydomonas reinhardtii*. *Plant Journal*, **31**, 149–160.

Goulding, S. E., Olmstead, R. G., Morden, C. W., and Wolfe, K. H. (1996). Ebb and flow of the chloroplast inverted repeat. *Molecular and General Genetics*, **252**, 195–206.

Hasebe, M. and Iwatsuki, K. (1990a). *Adiantum capillus-veneris* chloroplast DNA clone bank: as useful heterologous probes in the systematics of the leptosporangiate ferns. *American Fern Journal*, **80**, 20–25.

Hasebe, M. and Iwatsuki, K. (1990b). Chloroplast DNA from *Adiantum capillus-veneris* L., a fern species (Adiantaceae) – clone bank, physical map and unusual gene localization in comparison with angiosperm chloroplast DNA. *Current Genetics*, **17**, 359–364.

Hasebe, M. and Iwatsuki, K. (1992). Gene localization on the chloroplast DNA of the maiden hair fern: *Adiantum capillus-veneris*. *Botanical Magazine (Tokyo)*, **105**, 413–419.

Helfenbein, K. G. and Boore, J. L. (2004). The mitochondrial genome of *Phoronis architecta* – comparisons demonstrate that phoronids are lophotrochozoan protostomes. *Molecular Biology and Evolution*, **21**, 153–157.

Jansen, R. K., Raubeson, L. A., Boore, J. L., dePamphilis, C. W., Chumley, T. W., Haberle, R. C., Wyman, S. K., Alverson, A. J., Peery, R., Herman, S. J., Fourcade, H. M., Kuehl, J. V., McNeal, J. R., Leebens-Mack, J., and Cui, L. (2005). Methods for obtaining and analyzing whole chloroplast genome sequences. *Methods in Enzymology*, **395**, 348–384.

Kim, K.-J., Choi, K.-S., and Jansen, R. K. (2005). Two chloroplast DNA inversions originated simultaneously during the early evolution of the sunflower family (Asteraceae). *Molecular Biology and Evolution*, **22**, 1783–1792.

Kolodner, R. and Tewari, K. K. (1979). Inverted repeats in chloroplast DNA from higher plants. *Proceedings of the National Academy of Sciences of the United States of America*, **76**, 41–45.

Kugita, M., Yamamoto, Y., Fujikawa, T., Matsumoto, T., and Yoshinaga, K. (2003). RNA editing in hornwort chloroplasts makes more than half the genes functional. *Nucleic Acids Research*, **31**, 2417–2423.

Ledergerg, J. and McCray, A. T. (2001). "Ome sweet "omics – a genealogical treasury of words. *The Scientist*, **15**, 8–9.

Leister, D. (2003). Chloroplast research in the genomic age. *Trends in Genetics*, **19**, 47–56.

Martin, W. and Herrmann, R. G. (1998). Gene transfer from organelles to the nucleus: how much, what happens, and why? *Plant Physiology*, **118**, 9–17.

Martin, W. and Miller, M. (1998). The hydrogen hypothesis for the first eukaryote. *Nature*, **392**, 37–41.

Martin, W., Somerville, C. C., and Loiseaux-de Goel, S. (1992). Molecular phylogenies of plastid origins and algal evolution. *Journal of Molecular Evolution*, **35**, 385–404.

Martin, W., Stoebe, B., Goremyken, V., Hansmann, S., Hasegawa, M., and Kowallik, K. V. (1998). Gene transfer to the nucleus and the evolution of chloroplasts. *Nature*, **393**, 162–165.

McNeal, J. R., Leebens-Mack, J. H., Arumuganathan, K., Kuehl, J. V., Boore, J. L., and dePamphilis, C. W. (2006). Using partial genomic fosmid libraries for sequencing complete organellar genomes. *Biotechniques*, **41**, 69–72.

Nishiyama, T., Kugita, M., Sinclair, R. B., Sugita, M., Sugiura, C., Wakasugi, T., Wolf, P. G., Yamada, K., Yoshinaga, K., and Hasebe, M. (2004). Bryophytes are monophyletic and land plants comprise two extant lineages. *Molecular Biology and Evolution*, **21**, 1813–1819.

Ohyama, K., Fukuzawa, H., Kohchi, T., Shirai, H., Sano, T., Sano, S., Umesono, K., Shiki, Y., Takeuchi, M., Chang, Z., Aota, S., Inokuchi, H., and Ozeki, H. (1986).

Chloroplast gene organization deduced from complete sequence of liverwort *Marchantia polymorpha* chloroplast DNA. *Nature*, **322**, 572–574.

Olmstead, R. G. and Palmer, J. D. (1994). Chloroplast DNA systematics: a review of methods and data analysis. *American Journal of Botany*, **81**, 1205–1224.

Palmer, J. D. (1983). Chloroplast DNA exists in two orientations. *Nature*, **301**, 92–93.

Palmer, J. D. (1985a). Comparative organization of chloroplast genomes. *Annual Review of Genetics*, **19**, 325–354.

Palmer, J. D. (1985b). Evolution of cpDNA and mtDNA in plants and algae. In *Molecular and Evolutionary Genetics*, ed. R. J. MacIntyre. New York: Plenum Press, pp. 131–240.

Palmer, J. D. (1986). Isolation and structural analysis of chloroplast DNA. *Methods in Enzymology*, **118**, 167–186.

Palmer, J. D. (1987). Chloroplast DNA evolution and biosystematic uses of chloroplast DNA variation. *The American Naturalist*, **130**, 6–29.

Palmer, J. D. (1991). Plastid chromosomes: structure and evolution. In *Cell Culture and Somatic Genetics of Plant*, Vol. 7A, *Molecular Biology of Plastids*, ed. L. Bogorad and I. K. Vasil. San Diego, CA: Academic Press, pp. 5–53.

Palmer, J. D. and Stein, D. B. (1986). Conservation of chloroplast genome structure among vascular plants. *Current Genetics*, **10**, 823–833.

Palmer, J. D. and Thompson, W. F. (1981). Rearrangements in the chloroplast genomes of mung bean and pea. *Proceedings of the National Academy of Sciences of the United States of America*, **78**, 5533–5537.

Peltier, J., Friso, G., Kalume, D., Roepstorff, P., Nilsson, F., Adamska, I., van Wijk, K., and van Wijk, K. (2000). Proteomics of the chloroplast: systematic identification and targeting analysis of lumenal and peripheral thylakoid proteins. *Plant Cell*, **12**, 319–341.

Pryer, K. M., Schuettpelz, E., Wolf, P. G., Schneider, H., Smith, A. R., and Cranfill, R. (2004). Phylogeny and evolution of ferns (monilophytes) with a focus on the early leptosporangiate divergences. *American Journal of Botany*, **91**, 1582–1598.

Raubeson, L. A. and Jansen, R. K. (1992). Chloroplast DNA evidence on the ancient evolutionary split in vascular land plants. *Science*, **255**, 1697–2699.

Raubeson, L. A. and Stein, D. B. (1995). Insights into fern evolution from mapping chloroplast genomes. *American Fern Journal*, **85**, 193–204.

Robbens, S., Khadaroo, B., Camasses, A., Derelle, E., Ferraz, C., Inze, D., Van de Peer, Y., and Moreau, H. (2005). Genome-wide analysis of core cell cycle genes in the unicellular green alga *Ostreococcus tauri*. *Molecular Biology and Evolution*, **22**, 589–597.

Roper, J. M., Hansen, S. K., Wolf, P. G., Karol, K. G., Mandoli, D. F., Everett, K. D. E., Kuehl, J., and Boore, J. L. (2007). The complete plastid genome sequence of *Angiopteris evecta* (G. Forst.) Hoffm. *American Fern Journal*, **97**, 95–106.

Shinozaki, K., Ohme, M., Tanaka, M., Wakasugi, T., Hayashida, N., Matsubayashi, T., Zaita, N., Chunwongse, J., Obokata, J., Yamaguchi-Shinozaki, K., Ohto, C., Torazawa, K., Meng, B. Y., Sugita, M., Deno, H., Kamogashira, T., Yamada, K., Kusuda, J., Takaiwa, F., Kato, A., Tohdoh, N., Shimada, H., and Sugiura, M. (1986).

The complete nucleotide sequence of tobacco chloroplast genome: its gene organization and expression. *EMBO Journal*, **5**, 2043–2049.

Small, R. L., Lickey, E. B., Shaw, J., and Hauk, W. D. (2005). Amplification of noncoding chloroplast DNA for phylogenetic studies in lycophytes and monilophytes with a comparative example of relative phylogenetic utility from Ophioglossaceae. *Molecular Phylogenetics and Evolution*, **36**, 509–522.

Stein, D. B., Palmer, J. D., and Thompson, W. F. (1986). Structural evolution and flip-flop recombination of chloroplast DNA in the fern genus *Osmunda*. *Current Genetics*, **10**, 835–841.

Stein, D. B., Conant, D. S., Ahearn, M. E., Jordan, E. T., Kirch, S. A., Hasebe, M., Iwatsuki, K., Tan, M. K., and Thomson, J. A. (1992). Structural rearrangements of the chloroplast genome provide an important phylogenetic link in ferns. *Proceedings of the National Academy of Sciences of the United States of America*, **89**, 1856–1860.

Stoebe, B., Hansmann, S., Goremykin, V., Kowalik, K. V., and Martin, W. (1999). Proteins encoded in sequenced chloroplast genomes: an overview of gene content, phylogenetic information and endosymbiotic gene transfer to the nucleus. In *Molecular Systematics and Plant Evolution*, ed. P. M. Hollingsworth, R. M. Batesman, and R. J. Gornall. London: Taylor and Francis, pp. 327–352.

Sugiura, C., Kobayashi, Y., Aoki, S., Sugita, C., and Sugita, M. (2003). Complete chloroplast DNA sequence of the moss *Physcomitrella patens*: evidence for the loss and relocation of *rpoA* from the chloroplast to the nucleus. *Nucleic Acids Research*, **31**, 5324–5331.

van Wijk, K. J. (2000). Proteomics of the chloroplast: experimentation and prediction. *Trends in Plant Science*, **5**, 420–425.

Wolf, P. G., Rowe, C. A., Sinclair, R. B., and Hasebe, M. (2003). Complete nucleotide sequence of the chloroplast genome from a leptosporangiate fern, *Adiantum capillus-veneris* L. *DNA Research*, **10**, 59–65.

Wolf, P. G., Rowe, C. A., and Hasebe, M. (2004). High levels of RNA editing in a vascular plant chloroplast genome: analysis of transcripts from the fern *Adiantum capillus-veneris*. *Gene*, **339**, 89–97.

Wolf, P. G., Karol, K. G., Mandoli, D. F., Kuehl, J., Arumuganathan, K., Ellis, M. W., Mishler, B. D., Kelch, D. G., Olmstead, R. G., and Boore, J. L. (2005). The first complete chloroplast genome sequence of a lycophyte, *Huperzia lucidula* (Lycopodiaceae). *Gene*, **350**, 117–128.

Wolfe, K. H., Morden, C. W., Ems, S. C., and Palmer, J. D. (1992). Rapid evolution of the plastid translational apparatus in a nonphotosynthetic plant: loss or accelerated sequence evolution of tRNA and ribosomal protein genes. *Journal of Molecular Evolution*, **35**, 304–317.

Wu, M., Chang, C. H., Yang, J. M., Zhang, Y. L., Nie, Z. Q., and Hsieh, C. H. (1993). Regulation of chloroplast DNA replication in *Chlamydomonas reinhardtii*. *Botanical Bulletin of Academia Sinica*, **34**, 115–131.

Wyman, S. K., Boore, J. L., and Jansen, R. K. (2004). Automatic annotation of organellar genomes with DOGMA. *Bioinformatics*, **20**, 3252–3255.

7

Evolution of the nuclear genome of ferns and lycophytes

TAKUYA NAKAZATO, MICHAEL S. BARKER, LOREN H. RIESEBERG, AND GERALD J. GASTONY

7.1 Introduction

Analyses of gene expression and function, genetic networks, population polymorphisms, and genome organization at the whole genome level have enabled research on previously intractable questions (reviewed in Wolfe and Li, 2003). Among plant lineages, genomic approaches have been most widely applied in the angiosperms, where significant resources have been developed. Angiosperm studies utilizing genome scale analyses have made several important advances, including the identification of an extensive history of genome duplications (Blanc et al., 2003; Schlueter et al., 2004; Cui et al., 2006), progress in understanding flower development and evolution (Doust et al., 2005; Whibley et al., 2006), characterization of the genetics underlying speciation and adaptation (Bradshaw and Schemske, 2003; Rieseberg et al., 2003; Lai et al., 2005; Eyre-Walker, 2006), the identification and mapping of recombination hot spots (Drouaud et al., 2006), and the discovery and role of microRNAs (Bartel and Bartel, 2003; Bartel, 2004). Genomic analyses will undoubtedly continue to provide tests of longstanding questions and offer novel perspectives in biology. For example, modern genomic analyses are capable of explaining the origin of the exceptionally high chromosome numbers of homosporous ferns and lycophytes, a result that will shed light on eukaryotic genome organization and evolution.

Although there are rich biological and taxonomic resources for ferns and lycophytes, the genomics of these seed-free plants is still in its infancy, and the tools necessary for genomic studies lag behind those available for seed plants. The first homosporous fern linkage map was published only recently (Nakazato et al.,

Biology and Evolution of Ferns and Lycophytes, ed. Tom A. Ranker and Christopher H. Haufler. Published by Cambridge University Press. © Cambridge University Press 2008.

2006), whereas a large number of linkage maps for seed plants have accumulated since the 1980s. Only four modest Expressed Sequence Tags (EST) libraries are currently available for ferns and lycophytes (two ferns *Adiantum capillus-veneris* and *Ceratopteris richardii* and two lycophytes *Selaginella lepidophylla* and *S. moellendorffii*), and three Bacterial Artificial Chromosome (BAC) libraries (*C. richardii* and two *S. moellendorffii*) were recently constructed (www.greenbac.org). Whole genome sequences are also being developed, with three chloroplast genome sequences published recently (two from the ferns *Adiantum capillus-veneris* and *Psilotum nudum*, and one from the lycophyte *Huperzia lucidula*; see Chapter 6), and a nuclear genome sequencing effort is underway for *Selaginella moellendorffii*.

Development of fern and lycophyte genomic resources has been hindered primarily by two factors. First, genomic studies in these groups have been challenging because of their large genome sizes (mean 10 616 Mb, $N = 87$) compared to those of angiosperms (mean 6383 Mb, $N = 4427$, Plant DNA C-values Database, Royal Botanic Gardens, Kew; www.rbgkew.org.uk/cval/homepage.html). The development of low-cost, high-throughput molecular techniques and accumulating genetic resources will soon overcome this obstacle. A second hindrance to acquiring fern and lycophyte genomic resources is the limited funding opportunities attributable to the negligible economic importance of these plants. However, fully understanding the evolution of the economically important seed plants requires comparative data from related groups. For example, ferns are the second most diverse vascular plant group after angiosperms, including about 11 000 species (Smith *et al.*, 2006), and are the phylogenetic sister group to the entire clade of seed plants (Pryer *et al.*, 2001). Currently, our knowledge of plant genomes is based almost exclusively on a few groups of seed plants, and it is essential to explore diverse lineages before we can generalize about the organization and evolution of plant genomes. This issue is beginning to be addressed by funding agencies through projects whose goals are the development of diverse plant genome resources such as The Green Plant BAC Library Project (greenbac.org), Tree of Life Web Project (tolweb.org), and a number of ongoing whole genome sequencing projects involving phylogenetically diverse organisms.

Despite the currently limited genomic resources in ferns and lycophytes, the last decade has seen a significant increase in our understanding of their genomes. As reviewed in this chapter, the emerging data indicate that ferns and lycophytes share many genomic features with other plant groups. Still, some features such as the mode of evolution of chromosome number are unique to these lineages, suggesting that the biological characteristics of ferns and lycophytes and historical contingency have shaped their genome structure. Unfortunately, the available genomic data to date in ferns and lycophytes are mostly observational, and much more detailed data at the level of nucleotides, as well as careful hypothesis-driven studies, are necessary to understand fully the mechanisms of genome evolution in this group. Because genomes contain a tremendous amount of information and their evolution is influenced by a large number of factors, the resolution and accuracy of genomic knowledge are highly dependent on technological advances in molecular biology, genomics, and computational biology. In this chapter, we therefore first review historical advancements in our understanding of fern and lycophyte genomes based on several key technological innovations. We then integrate the information available today and attempt to interpret how fern and lycophyte genomes are organized and evolve. Finally, we discuss the shortcomings of our current knowledge and suggest how to improve our understanding of fern and lycophyte genomes.

7.2   Historical summary

Numerous chromosome counts accumulated over the past century provided the first view of fern and lycophyte genomes. Although these studies are descriptions of genomes in the broadest sense, they provided the fundamental hypotheses that continue to serve as the basis for much fern and lycophyte genome research. Comparison of chromosome counts among related species of many genera clearly showed that recognizable polyploids (neopolyploids) such as tetraploids and hexaploids are frequent throughout the fern and lycophyte lineages, eventually reaching the highest chromosome counts among the known living organisms in *Ophioglossum reticulatum* ($n > 600$, Khandelwal, 1990). These observations suggested that polyploidization is an ongoing process and a major component of fern and lycophyte evolution. In addition to the abundant occurrence of neopolyploids, it became clear that, with rare exceptions, even the lowest chromosome numbers in each genus are much higher than those of other plant groups. Klekowski and Baker (1966) estimated the average haploid chromosome numbers of homosporous ferns and lycophytes to be 57.05, compared to 15.99 for angiosperms. Haploid chromosome numbers higher than 14 are generally considered to be polyploids in angiosperms (Grant, 1981). If this rule were applied to homosporous ferns, 95% of the species would be considered polyploid. A significant exception to the generally high chromosome numbers in ferns are the heterosporous water ferns; the haploid chromosome number for heterosporous ferns averages 13.6, and 90% of these species have chromosome numbers less than 28. A similar situation is encountered in the lycophytes, where the homosporous Lycopodiaceae have significantly higher chromosome numbers than their sister group (Pryer et al., 2001), the heterosporous Selaginellaceae and Isöetaceae (Löve et al., 1977). Thus, homosporous ferns and lycophytes have significantly higher chromosome numbers compared to their close heterosporous relatives and seed plants.

Klekowski and Baker (1966) sought to explain the high chromosome numbers of homosporous ferns as a mechanism for maintaining heterozygosity in a putatively inbreeding group of plants. Because homosporous ferns are capable of producing hermaphroditic gametophytes, self-fertilization of these gametophytes (intragametophytic self-fertilization) results in completely homozygous sporophytes (see Chapter 2). Therefore, homosporous ferns potentially suffer severe losses of heterozygosity more frequently than do heterosporous plants. Based on these observations, Klekowski and Baker (1966) proposed that homosporous ferns acquired their high chromosome numbers in response to selection for increased heterozygosity via pairing of homologous chromosomes derived from polyploidization instead of conventional pairing of homologous chromosomes. According to Klekowski and Baker (1966), heterosporous ferns have lower chromosome numbers than homosporous ferns because they are obligately outcrossing and therefore cannot experience sharp reductions in heterozygosity through intragametophytic selfing. A variety of subsequent studies supported Klekowski's hypothesis. For example, Hickok and Klekowski (1974) and Hickok (1978) demonstrated homologous pairing in a small percentage (<1–3%) of self-fertilized offspring of the tetraploid *Ceratopteris thalictroides*. These studies demonstrated that homologous pairing can occur and may operate to promote heterozygosity in homosporous ferns.

Isozymes were an important tool in evaluating Klekowski's hypothesis and characterizing fern and lycophyte genomes. An early isozyme study allegedly showed support for the Klekowski and Baker hypothesis by demonstrating multiple bands inherited in non-Mendelian fashion (Chapman et al., 1979). However, a subsequent investigation demonstrated that a homosporous fern with the lowest (i.e., base) number for its genus (what is generally considered a diploid species) possessed a diploid, not polyploid, gene expression profile with Mendelian inheritance (Gastony and Gottlieb, 1982). Furthermore, it was demonstrated that the complex isozyme banding patterns of Chapman et al. (1979) were in part attributable to subcellularly compartmentalized isozymes (Gastony and Darrow, 1983; Wolf et al., 1987). Continued isozyme investigations supported the observation that homosporous ferns with base chromosome numbers for their genus have diploid expression profiles despite possessing relatively high chromosome numbers (Gastony and Gottlieb, 1985; Haufler and Soltis, 1986; Soltis, 1986; Haufler, 1987). These results differed from those in angiosperms, which showed that angiosperms with high chromosome numbers did indeed have duplicated sets of isozymes (reviewed in Gottlieb, 1982; Soltis and Soltis, 1990).

Analyses of natural fern populations also showed that they are predominantly outcrossing instead of engaging in intragametophytic selfing. Haufler and Soltis (1984) found that 83% of the examined populations of diploid *Bommeria hispida* had heterozygous enzyme banding patterns attributable to outcrossing between genetically different gametophytes. In a more extensive analysis of diploid populations of *Pellaea andromedifolia*, Gastony and Gottlieb (1985) determined that at least 81.3% of 145 sporophytes examined from natural populations arose through outcrossing between gametophytes carrying different alleles. Holsinger (1987) developed a statistical technique for estimating rates of self-fertilization in homosporous ferns and applied this to the data of Gastony and Gottlieb (1985). He found that there is no evidence of intragametophytic self-fertilization in the two populations with large enough sample sizes for his statistical analyses and concluded that intragametophytic self-fertilization occurs only a small fraction of the time, if at all. Thus Klekowski and Baker's (1966) original rationale for polyploid homologous heterozygosity, high levels of intragametophytic self-fertilization, was rejected.

To explain the paradox of high chromosome numbers and diploid isozyme expression in homosporous ferns, Haufler (1987) hypothesized that homosporous ferns may have acquired high chromosome numbers with diploid gene expression through repeated cycles of polyploidization and subsequent gene silencing without the loss of chromosomes. If such a mechanism is common, the genomes of extant homosporous ferns should contain multiple, silenced copies of each nuclear gene. In support of Haufler's hypothesis, a few studies have identified silenced nuclear genes in homosporous fern genomes. Gastony (1991) observed the process of gene silencing by documenting that the duplicated expression of a phosphoglucoisomerase isozyme in the neotetraploid homosporous fern *Pellaea rufa* progressively diminished to a diploid level in several populations. However, isozyme analyses can assess only the expression of enzyme-coding loci and not the presence of these genes. Therefore, this method cannot determine whether homosporous ferns with the lowest chromosome numbers in their genus have diploid expression of isozyme loci because they have never experienced polyploidization or because they have gone through the cycle of polyploidization and gene silencing, unless sequences of gene copies are examined (reviewed in Soltis and Soltis, 1987). Using DNA sequence based probes for chlorophyll a/b binding protein (CAB) genes in the homosporous fern *Polystichum munitum*, Pichersky et al. (1990) identified several copies of this high copy number gene and found that several gene copies have been pseudogenized, consistent with Haufler's hypothesis.

Subsequent studies evaluating gene copy number in homosporous ferns utilized restriction fragment length polymorphisms (RFLPs). RFLPs estimate the number of genes present in the genome by using DNA probes to detect length differences in restriction enzyme-digested genomic fragments. Using RFLPs with several cDNA probes, McGrath et al. (1994) found multiple copies of most genes tested in the genome of the diploid homosporous fern species *Ceratopteris richardii*, consistent with Haufler's hypothesis. Therefore, even diploid species of homosporous ferns seem to have multiple gene copies, although many may be inactive. McGrath et al. also compared the RFLP profiles of diploid *C. richardii* and tetraploid *C. thalictroides* and hypothesized that twice as many gene copies should be present in *C. thalictroides*. In contrast to their expectation, the tetraploid species had on average only 30% (marginally significant) more gene copies, and some of its genes were single-copy. Although some gene copies may have been undetected if they co-migrate on the gel or if their sequences are too diverged from the probe sequences, it seems reasonable to assume that at least in some cases, genes duplicated through polyploidization have been not only silenced but also completely deleted from the genome.

7.3 Review of critical recent advances

Recent advances in fern and lycophyte genomics have used new tools to overcome the limitations of earlier techniques and study the evolution of fern and lycophyte chromosome number. For example, RFLP studies cannot distinguish genome-wide polyploidization events from small-scale duplications because the number of detected RFLP fragments tends to overestimate the number of gene copies if restriction sites exist within the probe hybridizing sites. Furthermore, the relative positions of gene copies in the genome are unknown in RFLP studies without analysis of segregating populations. Some of these limitations can be overcome by fluorescent *in situ* hybridization (FISH) or "chromosome painting," which allows one to visualize the physical position of genes corresponding to the fluorescently labeled hybridizing probes on the fixed chromosomes. FISH using an rDNA probe in *C. richardii* showed that multiple rDNA gene copies exist on different chromosomes (McGrath and Hickok, 1999), supporting the hypothesis of paleopolyploidy in homosporous ferns. However, it is difficult to determine relative positions among different marker loci with FISH, and the approach is technically challenging and labor intense, especially when hybridizing many low copy markers. Other techniques can provide more information about fern and lycophyte genomes with greater reliability.

The most comprehensive view of the genome-wide distribution of duplicated gene copies presently available for fern and lycophyte genomes comes from genetic linkage mapping. Linkage maps describe the relative locations of polymorphic markers in genomes inferred from the segregation patterns of parental alleles in mapping populations. Inference of marker positions is based on a simple assumption: the frequency of recombination between two markers increases with the physical distance between these loci, although physical and genetic map distances are often not proportional because recombination rates are unevenly distributed across the genome. Linkage mapping is a powerful tool for investigating paleopolyploidy not only because gene copy number and relative marker positions can be estimated fairly accurately, but also because mapping a large number of duplicated genes allows us to infer large-scale duplication (e.g., polyploidy) if sets of duplicated loci occur in similar orders in different parts of the genome. Nakazato et al. (2006) constructed a genetic linkage map of the model homosporous fern *C. richardii* based on 729 markers (368 RFLPs, 358 amplified fragment length polymorphisms (AFLPs), and 3 isozymes) using a mapping population consisting of 488 doubled haploid lines (DHLs). Screening of 1037 RFLPs hybridized to fluorescently labeled cDNA probes showed that the majority of visualized genes (85%) occur in multiple copies, consistent with the results of previous RFLP studies. After correcting for potential multiple restriction sites within hybridizing fragments, 24% of genes were estimated to be single copy. This is one of the lowest proportions of single copy genes among plant species, at least as measured by RFLP linkage mapping. Local duplications such as tandem duplications are common, yet a large number of duplications were detected between rather than within chromosomes. Although these data favor the paleopolyploidy hypothesis as in previous studies, it is premature to reach a definite conclusion. First, the detected gene duplicates seemed to be scattered more or less haphazardly across the genome rather than occurring in recognizable homoeologous chromosomal segments (Figure 7.1). If there was a recent polyploidization event as in tetraploid cotton (1.1–1.9 million years ago (MYA), Wendel and Cronn, 2003) and maize (<4.8 MYA, Swigonova et al., 2004), syntenic chromosomal blocks should have been easily recognizable. Likewise, the genomes of *Brassica* species currently recognized as diploid are mainly composed of three sets of gene copies, suggesting that they are ancient hexaploids, although syntenic regions are shorter than in cotton or maize (Lagercrantz and Lydiate, 1996).

The absence of large syntenic regions consisting of duplicated gene copies in the mapping study of Nakazato et al. suggests that polyploidization has not occurred in the last several million years in the *C. richardii* lineage and perhaps not in diploid homosporous ferns in general. Statistical analyses of the distribution of gene copies, however, indicate that there is significant clustering in the *C. richardii* genome; that is, the copies of a set of markers in a given linkage group tend to co-occur on a different linkage group, although the pattern is not strong enough to be visually apparent. This result suggests that there may have been ancient polyploidization events but that the signals of such events have

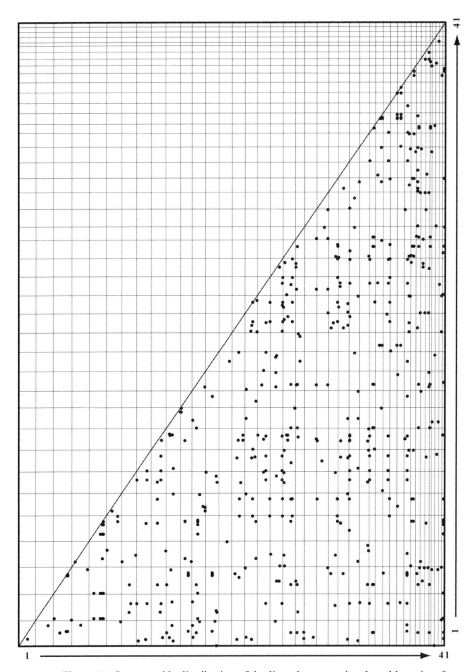

Figure 7.1 Genome-wide distribution of duplicated gene copies shared by pairs of linkage groups (indicated by the numbers). The width and the height of each block are proportional to the cM map length of the corresponding linkage group. (From Nakazato et al., 2006.)

been masked by chromosomal rearrangements and by ongoing duplications at the level of individual genes. Plant genomes are constantly altered by duplication, deletion, transposition, translocation, and transposable element insertion (reviewed in Bennetzen, 2002). In tetraploid cotton, for example, many gene copies have moved from homoeologous chromosomes in the less than 2 MY since polyploidization, indicating that plant genomes are surprisingly dynamic.

An alternative to map-based approaches to identify paleopolyploidy is to analyze duplicate gene pair age distributions. In these analyses, large-scale duplication events are identified when a significant fraction of duplications coalesce to a common time point. This approach has two attractive features that make it an economical approach for evaluating paleopolyploidy. First, ESTs developed for other purposes can be utilized to search for duplicate gene pairs. Second, assuming a clock-like rate of molecular evolution at synonymous sites, ages of duplication events can be inferred from their Ks value (the synonymous substitution rate), obviating the need for outgroup data to determine the age of duplication events (Lynch and Conery, 2000). Such an approach has been used successfully to detect ancient large-scale duplication events (putative paleopolyploidy) in *Arabidopsis thaliana* (Lynch and Conery, 2000; Blanc and Wolfe, 2004), diploid and tetraploid cotton, wheat, maize, potato, tomato, soybean, *Medicago truncatula* (Blanc and Wolfe, 2004), and seven species of poplar (Sterck et al., 2005). In these analyses, a generally exponential decline in the age of duplicate pairs was observed, as young duplications continually generated in the genome are lost over time. However, deviations from this exponentially declining distribution pattern were observed as significant peaks in the Ks value distribution and were interpreted as large-scale duplication events, putatively paleopolyploidization events.

Using this approach, Barker et al. (unpublished data) examined the age distribution of duplicate gene pairs in the homosporous fern *Ceratopteris richardii*. In the EST library of 6350 sequences, 631 duplicate gene pairs were identified, of which 387 had a Ks < 5.0. Barker et al. (unpublished data) identified a region that likely corresponds to a paleopolyploidization event in the history of *C. richardii* located at Ks 0.9–1.84, and dated between 78 and 151 MYA with a mean and median age of 115 MYA (Figure 7.2). Based on a SiZer analysis (Chaudhuri and Marron, 1999), which evaluates distribution data for significant features, this region is a broad but significant peak at $p < 0.05$ (Figure 7.3). These results suggest that *C. richardii* is likely a paleopolyploid that experienced its most recent duplication event in the ancient past, before the divergence of extant lineages of Pteridaceae and possibly before the divergence of the Pteridaceae and the eupolypods (Figure 7.2). Furthermore, these results are consistent with Nakazato et al.'s linkage map of *C. richardii*, where no recent genome-wide duplication

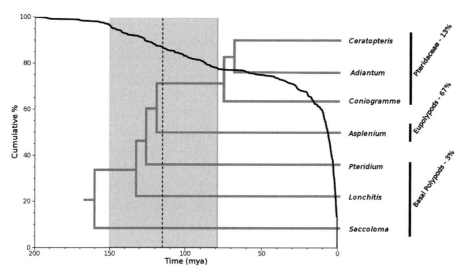

Figure 7.2 A cumulative distribution function plot (black line) of duplicate Ks values for *Ceratopteris richardii* superimposed over a phylogeny of polypod ferns (Pryer *et al.*, 2004). The gray box indicates the range of the duplication event, with the mean and median indicated by the vertical dashed line. The numbers on the right y-axis are the percentages of extant ferns (Schneider *et al.*, 2004).

events were detected. However, the statistical clustering observed in the linkage map is at least in part likely to be the remains of this ancient duplication event. Thus, duplicate gene age distributions and linkage mapping suggest that *C. richardii* is a paleopolyploid species.

The duplicate gene age distribution results are also consistent with a study of DNA methylation in vascular plant genomes. Rabinowicz *et al.* (2005) assessed gene numbers by examining a sample of the unmethylated fraction of nuclear genes. They found that the nuclear genome of *C. richardii* has an estimated 42 300 genes, similar to those of diploid angiosperms. However, *C. richardii* has significantly fewer genes than expected for its genome size and many fewer than expected if it is a recent polyploid. If *C. richardii* is an ancient polyploid that has retained much of its genome content with extensive methylation or loss of paralogs, these data are consistent with the linkage mapping and age distribution analyses of *C. richardii*. The emerging genome data from three different sources thus suggest that *C. richardii* is probably a paleopolyploid that has apparently retained much of its nuclear DNA during diploidization, a result consistent with Haufler's hypothesis.

In addition to the issue of gene duplications, the linkage map of Nakazato *et al.* revealed an interesting characteristic regarding the genome size of *C. richardii*. As expected, the total map length was greater than in crop plant species

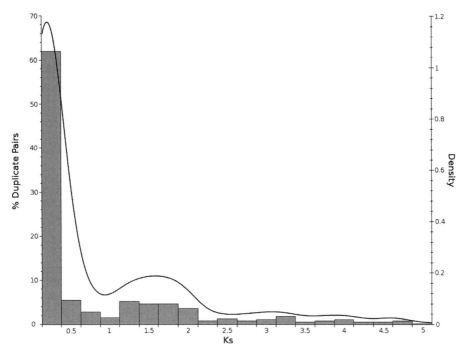

Figure 7.3 Histogram of *Ceratopteris richardii* duplicate Ks values demonstrating a peak in duplications at Ks 0.96–1.84. The bin size for the histogram was selected using the Sheather–Jones plugin method (Sheather and Jones, 1991). The solid line is a significant SiZer (Chaudhuri and Marron, 1999) kernel density estimate, with the peak significant at $p < 0.05$.

with high-density linkage maps, with the exception of tetraploid cotton. Thus, homosporous ferns may have more recombination per meiosis than other plants, mainly because of their high chromosome numbers. However, the map distance per linkage group (chromosome) ranged from 10.7 to 97.7 cM with an average of 53.1 in *Ceratopteris*. This is much lower than for the crop species listed in GenBank: soybean 158.0 cM, tomato 122.7 cM, oat 500.8 cM, barley 182.7 cM, bread wheat 170.4 cM, and corn 175.8 cM. This means that angiosperms typically have more than one recombination per chromosome per meiosis, whereas *Ceratopteris* has only about 0.5. Therefore, the homosporous fern genome seems to be characterized by a large number of chromosomes with tightly linked loci, as opposed to fewer more highly recombinant chromosomes found in seed plant groups.

The total map length of fern genomes is large, but does this correlate well with physical genome size (C-value estimates) from densitometry and flow cytometry measurements? C-values vary approximately 1000-fold across the land plants (reviewed in Leitch *et al.*, 2005) and approximately 450-fold across ferns

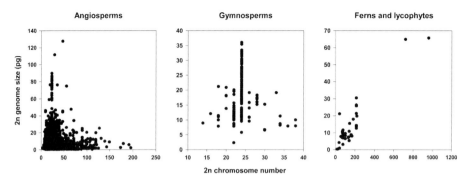

Figure 7.4 Pictograms showing the correlation between diploid chromosome numbers and genome size for angiosperms, gymnosperms, and ferns and lycophytes (data from the Plant DNA C-value database, Kew).

and lycophytes, from 0.16 pg in *Selaginella kraussiana* to 72.7 pg in *Psilotum nudum* var. *gasa* (Obermayer et al., 2002). The analysis of genome size estimates available in the Plant DNA C-values Database (Royal Botanic Gardens, Kew; www.rbgkew.org.uk/cval/homepage.html) shows that the C-values of ferns and lycophytes are much larger on average (mean 13.40, $N = 45$) than those of angiosperms (mean 6.36, $N = 3690$), but slightly and significantly smaller (2-tailed *t*-test, $p = 0.032$) than those of gymnosperms (mean 17.91, $N = 179$). Note that C-values are not sampled evenly from different taxonomic groups and may not accurately represent overall patterns. Despite their large genome size, the mean chromosome size (C-value/number of chromosomes) of ferns and lycophytes (mean 0.0465, $N = 45$) is only about one third that of angiosperms (mean 0.152, $N = 3690$), consistent with the small average map length of *C. richardii* chromosomes in the linkage map compared to those of crop species. The mean chromosome size of ferns and lycophytes is nearly an order of magnitude smaller than that of gymnosperms (mean 0.376, $N = 179$). Interestingly, whereas no correlation exists between genome size and chromosome number for angiosperms and gymnosperms ($r = -0.023$ and 0.106, respectively, $p > 0.1$), there is a highly significant positive correlation in ferns and lycophytes ($r = 0.913$, $p < 0.001$; Figure 7.4). This implies that chromosome size is more or less constant in ferns and lycophytes, but not in angiosperms and gymnosperms, consistent with the earlier observation in cytological studies that fern chromosomes are highly uniform in size and structure (reviewed in Wagner and Wagner, 1980).

Why is chromosome size seemingly constant in ferns, but flexible in seed plants? Recent genome analyses in angiosperms, particularly grasses, suggest that there was an increase in genome size in the last 10 million years largely due to transposable element (TE) expansion (SanMiguel et al., 1996, 1998; SanMiguel and Bennetzen, 1998; Vicient et al., 1999; Shirasu et al., 2000; Wicker et al., 2001; Hawkins et al., 2006). Further, genome size in angiosperms is positively correlated with numbers of retrotransposons, particularly long terminal repeat (LTR) transposons (Flavell, 1980). These elements comprise over 60% of large genomes like maize, wheat, and barley, but less than 50% in rice and about 10% in *Arabidopsis* (reviewed in Bennetzen, 2002). Although little is known about the numbers and kinds of TEs in ferns, reduced copy number of *Ty1-copia*-like retrotransposons were recently reported in the homosporous fern *Pteris cretica* compared to numbers in other plant species, despite the fact that it appears to be polyploid (Brandes et al., 1997). Thus one possible explanation for the constancy of chromosome sizes in ferns is suppression of TEs, but this is highly speculative and needs testing.

Superimposing the genome size on phylogenetic trees of angiosperms, gymnosperms, and ferns, presented in Leitch et al. (2005), showed that the mean and variance in genome size varies tremendously within and among taxonomic groups. Testing for phylogenetic independence of mean genome size (CACTUS 1.13, based on 10 000 randomizations; Schwilk and Ackerly, 2001) indicated that genome size evolved independently of phylogeny in angiosperms and gymnosperms ($p > 0.1$), but not in ferns and lycophytes ($p = 0.03$), mainly because of the unusually large genome sizes of Ophioglossaceae and Psilotaceae. These results suggest that genome size evolves fairly rapidly and more or less independently of phylogeny in vascular plants.

Aside from studies of genome evolution, recently there have been important technical advances in the area of functional fern and lycophyte genomics. A crucial tool for determining gene function is the ability to "knock out" a particular gene to assess its role in the organism. This is a requisite feature for a model organism, and until recently ferns and lycophytes did not have this capability. The successful targeted knock out of genes in *Marsilea vestita* via RNA interference (RNAi) by Klink and Wolniak (2000) provided the first demonstration of this technique in ferns. RNAi utilizes anti-sense or double stranded RNA that corresponds to a gene targeted for silencing. Subsequently, Stout et al. (2003) reported RNAi in *Ceratopteris richardii* and demonstrated RNAi silencing of genes selected from a *C. richardii* EST library. More recently, Kawai-Toyooka et al. (2004) developed a high throughput PCR based DNA interference (DNAi) approach for targeted gene silencing in *Adiantum capillus-veneris*. These tools will likely prove vital to understanding the roles of nuclear genes in ferns and lycophytes by facilitating reverse genetic approaches.

7.4 Synthesis of current perspectives

Thanks to recent advances in molecular technologies, we now have a substantial amount of information about fern and lycophyte nuclear genome organization and evolution. In particular, significant progress has been made toward understanding the origin of the remarkably high chromosome numbers of homosporous ferns and lycophytes, a major focus of research during the last half century. Despite this recent progress, there are many questions to be addressed.

Numerous studies over the last 50 years utilizing cytology (e.g., Manton, 1950; Klekowski and Baker, 1966), isozymes (reviewed in Haufler, 1987; Gastony, 1991), and DNA based methods (Pichersky et al., 1990; McGrath et al., 1994; McGrath and Hickok, 1999) indicated that homosporous ferns and lycophytes, with their high base chromosome numbers, are probably ancient polyploids, or paleopolyploids. However, the question of the origin of their high chromosome numbers has never been satisfactorily answered, primarily because of technical limitations. Recent genomic analyses (Rabinowicz et al., 2005; Nakazato et al., 2006; Barker et al., unpublished data) of the putatively diploid homosporous fern *Ceratopteris richardii* ($n = 39$) address this longstanding question and indicate that this species experienced an ancient duplication event, probably paleopolyploidy, between 78 and 151 MYA (Barker et al., unpublished data). These results suggest that the high chromosome numbers of homosporous ferns and lycophytes and many of the gene copies detected in their genomes (Pichersky et al., 1990; McGrath et al., 1994; McGrath and Hickok, 1999) originated via polyploidization. Furthermore, the recent genomic data indicate that these polyploidizations are probably ancient events with the genomes of "diploid" (but actually paleopolyploid) homosporous ferns and lycophytes fully diploidized, consistent with the body of isozyme work that demonstrated that homosporous fern and lycophytes with the lowest chromosome numbers in their genus possess diploid isozyme expression profiles (Gastony and Gottlieb, 1982, 1985; Gastony and Darrow, 1983; Haufler and Soltis, 1984, 1986; Haufler, 1987; Soltis and Soltis, 1987; Wolf et al., 1987). The linkage mapping results of Nakazato et al. also suggest that the chromosomes of homosporous ferns and lycophytes are extensively rearranged and diploidized with no strong visually syntenic evidence of homoeologs, a pattern consistent with previous cytological experiments (Rigby, 1975; Walker, 1985). Thus, the high chromosome numbered homosporous fern and lycophyte genomes appear to have been significantly influenced by polyploidy, a process that has left homosporous ferns and lycophytes a legacy of large chromosome numbers, multiple silenced gene copies, and probably large genome sizes.

Haufler's synthesis of fern and lycophyte genome organization and evolution developed a hypothesis that the high chromosome numbers of homosporous ferns and lycophytes result from repeated cycles of polyploidization with subsequent gene silencing and extinction of diploid progenitors. This model implies that multiple, independent polyploidization events have been responsible for the high chromosome numbers of homosporous ferns, with recurrent neopolyploidizations contributing chromosome numbers in an ever increasing cycle. Data presented by Barker *et al.* (unpublished data) suggest that there are fewer paleopolyploid events than predicted by Haufler's hypothesis. Under this new perspective, the high chromosome numbers appear to result from the survival of relatively few polyploid taxa that subsequently diversified to produce new lineages. This view of genome evolution is consistent with results from recent research in other organisms. Analyses of angiosperm genomes suggest that most angiosperms, even low chromosome numbered species such as *Arabidopsis thaliana*, have experienced multiple, shared paleopolyploidization events over the last several million years (Lynch and Conery, 2000; Blanc and Wolfe, 2004; Sterck *et al.*, 2005; Cui *et al.*, 2006). Data from animals (reviewed in Panopoulou and Poustka, 2005) also indicate two rounds of polyploidization in the vertebrate lineage (2R hypothesis), one before and one after the divergence of the jawless fishes and the rest of the vertebrates (500–430 MYA, evidence based mainly on the number of *Hox* gene copies). Thus, polyploidy has played a significant role in both plant and animal eukaryotic genome evolution.

Although paleopolyploidy is a common feature of eukaryotic genome evolution, homosporous ferns and lycophytes still possess large chromosome numbers and genome sizes relative to other groups of organisms. As suggested by the work of Barker *et al.* (unpublished data), paleopolyploidy does not appear to occur any more frequently in homosporous ferns and lycophytes than in other groups, suggesting a relatively high retention rate of chromosomes following polyploidy in this group. Homosporous ferns and lycophytes also appear to conserve the size of their chromosomes, as observed in the close correlation between genome size and chromosome number in ferns and lycophytes (presented above). Thus, it may be that diploidization occurs differently in homosporous ferns and lycophytes, with higher retention of chromosome number and size compared to other organisms.

7.5 Future goals and directions

Although the use of modern genomic tools has significantly advanced our knowledge of fern and lycophyte nuclear genome organization and evolution, new questions have surfaced that will fuel research for some time to come.

As with previous research, many of these questions focus on the high chromosome numbers of homosporous ferns. A recent analysis (Barker *et al.*, unpublished data) demonstrated that the homosporous fern *Ceratopteris richardii* is a paleopolyploid and that its large chromosome number ($n = 39$) is likely derived from polyploidization. Although this is probably a widespread phenomenon throughout ferns and lycophytes, as it is in angiosperms (Blanc and Wolfe, 2004), further EST library construction and analyses of diverse ferns and lycophytes are necessary to support this conclusion. In particular, sampling a mix of heterosporous and phylogenetically strategic homosporous ferns and lycophytes would be valuable. A good starting point for taxon selection would be those recommended by Pryer *et al.* (2001) for genomic studies. In addition, it would be useful to have EST libraries for lineages that are notable for their high chromosome numbers even among homosporous fern and lycophytes, such as a member of the Cyatheaceae. The construction of diverse EST data sets would also allow us to address questions about the timing and rate of paleopolyploidization, chromosome number evolution, and differences in these processes among disparate fern and lycophyte lineages. Currently, these questions cannot be addressed among these organisms because we lack the resources. Considering the growing collection of plant EST libraries, a community effort to produce a set of fern and lycophyte EST libraries should be possible because such data sets would facilitate research not only in genome evolution but also in phylogenetics, developmental genetics, and functional genomics.

In addition to examining further EST data sets, sequence analysis of multi-copy gene families is also needed to evaluate both silenced and expressed copies for evidence of paleopolyploidy. This would not only provide an alternative data set for testing paleopolyploidy, but would also reveal how gene copies were silenced. Furthermore, sequence analysis may also reveal how non-polyploid-derived duplicates are generated in the genome. Work on this is already underway for *Ceratopteris richardii* (Barker, unpublished data) with results forthcoming. A significant by-product of this research is that it will permit an estimate of a nuclear genome substitution per synonymous site per year (Ks/year) to be calculated with proper outgroup and fossil data. The calculation of this rate for *Ceratopteris* will provide the first estimate of the background nuclear genome mutation rate in homosporous ferns and will also facilitate more accurate dating of paleopolyploidization events in this group. Similar analyses in phylogenetically diverse ferns and lycophytes would generalize details on paleopolyploidization events and reveal the process of genic diploidization. Finally, calculation of a Ks/year rate for diverse fern and lycophyte taxa may also yield further insight into the rate heterogeneity observed in fern phylogenetic studies (Soltis *et al.*, 2002). Thus, our efforts in *Ceratopteris* should be viewed as only a starting point in these types of analyses because a number of significant questions require data from diverse taxa.

Although genomic changes accompanying polyploidization have been extensively investigated in synthetic polyploids of angiosperms (Song et al., 1995; Shaked et al., 2001; Madlung et al., 2002; Han et al., 2003; Liu and Wendel, 2003; Osborn et al., 2003; Pires et al., 2004), the long-term consequences of polyploidization in nature have not been extensively investigated. An exception is the study of homoeologous genes in natural tetraploid cotton, which experienced a polyploidization event <2 MYA (Adams et al., 2003). Those authors found that silencing or unequal expression of homoeologs from different diploid progenitors are common across tissue types. Because polyploidization plays an important role in fern and lycophyte evolution, similar investigations of the evolution of duplicated gene copies in ferns and lycophytes would be fruitful.

We also need information on gene copy numbers in neopolyploids and their diploid progenitors to estimate better the rate at which gene copies are deleted following polyploidization. Additional studies similar to that of allopatric polyploid populations of *Pellaea rufa* that showed progressive loss of duplicated isozyme expression (Gastony, 1991) would allow us to identify gene or genome copies that are more prone to deletion. Such studies may also shed light on whether polyploidization followed by gene silencing often contributes to reproductive isolation and speciation. Werth and Windham (1991) proposed that silencing of different gene copies in diverged polyploid populations (reciprocal silencing) can cause fertility reduction upon hybridization and can serve as a mechanism of speciation[1] (see Chapter 12).

In addition, mapping of additional fern and lycophyte genomes is needed to assess the generality of the linkage mapping results in *Ceratopteris*. Nakazato et al. (2006) identified a number of idiosyncratic features of the *C. richardii* genome, such as short chromosome length with relatively little recombination. Mapping of other *Ceratopteris* species, as well as phylogenetically more diverse taxa, would provide a strong test of the results reported by Nakazato et al. (2006). Furthermore, if common RFLP and isozyme markers were employed, major chromosomal rearrangements (e.g., inversions and translocations) could be detected (e.g.,

---

[1] After this chapter went to press, Nakazato et al. (2007) reported substantial reproductive barriers between the parental races of the hybrid mapping population used for their (Nakazato et al., 2006) linkage mapping study. Their data support the hypothesis that these barriers are a byproduct of divergence in allopatry and are attributable to a small number of genetic elements distributed throughout the genome. (Nakazato, T., Jung, M.-K., Housworth, E. A., Rieseberg, L. H., and Gastony, G. J. (2007). A genomewide study of reproductive barriers between allopatric populations of a homosporous fern, *Ceratopteris richardii*, Genetics, **177**, 1141–1150.)

Ahn and Tanksley, 1993; Lagercrantz and Lydiate, 1996), which would provide a preliminary estimate of rates of chromosomal structural evolution in ferns.

Despite genic diploidization and extensive rearrangements (Nakazato et al., 2006), the large chromosome numbers of homosporous ferns and lycophytes have probably been maintained by a combination of forces rather than continually produced by recurrent paleopolyploidization (Barker et al., unpublished data). Possibly involved in the maintenance of these chromosomes is another peculiar fern and lycophyte trait, the strong bivalent pairing of chromosomes (Wagner and Wagner, 1980; Gastony and Windham, 1989; Gastony, 1990). Whether high chromosome numbers and stringent bivalent pairing are related is still an open question. Further work is needed to study this relationship and to determine whether it is characteristic of other high chromosome numbered taxa (such as fishes). In addition, research addressing how homologous chromosomes recognize each other in ferns and lycophytes might be rewarding. A starting point for this work would be to seek genetically tractable species or laboratory lines with segregating variation for pairing behavior.

Additional C-value estimates for ferns and lycophytes would also help identify factors that influence evolution of the genome size of ferns and lycophytes. In particular, underrepresented fern and lycophyte lineages need to be added to the C-value database to determine whether current C-value estimates are representative. Also, comparisons of C-values between neopolyploids, hybrids, and progenitor taxa would elucidate whether genome size is conserved during these transitions. Another approach that would contribute to our understanding of the evolution of genome size and paleopolyploidy would be to examine the guard cell sizes of fossil taxa to infer ancient genome sizes. As demonstrated by Barrington et al. (1986), fern and lycophyte guard cell sizes are correlated with genome size and ploidy level, and have been used to infer relative ploidy level among extant homosporous ferns (Tryon, 1968; Barker and Hickey, 2006). Masterson (1994) utilized this approach among selected angiosperm taxa and demonstrated an increase in guard cell size over time, implicating polyploidy in the ancestry of most angiosperms. Such an approach could be used in ferns and lycophytes to assess how genome size has changed over time and whether the changes are consistent with polyploidy. Furthermore, Barker and Hickey (2006) demonstrated that spore size and indurated sporangial arcus cell sizes also appear to be directly related to ploidy level, extending the fossil sources for measurements in ferns and lycophytes. Expanding our knowledge of both extant and ancient genome sizes would provide crucial information on the rate of genome size change in ferns and lycophytes, the relationship of polyploidy and genome size, and the predominant pattern of fern and lycophyte genome size change.

Perhaps the ultimate tool for contemporary genomic research is a whole genome sequence for a representative taxon. In ferns and lycophytes, a whole genome sequencing effort is underway for the lycophyte *Selaginella moellendorffii*. This will likely yield a wealth of information on lycophyte genomes and provide a comparison to the sequenced genomes of mosses and seed plants. With the sequencing of *Selaginella moellendorffii*, ferns will be the only major lineage of land plants without a sequenced genome. However, given the large genome sizes of most ferns, the sequencing of a whole nuclear genome is not financially feasible with current technology. This obstacle is guaranteed to be short lived, because innovations in DNA sequencing will make a whole genome sequence for a fern feasible in the near future. With whole genome data, we could thoroughly assess the duplication history of fern and lycophyte genomes and the role and activity of transposable elements, measure gene density and distribution, and have a sister group genome for comparison with seed plants. An obvious candidate for whole genome or substantial BAC sequencing would be *Ceratopteris richardii*, especially considering the accumulating genetic resources for this model fern. Ultimately, multiple whole genome sequences from phylogenetically diverse taxa will be available, and these data will answer many of today's questions and some that have not yet been asked.

References

Adams, K. L., Cronn, R., Percifield, R., and Wendel, J. F. (2003). Genes duplicated by polyploidy show unequal contributions to the transcriptome and organ-specific reciprocal silencing. *Proceedings of the National Academy of Sciences of the United States of America*, **100**, 4649–4654.

Ahn, S. and Tanksley, S. D. (1993). Comparative linkage maps of the rice and maize genomes. *Proceedings of the National Academy of Sciences of the United States of America*, **90**, 7980–7984.

Barker, M. S. and Hickey, R. J. (2006). A taxonomic revision of Caribbean *Adiantopsis* (Pteridaceae). *Annals of the Missouri Botanical Garden*, **93**, 371–401.

Barrington, D. S., Paris, C. A., and Ranker, T. A. (1986). Systematic inferences from spore and stomate size in the ferns. *American Fern Journal*, **76**, 149–159.

Bartel, B. and Bartel, D. P. (2003). MicroRNAs: at the root of plant development? *Plant Physiology*, **132**, 709–717.

Bartel, D. P. (2004). MicroRNAs: genomics, biogenesis, mechanism, and function. *Cell*, **116**, 281–297.

Bennetzen, J. L. (2002). Mechanisms and rates of genome expansion and contraction in flowering plants. *Genetica*, **115**, 29–36.

Blanc, G. and Wolfe, K. (2004). Widespread paleopolyploidy in model plant species inferred from age distributions of duplicate genes. *Plant Cell*, **16**, 1667–1678.

Blanc, G., Hokamp, K., and Wolfe, K. (2003). A recent polyploidy superimposed on older large-scale duplications in the *Arabidopsis* genome. *Genome Research*, **13**, 137–144.

Bradshaw, H. D. and Schemske, D. W. (2003). Allele substitution at a flower colour locus produces a pollinator shift in monkeyflowers. *Nature*, **426**, 176–178.

Brandes, A., Heslop-Harrison, J. S., Kamm, A., Kubis, S., Doudrick, R. S., and Schmidt, T. (1997). Comparative analysis of the chromosomal and genomic organization of Ty1-copia-like retrotransposons in pteridophytes, gymnosperms and angiosperms. *Plant Molecular Biology*, **33**, 11–21.

Chapman, R. H., Klekowski, E. J. J., and Selander, R. K. (1979). Homoeologous heterozygosity and recombination in the fern *Pteridium aquilinum*. *Science*, **204**, 1207–1209.

Chaudhuri, P. and Marron, J. S. (1999). SiZer for exploration of structures in curves. *Journal of the American Statistical Association*, **94**, 807–823.

Cui, L. Y., Wall, P. K., Leebens-Mack, J. H., Lindsay, B. G., Soltis, D. E., Doyle, J. J., Soltis, P. S., et al. (2006). Widespread genome duplications throughout the history of flowering plants. *Genome Research*, **16**, 738–749.

Doust, A. N., Devos, K. M., Gadberry, M. D., Gale, M. D., and Kellogg, E. A. (2005). The genetic basis for inflorescence variation between foxtail and green millet (Poaceae). *Genetics*, **169**, 1659–1672.

Drouaud, J., Camilleri, C., Bourguignon, P. Y., Canaguier, A., Berard, A., et al. (2006). Variation in crossing-over rates across chromosome 4 of *Arabidopsis thaliana* reveals the presence of meiotic recombination "hot spots." *Genome Research*, **16**, 106–114.

Eyre-Walker, A. (2006). The genomic rate of adaptive evolution. *Trends in Ecology and Evolution*, **21**, 569–575.

Flavell, R. (1980). The molecular characterization and organization of plant chromosomal DNA sequences. *Annual Review of Plant Physiology and Plant Molecular Biology*, **31**, 569–596.

Gastony, G. J. (1990). Electrophoretic evidence for allotetraploidy with segregating heterozygosity in South African *Pellaea rufa* Tryon, A. F. (Adiantaceae). *Annals of the Missouri Botanical Garden*, **77**, 306–313.

Gastony, G. J. (1991). Gene silencing in a polyploid homosporous fern: paleopolyploidy revisited. *Proceedings of the National Academy of Sciences of the United States of America*, **88**, 1602–1605.

Gastony, G. J. and Darrow, D. C. (1983). Chloroplastic and cytosolic isozymes of the homosporous fern *Athyrium filix-femina* L. *American Journal of Botany*, **70**, 1409–1415.

Gastony, G. J. and Gottlieb, L. D. (1982). Evidence for genetic heterozygosity in a homosporous fern. *American Journal of Botany*, **69**, 634–637.

Gastony, G. J. and Gottlieb, L. D. (1985). Genetic variation in the homosporous fern *Pellaea andromedifolia*. *American Journal of Botany*, **72**, 257–267.

Gastony, G. J. and Windham, M. D. (1989). Species concepts in pteridophytes: the treatment and definition of agamosporous species. *American Fern Journal*, **79**, 65–77.

Gottlieb, L. D. (1982). Conservation and duplication of isozymes in plants. *Science*, **216**, 373–380.

Grant, V. (1981). *Plant Speciation*. New York: Columbia University Press.

Han, F. P., Fedak, G., Ouellet, T., and Liu, B. (2003). Rapid genomic changes in interspecific and intergeneric hybrids and allopolyploids of Triticeae. *Genome*, **46**, 716–723.

Haufler, C. H. (1987). Electrophoresis is modifying our concepts of evolution in homosporous pteridophytes. *American Journal of Botany*, **74**, 953–966.

Haufler, C. H. and Soltis, D. E. (1984). Obligate outcrossing in a homosporous fern: field confirmation of a laboratory prediction. *American Journal of Botany*, **71**, 878–881.

Haufler, C. H. and Soltis, D. E. (1986). Genetic evidence suggests that homosporous ferns with high chromosome numbers are diploid. *Proceedings of the National Academy of Sciences of the United States of America*, **83**, 4389–4393.

Hawkins, J. S., Kim, H., Nason, J. D., Wing, R. A., and Wendel, J. F. (2006). Differential lineage-specific amplification of transposable elements is responsible for genome size variation in *Gossypium*. *Genome Research*, **16**, 1252–1261.

Hickok, L. G. (1978). Homoeologous chromosome pairing and restricted segregation in the fern *Ceratopteris*. *American Journal of Botany*, **65**, 516–521.

Hickok, L. G. and Klekowski, E. J., Jr. (1974). Inchoate speciation in *Ceratopteris*: an analysis of the synthesized hybrid *C. richardii* × *C. pteridoides*. *Evolution*, **28**, 439–446.

Holsinger, K. E. (1987). Gametophytic self-fertilization in homosporous plants: development, evaluation, and application of a statistical method for evaluating its importance. *American Journal of Botany*, **74**, 1173–1183.

Kawai-Toyooka, H., Kuramoto, C., Orui, K., Motoyama, K., Kikuchi, K., Kanegae, T., and Wada, M. (2004). DNA interference: a simple and efficient gene-silencing system for high-throughput functional analysis in the fern *Adiantum*. *Plant and Cell Physiology*, **45**, 1648–1657.

Khandelwal, S. (1990). Chromosome evolution in the genus *Ophioglossum* L. *Botanical Journal of the Linnean Society*, **102**, 205–217.

Klekowski, E. J., Jr. and Baker, H. G. (1966). Evolutionary significance of polyploidy in the Pteridophyta. *Science*, **153**, 305–307.

Klink, V. P. and Wolniak, S. M. (2000). The efficacy of RNAi in the study of the plant cytoskeleton. *Journal of Plant Growth Regulation*, **19**, 371–384.

Lagercrantz, U. and Lydiate, D. J. (1996). Comparative genome mapping in *Brassica*. *Genetics*, **144**, 1903–1910.

Lai, Z., Nakazato, T., Salmaso, M., Burke, J. M., Tang, S. X., Knapp, S. J., and Rieseberg, L. H. (2005). Extensive chromosomal repatterning and the evolution of sterility barriers in hybrid sunflower species. *Genetics*, **171**, 291–303.

Leitch, I. J., Soltis, D. E., Soltis, P. S., and Bennett, M. D. (2005). Evolution of DNA amounts across land plants (Embryophyta). *Annals of Botany*, **95**, 207–217.

Liu, B. and Wendel, J. F. (2003). Epigenetic phenomena and the evolution of plant allopolyploids. *Molecular Phylogenetics and Evolution*, **29**, 365–379.

Löve, Á., Löve, D., and Pichi-Sermolli, R. E. G. (1977). *Cytotaxonomical Atlas of the Pteridophyta*. Vaduz: J. Cramer.

Lynch, M. and Conery, J. S. (2000). The evolutionary fate and consequences of duplicate genes. *Science*, **290**, 1151–1155.

Madlung, A., Masuelli, R. W., Watson, B., Reynolds, S. H., Davison, J., and Comai, L. (2002). Remodeling of DNA methylation and phenotypic and transcriptional changes in synthetic *Arabidopsis* allotetraploids. *Plant Physiology*, **129**, 733–746.

Manton, I. (1950). *Problems of Cytology and Evolution in the Pteridophyta*. Cambridge: Cambridge University Press.

Masterson, J. (1994). Stomatal size in fossil plants: evidence for polyploidy in majority of angiosperms. *Science*, **264**, 421–424.

McGrath, J. M. and Hickok, L. G. (1999). Multiple ribosomal RNA gene loci in the genome of the homosporous fern *Ceratopteris richardii*. *Canadian Journal of Botany*, **77**, 1199–1202.

McGrath, J. M., Hickok, L. G., and Pichersky, E. (1994). Assessment of gene copy number in the homosporous ferns *Ceratopteris thalictroides* and *C. richardii* (Parkeriaceae) by restriction fragment length polymorphisms. *Plant Systematics and Evolution*, **189**, 203–210.

Nakazato, T., Jung, M.-K., Housworth, E. A., Rieseberg, L. H., and Gastony, G. J. (2006). Genetic map-based analysis of genome structure in the homosporous fern *Ceratopteris richardii*. *Genetics*, **173**, 1585–1597.

Obermayer, R., Leitch, I. J., Hanson, L., and Bennett, M. D. (2002). Nuclear DNA C-values in 30 species double the familial representation in pteridophytes. *Annals of Botany*, **90**, 209–217.

Osborn, T. C., Pires, J. C., Birchler, J. A., Auger, D. L., Chen, Z. J., Lee, H. S., Comai, L., et al. (2003). Understanding mechanisms of novel gene expression in polyploids. *Trends in Genetics*, **19**, 141–147.

Panopoulou, G. and Poustka, A. J. (2005). Timing and mechanism of ancient vertebrate genome duplications: the adventure of a hypothesis. *Trends in Genetics*, **21**, 559–567.

Pichersky, E., Soltis, D. E., and Soltis, P. S. (1990). Defective chlorophyll a/b-binding protein genes in the genome of a homosporous fern. *Proceedings of the National Academy of Sciences of the United States of America*, **87**, 195–199.

Pires, J. C., Zhao, J. W., Schranz, M. E., Leon, E. J., Quijada, P. A., Lukens, L. N., and Osborn, T. C. (2004). Flowering time divergence and genomic rearrangements in resynthesized *Brassica* polyploids (Brassicaceae). *Biological Journal of the Linnean Society*, **82**, 675–688.

Pryer, K. M., Schneider, H., Smith, A. R., Cranfill, R., Wolf, P. G., Hunt, J. S., and Sipes, S. D. (2001). Horsetails and ferns are a monophyletic group and the closest living relatives to seed plants. *Nature*, **409**, 618–622.

Pryer, K. M., Schuettpelz, E., Wolf, P. G., Schneider, H., Smith, A. R., and Cranfill, R. (2004). Phylogeny and evolution of ferns (Monilophytes) with a focus on the early leptosporangiate divergences. *American Journal of Botany*, **91**, 1582–1598.

Rabinowicz, P. D., Citek, R., Budiman, M. A., Nunberg, A., Bedell, J. A., Lakey, N., O'Shaughnessy, A. L., et al. (2005). Differential methylation of genes and repeats in land plants. *Genome Research*, **15**, 1431–1440.

Rieseberg, L. H., Raymond, O., Rosenthal, D. M., Lai, Z., Livingstone, K., Nakazato, T., Durphy, J. L., et al. (2003). Major ecological transitions in wild sunflowers facilitated by hybridization. *Science*, **301**, 1211–1216.

Rigby, S. J. (1975). Meiosis and sporogenesis in a haploid plant of *Pellaea glabella* var. *occidentalis*. *Canadian Journal of Botany*, **53**, 894–900.

SanMiguel, P. and Bennetzen, J. L. (1998). Evidence that a recent increase in maize genome size was caused by the massive amplification of intergene retrotransposons. *Annals of Botany*, **82**, 37–44.

SanMiguel, P., Gaut, B. S., Tikhonov, A., Nakajima, Y., and Bennetzen, J. L. (1998). The paleontology of intergene retrotransposons of maize. *Nature Genetics*, **20**, 43–45.

SanMiguel, P., Tikhonov, A., Jin, Y. K., Motchoulskaia, N., Zakharov, D., MelakeBerhan, A., Springer, P. S., et al. (1996). Nested retrotransposons in the intergenic regions of the maize genome. *Science*, **274**, 765–768.

Schlueter, J. A., Dixon, P., Granger, C., Grant, D., Clark, L., Doyle, J. J., and Shoemaker, R. C. (2004). Mining EST databases to resolve evolutionary events in major crop species. *Genome*, **47**, 868–876.

Schneider, H., Schuettpelz, E., Pryer, K. M., Cranfill, R., Magallon, S., and Lupia, R. (2004). Ferns diversified in the shadow of angiosperms. *Nature*, **428**, 553–557.

Schwilk, D. W. and Ackerly, D. D. (2001). Flammability and serotiny as strategies: correlated evolution in pines. *Oikos*, **94**, 326–336.

Shaked, H., Kashkush, K., Ozkan, H., Feldman, M., and Levy, A. A. (2001). Sequence elimination and cytosine methylation are rapid and reproducible responses of the genome to wide hybridization and allopolyploidy in wheat. *Plant Cell*, **13**, 1749–1759.

Sheather, S. J. and Jones, M. C. (1991). A reliable data-based bandwidth selection method for kernel density estimation. *Journal of the Royal Statistical Society, Series B*, **53**, 683–690.

Shirasu, K., Schulman, A. H., Lahaye, T., and Schulze-Lefert, P. (2000). A contiguous 66-kb barley DNA sequence provides evidence for reversible genome expansion. *Genome Research*, **10**, 908–915.

Smith, A. R., Pryer, K. M., Schuettpelz, E., Korall, P., Schneider, H., and Wolf, P. C. (2006). A classification for extant ferns. *Taxon*, **55**, 705–731.

Soltis, D. E. (1986). Genetic evidence for diploidy in *Equisetum*. *American Journal of Botany*, **73**, 908–913.

Soltis, D. E. and Soltis, P. S. (1987). Polyploidy and breeding systems in homosporous Pteridophyta: a reevaluation. *American Naturalist*, **130**, 219–232.

Soltis, D. E. and Soltis, P. S. (1990). Isozyme evidence for ancient polyploidy in primitive angiosperms. *Systematic Botany*, **15**, 328–337.

Soltis, P. S., Soltis, D. E., Savolainen, V., Crane, P. R., and Barraclough, T. G. (2002). Rate heterogeneity among lineages of tracheophytes: integration of molecular and fossil data and evidence for molecular living fossils. *Proceedings of the National Academy of Sciences of the United States of America*, **99**, 4430–4435.

Song, K., Lu, P., Tang, K., and Osborn, T. (1995). Rapid genome change in synthetic polyploids of *Brassica* and its implications for polyploid evolution. *Proceedings of the National Academy of Sciences of the United States of America*, **92**, 7719–7723.

Sterck, L., Rombauts, S., Jansson, S., Sterky, F., Rouze, P., and Van de Peer, Y. (2005). EST data suggest that poplar is an ancient polyploid. *New Phytologist*, **167**, 165–170.

Stout, S. C., Clark, G. B., Archer-Evans, S., and Roux, S. J. (2003). Rapid and efficient suppression of gene expression in a single-cell model system, *Ceratopteris richardii*. *Plant Physiology*, **131**, 1165–1168.

Swigonova, Z., Lai, J. S., Ma, J. X., Ramakrishna, W., Llaca, V., Bennetzen, J. L., and Messing, J. (2004). Close split of sorghum and maize genome progenitors. *Genome Research*, **14**, 1916–1923.

Tryon, A. F. (1968). Comparison of sexual and apogamous races in the fern genus *Pellaea*. *Rhodora*, **70**, 1–24.

Vicient, C. M., Suoniemi, A., Anamthamat-Jonsson, K., Tanskanen, J., Beharav, A., Nevo, E., and Schulman, A. H. (1999). Retrotransposon BARE-1 and its role in genome evolution in the genus *Hordeum*. *Plant Cell*, **11**, 1769–1784.

Wagner, W. H., Jr. and Wagner, F. S. (1980). Polyploidy in Pteridophytes. In *Polyploidy, Biological Relevance: Proceedings of the International Conference on Polyploidy, Biological Relevance*, ed. W. H. Lewis. New York: Plenum Press, pp. 199–214.

Walker, T. G. (1985). Cytotaxonomical studies of the ferns of Trinidad 2. The cytology and taxonomic implications. *Bulletin of the British Museum of Natural History (Botany)*, **13**, 149–249.

Wendel, J. F. and Cronn, R. C. (2003). Polyploidy and the evolutionary history of cotton. *Advances in Agronomy*, **78**, 139–186.

Werth, C. R. and Windham, M. D. (1991). A model for divergent, allopatric speciation of polyploid pteridophytes resulting from silencing of duplicate-gene expression. *American Naturalist*, **137**, 515–526.

Whibley, A. C., Langlade, N. B., Andalo, C., Hanna, A. I., Bangham, A., Thebaud, C., and Coen, E. (2006). Evolutionary paths underlying flower color variation in *Antirrhinum*. *Science*, **313**, 963–966.

Wicker, T., Stein, N., Albar, L., Feuillet, C., Schlagenhauf, E., and Keller, B. (2001). Analysis of a contiguous 211 kb sequence in diploid wheat (*Triticum monococcum* L.) reveals multiple mechanisms of genome evolution. *Plant Journal*, **26**, 307–316.

Wolf, P. G., Haufler, C. H., and Sheffield, E. (1987). Electrophoretic evidence for genetic diploidy in the bracken fern (*Pteridium aquilinum*). *Science*, **236**, 947–949.

Wolfe, K. H. and Li, W.-H. (2003). Molecular evolution meets the genomics revolution. *Nature Genetics*, **33**, 255–265.

PART III ECOLOGY

8

Phenology and habitat specificity of tropical ferns

KLAUS MEHLTRETER

8.1 Introduction

The focus of this chapter is two aspects of fern sporophyte ecology: phenology and habitat specificity. I define phenology as the study of the periodicity of biological processes caused by intrinsic factors (hormones, circadian clock) or triggered by extrinsic, environmental factors, mainly rainfall, temperature, and photoperiod, or some combination of those elements. Habitat specificity is defined as the biotic and abiotic conditions that favor the development and, consequently, the presence and abundance of fern species on a spatial scale.

8.2 Historical summary

Descriptive treatments considering ecological aspects of ferns and lycophytes have been organized geographically (Christ, 1910) and by vegetation types and/or growth forms (Holttum, 1938; Tryon, 1964; Page, 1979a). The latter organization is followed for the two ecological issues treated within this chapter, starting with terrestrial species, followed by rheophytes (fluvial plants), lithophytes (rock plants), epiphytes, and climbers. All other growth forms (e.g., hemi-epiphytes, mangrove ferns) are either treated marginally within the nearest group (e.g., tree ferns within terrestrial ferns, mangrove ferns within rheophytes) or omitted because of lack of information.

Holttum (1938) observed that ferns and lycophytes are rarely dominant in any plant community. His statement that most vegetation types would not be greatly

Biology and Evolution of Ferns and Lycophytes, ed. Tom A. Ranker and Christopher H. Haufler. Published by Cambridge University Press. © Cambridge University Press 2008.

modified if all ferns were removed reflects the low importance he accorded ferns in a functional context within tropical forest ecosystems. In fact, we simply do not understand the ecological importance of ferns, because few studies have addressed this issue. Page (1979a) presented an opposite point of view. In his opinion, ferns play an important ecological role in a variety of vegetation types, partly because of their diversity of life forms. Ferns and lycophytes represent about 2–5% of the species diversity of vascular plants. The majority of that diversity is concentrated in the tropics (~85%; Tryon, 1964), especially in cloud forests at mid-elevation and on oceanic islands (e.g., Hawaiian Islands, La Réunion Island, Tristan da Cunha; see Chapter 14) where they comprise 16–60% of the flora of vascular plants (Kramer *et al.*, 1995).

Holttum (1938) was possibly the first to study the phenology of tropical ferns, because he was aware of the seasonality of several tropical ferns. He stated that *Drynaria fortunei* is an obligately deciduous epiphyte, which loses its leaves seasonally even under greenhouse conditions. He reported on the coexistence of seasonal (e.g., *Platycerium grande*) and aseasonal (*P. coronarium*) fern epiphytes on the Malayan Peninsula. Interestingly, these observations have not been confirmed by more detailed studies. Kornás (1977) observed seasonal phenological patterns of ferns in Zambia and distinguished between species that tolerated the climatic conditions of the dry season and the deciduous species that entered into a dormant stage. Seiler (1981) initiated the first detailed phenological study of a tree fern in El Salvador. He recognized that *Alsophila salvinii* produced leaves asynchronously during the entire year, but exhibited a strong seasonal emergence of leaves at the start of the rainy season. This somewhat unexpected result was the first to show that seasonal patterns in tropical ferns may occur in spite of the humid conditions at the cloud forest site (2250 mm annual rainfall), apparently because of the drier winter season with less rainfall. However, leaf emergence in spring could be a consequence of rising temperatures or increasing photoperiod, and these hypotheses of correlation between climatic parameters and plant phenology should be tested. Seiler (1981) also used the quotient of mean leaf number (6) and annual leaf production (3) to calculate a mean leaf life span for this species of 2 years. Tanner (1983) identified a correlation between height and age of the tree fern trunk in his study of *Alsophila auneae* (syn. *Cyathea pubescens*) in Jamaica. Although the correlation between leaf number and height was significant, leaf production and the number of leaf scars did not correlate well with trunk height and were consequently a poor predictor of trunk age. However, his results confirmed that leaf production was higher after periods of heavy rainfall. These few studies were the starting point for an increasing interest in fern phenology in the 1990s.

Holttum (1938) presented some controversial ideas on habitat specificity (also known as habitat preference) based on his observations and the scarce information available in his time. He described dozens of habitat specialists, e.g., *Dipteris lobbiana* and *Tectaria semibipinnata* as rheophytes which can tolerate periodic flooding, and *Davallia parvula* as a trunk epiphyte of mangroves, but also stated that most rock ferns are not specialized to their substrate. Page (1979a) reviewed the apparent habitat specificities of ferns and some of their noteworthy adaptations, but no quantitative data were published until Moran *et al.* (2003) and Mehltreter *et al.* (2005). However, tree ferns have been recognized as a specific substrate for epiphytes, especially some species of mosses (Oliver, 1930; Johansson, 1974; Pócs, 1982; Beever, 1984). Based on such observations, it appeared that some epiphytic ferns might have specific host preferences. For example, Oliver (1930) reported that *Trichomanes ferrugineum* and *Crepidomanes venosum* from New Zealand grow exclusively on tree ferns, and Copeland (1947) found that *Stenochlaena areolaris* selectively develops in the leaf axils of *Pandanus* trees. Rock ferns have been reported to be somewhat habitat specific. For example, Page (1979a) listed *Adiantum capillus-veneris*, *A. philippense*, and *A. reniforme* as limestone ferns and *Cheilanthes* and *Notholaena* species as volcanic rock ferns. Today we know that the latter two genera have species with a variety of substrate preferences.

8.3   Review of critical recent advances

Although the number of ecological studies of ferns has grown significantly over the last 20 years, few researchers contributed to these advances and consequently the results were restricted to a few sites. Phenological studies were undertaken for dozens of species in Costa Rica (Sharpe and Jernstedt, 1990, 1991; Sharpe, 1993; Bittner and Breckle 1995), Puerto Rico (Sharpe, 1997), Mexico (Mehltreter and Palacios-Rios, 2003; Hernández, 2006; Mehltreter, 2006), Brazil (Schmitt and Windisch, 2005, 2006), Hawaiian Islands (Durand and Goldstein, 2001), Fiji (Ash, 1986, 1987), and Taiwan (Chiou *et al.*, 2001). Most studies focused on tree ferns, with the exception of the mangrove fern *Acrostichum danaeifolium* (Mehltreter and Palacios-Rios, 2003), the herbaceous forest understory fern *Danaea wendlandii* (Sharpe and Jernstedt, 1990, 1991; Sharpe, 1993), the climbing fern *Lygodium venustum* (Mehltreter, 2006), and the rheophyte *Thelypteris angustifolia* (Sharpe, 1997). The focus of these studies was relating plant age to the correlation between climatic triggers and phenological responses, the measurement of growth rates of leaves and trunks, and the periodicity of leaf production and leaf fertility.

8.3.1 Seasonality in tropical ferns

Seasonality or periodicity of growth and reproduction is typical for plants of temperate zones because during the cold winter, soil water freezes and becomes unavailable for plant roots. In tropical environments periodic dry seasons can cause seasonal variation in plant growth, but even without a definite dry season some plants express seasonal phenological patterns because of the periodic appearance of pollinators, animal dispersers, or herbivores. Given that ferns do not interact with pollinators or specialized dispersers (Barrington, 1993), it had been assumed that they would have an aseasonal phenology (Tryon, 1960). Most recent phenological studies of tropical ferns, however, have shown predominantly seasonal patterns (Table 8.1), similar to those of woody angiosperms. In tropical deciduous forests, woody angiosperms are leafless during the dry season and produce a flush of leaves at the end of this period (Rivera *et al.*, 2002) or during the wet season (Lieberman and Lieberman, 1984; Bullock and Solis-Magallanes, 1990). In semideciduous lowland forests, seasonal woody understory plants flush at the beginning of the dry or at the beginning of the wet season (Aide, 1993). In wet tropical climates of montane regions, deciduous trees flush early in the dry season (Williams-Linera, 1997). Although most ferns do not drop their leaves, new leaves are produced mainly during the wet season or rarely earlier at the end of the dry season. Even the evergreen mangrove fern *Acrostichum danaeifolium*, which stands the entire year with its roots in the water, produces larger leaves and grows faster during the hot rainy summer season, when it produces 1–3 fertile leaves (Mehltreter and Palacios-Rios, 2003). This species has slightly dimorphic leaves, with fertile leaves up to one third larger than sterile leaves and the pinnae of fertile leaves about one third narrower than those of sterile leaves. Leaf dimorphism is generally considered to be a functional adaptation to spore dispersal, with larger fertile leaves favoring wind dispersal of spores (Wagner and Wagner, 1977). If this is true, we might expect fertile leaves to be produced during the dry season, when deciduous canopy species are leafless and wind speeds should be higher. We might also expect that fertile leaves would be shorter lived than sterile leaves because their reduced laminar surface makes them less photosynthetically efficient than sterile leaves. The first hypothesis is probably erroneous, because storms with high winds causing tree falls are frequent during the rainy season (Brokaw, 1996), and may promote spore dispersal. There is some support for the second hypothesis in dimorphic species (Table 8.2), but a larger sample size is needed to test this rigorously. In *Cibotium taiwanense*, a monomorphic species, fertile leaves have a longer life span than sterile leaves, and shed their spores in the second year after formation (Chiou *et al.*, 2001). The reason for this delay in spore dispersal

Table 8.1 *Tropical ferns of different vegetation types with seasonal leaf traits; seasonal growth (and leaf production) is more common than seasonal fertility*

Species	Life form	Seasonality of leaf traits		Vegetation type
		Growth	Fertility	
Acrostichum danaeifolium[7]	herbaceous	yes	yes	mangrove
Alsophila auneae[2]	arborescent	yes	?	cloud forest
Alsophila polystichoides[4]	arborescent	yes	?	cloud forest
Alsophila salvinii[1]	arborescent	yes	?	cloud forest
Alsophila setosa[10]	arborescent	yes	?	lower montane forest
Botrychium virginianum[8]	herbaceous	yes	yes	lower montane forest
Cibotium taiwanense[6]	arborescent	no	yes	subtropical forest
Ctenitis melanosticta[8]	herbaceous	yes	no	lower montane forest
Cyathea nigripes[4]	arborescent	yes	?	cloud forest
Danaea wendlandii[3]	herbaceous	yes	yes	lowland rain forest
Lygodium venustum[9]	climber	yes	?	semideciduous forest
Pteris orizabae[8]	herbaceous	yes	no	lower montane forest
Pteris quadriaurita[8]	herbaceous	yes	no	lower montane forest
Thelypteris angustifolia[5]	rheophyte	yes	no	lowland rain forest
Woodwardia semicordata[8]	herbaceous	yes	no	lower montane forest

[1] Seiler, 1981;
[2] Tanner, 1983;
[3] Sharpe, 1993;
[4] Bittner and Breckle, 1995;
[5] Sharpe, 1997;
[6] Chiou et al., 2001;
[7] Mehltreter and Palacios-Rios, 2003;
[8] Hernández, 2006;
[9] Mehltreter, 2006;
[10] Schmitt and Windisch, 2006.

remains unclear. Production of large fertile leaves may be too costly, causing a decreased rate of sporangial development, or delayed spore release may be considered an adaptation to await better dispersal conditions or improve opportunities for gametophyte development. In monomorphic species, fertile leaves shed their spores earlier but they also have longer life spans than sterile leaves (Table 8.2). This might be a consequence of an ontogenetic effect, when young plants have only sterile leaves and mature plants only fertile leaves, which is typical for *Macrothelypteris torresiana* (Mehltreter, personal observation).

Aseasonal species in the tropics encounter favorable growth conditions throughout the year and are characterized by continuous leaf turnover. Because

Table 8.2 *Leaf life span in months of monomorphic and dimorphic ferns in ascending order of the life span of fertile leaves*

Species	Sterile leaves	Fertile leaves	Dimorphism
Danaea wendlandii[1]	39.6	4.0	yes
Acrostichum danaeifolium[4]	7.7	4.1	yes
Thelypteris angustifolia[2]	11.0	9.6	yes
Macrothelypteris torresiana[5]	5.5	14.5	no
Cibotium taiwanense[3]	16.0	25.0	no
Ctenitis melanosticta[5]	11.5	30.1	no

[1] Sharpe and Jernstedt, 1990; Sharpe 1993;
[2] Sharpe, 1997;
[3] Chiou *et al.*, 2001;
[4] Mehltreter and Palacios-Rios, 2003;
[5] Hernández, 2006.

their leaf life span is not restricted by a cold winter or a dry season, it may vary from several months to several years (Westoby *et al.*, 2000). Most terrestrial ferns have mean leaf life spans of 6 months (e.g., *Sphaeropteris cooperi*; Durand and Goldstein, 2001) to 24 months (e.g., *Alsophila salvinii*; Seiler, 1981), exceptionally up to 40 months (Table 8.2). Aseasonal epiphytic ferns possess longer leaf life spans between 20 and 30 months (Mehltreter *et al.*, 2006). These longer-lived leaves may invest more in biochemical defenses against herbivores and consequently have a higher specific leaf weight (defined as dry weight per cm^2 leaf surface).

Terrestrial ferns, especially tree ferns

Danaea wendlandii is an herbaceous understory fern growing in the lowland rain forest of Costa Rica. Leaf growth is slow, only 1.6 leaves per year being produced, because of the low light level in the forest understory (Sharpe and Jernstedt, 1990; Sharpe, 1993). However, despite the constantly high humidity within the understory, leaf production and leaf fertility are seasonal, occurring mainly during the wetter summer. *Alsophila firma*, a Mexican tree fern of the montane forest, is even more seasonal. This species is short-deciduous and surprisingly drops its leaves during the wetter and hotter summer season, replacing them synchronously in one flush after 1–2 months of leaflessness (Mehltreter and Garcia-Franco, in press, Figure 8.1a). Some plants grow up to 50 cm per year (Figure 8.1c, 8.1d). Synchronous leaf production is thought to satiate herbivores during the leaf production peak (Aide, 1993) and may allow for a reduction in the costs of biochemical defense. This explanation seems to apply

Phenology and habitat specificity of tropical ferns 207

Figure 8.1 (a, b) Leaf emergence and (c, d) annual trunk growth of Mexican tree ferns. (a) Synchronous emergence of all new crosiers in *Alsophila firma*.
(b) Asynchronous leaf emergence with one new crosier at a time in *Cyathea bicrenata*.
(c, d) Trunk height of a plant of *Alsophila firma* in (c) May 2005 and (d) June 2006, illustrating an exceptional annual trunk growth of up to 50 cm.

to *Alsophila firma*, because its leaf flush falls in the middle of the wet season when herbivore pressure is very high. At the same site different fern species may follow distinct phenological strategies when these have the same cost benefit. *Cyathea bicrenata*, another tree fern growing at the same site as *A. firma*, produces leaves asynchronously throughout the year and considerable herbivore damage has been observed (Mehltreter, personal observation, Figure 8.1b). The cost benefit may be the same for this species, because it saves the costs of storage and recycling of nutrients of seasonal species, perhaps invests in more biochemical defenses, and may compensate for herbivore damage by a stronger continuous growth. *Alsophila setosa*, another tree fern species of Southern Brazil (Schmitt and Windisch, 2006), seems to be seasonal because of exposure to occasional frost during the winter that damages all older leaves. Plants recover in spring with a strong leaf flush.

Rheophytes

van Steenis' (1981, 1987) list of rheophytes includes 40 ferns and two species of Isoëtes. Rheophytes grow on rocks and boulders or along the stream bank, where they are exposed to periodic flooding (van Steenis, 1981). Leaves are often pinnate and possess narrow, long pinnae (e.g., *Dipteris lobbiana*; Holttum, 1938), which do not resist the water currents when the plant is submersed during periods of flooding. Spores might be expected to be dispersed by both wind and water currents. Only one rheophytic fern has been studied phenologically, *Thelypteris angustifolia* from a Puerto Rican rain forest (Sharpe, 1997). It has a branching rhizome that can break apart; these pieces can serve as vegetative propagules. Fertility is low, with most plants producing one or no fertile leaves per year. Although it is an evergreen plant primarily found in humid conditions, leaf development occurs mainly in the rainy season.

Climbing ferns and epiphytes

Phenological patterns vary across different life forms. Climbing ferns must have long-lived leaves to become established in the canopy. In the case of *Lygodium venustum*, the lamina is replaced after about 1 year, because the same leaf can sprout repeatedly by lateral dormant petiole buds (Mehltreter, 2006). These lateral axes renew the lamina in a way similar to that found for new leaves in woody angiosperms, differing only in that *Lygodium venustum* has a leaf petiole and rachis that persist rather than the shoot branches. Because there is no secondary stem growth in ferns, all water needs to be transported through the narrow rachis (diameter of 1–2 mm), and consequently collapsed vessels cannot be replaced by new tissues. Ewers *et al.* (1997) showed that roots of *L. venustum* hold a positive xylem pressure of up to 66 kPa, which allows refilling of embolized tracheids up to 7 m of plant height. Problems of water conduction within the rhizome may limit the height of most hemi-epiphytic ferns (e.g., *Bolbitis* and *Lomariopsis*) to 2–5 m.

Epiphytes do not root in the ground, but may have some humus substrate accumulated in tree branches, which allows for nutrient and water storage. However, most epiphytes grow on the bark surface and are exposed directly to daily changes in humidity. During the rainy season they experience strong solar radiation and dryness around midday, but recover quickly with daily rainfall events or the appearance of fog, especially on mountain ridges, and during the night when relative humidity increases.

For epiphytes the dry season is more challenging because they cannot rely on water stored within the soil. Species with glabrous, thin textured leaves drop them in response to drying (e.g., *Polypodium rhodopleuron*). Other epiphytic ferns tolerate the drought, roll their pinnae inwards, exposing the lower scaly leaf surface (e.g., *Pleopeltis furfuraceum*, Figure 8.2c), or fold their thicker-textured pinnae accordion-like (e.g., *Asplenium praemorsum*, Figure 8.2d). It is unknown for how long the leaves of these species can remain in this dormant stage and completely recover afterwards.

8.3.2 Habitat specificity

Habitat specificities were often reported without any quantitative measurements (Holttum, 1938; Page 1979a, 1979b; Cortez, 2001), with the exception of recent studies on epiphytes (Moran *et al.*, 2003; Mehltreter *et al.*, 2005) and terrestrial species (Poulsen *et al.*, 2006). The role of mycorrhizae in ferns for mediating habitat specificity has been poorly studied. Despite the obligatory mycorrhizae of some fern groups (e.g., *Lycopodium*, *Ophioglossum*, *Psilotum*, *Tmesipteris*), only facultative mycorrhizae were reported for the sporophytes in a study of Hawaiian ferns (Gemma *et al.*, 1992).

Terrestrial species

Poulsen *et al.* (2006) studied the floristic diversity of a plot of 1 ha of lowland Amazonian rain forest in Ecuador with an acidic mean soil pH of 3.33. The distribution and abundance of 29 fern species were correlated with soil calcium and sand content, and to a lesser degree with aluminum content. More studies of this type are needed on ferns at this geographical scale to allow for generalizations; the results of Poulsen *et al.* (2006) may not be applicable to sites with higher soil pH values, and where the calcium and aluminum contents could be less important to fern distribution. Edaphic niches of *Polybotrya* spp. in Northwestern Amazonia were best described by differences in soil texture, and cation content (Tuomisto, 2006).

Figure 8.2 (a, b) Vegetative reproduction and (c, d) leaf desiccation in some Mexican ferns. (a) Apical leaf buds of *Asplenium alatum* with young plant rooted in the ground. (b) *Asplenium sessilifolium* with completely developed plantlet at the leaf tip. (c) Desiccating leaf of *Polypodium furfuraceum*, bending pinnae inwards, exposing the lower scaly surface. (d) Desiccating leaf of *Asplenium praemorsum* folding pinnae accordion-like starting at the leaf tip.

Mangrove species

The three species of mangrove ferns in the genus *Acrostichum* are exceptionally tolerant to salt stress. They root in the soil and are often flooded, even though never completely submersed. The largest species, *A. danaeifolium*, is restricted to the Neotropics and is the least salt tolerant (Lloyd and Buckley, 1986; Mehltreter and Palacios-Rios, 2003). The smallest species, *A. speciosum*, from the Paleotropics seems to be the only obligately halophytic fern species (Kramer et al., 1995). The pantropically distributed *A. aureum* is ecologically intermediate

Table 8.3 *Number of species with substrate specificity of four genera of Mexican rock ferns*

Genus	Limestone	Gypsum	Igneous rocks	No preference
Argyrochosma	5	0	2	5
Cheilanthes	12	2	11	35
Notholaena	4	2	5	13
Pellaea	2	1	4	7

Data from Mickel and Smith (2004).

between the other two species. Other species that may grow in the coastal zones adjacent to the mangroves are *Stenochlaena palustris* in southeast Asia and *Ctenitis maritima* on La Réunion Island (Mehltreter, personal observation).

Lithophytes

Lithophytes (petrophytes) are plants that grow primarily on rocks and boulders. Species that grow on rocks along rivers and stream banks often reproduce vegetatively by apical leaf buds, e.g., *Asplenium* (Figure 8.2a, 8.2b). Rock ferns are often good indicators for the underlying chemistry of the substrate. For temperate zones, there are dozens of examples of ferns that only grow on igneous rocks (e.g., granite), metamorphic rocks (e.g., serpentine), or sedimentary rocks (e.g., sandstone, gypsum, and limestone). In some genera such as *Asplenium* these substrate specificities change in newly formed hybrids compared to their parental taxa, e.g. *A. adulterinum* is restricted to serpentine rocks, although it is the allopolyploid hybrid between calciphilous *A. viride* and *A. trichomanes* which grows on silicate rocks (Kramer *et al.*, 1995). Although in tropical humid zones limestone is scarcely found because it erodes quickly, some species are typically restricted to this habitat, especially those in the genera *Argyrochosma* and *Cheilanthes* (Table 8.3). Mickel and Smith (2004) report rock substrate specificities for 50 out of 110 species of *Argyrochosma*, *Cheilanthes*, *Notholaena*, and *Pellaea* with more or less equal proportions occurring on limestone (23 species) or igneous rocks (22 species). Consequently, these species are restricted to mountain ranges with these substrates and may be less common and more frequently endangered than more widespread, substrate-generalist species. Over the last twenty years, some ferns have been reported to hyper-accumulate heavy metals, especially arsenic, cadmium, copper, and zinc (Table 8.4) and, thus, may be good indicators of soil type. Recently, some fern species, especially *Athyrium yokoscense* (Nishizono *et al.*, 1987) and *Pteris vittata* (Ma *et al.*, 2001), have been used as phytoremediators to manage heavy-metal contaminated soils.

Table 8.4 *Heavy metal hyper-accumulating fern species*

Species	Life form	Accumulated heavy metals
Pityrogramma calomelanos[5]	terrestrial	As
Azolla caroliniana[6]	aquatic	Cr, Hg
Azolla filiculoides[2]	aquatic	Cd, Cr, Cu, Ni, Zn
Marattia spp.[3]	terrestrial	Al
Pteris vittata[4]	terrestrial and lithophytic	As
Athyrium yokoscense[1]	terrestrial	Cd, Cu, Zn

[1] Nishizono *et al.*, 1987;
[2] Sela *et al.*, 1989;
[3] Kramer *et al.*, 1995;
[4] Ma *et al.*, 2001; Barger *et al.*, 2007;
[5] Francesconi *et al.*, 2002;
[6] Bennicelli *et al.*, 2004.

Epiphytes

Ferns are the third most species-rich group of epiphytes after orchids and bromeliads in the New World. The three fern families with the greatest number of epiphytic species are Polypodiaceae, Hymenophyllaceae, and Aspleniaceae, all with more than 50% of the family occurring as epiphytes. Vittarioid ferns (Pteridaceae) are entirely epiphytic (Gentry and Dodson, 1987).

In comparison to terrestrial ferns, epiphytes must grow in substrates that have a lower nutrient availability and water retention capacity (Benzing, 1990, 1995), and they must cope with large seasonal and daily changes in humidity, as well as an enormous variety of potential hosts (Kramer *et al.*, 1995). Apparent adaptations to prevent water loss include such characteristics as simple or pinnate leaves, thick leaf texture (e.g., *Niphidium*), an often dense cover of scales (e.g., *Polypodium, Elaphoglossum*), water storing rhizomes (e.g., *Davallia*), and leaf succulence together with crassulacean acid metabolism (e.g., *Pyrrosia*). For humus accumulation and water storage, some ferns have large nest-forming rosettes (e.g., *Asplenium*) or specific niche-forming leaves (e.g., *Drynaria, Platycerium*) or leaf bases (e.g., *Aglaomorpha*). All of the latter group are restricted to the Paleotropics, with the exception of one species of *Platycerium* (*P. andinum*) and perhaps *Niphidium* spp. with large, but morphologically undifferentiated leaves. This nearly complete restriction of humus accumulating fern life forms to the Paleotropics has been interpreted as competitive exclusion by bromeliads, which are absent in the Paleotropics (Kramer *et al.*, 1995). Some other fern species (*Solanopteris* spp. in the Neotropics, *Lecanopteris* spp. in the Paleotropics) have ants collecting nutrients for them. These ferns have thick hollow rhizomes that are inhabited by ants. The plants form roots within the interior of the rhizome to take up the nutrients that the ants bring in (Wagner, 1972; Gómez, 1974; Walker, 1986; Gay, 1991).

The concentration of gemmae-forming gametophytes in mainly epiphytic groups such as grammitids (Polypodiaceae), Hymenophyllaceae, and vittarioids (Pteridaceae) is often interpreted as an adaptation of epiphytism (Page, 1979b; Farrar, 1990; Dassler and Farrar, 2001).

Within the understory of wet tropical forests there is a fairly constant air humidity (i.e., from the base of the trunk to about 3 m above ground). Epiphytic ferns abound as lower trunk epiphytes, especially Hymenophyllaceae, and may be the dominant plant group (Zotz and Büche, 2000; Mehltreter et al., 2005). In the drier and exposed canopy, bromeliads and orchids are apparently better adapted to the extreme changes of temperature and humidity (Johansson, 1974; Kelly, 1985; Benzing, 1990).

The first fern epiphytes (e.g., *Botryopteris forensis*) are known from the Carboniferous, where they grew on marattiaceous tree ferns of the genus *Psaronius* (Rothwell, 1991). For contemporary epiphytes, three main host groups are available as substrate: tree ferns, gymnosperms (e.g., pines), and woody angiosperms. One species usually found growing on tree ferns is *Polyphlebium capillaceum* (syn. *Trichomanes capillaceum*) (Mickel and Beitel, 1988; Moran et al., 2003; Mehltreter et al., 2005; Ebihara et al., 2006; (see Table 8.5, Figure 8.3b). It develops especially on the trunk bases when the tree fern develops its adventitious roots forming a continuous root mantle up to half way to the top. This root mantle has a high water retention capacity and porosity because of its entangled roots, which makes it a favorable substrate for many epiphyte species (Heatwole, 1993; Medeiros et al., 1993; Mehltreter et al., 2005), including recently derived fern clades (e.g., *Elaphoglossum*, *Polypodium*) that diversified along with the angiosperms (Schneider et al., 2004). However, most modern epiphytes are not restricted to tree ferns and also grow on a wide range of angiosperm hosts. Nevertheless they may be very abundant on tree fern trunks (Table 8.5), which they can cover entirely, for example *Asplenium harpeodes*, *Blechnum fragile* (Figure 8.3c), *Elaphoglossum lonchophyllum* (Figure 8.3a), and *Terpsichore asplenifolia* (Figure 8.3a) in Mexican cloud forest (Mehltreter, personal observation). On the other hand, some epiphytic ferns do not occur on tree ferns, perhaps because of the acidic pH and high tannin content of the root mantle (Frei and Dodson, 1972), e.g., *Elaphoglossum peltatum* (Table 8.5, Figure 8.3d). Another example is *Vittaria isoetifolia* on La Réunion Island, which is only found on angiosperm trees. Its short rhizome may have problems attaching to the surface of the root mantle or even may be overgrown by the latter. Horizontally extended branches of angiosperm trees allow the large leaves of *V. isoetifolia* to hang down freely and avoid touching the

Table 8.5 *Epiphytic ferns with significant host specificities for tree ferns or angiosperm trees in order of their frequency on tree ferns; data are percentages of presence on host trunks*

	Tree ferns	Angiosperm trees
Elaphoglossum decursivum[1]	95	30
Blechnum fragile[1]	90	0
Trichomanes capillaceum[2]	89	0
Pecluma eurybasis[1]	65	20
Dryopteris patula[1]	50	0
Elaphoglossum stenoglossum[1]	50	10
Blechnum attenuatum[3]	41	4
Campyloneurum sphenodes[1]	40	0
Hymenophyllum elegans[1]	35	0
Asplenium auriculatum[1]	35	5
Asplenium serratum[1]	30	0
Elaphoglossum petiolatum[2]	29	4
Trichomanes reptans[2]	18	67
Asplenium nitens[3]	2	24
Vittaria isoetifolia[3]	0	16
Elaphoglossum peltatum[2]	0	19

[1] Moran *et al.*, 2003;
[2] Mehltreter *et al.*, 2005;
[3] Mehltreter, unpublished data from La Réunion Island.

vertical trunk surface where they would be exposed to competition with other epiphytes.

In conclusion, dozens of epiphytic ferns exhibit host specificity but few are restricted to one host group. Tree ferns are generally a better substrate for trunk epiphytes than angiosperm trees, but there are important exceptions, for which we only have speculative explanations. Specific needs of the gametophyte, the life form of the sporophyte, pH, content of tannins, and structural differences of the root mantle of tree ferns may be involved. Moreover, fern epiphytes specific to angiosperm trees may have been overlooked (Callaway *et al.*, 2002), because these may be restricted to regions where no tree ferns occur. For example, palm trees are especially good substrates when their leaf bases stay attached to the trunk (Zotz and Vollrath, 2003). Their epiphyte communities have been studied mainly in cultivated species. In Costa Rican oil palm fields, *Nephrolepis* spp. and *Phlebodium* spp. commonly occur at palm leaf bases (Mehltreter, personal observation). Interestingly, tree fern skirts formed by old leaves that stay attached to the

Phenology and habitat specificity of tropical ferns 215

Figure 8.3 Habitat specificities of Mexican trunk epiphytes on tree ferns (a–c) or on angiosperm trees (d). (a) *Elaphoglossum lonchophyllum* with entire leaves and *Terpsichore asplenifolia* with pinnate leaves on *Alsophila firma*. (b) *Trichomanes capillaceum* on *Alsophila firma*. (c) *Blechnum fragile* on the root mantle of *Dicksonia sellowiana*. (d) *Elaphoglossum peltatum* on the tree stem of *Quercus* spp.

trunk were found to inhibit the colonization by larger epiphytes and climbers (Page and Brownsey, 1986).

8.4 Synthesis of current perspectives

Unexpected phenological patterns suggest that we cannot easily draw general conclusions because of limited observations and little quantitative data. We need more detailed quantitative field data across wider geographical and taxonomic scales to understand the fascinating phenology and habitat requirements of ferns. For horticultural purposes, both issues should be of some concern, when species are known to be difficult to cultivate (for example, grammitids (Polypodiaceae), Hymenophyllaceae, and Gleicheniaceae). For conservation purposes, specific habitat requirements may restrict some fern species geographically, and may be responsible in part for their endangered status. Hyper-accumulators of heavy metals may be more common in ferns than known until recently and can play an important role in the future as bioindicators and for phytoremediation.

8.5   Future goals and directions

Within the field of ecological research of ferns and lycophytes, there are at least three areas on which future studies should be focused.

(1) Long-term research, especially in the tropics where ferns are most diverse and abundant.
(2) Broad-scale studies within a wide range of species at different latitudes, altitudes, and habitats.
(3) Multi-disciplinary approaches of ecology, physiology, biochemistry, morphology, systematics, and genetics, which combine molecular methods and field experiments.

Whatever approach is selected, quantitative approaches are preferable over observational and merely qualitative studies. Field studies should be comparative or experimental to answer specific questions, for example temperature and soil optima for new species of horticultural interest.

8.6   Importance of long-term studies

Most results of phenological studies depend heavily on climatic conditions at the study site during the years of observation. If these have been exceptional, extrapolations and general conclusions cannot be drawn without the risk of committing significant errors.

For this reason long-term studies are particularly important. Moreover, for conservation purposes it is critical to document complete life cycles of ferns to improve our understanding of demographic processes (see Chapter 9). Long-term growth measurements allow calculation of the ages of individual plants, and determination of quantified survival rates for better estimations of population turnover, which are fundamental values for conservation management.

Our actual understanding of fern phenology is restricted to a few species and even fewer locations. Only studies on a wider geographical scale will allow us to understand, for example, how phenological patterns change within and among species at different latitudes and altitudes. This knowledge is fundamental for addressing future challenges, such as understanding the possible consequences of global warming on ferns. Will tropical fern communities benefit from a warmer climate or will they decline because of a possibly longer dry season? Some species that now produce only one or two fertile leaves per year may discontinue doing so when climatic changes affect their development (Sharpe, 1997; Mehltreter and Palacios-Rios, 2003). Over the long term, this will significantly reduce the reproductive success and increase the chance of extinction of local populations, when these cannot recover from a spore bank.

References

Aide, T. M. (1993). Patterns of leaf development and herbivory in a tropical understorey community. *Ecology*, **74**, 455–466.

Ash, J. (1986). Demography and production of *Leptopteris wilkesiana* (Osmundaceae), a tropical tree fern from Fiji. *Australian Journal of Botany*, **34**, 207–215.

Ash, J. (1987). Demography of *Cyathea hornei* (Cyatheaceae), a tropical tree-fern from Fiji. *Australian Journal of Botany*, **35**, 331–342.

Barger, T. W., Durham, T. J., Andrews, H. T., and Wilson, M. S. (2007). Gametophytic and sporophytic responses of *Pteris* spp. to arsenic. *American Fern Journal*, **97**, 30–45.

Barrington, D. S. (1993). Ecological and historical factors in fern biogeography. *Journal of Biogeography*, **20**, 275–280.

Beever, J. E. (1984). Moss epiphytes of tree-ferns in a warm-temperate forest, New Zealand. *Journal of the Hattori Botanical Laboratory*, **56**, 89–95.

Bennicelli, R., Stepniewska, Z., Banach, A., Szajnocha, K., and Ostrowski, J. (2004). The ability of *Azolla caroliniana* to remove heavy metals (Hg(II), Cr(III), Cr(VI)) from municipal waste water. *Chemosphere*, **55**, 141–146.

Benzing, D. H. (1990). *Vascular Epiphytes*. Cambridge: Cambridge University Press.

Benzing, D. H. (1995). Vascular epiphytes. In *Forest Canopies*, ed. M. D. Lowman and N. M. Nadkarni. San Diego, CA: Academic Press, pp. 225–254.

Bittner, J. and Breckle, S. W. (1995). The growth rate and age of tree fern trunks in relation to habitats. *American Fern Journal*, **85**, 37–42.

Brokaw, N. V. L. (1996). Treefalls: frequency, timing and consequences. In *The Ecology of a Tropical Forest: Seasonal Rhythms and Long-term Changes*, ed. E. G. Leigh, A. S. Rand, and D. M. Windsor, 2nd edn., Washington, DC: Smithsonian Institution Press, pp. 101–108.

Bullock, S. H. and Solis-Magallanes, J. A. (1990). Phenology of canopy trees of a tropical deciduous forest in Mexico. *Biotropica*, **22**, 22–35.

Callaway, R. M., Reinhart, K. O., Moore, G. W., Moore, D. J., and Pennings, S. C. (2002). Epiphyte host preferences and host traits: mechanisms for species-specific interactions. *Oecologia*, **132**, 221–230.

Chiou, W.-L., Lin, J. C., and Wang, J. Y. (2001). Phenology of *Cibotium taiwanense* (Dicksoniaceae). *Taiwan Journal of Forestry Science*, **16**, 209–215.

Christ, H. (1910). *Die Geographie der Farne*. Jena: Fischer.

Copeland, E. B. (1947). *Genera Filicum*. Waltham, MA: Chronica Botanica.

Cortez, L. (2001). Pteridofitas epífitas encontradas en Cyatheaceae y Dicksoniaceae de los bosques nublados de Venezuela. *Gayana Botanica* **58**, 13–23.

Dassler, C. L. and Farrar, D. R. (2001). Significance of gametophyte form in long distance colonization by tropical, epiphytic ferns. *Brittonia*, **53**, 352–369.

Durand, L. Z. and Goldstein, G. (2001). Photosynthesis, photoinhibition, and nitrogen use efficiency in native and invasive tree ferns in Hawaii. *Oecologia*, **126**, 345–354.

Ebihara, A., Dubuisson, J.-Y., Iwatsuki, K., Hennequin, S., and Ito, M. (2006). A taxonomic revision of Hymenophyllaceae. *Blumea*, **51**, 221–280.

Ewers, F. W., Cochard, H., and Tyree, M. T. (1997). A survey of root pressures in vines of a tropical lowland forest. *Oecologia*, **110**, 191–196.

Farrar, D. R. (1990). Species and evolution in asexually reproducing independent fern gametophytes. *Systematic Botany*, **15**, 98–111.

Francesconi, K., Visoottiviseth, P., Sridokchan, W., and Goessler, W. (2002). Arsenic species in an arsenic hyperaccumulating fern, *Pityrogramma calomelanos*: a potential phytoremediator of arsenic-contaminated soils. *Science of the Total Environment*, **284**, 27–35.

Frei, J. K. and Dodson, C. H. (1972). The chemical effect of certain bark substrates on the germination and early growth of epiphytic orchids. *Bulletin of the Torrey Botanical Club*, **99**, 301–307.

Gay, H. (1991). Ant-houses in the fern genus *Lecanopteris*: the rhizome morphology and architecture of *L. sarcopus* and *L. darnaedii*. *Botanical Journal of the Linnean Society*, **106**, 199–208.

Gemma, J. N., Koske, R. E., and Flynn, T. (1992). Mycorrhizae in Hawaiian Pteridophytes: occurence and evolutionary significance. *American Journal of Botany*, **79**, 843–852.

Gentry, A. H. and Dodson, C. H. (1987). Diversity and biogeography of neotropical vascular epiphytes. *Annals of the Missouri Botanical Garden*, **74**, 205–233.

Gómez, L. D. (1974). Biology of the potato-fern, *Solanopteris brunei*. *Brenesia*, **4**, 37–61.

Heatwole, H. (1993). Distribution of epiphytes on trunks of the arborescent fern *Blechnum palmiforme*, at Gough Island, South Atlantic. *Selbyana*, **14**, 46–58.

Hernández, A. C. (2006). Fenología foliar de helechos terrestres en un fragmento de bosque mesófilo de montaña en Xalapa, Veracruz, México. Tesis de Licenciatura en Biología, Universidad Veracruzana, Xalapa.

Holttum, R. E. (1938). The ecology of tropical pteridophytes. In *Manual of Pteridology*, ed. F. Verdoorn. The Hague: M. Nijhoff, pp. 420–450.

Johansson, D. (1974). Ecology of vascular epiphytes in West African rain forest. *Acta Phytogeographica Suecica*, **59**, 1–130.

Kelly, D. L. (1985). Epiphytes and climbers of a Jamaican rain forest: vertical distribution, life forms and life histories. *Journal of Biogeography*, **12**, 223–241.

Kornás, J. (1977). Life-forms and seasonal patterns in the pteridophytes of Zambia. *Acta Societatis Botanicorum Poloniae*, **46**, 668–690.

Kramer, K. U., Schneller, J. J., and Wollenweber, E. (1995). *Farne und Farnverwandte*. Stuttgart: Thieme.

Lieberman, D. and Lieberman, M. (1984). The causes and consequences of synchronous flushing in a tropical dry forest. *Biotropica*, **16**, 193–201.

Lloyd, R. M. and Buckley, D. P. (1986). Effects of salinity on gametophyte growth of *Acrostichum aureum* and *Acrostichum danaeifolium*. *Fern Gazette*, **13**, 97–102.

Ma, L. Q., Komar, K. M., Tu, C., Zhang, W., Cai, Y., and Kennelley, E. D. (2001). A fern that hyperaccumulates arsenic. *Nature*, **409**, 579.

Medeiros, A. C., Loope, L. L., and Anderson, S. J. (1993). Differential colonization by epiphytes on native (*Cibotium* spp.) and alien (*Cyathea cooperi*) tree ferns in a Hawaiian rain forest. *Selbyana*, **14**, 71–74.

Mehltreter, K. (2006). Leaf phenology of the climbing fern *Lygodium venustum* in a semi-deciduous lowland forest on the Gulf of Mexico. *American Fern Journal*, **96**, 21–30.

Mehltreter, K. and García-Franco, J. G. (in press). Leaf phenology and trunk growth of the deciduous tree fern *Alsophila firma* in a Mexican lower montane forest. *American Fern Journal*.

Mehltreter, K. and Palacios-Rios, M. (2003). Phenological studies of *Acrostichum danaeifolium* (Pteridaceae, Pteridophyta) at a mangrove site on the Gulf of Mexico. *Journal of Tropical Ecology*, **19**, 155–162.

Mehltreter, K., Flores-Palacios, A., and García-Franco, J. G. (2005). Host preferences of vascular trunk epiphytes in a cloud forest of Veracruz, México. *Journal of Tropical Ecology*, **21**, 651–660.

Mehltreter, K., Hülber, K., and Hietz, P. (2006). Herbivory on epiphytic ferns of a Mexican cloud forest. *Fern Gazette*, **17**, 303–309.

Mickel, J. T. and Beitel, J. M. (1988). *Pteridophyte Flora of Oaxaca, Mexico*. New York: New York Botanical Garden.

Mickel, J. T. and Smith, A. R. (2004). *The Pteridophytes of Mexico*. New York: New York Botanical Garden.

Moran, R. C., Klimas, S., and Carlsen, M. (2003). Low-trunk epiphytic ferns on tree ferns versus angiosperms in Costa Rica. *Biotropica*, **35**, 48–56.

Nishizono, H., Suzuki, S., and Ishii, F. (1987). Accumulation of heavy metals in the metal-tolerant fern *Athyrium yokoscense*, growing on various environments. *Plant and Soil*, **102**, 65–70.

Oliver, W. R. B. (1930). New Zealand epiphytes. *Journal of Ecology*, **18**, 1–50.

Page, C. N. (1979a). The diversity of ferns. An ecological perspective. In *The Experimental Biology of Ferns*, ed. A. F. Dyer. London: Academic Press, pp. 10–56.

Page, C. N. (1979b). Experimental aspects of fern ecology. In *The Experimental Biology of Ferns*, ed. A. F. Dyer. London: Academic Press, pp. 552–589.

Page, C. N. and Brownsey, P. J. (1986). Tree-fern skirts: a defense against climbers and large epiphytes. *Journal of Ecology*, **74**, 787–796.

Pócs, T. (1982). Tropical forest bryophytes. In *Bryophyte Ecology*, ed. A. J. E. Smith. London: Chapman and Hall, pp. 59–104.

Poulsen, A. D., Tuomisto, H., and Balslev, H. (2006). Edaphic and floristic variation within a 1-ha plot of lowland Amazonian rain forest. *Biotropica*, **38**, 468–478.

Rivera, G., Elliott, S., Caldas, L. S., Nicolossi, G., Coradin, V. T. R., and Borchert, R. (2002). Increasing day-length induces spring flushing of tropical dry forest trees in the absence of rain. *Trees*, **16**, 445–456.

Rothwell, G. W. (1991). *Botryopteris forensis* (Botryopteridaceae), a trunk epiphyte of the tree fern Psaronius. *American Journal of Botany*, **78**, 782–788.

Schmitt, J. L. and Windisch, P. G. (2005). Aspectos ecológicos de *Alsophila setosa* Kaulf. (Cyatheaceae, Pteridophyta) no Rio Grande do Sul, Brasil. *Acta Botanica Brasilica*, **19**, 859–865.

Schmitt, J. L. and Windisch, P. G. (2006). Phenological aspects of frond production in *Alsophila setosa* (Cyatheaceae, Pteridophyta) in Southern Brazil. *Fern Gazette*, **17**, 263–270.

Schneider, H., Schuettpelz, E., Pryer, K. M., Cranfill, R., Magallón, S., and Lupia, R. (2004). Ferns diversified in the shadow of angiosperms. *Nature*, **428**, 553–557.

Seiler, R. L. (1981). Leaf turnover rates and natural history of the Central American tree fern *Alsophila salvinii*. *American Fern Journal*, **71**, 75–81.

Sela, M., Garty, J., and Tel-Or, E. (1989). The accumulation and the effect of heavy metals on the water fern *Azolla filiculoides*. *New Phytologist*, **112**, 7–12.

Sharpe, J. M. (1993). Plant growth and demography of the neotropical herbaceous fern *Danaea wendlandii* (Marattiaceae) in a Costa Rican rain forest. *Biotropica*, **25**, 85–94.

Sharpe, J. M. (1997). Leaf growth and demography of the rheophytic fern *Thelypteris angustifolia* (Willdenow) Proctor in a Puerto Rican rainforest. *Plant Ecology*, **130**, 203–212.

Sharpe, J. M. and Jernstedt, J. A. (1990). Leaf growth and phenology of the dimorphic herbaceous layer fern *Danaea wendlandii* (Marattiaceae) in a Costa Rican rain forest. *American Journal of Botany*, **77**, 1040–1049.

Sharpe, J. M. and Jernstedt, J. A. (1991). Stipular bud development in *Danaea wendlandii* (Marattiaceae). *American Fern Journal*, **81**, 119–127.

Tanner, E. V. J. (1983). Leaf demography and growth of the tree-fern *Cyathea pubescens* Mett. ex Kuhn in Jamaica. *Botanical Journal of the Linnaean Society*, **87**, 213–227.

Tuomisto, H. (2006). Edaphic niche differentiation among *Polybotrya* ferns in western Amazonia: implications for coexistence and speciation. *Ecography*, **29**, 273–284.

Tryon, R. M. (1960). The ecology of Peruvian ferns. *American Fern Journal*, **50**, 46–55.

Tryon, R. M. (1964). Evolution in the leaf of living ferns. *Bulletin of the Torrey Botanical Club*, **21**, 73–85.

van Steenis, C. G. G. J. (1981). *Rheophytes of the World*. Alpen an den Rijn: Sijthoff and Noordhoff.

van Steenis, C. G. G. J. (1987). Rheophytes of the world: supplement. *Allertonia*, **4**, 267–330.

Wagner, W. H. (1972). *Solanopteris brunei*, a little known fern epiphyte with dimorphic stems. *American Fern Journal*, **62**, 33–43.

Wagner, W. H., Jr. and Wagner, F. S. (1977). Fertile-sterile leaf dimorphy in ferns. *Gardens Bulletin Singapore*, **30**, 251–267.

Walker, T. G. (1986). The ant-fern *Lecanopteris mirabilis*. *Kew Bulletin*, **41**, 533–545.

Westoby, M., Warton, D., and Reich, P. B. (2000). The time value of leaf area. *American Naturalist*, **155**, 649–656.

Williams-Linera, G. (1997). Phenology of deciduous and broadleaved-evergreen tree species in a Mexican tropical lower montane forest. *Global Ecology and Biogeography Letters*, **6**, 115–127.

Zotz, G. and Büche, M. (2000). The epiphytic filmy ferns of a tropical lowland forest – species occurrence and habitat preferences. *Ecotropica*, **6**, 203–206.

Zotz, G. and Vollrath, B. (2003). The epiphyte vegetation of the palm *Socratea exorrhiza* – correlations with tree size, tree age and bryophyte cover. *Journal of Tropical Ecology*, **19**, 81–90.

9

Gametophyte ecology

DONALD R. FARRAR, CYNTHIA DASSLER, JAMES E. WATKINS, JR., AND CHANDA SKELTON

9.1 Introduction

Seed plant ecologists would find incredulous a proposal to study the ecology of a species that did not include critical examination of all aspects of recruitment in the field, relying instead on laboratory studies of seed germination, seedling growth and mortality, etc. Yet students of fern and lycophyte ecology are limited to data on gametophyte growth and reproduction collected almost exclusively from laboratory studies. They are expected to assume that these studies accurately reflect growth and reproduction in nature. Field investigations of gametophyte biology are minimal on such critical topics as: the role of morphological and physiological diversity among gametophyte taxa in habitat selection; the time frame and method of gametophyte development, maturation, sexual differentiation, and sporophyte production; the breeding systems and habitats effectively contributing new recruits to sporophyte populations; or the number and frequency of recruits. In addition to producing a decidedly unbalanced view of fern and lycophyte ecology, the absence of ecological data on the gametophytic phase of fern biology has left science with important misconceptions regarding this critical phase of the life cycle. The gametophyte not only provides the opportunity for sexual reproduction (and thus controls genetic diversity) but also determines (along with vegetative reproduction) recruitment, species habitat selection, species migration, and, ultimately, fern and lycophyte evolution.

Reasons for the dearth of field studies on gametophyte ecology are both perceived and real. It is much more difficult to find and to identify gametophyte

Biology and Evolution of Ferns and Lycophytes, ed. Tom A. Ranker and Christopher H. Haufler. Published by Cambridge University Press. © Cambridge University Press 2008.

plants than to do the same with sporophytes. Following the fate of gametophyte populations and newly produced sporophytes requires continuous access to study sites and dedicated observation for long periods. Although difficult, such studies are possible and the rewards, in the context of current knowledge, are immense. In this chapter we will examine the problems posed by field studies and will

- discuss why field studies are important to understanding fern biology,
- describe methods that have been successfully employed, and
- present recent advances in our understanding of the gametophyte generation that promise deeper insight into the role of that generation in determining the ecology and evolution of fern and lycophyte species.

Ferns and lycophytes are vascular land plants composed of two alternating and morphologically very different growth forms, each of which is free-living, i.e., the majority of the life span of each is nutritionally and physically independent of the other. We refer to these two forms as "sporophyte," the relatively large and familiar spore-producing plant and "gametophyte," the generally obscure and relatively featureless gamete-producing plant (see Chapter 2). Even apogamous species, which avoid complications of sexual reproduction, have a gametophytic phase, and most ecological considerations apply to their prothalli.

With few exceptions, the sporophyte phase of ferns and lycophytes is a perennial plant of potentially indeterminate growth. As long as the local environment is suitable, an individual plant or clone can exist indefinitely. It is likely that terrestrial and epipetric ferns and lycophytes are the oldest members of their community, having persisted vegetatively in their local area for centuries or perhaps millennia.

Gametophytes are usually not considered to be perennial, and for many, especially terrestrial species with photosynthetic gametophytes, this may be the rule. However, a large and underappreciated number of epiphytic and epipetric (and some terrestrial) species do have gametophytes that persist indefinitely through vegetative growth and proliferation. Furthermore, a large number of these species are capable of local dispersion through production of vegetative propagules (gemmae). Thus both gametophyte and sporophyte phases of ferns and lycophytes can persist indefinitely in a local supportive habitat. Regardless, the initial introduction of a species into a local area is through successful spore dispersal, spore germination, gametophyte development, gametangia formation, fertilization, and production of a viable sporophyte. From this it follows that studies of species distributions, species migration, habitat selection, etc. must take into account the particular growth form *and* the ecology of the gametophyte generation.

Important historical studies on the gametophyte generation of ferns and lycophytes began with the elaboration of the fern life cycle by Hofmeister (1862). Made aware of the vital role of the gametophyte, natural historians of the late nineteenth and early twentieth centuries sought and illustrated gametophytes of a number of fern and lycophyte species (e.g., Campbell, 1905; Goebel, 1905; Bower, 1923). It was not until the mid twentieth century, however, after detailed description of a wide diversity of taxa by Stokey (1951), Atkinson and Stokey (1964), Nayar and Kaur (1971), Atkinson (1973), and others that important patterns in development and mature morphologies were detected within the diverse array of gametophyte forms. Similarly, Nayar and Kaur (1969) detected four fundamentally different pathways of early development in fern gametophytes that segregate along genus and family lines. In addition to describing characters useful in phylogenetic classification, these studies also formed the basis for field identification of gametophytes, especially to the level of genus, and laid the groundwork for future gametophyte-based ecological studies.

Development of culture methods (e.g., Andersson, 1923) and recognition of systematic differences among taxa also led to more elaborate studies on the developmental response of gametophytes to environmental stimuli. Effects of light quantity, quality, and direction, as well as reponses to chemical signals, revealed a highly complex physiology within these morphologically simple plants (Miller, 1968; Raghavan, 1989). Of special importance was the elaboration of an antheridiogen system by which older or developmentally more advanced plants within a population secrete a gibberellin-like chemical into the substrate that promotes antheridium formation in younger or developmentally less advanced plants (Döpp, 1962; Näf, 1979; Chapter 5). This "antheridiogen" also promoted dark germination of spores of species otherwise requiring light for germination (reviewed in Voeller and Weinberg, 1969). Thus sexual differentiation in gametophyte populations could involve not only gametophytes developed from recently deposited spores, but also gametophytes developed from spores that had arrived earlier and were dormant in a shallow but dark spore bank.

Detailed field studies of natural populations of fern gametophytes began in the 1970s (e.g., Lloyd, 1974; Farrar and Gooch, 1975; Cousens, 1981, 1988; Peck et al., 1990). These studies greatly elevated holistic ecological studies by (1) analyzing gametophyte establishment and persistence/mortality through time, (2) considering the nature of "safe sites" that can support new sporophyte recruitment, and (3) recognizing the interplay of genetic variability, breeding systems, and gametophyte density in governing the migratory ability of species. In this regard the "isolate potential," the potential for viable sporophyte production by individual gametophytes developing in isolation following long-distance dispersal,

Table 9.1 *Classification of leptosporangiate fern gametophytes into five types in three ecologically functional groups based on form, type of meristem, type of proliferation and longevity*

Type	Shape	Meristem	Description	Functional group
I	cordiform	notch	non-proliferating	annual
II	strap	notch	regenerative branching	perennial
III	ribbon	marginal	apical branching	perennial
IV	gemmiferous strap	notch	regenerative branching and gemmiferous	gemmiferous
V	gemmiferous ribbon	marginal	apical branching and gemmiferous	gemmiferous

was shown to be as important as their development and differentiation within populations (Peck *et al.*, 1990). These pioneering field studies also demonstrated the feasibility of gametophyte identification and study in the field and promoted integrative studies combining gametophyte development, morphology, and breeding systems to understand the ecology and distribution of species (Dassler and Farrar, 1997; Chiou *et al.*, 1998, 1999; Chiou and Farrar, 2002).

In this chapter, along with brief discussions of relevant literature, especially reviews of studies on the gametophyte generation, we describe recent advances in gametophyte study by the authors in four areas. In Section 9.2, we consider how the form and growth patterns of fern gametophytes affect and reflect habitats, as well as survival and reproductive strategies. Section 9.3 describes studies that relate gametophytic physiological differentiation among species to both habitat and phylogenetic groups. In Section 9.4, we describe investigations of the relevancy of laboratory cultures of fern gametophytes to natural systems.

9.2 Ecomorphology

As noted above, examination of the many published descriptions of fern gametophytes reveals a diverse set of morphologies, as might be expected of a large group of taxa adapted to a diversity of habitat types and survival strategies. For purposes of exploring survival strategies and evolutionary trends among photosynthetic fern gametophytes, gametophyte form can be grouped into five morphological types within three functional forms (Table 9.1). Our descriptions of these forms are primarily taken from the works of Stokey and Atkinson (1958), Atkinson (1973), Atkinson and Stokey (1964), Nayar and Kaur (1971), and personal

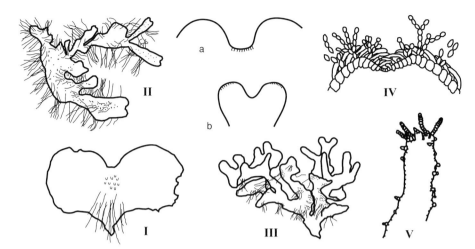

Figure 9.1 Growth forms of five basic types of leptosporangiate fern gametophytes: I, cordiform; II, strap; III, ribbon; IV, gemmiferous strap; V, gemmiferous ribbon. A recessed apical notch meristem (a) is characteristic of types I, II and IV. A discontinuous marginal meristem is characteristic of types III and V. See Table 9.1 for additional description.

observations. Each of these types can be related to the habitat types frequented by the sporophyte generation and to species' dispersal capabilities.

Gametophytes of type 1 morphology are "cordiform", butterfly or heart shaped, and of the functional form we classify as annual (Figure 9.1, I). At maturity these gametophytes consist of a single apical meristem that typically resides in an apical notch. New cells derived from the meristem expand more in width than in length. Gametophytes grow rapidly and do not branch or proliferate over a usual life span of about 1 year or less. Archegonia are produced on a central multilayer cushion behind the apex. Antheridia are produced variously on the same or on different, sexually differentiated thalli. Following production of a single sporophyte, the gametophyte blanches and dies. Annual cordiform gametophytes are typical of most terrestrial fern species in the Polypodiales (e.g., *Dryopteris*, *Pteris*). (Order and family concepts in this chapter follow Smith *et al.*, 2006 and Chapter 16.)

Gametophytes of type II morphology are strap shaped and are one of two types that we classify as perennial. Strap-shaped gametophytes also have their meristem residing in an apical notch, giving the thallus a cordate apex (Figure 9.1, II), but cells derived from the meristem expand in length equally or more than in width. Perennial strap gametophytes differ from annual cordiform gametophytes in growing more slowly, gradually elongating, and, most importantly, proliferating through production of new gametophyte thalli from the original thallus surface or margins. These proliferations arise well behind the growing principal apex through dedifferentiation of mature cells, and thus can be considered as regenerative branching. Gametangia are produced as in annual gametophytes with young proliferations often passing through an initial antheridiate phase. Sporophyte production may result in death of the apex producing it, but due to its proliferations, multiple, sequential sporophytes can be produced by a gametophyte clone that originated from a single spore. Gametophyte clones continue to grow as long as local habitat conditions are supportive. Gametophytes of epiphytic species of Polypodiaceae and Elaphoglossaceae are representative of perennial strap forms. Terrestrial gametophytes of Osmundaceae and most basal leptosporangiate families are also strap shaped and perennial, but usually not as regularly proliferating as those of epiphytic species.

Perennial gametophytes of type III morphology are ribbon-like. They differ from annual cordiform and perennial strap gametophytes in the basic structure of their apical meristems and their method of branching (Figure 9.1, III). Rather than residing in an apical notch, a marginal meristem extends across the leading edge of the thallus. Cell expansion behind the meristem is primarily in length. Branching occurs by cessation of activity at points along the meristem resulting in a discontinuous meristem, with each section of meristem producing an independent axis that can further rebranch. Early and frequent apical branching yields an expanding colonial network of overlapping ribbon thalli, most of which are uniformly one cell thick. Archegonia are produced on multilayered cushions on robust branches in contact with moist substrate. Antheridia are produced variously over the gametophyte thallus, but usually on branches distant from those bearing archegonia, suggesting antheridiogen activity. As with perennial strap gametophytes, multiple, sequential sporophytes can be produced from a single gametophyte clone, which can grow indefinitely. Perennial ribbon gametophytes are typified by basal members of the Hymenophyllaceae (*Hymenophyllum s.s.*) and the vittarioids (*Ananthacorus*), both of whose derived taxa produce gametophytes of type V. A few species (e.g., of *Lomariopsis*) produce either strap or ribbon gametophytes depending on available light and nutrients (Farrar, personal observation). In such species, archegonia and sporophytes are produced only on thalli achieving strap growth.

Gametophytes of type IV and type V morphologies are strap shaped and ribbon-like respectively, perennial, *and* produce gemmae. We classify these in the functional group gemmiferous. Gemmae are vegetative units consisting of small strings or plates of cells that are produced by, and at maturity abscise from, the parent thallus. Gemmae can be dispersed locally by wind, water, animal traffic, and gravity. Upon deposit on a supportive substrate, a gemma develops either into a new gametophyte thallus or, if in the vicinity of already mature gametophytes, it can immediately produce antheridia with little or no intervening vegetative growth (Emigh and Farrar, 1977; Dassler and Farrar, 1997, 2001). Gemmiferous strap gametophytes are produced by many species of grammitids (Polypodiaceae; Stokey and Atkinson, 1958) (Figure 9.1, IV). Gemmiferous ribbon gametophytes are typical of most species of Hymenophyllaceae and the vittarioids, which are otherwise of type III morphology (Figure 9.1, V).

9.2.1 Ecological adaptations of gametophyte types

Fast-growing, annual, cordiform gametophytes appear well adapted to short-lived habitats produced by disturbance on the forest floor by erosion, tip-up mounds, trail and roadsides, etc. For a period of time, gametophytes in these sites would be largely free of competition and free from shading effects of dense forest litter. Annual cordiform gametophytes tend to grow in large populations sufficiently dense for intergametophytic interactions, including responses induced by antheridiogens (e.g., Tryon and Vitale, 1977; Hamilton and Lloyd, 1991). Successful populations would need to produce sporophytes before litter deposition or competing plant growth renders the site unsuitable for their continued growth.

More slowly growing, perennial, strap and ribbon gametophytes are ill equipped for reproduction in the fast lane of temporary habitats. Through their longevity and branching habit, they appear better suited to more or less mature habitats already occupied by competing vegetation, especially in bryophyte mats. Gametophytes of these types can grow intertwined among bryophyte stems, which may allow them to survive until favorable conditions for sporophyte production arise and, through elongation and branching, explore their habitat for more supportive microsites.

A putatively important capability of perennial strap and ribbon gametophytes is to provide increased opportunity for sporophyte production via outcrossing. For species carrying high genetic variability that includes recessive deleterious alleles, outcrossing may be necessary for production of viable sporophytes (see Chapter 4 and references therein). To this end, long-lived gametophytes and gametophyte clones may provide increased opportunity in both space and time for breeding between different genotypes arising from spores that do not arrive in the spatial and temporal proximity that is required by annual gametophytes. This hypothesis becomes especially relevant in considering long-distance migration.

Perennial ribbon gametophytes grow and branch more rapidly than do strap-shaped gametophytes (Farrar, personal observations). Ribbon gametophytes begin branching almost immediately upon germination, and, in their narrow, one-cell-thick thalli, seem to invest fewer resources per linear unit than do strap-shaped gametophytes. This reduction of morphological complexity reaches the extreme in the permanently filamentous gametophytes of Hymenophyllaceae and Schizaeaceae.

Gemma production by both strap-shaped and ribbon-like gametophytes would seem to expand their capacity for habitat exploration. Propagules derived from a single founding spore can be dispersed to supportive microsites throughout the local area, including different branches of different trees, a feat not achievable by perennial but non-gemmiferous gametophytes. Continuous production and dispersal of gemmae may make possible metapopulation dynamics in which newly supportive habitats can be colonized while older sites become unsuitable, thus accommodating individual population extinction.

The indefinite longevity of all perennial gametophytes, with or without gemmae, may be of special significance in colonization of new habitats that are kilometers, or thousands of kilometers, distant from the spore source (e.g., Rumsey et al., 1998). For species requiring mating between different genotypes, absence of a second genotype in close proximity following long-distance dispersal may severely restrict migratory ability. Indefinite persistence may allow gametophyte clones to avoid this impediment by acting in a manner similar to seed and spore banks. Banks of perennial gametophytes would permit interaction among genotypes (spores) arriving at different times, thus greatly enhancing colonization by outbreeding species of sites infrequently receiving spores. It is important in this regard to recognize the dual role of gemmae as both vegetative propagules *and* dispersible male gametophytes that may promote breeding between spatially separated colonies.

Gametophyte banks may also permit species' survival in areas that are unable to support sporophyte growth during unfavorable climatic periods. This is possible because of the gametophyte's ability to grow in smaller and more moderated microsites, and because of their possible greater tolerance of desiccation and/or freezing than their sporophyte counterparts (see Section 9.3). Extreme examples are the independent (non-sporophyte producing) gametophytes of several tropical species in temperate regions of the eastern USA (Farrar, 1998).

9.2.2 *Support for hypotheses of ecomorphological differentiation*

Based on variations in the morphology of the several hundred gametophytes that have been described, and through extrapolation based on published and personal observations of similarities within genera and families, it is possible to find support for the hypotheses generated in the foregoing discussions of gametophyte types. Most terrestrial Polypodiales (derived ferns) produce annual cordiform gametophytes, whereas most epiphytic Polypodiales produce perennial gametophytes, many of which are also gemmiferous. In the broader context of taxa noted for possessing perennial gametophytes, grammitids

(Polypodiaceae), Elaphoglossaceae, vittarioids (Pteridaceae), and Hymenophyllaceae, at least 80% of the species grow in epiphytic habitats (Dassler and Farrar, 1997). Thus the correlation appears high between gametophyte type and the habitats where successful gametophyte growth and sporophyte recruitment occur.

In tropical rainforests where epiphytic species abound, one seldom sees a primarily terrestrial species (e.g., *Thelypteris, Pteris*, etc.) growing as an epiphyte, even as a juvenile sporophyte (Watkins et al., 2006a). This correlation indicates that the reproductive strategies of terrestrial species with annual cordiform gametophytes are ill suited for epiphytic habitats. The opposite observation of primarily epiphytic species occurring on terrestrial habitats is more common, but such occurrences are usually in mature, bryophyte dominated sites more similar to epiphytic habitats than to the disturbed soil occurrences of annual cordiform gametophytes. Further, soil-dwelling species of otherwise epiphytic genera (e.g., *Trichomanes rigidum* and *T. osmundoides*) retain the gametophyte form of their epiphytic relatives (Farrar, personal observation; Dassler and Farrar, 1997).

Comparisons of physiological functions indicate that adaptations in the gametophytes of terrestrial and epiphytic ferns are not limited to growth form. See Watkins *et al.* (2007b) and Section 9.3 for a more detailed discussion of this issue. To test the role of gametophyte form in species' migration, Dassler and Farrar (2001) examined the composition of fern floras of continents versus islands in the same floristic zone. They assumed that significant differences in percentage of species with gametophytes of different types would reflect contributions of gametophyte form to successful long-distance migration. Included in the comparison were 32 floras across five continent–island sets (north Pacific, equatorial Pacific, south Pacific, Caribbean, and Indian Ocean). The average total epiphyte percentage was nearly unchanged between continents and islands. Island floras displayed a significant increase in species with gemmiferous gametophytes and a decrease in species with perennial gametophytes lacking gemmae (Figure 9.2). These observations might indicate that the ability to multiply and disperse locally via gemmae, following initial introduction by spores, significantly enhances the probability of successful migration when spore dispersal is limited. It should be noted that these estimates did not correct for phylogenetic history, thus the ability to produce gemmae may also lead to greater levels of lineage diversification post-migration. It is inappropriate, however, to conclude from these analyses that perennial but non-gemmiferous gametophytes are maladapted to long-distance migration. Neither is it informative to compare across habitat types, but we can speculate that epiphytic species with perennial gametophytes are better adapted for long-distance migration than they would be if they possessed annual gametophytes.

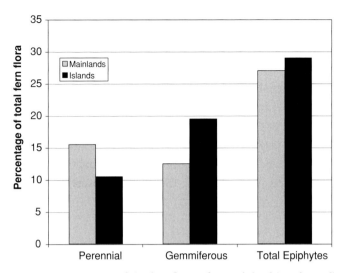

Figure 9.2 Percentage of the fern floras of 12 mainland (continental) regions and 20 islands in five oceanic regions composed of epiphytic species with perennial gametophytes and perennial gemmiferous gametophytes and of total epiphytes. Perennial gametophytes are Polypodiaceae (minus grammitids) and Elaphoglossaceae. Gemmiferous gametophytes are vittarioids, grammitids and Hymenophyllaceae. Means of mainlands and islands are significantly different for species with perennial ($p = 0.0003$) and gemmiferous ($p = 0.0002$) gametophytes as determined by one-tailed t-tests following arcsine transformation of percentages to give the data a normal distribution (see Dassler and Farrar, 2001).

Outbreeding species with annual gametophytes reproduce successfully only if multiple spores arrive at a safe site at nearly the same time and in nearly the same place (e.g., Cousens et al., 1988). Perennial, non-gemmiferous gametophytes may free species from the temporal constraint, but much less so from the spatial constraint, i.e., they are still dependent upon the deposition of several spores in nearly the same place. By providing a mechanism for much greater local dispersion of gametophytes, gemmae may alleviate both temporal and spatial constraints.

When the gametophyte form and the breeding system of a sufficient number of species have been characterized, comparisons such as these may yield even greater insight into the influence exerted by the gametophyte generation on species dispersal. Published descriptive data suggest the existence of some variation in the form of gametophytes even within genera and families. Initially, some species appear to assume the annual gametophyte strategy, then, if not successful in sporophyte production, convert to the strategy of perennial gametophytes. Examination of such variation for correlations with habitat types may be instructive in understanding the evolution and plasticity of growth form.

9.2.3 Evolution of growth form in fern gametophytes

A detailed analysis of gametophyte form versus phylogenetic position is beyond the scope of this chapter, but a brief look at evolution of form is instructive. No fossil forms attributable to early ferns are known, but several unquestionable gametophytic forms have been described from the Devonian Rhynie Chert (Remy and Remy, 1980; Remy et al., 1993) and may be relevant to tracing the evolution of gametophyte habitat selection (Taylor et al., 2005). All of these appear to have been thickened, possibly radially symmetric, thalli. They also had massive antheridia similar to those of eusporangiate ferns. Gametophytes of extant eusporangiate ferns vary greatly, from the underground tuberous forms of the Psilotales (Bierhorst, 1953) and Ophioglossales (Mesler, 1975a, 1975b, 1976), to the multi-lobed green thalli of the Equisetales (Mesler and Lu, 1977), to the massively thickened strap-like gametophytes of the Marattiales. Gametophytes in Osmundales are strap shaped with a very thick midrib (Hiyama et al., 1992), and thus most closely resemble, among eusporangiate ferns, those of Marattiales, although without the embedded antheridia typical of eusporangiate ferns. Gleicheniales (Haufler and Adams, 1982) and Cyatheales also possess perennial strap-like gametophytes with a thick midrib similar to those of Osmundales. Though terrestrial and not always highly proliferous, all of these are more similar to the perennial strap gametophytes of epiphytic Polypodiales than to the annual, cordiform gametophytes of terrestrial Polypodiales.

The Hymenophyllaceae (Dassler and Farrar, 2001; Makgomol and Sheffield, 2001), the only epiphytes among basal leptosporangiate ferns, have perennial ribbon and gemmiferous ribbon gametophytes. Annual cordiform gametophyte morphology is first encountered in the Schizaeales (Twiss, 1910; Nester and Schedlbauer, 1981). These are within an array of gametophyte diversity that includes filamentous (e.g., Kiss and Swatzell, 1996) and subterranean gametophytes (Bierhorst, 1966) and are phylogenetically isolated from the remainder of annual, cordiform gametophytes. Families with annual cordiform gametophytes are common in the Polypodiales. Most of the polypod radiation (Pteridoids and Eupolypods) appears in concert with the late Mesozoic/early Cenozoic development of angiosperm forests, these producing an abundance of both epiphytic and deeply shaded terrestrial habitats (Lovis, 1977; Schneider et al., 2004).

We present here a testable hypothesis of gametophyte evolution. Gametophytes of early diverging clades were probably terrestrial, but relatively large and long lived; the thallus possessed a cordate apex and a thick midrib as postulated by Atkinson and Stokey (1964). These are basically perennial, strap-shaped gametophytes with the potential, if not always the propensity of epiphytic species, for vegetative proliferation. An early transformation to ribbon and to gemmiferous ribbon forms occurred with adoption of the epiphytic habit by members of the Hymenophyllaceae, resulting in gametophytes adapted to grow in the dense, moist forests of the late Paleozoic era. At about the same time, an early and probably independent derivation of annual, cordiform gametophytes occurred in the Schizaeales, but the perennial strap gametophyte form dominated until the origin of basal Polypodiales when annual cordiform gametophytes became common in the terrestrial Pteridaceae, Dennstaedtiaceae, and Eupolypod families. Ribbon and gemmiferous ribbon forms arose again in the vittarioids (Pteridaceae), and gemmiferous strap forms arose in the grammitids (Polypodiaceae), while the remainder of the epiphytic Polypodiaceae retained perennial strap gametophyte morphology, often with a high degree of regenerative branching.

9.3   Ecophysiology

Knowledge of the ecophysiology of ferns is limited, but we do know that the fern sporophyte leaf essentially acts as a typical vascular plant leaf (Nobel, 1978; Hietz and Briones, 1998; Brodribb and Holbrook, 2004), whereas the gametophyte may be more profitably compared to a bryophyte (Hagar and Freeberg, 1980; Ong et al., 1998). Perhaps more than any other vascular plant lineages, ferns and lycophytes exhibit a system dominated by dual ecologies. While both gametophytes and sporophytes face similar physiological needs, they have diverged along radically different biological paths to ensure that these needs are met.

This section reviews past research and presents new data on the ecophysiology of fern gametophytes. It emphasizes the water relations of gametophytes to understand better how organisms with limited internal water storage, no stomata, and little or no cuticle respond to a water-limited environment. The influence of light stress on the carbon gain capacity of gametophytes is also examined to develop a better concept of how species from different naturally occurring light regimes respond to excess light. In Holttum's (1938) seminal work on "The ecology of tropical pteridophytes," he combined intimate knowledge of ferns with careful observation and questioning to draw a big-picture conclusion about the ecology of the group. He included this perspective on the ecology of the gametophyte generation:

> We are accustomed to see and to marvel at the great varied form and adaptation of the sporophytes, which are the ferns as we know them, but indeed there must be nearly as much variety of adaptation among the gametophytes. It is true that if the prothallus of *Platycerium* grew upon the forest floor, the resulting sporophyte, if produced, would find

itself in uncongenial surroundings, and would not develop very far; but it is also true that the *Platycerium* prothallus must be able to develop in relatively exposed position on the tree trunk in which prothalli of many ferns would be unable to exist (Holttum, 1938, pp. 421–422).

One of Holttum's key observations was the recognition of gametophyte-mediated controls on fern recruitment. *Platycerium*, commonly known as staghorn ferns, are Old World epiphytes with diverse ecology; many species grow on highly exposed emergent canopy tree trunks, but they are never found on the forest floor. A more recent study has made similar observations for hundreds of species to suggest that there is extreme habitat fidelity between epiphytic and terrestrial species (Watkins *et al.*, 2006a). This observation, combined with the copious production of wind-dispersed spores, demonstrates that whereas dispersal is important in the distribution of ferns, gametophytes play a critical role in recruitment. Unfortunately, Holttum's gametophytic call to arms was largely ignored and we have progressed but little since the publication of his work.

9.3.1 Gametophyte autecology

Multiple factors clearly act to control the distributions of fern gametophytes and studies have demonstrated the importance of disturbance and edaphic factors (Farrar and Gooch, 1975; Cousens, 1979, 1981; Cousens *et al.*, 1985, 1988; Peck, 1980; Peck *et al.*, 1990; Greer and McCarthy, 1999) and plasticity (Greer and McCarthy, 1999) on the distribution and survival of temperate gametophytes. Data on tropical fern gametophytes suggest that there may be considerable differences among different functional groups. For example, both gametophyte ontogeny and morphology vary tremendously among species of different functional groups (e.g., epiphytic and terrestrial) (Atkinson and Stokey, 1964; Nayar and Kaur, 1971). A common observation is that there are fundamental differences in morphology and potential longevity between the gametophytes of epiphytic and terrestrial species (Dassler and Farrar, 1997, 2001; Watkins *et al.*, 2007a). Thus, there is likely to be considerable variation in gametophyte ecology especially as it relates to different fern functional types.

Fern gametophytes are not simply transient members of their environment. Early studies have produced either anecdotal or laboratory-based data to suggest that if fertilization is prevented, fern gametophytes can be extremely long lived. Longevities of 12 years have been reported in some laboratory-reared gametophytes (Walp, 1951). Watkins *et al.* (2007a) have shown that the gametophytes of tropical ferns, especially those of epiphytic taxa, can live for years in natural

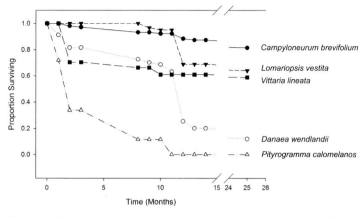

Figure 9.3 Kaplan–Meier survivorship curves of the gametophytes of five species over a 25 month study period. No data were collected during months 4–7 and 15–24. Open symbols indicate terrestrial species and closed symbols indicate epiphytic and hemi-epiphytic species. *Campyloneurum brevifolium* (Lodd. ex Link) Link and *Vittaria lineata* (L.) Sm. are epiphytes that commonly occur in the inner canopy and in high-light understory habitats. *Lomariopsis vestita* E. Fmyn. is an abundant hemi-epiphyte that occurs in the understory of both primary and secondary forests throughout the study site. *Danaea wendlandii* Rchb. f. is a terrestrial species typical of more stable understory habitats of primary forests. *Pityrogramma tartarea* (Cav.) Maxon is an abundant species often found in disturbed sites such as trail and road sides and on the root balls of recently fallen trees.

settings. Provided that fertilization and more importantly, disturbance, is limited, many terrestrial species can also live for years (Figure 9.3).

Increased longevity generally results in increased gametophyte size. Gametophytes of many square centimeters that can produce dozens of sporophytes in space and time are not an uncommon occurrence in tropical forests (Watkins, personal observations). Even in the case of many temperate species, gametophytes have been shown to over-winter (Pickett, 1914; Sakai, 1980; Farrar and Gooch, 1975; Sato and Sakai, 1981). Such observations call for fundamental revision of the traditional view of gametophyte longevity and sporophyte production. Although the view that gametophytes are short lived and produce a single sporophyte may be applicable to many temperate and tropical terrestrial species, this perspective should be reconsidered in the broader context of fern and lycophyte diversity and evolution.

The role of habitat disturbance provides another example of differences between epiphytic and terrestrial gametophyte species. Recent studies by Watkins *et al.* (2007a) have confirmed earlier studies (Peck, 1980; Peck *et al.*, 1990) showing the general dependence of terrestrial gametophytes upon disturbance for the creation of new habitats whereas epiphytic species appear to be negatively impacted by disturbance. Epiphytic gametophytes maintain significantly lower densities in more highly disturbed sites than do terrestrial species (Watkins et al., 2007a). In contrast, terrestrial species rely on disturbance that produces a bare, litter-free habitat for gametophyte establishment and growth (Cousens et al., 1985), yet such sites are inherently unstable and likely to be short lived. Mortality rates of most terrestrial gametophytes are directly linked to microsite instability that frequently results in habitat collapse and wholesale loss of gametophyte populations (Peck et al., 1990; Watkins et al., 2007a). Epiphytic gametophytes produced significantly lower densities in more highly disturbed sites compared to terrestrial species. These observations suggest that the gametophytes of epiphytic and terrestrial life forms have evolved radically different ecological strategies. It is important that these two functional groups be considered separately in attempts to understand species ecology.

9.3.2 Water relations and desiccation tolerance

Even in tropical rainforests, there is considerable variation in vapor pressure deficit throughout the day and especially throughout the year. Thus, gametophytes that live for years are faced with frequent changes in the water status of their environment. Such changes are even more pronounced when one compares epiphytic and terrestrial environments. Epiphytic habitats in both the exposed rainforest canopy and in the more protected understory tend to have higher light levels and dry more quickly than buffered understory terrestrial habitats (unpublished data). The low relative water content and exposure to high light that is typical of canopy habitats can be a deadly combination for plants.

The vast majority of vascular plants, including ferns and lycophytes, cope with periods of drought by avoiding it. Sporophytes do this by having waxy cuticles, closing stomata when low relative water contents are sensed, and producing relatively deep roots to tap additional water sources. Species of some of the most drought prone habitats combine initial avoidance with a more radical form of drought tolerance: desiccation tolerance. A desiccation tolerant (DT) plant is defined as one with the ability to lose all vegetative internal water (Bewley, 1979) and to recover from anhydrobiosis (i.e., the cessation of metabolic activity due to low intracellular water content). An environment that brings a plant to an air-dried anhydrobiotic state is sufficiently lethal to kill all modern agricultural crops and >99% of all vascular land plants (Alpert, 2000; Alpert and Oliver, 2002).

True desiccation tolerance in the vegetative sporophyte stage of ferns is known from and likely exists in relatively few species (Gaff, 1987; Porembski and Barthlott, 2000). In a recent review on the subject, Proctor and Pence (2002) recorded that <1% (64 species) of the fern sporophytes studied exhibited DT and, of those, 40 were cheilanthoid taxa that are commonly associated with desert-like habitats. Much less is known of species from tropical habitats, but DT has been recorded in genera as phylogenetically disparate as *Asplenium* and *Polypodium* (Kappen, 1964; Gaff, 1987; Proctor and Pence, 2002). Thus, in spite of well-known occurrences of "resurrection" ferns of dry habitats, DT remains rare in fern sporophytes.

Desiccation tolerance is the principal mechanism employed by bryophytes to cope with life in a water limited environment. Fern gametophytes lack the morphological machinery to utilize the well-developed avoidance mechanisms of the sporophyte and are consequently left with two options: survive only in perpetually wet habitats *or* evolve effective desiccation tolerance. Given the similarity in structure of fern and bryophyte gametophytes a logical starting hypothesis is that fern gametophytes also employ DT. Ferns are certainly not limited to life in mesic environments and have radiated into the same drought prone habitats as seed plants.

Although anecdotal observations of gametophytes surviving desiccation *in situ* were published as early as 1904 (Campbell, 1905), the first experimental evidence was generated by an elegant series of studies by Pickett (1913, 1914, 1931) who, through a series of desiccation experiments, was the first to show clearly that the gametophytes of *Asplenium rhizophyllum* and *A. platyneuron* could recover growth following extreme desiccation. He also noted a greater degree of desiccation tolerance in sporophytes of *A. rhizophyllum*, a species of more exposed and drier epipetric habitats, relative to *A. platyneuron*, which is generally confined to more mesic terrestrial sites. This was the first documented link of desiccation tolerance in the gametophyte generation to sporophyte distributions and species ecology. Ong and Ng (1998) produced the first evidence of DT in gametophytes in a study of the tropical epiphyte *Pyrrosia piloselloides*.

Watkins (2006) and Watkins *et al.* (2007b) examined the degree of tolerance in the gametophytes of a number of tropical species. This work examined the gametophytes of three tropical species (*Nephrolepis biserrata*, *Phlebodium pseudoaureum*, and *Microgramma reptans*) from epiphytic habitats and nine species (*Adiantum latifolium*, *Cyclopeltis semicordata*, *Dennstaedtia bipinnata*, *Diplazium subsilvaticum*, *Pityrogramma tartarea*, *Pteris altissima*, *Thelypteris balbisii*, *Thelypteris curta*, and *Thelypteris nicaraguensis*) from terrestrial habitats. All species were collected in Costa Rica and were chosen to represent degrees of natural drought exposure. An initial survey of these species revealed a surprising degree of desiccation tolerance across all species. For example, after gametophytes were exposed to and rehydrated from a vapor pressure deficit (VPD) of 1.3 kPa (extremely dry air for tropical wet forest environments) for a period of 24 hours, all 12 species exhibited greater than 50% recovery of the pre-treatment chlorophyll fluorescence (F_v/F_m) values and the majority had recovered more than 70% of this value with only slight differentiation among species from different habitats. Chlorophyll fluorescence is widely used as a proxy for photosynthetic ability and general physiological health of green tissue.

To develop perspectives on the dynamics of the desiccation response, Watkins (2006) and Watkins et al. (2007b) also examined the influence of different desiccation intensities and multiple desiccation cycles on gametophyte recovery ability. The influence of desiccation intensity (i.e., increasingly dry air) on physiological recovery, was studied by exposing species to three different desiccation intensities: (1) the typical daily VPD in nature (0.53 kPa); (2) the VPD representative of a typical drought event (1.32 kPa); and (3) an extremely low VPD that species in this site rarely if ever experience (2.12 kPa). Recovery of physiological function following rehydration of the gametophytes was measured using the modified chlorophyll fluorescence method of Bjorkman and Demmig (1987). The results from this experiment (a subset of which are presented in Figure 9.4) demonstrated remarkable tolerance to desiccation intensity that is tightly linked to species ecology. *Diplazium subsilvaticum*, a low-light mesophyte, had little tolerance of desiccation intensities imposed by a relative humidity of approximately 50%. On the other hand, the epiphyte, *Microgramma reptans*, which often occurs on the distal twigs of emergent canopy trees, exhibited essentially no sensitivity to desiccation of this intensity (Figure 9.4a). A similar pattern was observed when these same species were subjected to multiple desiccation cycles (i.e., one, two, or three dehydration/rehydration cycles). The more mesic *Diplazium subsilvaticum* was desiccation sensitive to one cycle and showed little recovery after two and three cycles, whereas the more xeric *Microgramma reptans* exhibited strong recovery even after three desiccation cycles (Figure 9.4b).

Another critical finding of these experiments was the considerable variation in drying rates among the different species. Given the relative simplicity of gametophyte anatomy and morphology, such variation was surprising. Is this indicative of some ability of the gametophyte to control water loss? If so, what are the mechanisms behind such abilities? When Watkins (2006) and Watkins et al. (2007b) compared gametophyte drying rate to a number of morphological parameters, he found this rate was linked to gross gametophyte morphology. Gametophytes with more complex three-dimensional morphologies dried more slowly than simple planar gametophytes.

Such observations may be important in understanding the radiation of ferns from purported mesic understory species to modern xeric species of exposed epiphytic habitats. For example, the derived Eupolypod clade is composed of two lineages: one that is ~90% epiphytic and the other that is ~90% terrestrial.

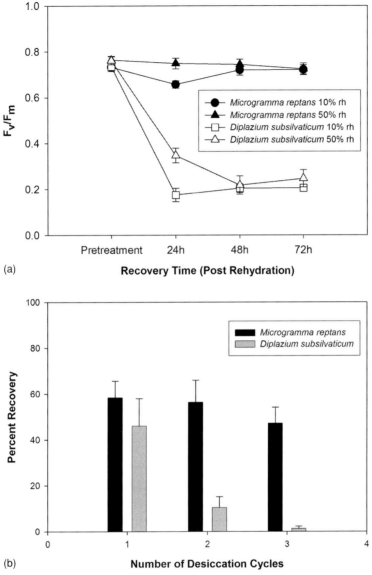

Figure 9.4 (a) F_v/F_m recovery graphs for gametophytes of two species exposed to two different desiccation intensities. Gametophytes of *Diplazium subsilvaticum* typically occur in the understory, whereas those of *Microgramma reptans* occur in exposed canopy. Gametophytes were kept at vapor pressure deficits equivalent to 10% or 50% relative humidity for 48 hours then rehydrated. Measurements of chlorophyll fluorescence (F_v/F_m) were taken at 24, 48, and 72 hours post rehydration.
(b) Chlorophyll fluorescence (F_v/F_m) recovery results for gametophytes exposed to 1, 2, or 3 desiccation cycles at a vapor pressure deficit equivalent to ~10% relative humidity. In a cycle, gametophytes were desiccated for 48 hours then rehydrated and measured at 72 hours post rehydration.

Gametophyte morphology generally tracks this split. As discussed elsewhere in this chapter, most epiphytic species exhibit morphologies more complex than the simple heart-shaped thallus typical of terrestrial species. Gametophyte colonies of epiphytic species also tend to develop an imbricated, three-dimensional structure more typical of bryophytes. If complex morphology plays the role of slowing (or at least controlling) water loss, could variation in morphology have been important in the radiation of ferns into drier habitats?

Whereas morphology could act to slow desiccation, there are also anatomical specializations that can both retard the rate of desiccation and protect desiccating cells. Extreme drying can often result in cell wall collapse, and recovery from such collapse is thought to be difficult. Thus, it has been hypothesized that cells with thicker walls will lose water more slowly and can resist collapse more effectively than cells with thinner cell walls. Indeed, some xerophytic species tend to have thicker cell walls when compared to mesophytic species (Oertli et al., 1990; Oliver, 1996; Moore et al., 2006). Whereas direct measurements of cell wall thickness have not been made across a diversity of fern gametophytes, indirect evidence indicates that epiphytic gametophytes tend to have thicker cell walls than terrestrial species. When mounted in an initially desiccating medium (i.e., Hoyer's), gametophytes of most epiphytic species maintain their rigid structure whereas those of terrestrial species usually become limp and nearly invisible (Farrar, personal observations). Such observations suggest that there are also significant differences at anatomical levels between the gametophytes of epiphytic and terrestrial species.

9.3.3 Gametophyte carbon and light stress relations

Fern gametophytes have intrinsic limitations for taking up and storing water, therefore, water relations may be an important area of gametophyte physiological ecology. As discussed above, and as will probably emerge from future studies, fern gametophytes are true poikilohydric plants. Poikilohydric taxa have evolved complicated morphological, biochemical, and anatomical mechanisms to cope with limited water and are robust in tolerating periods of drought. These plants face the additional complication of stress derived from exposure to excess light when their vegetative tissues experience reduced water contents. Light in excess of that which the photosynthetic machinery can transfer can result in photoinhibition (the light mediated depression of photosynthesis) and eventually significant photodamage (wholesale destruction of photosystems). Such depression can severely limit plant fitness. Low water content combined with excess light can be a deadly mix. Thus, even the most advanced poikilohydric species is faced with special challenges when presented with such combinations.

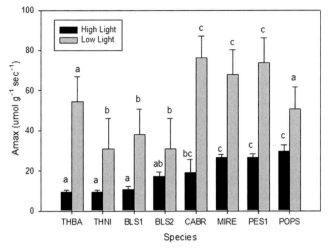

Figure 9.5 Maximum photosynthetic rates of terrestrial and epiphytic fern gametophytes from laboratory cultures in high (500 µmol m^{-2} s^{-1}) and low (100 µmol m^{-2} s^{-1}) light conditions. Maximum photosynthetic rates were measured using a LICOR 6400 infrared gas analyzer on three groups of 10–20 gametophytes. Error bars represent standard errors; differences among species at the same light levels were calculated by *post hoc* Tukey tests ($p < 0.000$ for high light; $p = 0.0015$ for low light); species sharing the same letter *within* epiphytes or terrestrial species at the same light level are not significantly different. THBA, *Thelypteris balbisii* (Spreng.) Ching; THNI, *Thelypteris nicaraguensis* (E. Fourn.); BLS1, *Blechnum* species 1; BLS2, *Blechnum* species 2; CABR, *Campyloneurum brevifolium* (Lodd. ex Link) Link; MIRE, *Microgramma reptans* (Cav.) A. R. Sm.; PES1, *Pecluma* species 1, POPS, *Polypodium pseudoaureum* Cav. L.

The fern sporophyte contains numerous morphological (Watkins *et al.*, 2006b) and biochemical mechanisms (Tausz *et al.*, 2001) to meet the challenge of excess light. Watkins *et al.* (2006b) showed that epiphytic species exhibited less photoinhibition at supersaturating light intensities and higher photosynthesis rates at subsaturating light intensities than the terrestrial species (Figure 9.5). These results indicate that there are distinct physiological/biochemical differences between the two groups and provide evidence for gametophytic adaptation to high light.

Because of limited research on gametophyte physiological ecology, discussions here have focused on comparisons of epiphytic and terrestrial taxa. However, the differences demonstrated between these two functional groups substantiate the need to consider gametophyte ecology on a species and habitat basis in order to understand the ecology of individual species. These studies also yield the promise of deeper insight into species ecology when conducted on finer taxon and ecological scales.

Ferns are assumed to have radiated from a terrestrial understory ancestor. As ferns diversified beyond this environment they would have needed several innovations to cope with life in more drought prone and higher light habitats. These innovations would have involved drought and desiccation and light stress tolerance and would have also been necessary in the gametophyte generation. The discussion above demonstrates that variation within such characters exists in modern species and likely plays a significant role in species evolution.

Are there advantages to maintaining an independent gametophyte? Fern gametophytes have historically been viewed as short-lived mesophytes that are limited to wet environments. Whereas this is the case for many species, there is a significant number of species that deviate from such characteristics. Tremendous variation exists in sporophyte ecology. Should we expect less in the gametophyte stage? Gametophytes can be tough, long-lived individuals (and clones) that produce multiple sporophytes over space and time, grow in areas that are uninhabitable to sporophytes (Farrar, 1967; Peck, 1980; Peck et al., 1990), and exhibit a greater degree of stress tolerance than sporophytes (Sakai, 1980; Sato and Sakai, 1981; Watkins, 2006). An underappreciated role of the gametophyte generation may be its exploratory role in continually testing habitat suitability for sporophytes. Another critical aspect may be the gametophyte's ability to survive harsh times when sporophytes, and thus the species, might otherwise perish from a given area (Farrar, 1985, 1998).

9.4 Ecovalidation

Most of what we know about fern gametophyte development, morphology, physiology, and sexuality is based primarily on laboratory grown populations. Few studies, however, have tested the assumption that laboratory populations mirror natural conditions. Consequently the possibility that laboratory-derived data imperfectly reflect natural systems persists (e.g., Ranker and Houston, 2002). This section reviews relevant literature, then presents new data to address this critical question and to discuss whether differences between laboratory and field systems can be explained. We also explore whether laboratory conditions can be modified to reflect the conditions in nature more accurately.

9.4.1 Morphological comparison of laboratory and field grown gametophytes

By comparing field and laboratory grown gametophytes of the same species, we can consider which morphological characters remain constant across growth conditions. Such characters are presumably genetically determined, they form the basis for field identification of gametophyte taxa, and may be used for phylogenetic analyses. If culture conditions are altered, it may be possible to determine the causes of differences we observe in field grown gametophytes, especially with regard to sexual development and sex ratios. Results of these studies are relevant to determination of breeding systems.

The work of Stokey (1951), Atkinson (1973), Nayar and Kaur (1971), and others identified developmental and morphological characters that remained constant within genera and higher taxa. These characters, along with juvenile sporophytes, enabled later workers (e.g., Peck et al., 1990) to identify the species that produce natural populations of fern gametophytes. However, no one has determined the constancy/non-constancy of morphological and developmental characteristics between field and laboratory grown gametophytes of any species.

In our studies (Skelton, 2007), in addition to taxon specific gametophyte characters, identification of most fern gametophyte populations in the field was aided because young sporophytes at various ages were found associated with gametophytes and could be traced through development to stages old enough to exhibit mature frond characters. Comparisons between laboratory and field gametophytes of 15 species from a range of eusporangiate and leptosporangiate fern families showed far more similarities than differences (Table 9.2). Basic growth form was conserved, as well as structure and position of rhizoids, hairs, and gametangia. Field gametophytes of epiphytic species tended to be more elongated and highly branched, possibly due to bryophyte competition that was present in field populations but absent from laboratory cultures. Field plants may also have been older than the cultured plants. The most frequent difference was the common production by laboratory gametophytes of some archegonia on the dorsal side of the thallus, instead of exclusively on the ventral side as in field samples.

Other differences included density and intensity of color of rhizoids and hairs in some taxa, but other than the degree of rhizoid branching these were quantitative differences and not consistently characteristic of field plants. Alteration of densities of hairs and rhizoids could have resulted from inadvertent removal in the process of cleaning field gametophytes. Hairs could also have been removed in the field by invertebrate activity or other disturbances.

In summary, the basic morphological structure of fern gametophytes appears to be highly conservative across growth environments. There is little suggestion of high plasticity in development or in mature morphology that would preclude use of morphological characteristics for gametophyte identification in the field or in use of these characters in phylogenetic classification, especially at the level of genus.

Table 9.2 Comparison of morphological characters in laboratory and field grown gametophytes of the same species

Blank entries indicate characters that were consistent between laboratory and field samples, others indicate differences. A. Some cultured gametophytes produced elongate projections from the apical meristem (percent observed ~33%). B. Field gametophytes were more elongate, strap-like, and branched (percent observed ~82%). C. Some field gametophytes were more elongate (percent observed ~57%). D. Field gametophytes were more strap-like (percent observed ~66%). E. Rhizoids were darker in laboratory gametophytes (percent observed ~85%). F. Rhizoids were darker in field gametophytes (percent observed ~80%). G. Some field gametophytes had branched rhizoids (percent observed ~66%). H. Hairs were darker in field gametophytes (percent observed ~80%). I. Fewer hairs in field gametophytes (percent observed ~90%). J. Fewer hairs in cultured gametophytes (percent observed ~90%). K. Antheridia on the "wings" of field gametophytes (percent observed ~16%); at base of thallus among rhizoids in the cultured gametophytes. L. Archegonia on dorsal surface of cultured gametophytes (percent observed ranged from ~25% to 50%).

Species	Field gametophytes total / number of populations	Laboratory gametophytes total / number of populations	Gametophyte outline	Rhizoid structure and color	Rhizoid position	Presence/ absence of hairs	Hair structure and color	Abundance of marginal hairs	Abundance of suficial hairs	Gametangial structure	Antheridia position	Archegonia position
Angiopteris lygodiifolia	23/1	4/1										
Dicranopteris linearis	87/1	38/1										
Adiantum latifolium	101/1	82/2										L
Adiantum petiolatum	82/2	74/1		E								L
Cyathea podophylla	15/1	39/2										
Histiopteris incisa	166/2	59/4		F								
Dictyocline griffithii var. wilfordii	106/3	102/5									K	L
Diplazium cristatum	35/1	22/1		G								
Dryopteris subexaltata	81/1	27/1	A									L
Elaphoglossum peltatum	38/1	11/2	B									
Tectaria subebeana	96/1	27/1							I			L
Asplenium antiquum	88/1	53/2	C									
Blechnum orientale	88/1	76/2					H	I				
Blechnum gracile	41/1	63/2										L
Campyloneuron brevifolia	92/3	42/1	D					J				

9.4.2 Culture conditions and sexual development

Ranker and Houston (2002) reported that laboratory cultured gametophyte populations of *Sadleria* spp. exhibited significantly different sex ratios than those observed in natural gametophyte populations, with the wild populations containing a significantly higher proportion of male gametophytes. Because ratios of male to female gametophytes influence breeding systems, these results suggest that laboratory studies may yield misleading predictions of mating systems in nature. A number of differences exist between the growth conditions of laboratory and field that could influence sexual development, including the differential impacts of mineral nutrient agar versus natural soil as a culture substrate. A less obvious difference is the process of spore introduction onto the substrate. Skelton (2007) hypothesized that discrepancies in sex ratios between laboratory and field populations could result from a greater duration of spore deposit and germination in natural populations relative to laboratory cultures. Ranker and Houston (2002) also commented on the possible impact of mixed-age populations on sex ratios.

In laboratory cultures, all spores are typically sown onto a uniform surface of agar-solidified media in a single sowing event. These spores germinate at approximately the same time and, because the medium surface is homogeneous, all gametophytes have approximately the same initial growth rate. In natural populations spores are released from source plants and even individual fronds over an extended a period of time (Peck *et al.*, 1990; Farrar, 1976), consequently spores contributing to a natural population can produce uneven aged populations. This may be especially true in tropical systems where sporophytes of many species release spores nearly continuously over periods of months. The resulting populations can contain a mixture of gametophytes of different ages, the younger subject to influences from the older. Further contributing to heterogeneity in natural populations is the heterogeneity of natural substrates, both physically and nutritionally, which may differentially influence growth rates and development.

Many species of homosporous ferns have an antheridiogen system that promotes the production of sporophytes via outcrossing (see Chapter 5). Late arrival of spores, as well as delayed growth due to substrate heterogeneity, could result in an increase in male gametophytes in natural populations due to the effects of antheridiogen being produced by older, established gametophytes. To test this hypothesis we performed two sequential spore sowings of *Pteridium aquilinum* on mineral nutrient agar cultures, then examined sex ratios in the cultured populations. Results showed a nearly two-fold increase in the percentage of male gametophytes in mixed age populations when the second sowing followed the first by 6 days, compared to single-day sowings (Figure 9.6). In experiments where

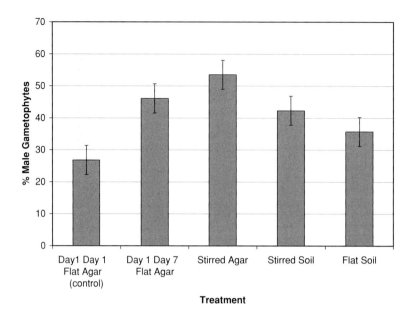

Figure 9.6 Effect of various treatments on the percentage of male gametophytes in laboratory cultures of *Pteridium aquilinum*. Day 1 Day 7 indicates that a second sowing of spores was made in the same culture 6 days after the first sowing. Approximately the same total number of spores was sown in each replicate of each treatment. Results for the flat agar treatments are an average of six replicates for each treatment; the number per replicate is 24 to 30 individuals. Results for the stirred agar and soil treatments are an average of three replicates per treatment; the number per replicate is 20 individuals. Bars are means ±1 standard error.

the first and second sowings were in adjacent strips (so that they could be analyzed separately), male gametophyte percentages in the second sowing reached 88%, indicating that the increase in percentage of males in the mixed-age populations was due primarily to effects on younger gametophytes derived from the second sowing.

To determine whether a similar effect could result from uneven developmental stages induced by a heterogeneous substrate surface, we compared even-aged populations developed on flat and uneven agar surfaces, and compared these to development on flat and uneven (stirred) soil surfaces in Petri dish cultures. Uneven agar surfaces produced an even higher increase in male gametophytes than did sequential sowings (Figure 9.6). Likewise uneven soil surfaces produced an elevated male-to-female ratio, though not as great as that observed on uneven agar. However, flat soil cultures contained significantly higher male-to-female ratios than did flat agar cultures. Thus it seems probable that uneven-aged populations resulting from either heterogeneous substrates or from sequential spore introduction, both of which might be expected in nature, could produce elevated male-to-female ratios relative to even-aged laboratory cultures.

9.4.3 *Effects of light, gravity and humidity on archegonial development*

The reduced effect of surface heterogeneity in soil cultures versus agar cultures, the increase in percentage of males in flat soil versus flat agar cultures, and our observations of the anomalous position of archegonia on the dorsal surface of gametophytes in agar cultures suggest that there may be additional differences between natural and laboratory cultured gametophyte populations.

Humidity, light, and gravity orientation are environmental parameters that can also differ sharply in nature from the environment of standard laboratory cultures. Wild gametophytes grow in diverse orientations with respect to gravity, and to some extent with respect to light. An additional significant consideration may be the degree of difference in light received between the two gametophyte surfaces. Gametophytes growing on a dark, non-reflective surface, as is the usual case in nature, experience a much sharper light-intensity gradient from one surface to the other than do gametophytes growing on a translucent agar substrate.

To investigate the effects of light and gravity orientation, gametophytes of *Pteridium aquilinum* were grown on an agar substrate in Petri dishes positioned above and below a fixed light source and in both normal (right-side-up) and inverted (upside-down) orientations with respect to gravity. Gametophytes in all orientations continued to produce some archegonia on their morphologically dorsal surface (air side), but significantly fewer when that surface was oriented toward the light source (Table 9.3). Regardless of orientation, the gametophyte surface facing away from the light source produced the greater total number of archegonia, and, on dorsal surfaces facing away from the light source, more archegonia were produced on that surface whether in normal or inverted orientation with respect to gravity. Gametophytes cultured on soil in normal orientation with respect to light and gravity produced very few archegonia on the dorsal surface, mimicking those observed in wild soil-grown gametophytes. Production of archegonia on the dorsal surface in agar grown cultures likely results from a reduced differential in light incidence due to the translucence of the medium, allowing much greater light incidence on the ventral surface of the gametophyte than occurs in field grown gametophytes.

Gravitational orientation did not exert an overriding effect on archegonial position although gametophytes receiving normal orientation of gravity and light produced slightly fewer archegonia on the dorsal surface and fewer archegonia overall than did cultures with reversed gravity and normal light orientation. Because gametophytes in nature grow on substrates of all aspects, the anomalous minimal effects of gravitational orientation are not surprising.

Table 9.3 *Effect of light orientation and gravity on the position and number of archegonia per gametophyte of Pteridium aquilinum*

Air side, morphologically dorsal surface. ASU-B, air side up, below light; ASD-B, air side down, above light; ASU-A, air side up, above light; ASD-B, air side down, below light. Results are an average of three replicates; the number per replicate ranged from 11 to 40 individuals (20 individuals per plate were sampled where possible, but plates with an atypical gravity/light orientation produced fewer archegoniate gametophytes). Mean values for % of archegonia on the dorsal surface that do not share a letter in common are significantly different at $P < 0.05$. Differences in total number of archegonia were not significant among treatments.

Treatment	Average number of archegonia on air side	Average number of archegonia on substrate side	% of total archegonia on air side	% gametophytes with archegonia only on air side	% gametophytes with archegonia only on substrate side	Total number of archegonia
Light normal, gravity normal (ASU-B)	2.98	18.95	13.6 (a)	0	19.2	21.93
Light normal, gravity reversed (ASD-A)	6.58	18.08	26.7 (b)	0	3.1	24.66
Light reversed, gravity normal (ASU-A)	13.79	5.33	72.1 (c)	38.5	0	19.13
Light reversed, gravity reversed (ASD-B)	17.31	6.46	72.8 (c)	42.6	4.2	23.77

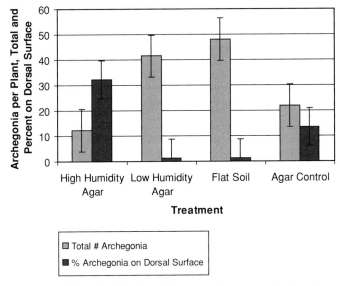

Figure 9.7 Effect of relative humidity on archegonia production in gametophytes of *Pteridium aquilinum*. High humidity, 100%; low humidity, approximately 90%. Results are averages of three replicates for each treatment; the number per replicate is 20 individuals. Bars are means ±1 standard error.

Experiments indicated that gametophytes in high humidity cultures produce a large percentage of archegonia (48.7%) on the dorsal side of the thallus, whereas the low humidity cultures produced very few archegonia (1.4%) on the dorsal surface (Figure 9.7). Perhaps of greater significance, the total numbers of archegonia produced increased nearly four-fold in low relative humidity cultures, compared to high humidity cultures, and nearly two-fold relative to typical flat agar cultures (grown in small culture boxes not within terraria). The total number of archegonia produced by low humidity cultures mimicked the results of cultures grown on soil. Since it was necessary to add water to the soil cultures periodically to prevent them from dying (whereas it was not necessary to add water to agar cultures) it may be that soil cultures experienced lower relative humidities at least part of the time.

A unifying theme may explain both the anomalous production of archegonia on the dorsal surface and the lower total numbers of archegonia produced in standard agar cultures. Both appear to be induced by reduced light and humidity differentials across the gametophyte's two surfaces relative to that experienced by gametophytes in nature. Though relatively simple anatomically and morphologically, fern gametophytes do have complex differentiation patterns including establishment of polarities in development (Raghavan, 1989). It may be that the strong differentials of light and moisture between dorsal and ventral surfaces experienced by gametophytes in natural habitats provide environmental signals critical to establishment of normal polarity. Incomplete establishment of this polarity in agar cultures may disrupt normal differentiation such that archegonia are produced in abnormal positions and overall in smaller numbers.

Additional experiments may elucidate further the environmental signals and physiological responses involved, but it is clear from these experiments that gametophytes grown in standard agar cultures in Petri dishes may yield seriously misleading data relative to sexual development of natural populations of fern gametophytes. Analysis of breeding systems of fern species based on laboratory data should take these discrepancies into consideration.

9.5 Summary

Every fern and lycophyte population owes its beginnings to the gametophyte generation and the simple gametophyte plant contains the same genetic information that produces the highly complex sporophyte plant. We should not be surprised to find that sophisticated differentiation of form and physiology also exists among gametophytes of different taxa. Studies described in this chapter have made it clear that combinations of growth rate, mature form, sexual differentiation, longevity, and differential physiological interactions with environmental parameters lead to functional gametophyte strategies that initiate the sporophyte phase of each species in its preferred habitat. Without knowledge of these gametophyte-level strategies we cannot comprehend habitat compartmentalization within fern and lycophyte communities or migration, distribution, and evolution of species.

The simplistic "one size fits all" depiction of the cordiform, bisexual fern gametophyte in standard textbooks does a great disservice not only to ferns and lycophytes, but to the disciplines of ecology, genetics, and evolution as well. The thoughtful student is left to conclude, incorrectly, that ferns and lycophytes do not obey principles of variation and selection and can be set aside as inexplicable curiosities in the broader context of plant evolution.

Data derived from laboratory cultures provide important starting points for ecological studies but must not be surrogate to field investigations. Laboratory cultures do not provide information on ecological safe sites, on interaction with competing vegetation, or on conditions necessary for successful sporophyte recruitment. Data from culture studies may, in fact, yield misleading data. Absence of field data on recruitment leaves the haunting prospect, as expressed by Peck (1980) in his field studies, that gametophytes we study in the field could be nothing more than "reproductive noise." Determining which gametophytic ecological and reproductive strategies result in sporophytes capable of completing the life cycle remains a challenge that must be accepted.

Both short-term studies of habitat characterization and population demographics, including sexual structure, and long-term studies on sporophyte recruitment and population dynamics of perennial gametophytes need to be conducted on a broad scale. Functional classes of ecological strategies will likely be redefined from such studies. However, we must be cautious in extrapolation. We must not assume a priori that the gametophyte ecology of a given species is representative of another in the same genus or habitat any more so than is the ecology of the sporophyte generation. These studies will challenge the next generation of fern and lycophyte biologists, but they are sure to increase immensely our insight into the occurrence, survival, and evolution of species.

References

Alpert, P. (2000). The discovery, scope, and puzzle of desiccation tolerance in plants. *Plant Ecology*, **151**, 5–17.

Alpert, P. and Oliver, M. J. (2002). Drying without dying. In *Desiccation and Survival in Plants*, ed. M. Black and H. W. Pritchard. Wallingford: CAB International, pp. 3–43.

Andersson, I. (1923). The genetics of variegation in a fern. *Journal of Genetics*, **13**, 1–11.

Atkinson, L. R. (1973). The gametophyte and family relationships. In *The Phylogeny and Classification of Ferns*, ed. A. C. Jermy, J. A. Crabbe, and B. A. Thomas. New York: Academic Press, pp. 73–90.

Atkinson, L. R. and Stokey, A. G. (1964). Comparative morphology of the gametophyte of homosporous ferns. *Phytomorphology*, **14**, 51–70.

Bewley, J. D. (1979). Physiological aspects of desiccation tolerance. *Annual Review of Plant Physiology and Plant Molecular Biology*, **30**, 195–238.

Bierhorst, D. W. (1953). Structure and development of the gametophyte of *Psilotum nudum*. *American Journal of Botany*, **40**, 649–658.

Bierhorst, D. W. (1966). Fleshy cylindrical subterranean gametophyte of *Schizaea melanesica*. *American Journal of Botany*, **53**, 123–133.

Bjorkman, O. and Demmig, B. (1987). Photon yield of O_2 evolution and chlorophyll fluorescence characteristics at 77-K among vascular plants of diverse origins. *Planta*, **170**, 489–504.

Bower, F. O. (1923). *The Ferns*, Vol. I. Cambridge: Cambridge University Press.

Brodribb, T. J. and Holbrook, N. M. (2004). Stomatal protection against hydraulic failure: a comparison of coexisting ferns and angiosperms. *New Phytologist*, **162**, 663–670.

Campbell, D. H. (1905). *The Structure and Development of the Mosses and Ferns*. New York: Macmillan.

Chiou, W.-L. and Farrar, D. R. (2002). The mating systems of some epiphytic Polypodiaceae. *American Fern Journal*, **92**, 65–79.

Chiou, W. L., Farrar, D. R., and Ranker, T. A. (1998). Reproductive biology of some species of *Elaphoglossum*. *Canadian Journal of Botany*, **76**, 1967–1977.

Chiou, W.-L., Farrar, D. R., and Ranker, T. A. (1999). Gametophyte growth and sexual reproduction of some epiphytic ferns. In *Ching Memorial Volume – A Collection of Pteridological Papers Published to Commemorate the Centenary of the Birth of Professor Ren-Change Ching*, ed. X.-C. Zang and K.-H. Shing. Beijing: China Forestry Publishing House, pp. 303–315.

Cousens, M. I. (1979). Gametophyte ontogeny, sex expression, and genetic load as measures of population divergence in *Blechnum spicant*. *American Journal of Botany*, **66**, 116–132.

Cousens, M. I. (1981). *Blechnum spicant*: habitat and vigor of optimal, marginal, and disjunct populations, and field observations of gametophytes. *Botanical Gazette*, **142**, 251–258.

Cousens, M. I. (1988). Reproductive strategies of pteridophytes. In *Plant Reproductive Ecology: Patterns and Strategies*, ed. J. Lovett-Doust and L. Lovett-Doust. Oxford: Oxford University Press, pp. 307–328.

Cousens, M. I., Lacey, D. G., and Kelly, E. M. (1985). Life-history studies of ferns – a consideration of perspective. *Proceedings of The Royal Society of Edinburgh Section B, Biological Sciences*, **86**, 371–380.

Cousens, M. I., Lacey, D. G., and Scheller, J. M. (1988). Safe sites and the ecological life history of *Lorinseria areolata*. *American Journal of Botany*, **75**, 797–807.

Dassler, C. L. and Farrar, D. R. (1997). Significance of form in fern gametophytes: clonal, gemmiferous gametophytes of *Callistopteris baueriana* (Hymenophyllaceae). *International Journal of Plant Sciences*, **158**, 622–639.

Dassler, C. L. and Farrar, D. R. (2001). Significance of gametophyte form in long-distance colonization by tropical, epiphytic ferns. *Brittonia*, **53**, 352–369.

Döpp, W. (1962). Eine die Antheridienbildung bei Farnen fördernde Substanz in den Prothallien von *Pteridium aquilinum* (L.) Kuhn. *Berichte der Deutschen Botanischen Gesellschaft*, **63**, 139–147.

Emigh, V. D. and Farrar, D. R. (1977). Gemmae: a role in sexual reproduction in the fern genus *Vittaria*. *Science*, **198**, 297–298.

Farrar, D. R. (1967). Gametophytes of four tropical fern genera reproducing independently of their sporophytes in the southern Appalachians. *Science*, **155**, 1266–1267.

Farrar, D. R. (1976). Spore retention and release from overwintering fern fronds. *American Fern Journal*, **66**, 49–52.

Farrar, D. R. (1985). Independent fern gametophytes in the wild. *Proceedings of the Royal Society of Edinburgh*, **86**, 361–369.

Farrar, D. R. (1998). The tropical flora of rockhouse cliff formations in the eastern United States. *Journal of the Torrey Botanical Club*, **125**, 91–108.

Farrar, D. R. and Gooch, R. D. (1975). Fern reproduction in Woodman Hollow, central Iowa: preliminary observations and a consideration of the feasibility of conducting studies on fern reproductive biology in nature. *Proceedings of the Iowa Academy of Science*, **82**, 119–122.

Gaff, D. F. (1987). Desiccation tolerant plants in South America. *Oecologia*, **74**, 133–136.

Goebel, K. (1905). *Organography of Plants*. Oxford: Clarendon Press.

Greer, G. K. and McCarthy, B. C. (1999). Gametophytic plasticity among four species of ferns with contrasting ecological distributions. *International Journal of Plant Sciences*, **160**, 879–886.

Hagar, W. G. and Freeberg, J. A. (1980). Photosynthetic rates of sporophytes and gametophytes of the fern, *Todea barbara*. *Plant Physiology*, **65**, 584–586.

Hamilton, R. G., Lloyd, R. M. (1991). Antheridiogen in the wild. The development of fern gametophyte communities. *Functional Ecology*, **6**, 804–809.

Haufler, C. H. and Adams, W. W. (1982). Early gametophyte ontogeny of *Gleichenia bifida* (Willd.) Spreng. – phylogenetic and ecological implications. *American Journal of Botany*, **69**, 1560–1565.

Hietz, P. and Briones, O. (1998). Correlation between water relations and within-canopy distribution of epiphytic ferns in a Mexican cloud forest. *Oecologia*, **114**, 305–316.

Hiyama, T., Imaichi, R., and Kato, M. (1992). Comparative development of gametophytes of *Osmunda lancea* and *Osmunda japonica* (Osmundaceae) – adaptation of rheophilous fern gametophyte. *Botanical Magazine (Tokyo)*, **105**, 215–225.

Hofmeister, W. (1862). *On the Germination, Development, and Fructification of Higher Cryptogamia*. London: Ray Society.

Holttum, R. E. (1938). The ecology of tropical pteridophytes. In *Manual of Pteridology*, ed. F. Verdoorn. The Hague: Martinus Nijhoff, pp. 420–450.

Kappen, L. (1964). Untersuchungen uber die Jahreslauf der Frost-, Hitze-und Austrocknungsresistenz vin sporophyten einheimischer Polypodiaceen. *Flora*, **155**, 123–166.

Kiss, J. Z. and Swatzell, L. J. (1996). Development of the gametophyte of the fern *Schizeae pusilla*. *Journal of Microscopy, Oxford*, **181**, 213–221, Part 2.

Lloyd, R. M. (1974). Mating systems and genetic load in pioneer and non-pioneer Hawaiian pteridophyta. *Botanical Journal of the Linnean Society*, **69**, 23–35.

Lovis, J. D. (1977). Evolutionary patterns and processes in ferns. In *Advances in Botanical Research*, ed. R. D. Preston and H. W. Woolhouse. London: Academic Press, pp. 229–440.

Makgomol, K. and Sheffield, E. (2001). Gametophyte morphology and ultrastructure of the extremely deep shade fern, *Trichomanes speciosum*. *New Phytologist*, **151**, 243–255.

Mesler, M. R. (1975a). Gametophytes of *Ophioglossum palmatum* L. *American Journal of Botany*, **62**, 97–98.

Mesler, M. R. (1975b). Mature gametophytes and young sporophytes of *Ophioglossum nudicaule*. *Phytomorphology*, **25**, 156–166.

Mesler, M. R. (1976). Gametophytes and young sporophytes of *Ophioglossum crotalophoroides* Walt. *American Journal of Botany*, **63**, 443–448.

Mesler, M. R. and Lu, K. L. (1977). Large gametophytes of *Equisetum hyemale* in northern California. *American Fern Journal*, **67**, 97–98.

Miller, J. H. (1968). Fern gametophytes as experimental material. *Botanical Review*, **34**, 361–440.

Moore, J. P., Nguema-Ona, E., Chevalier, L., Lindsey, G. G., Brandt, W. F., Lerouge, P., Farrant, J. M., and Driouich, A. (2006). Response of the leaf cell wall to desiccation in the resurrection plant *Myrothamnus flabellifolius*. *Plant Physiology*, **141**, 651–662.

Näf, U. (1979). Antheridiogens and antheridia development. In *The Experimental Biology of Ferns*, ed. A. F. Dyer. London: Academic Press, pp. 435–470.

Nayar, B. K. and Kaur, S. (1969). Types of prothallial development in homosporous ferns. *Phytomorphology*, **19**, 179–188.

Nayar, B. K. and Kaur, S. (1971). Gametophytes of homosporous ferns. *Botanical Review*, **37**, 295–396.

Nester, J. E. and Schedlbauer, M. D. (1981). Gametophyte development in *Anemia mexicana* Klotzsch. *Botanical Gazette*, **142**, 242–250.

Nobel, P. S. (1978). Microhabitat, water relations, and photosynthesis of a desert fern, *Notholaena parryi* [Pteridophyta]. *Oecologia*, **31**, 293–309.

Oertli, J. J., Lips, S. H., and Agami, M. (1990). The strength of sclerophyllous cells to resist collapse due to negative turgor pressure. *Acta Oecologica, International Journal of Ecology*, **11**, 281–289.

Oliver, M. J. (1996). Desiccation tolerance in vegetative plant cells. *Physiologia Plantarum*, **97**, 779–787.

Ong, B. L. and Ng, M. L. (1998). Regeneration of drought-stressed gametophytes of the epiphytic fern, *Pyrrosia piloselloides* (L.) Price. *Plant Cell Reports*, **18**, 125–228.

Ong, B. L., Koh, C. K. K., and Wee, Y. C. (1998). Effects of CO_2 on growth and photosynthesis of *Pyrrosia piloselloides* (L.) Price gametophytes. *Photosynthetica*, **35**, 21–27.

Peck, J. H. (1980). The ecology and reproductive biology of ferns in Woodman Hollow, Iowa. Unpublished Ph.D. Thesis, Iowa State University, Ames, IA.

Peck, J. H., Peck, C. J., and Farrar, D. R. (1990). Comparative life history studies and the distribution of pteridophyte populations. *American Fern Journal*, **80**, 126–142.

Pickett, F. L. (1913). Resistance of the prothallia of *Camptosorus rhizophyllus* to desiccation. *Bulletin of the Torrey Botanical Club*, **40**, 641–645.

Pickett, F. L. (1914). Some ecological adaptations of certain fern prothallia – *Camptosorus rhizophyllus* Link., *Asplenium platyneuron* Oakes. *American Journal of Botany*, **1**, 477–498.

Pickett, F. L. (1931). Notes on xerophytic ferns. *America Fern Journal*, **21**, 49–57.

Porembski, S. and Barthlott, W. (2000). Granitic and gneissic outcrops (inselbergs) as centers of diversity for desiccation-tolerant vascular plants. *Plant Ecology*, **151**, 19–28.

Proctor, M. C. F. and Pence, V. (2002). Vegetative tissues: bryophytes, vascular resurrection plants and vegative propagules. In *Desiccation and Survival in Plants:*

Drying Without Dying, ed. M. Black and H. W. Prichard. Wallingford: CAB International, pp. 207–237.

Raghavan, V. (1989). *Developmental Biology of Fern Gametophytes*. New York: Cambridge University Press.

Ranker, T. A. and Houston, H. A. (2002). Is gametophyte sexuality in the lab a good predictor of sexuality in nature? *Sadleria* as a case study. *American Fern Journal*, **92**, 112–118.

Remy, R. and Remy, W. (1980). Devonian gametophytes with anatomically preserved gametangia. *Science*, **208**, 295–296.

Remy, R., Gensel, P. G., and Hass, H. (1993). The gametophyte generation of some early Devonian land plants. *International Journal of Plant Science*, **154**, 35–58.

Rumsey, F. J., Vogel, J. C., Russell, S. J., Barrett, J. A., and Gibby, M. (1998). Climate, colonisation and celibacy: population structure in central European *Trichomanes speciosum* (Pteridophyta). *Botanica Acta*, **111**, 481–489.

Sakai, A. (1980). Freezing resistance of gametophytes of the temperate fern, *Polystichum retroso-paleaceum*. *Canadian Journal of Botany*, **58**, 1144–1148.

Sato, T. and Sakai, A. (1981). Cold tolerance of gametophytes of some cool temperare ferns native to Hokkaido. *Canadian Journal of Botany*, **59**, 604–608.

Schneider, E., Schuettpelz, E., Pryer, K. M., Cranfill, R., Magallon, S., and Lupia, R. (2004). Ferns diversified in the shadow of angiosperms. *Nature*, **428**, 553–557.

Skelton, C. L. (2007). Investigations into gametophyte morphology and population sex ratios through direct comparisons between laboratory grown and field-grown fern gametophytes. Unpublished MS Thesis, Iowa State University, Ames, IA.

Smith, A. R., Pryer, K. M., Schuettpelz, E., Korall, P., Schneider, H., and Wolf, P. G. (2006). A classification for extant ferns. *Taxon*, **55**, 705–731.

Stokey, A. G. (1951). The contribution of the gametophyte to the classification of the homosporous ferns. *Phytomorphology*, **1**, 39–58.

Stokey, A. G. and Atkinson, L. R. (1958). The gametophyte of the Grammitidaceae. *Phytomorphology*, **8**, 391–403.

Tausz, M., Hietz, P., and Briones, O. (2001). The significance of carotenoides and tocopherols in photoprotection of seven epiphytic fern species of a Mexican cloud forest. *Australian Journal of Plant Physiology*, **28**, 775–783.

Taylor, T. N., Kerp, H., and Hass, H. (2005). Life history biology of early land plants: deciphering the gametophyte phase. *Proceedings of the National Academy of Sciences of the United States of America*, **102**, 5892–5897.

Tryon, R. M and Vitale, G. (1977). Evidence for antheridiogen production and its mediation of a mating system in natural populations of fern gametophytes. *Botanical Journal of the Linnean Society*, **74**, 243–249.

Twiss, E. M. (1910). The prothallia of *Anemia* and *Lygodium*. *Botanical Gazette*, **49**, 168–181.

Voeller, B. and Weinberg, E. S. (1969). Evolutionary and physiological aspects of antheridium induction in ferns. In *Current Topics in Plant Science*, ed. J. E. Gunckel. London: Academic Press, pp. 77–93.

Walp, R. L. (1951). Fern prothallia under cultivation for 12 years. *Science*, **113**, 128–129.

Watkins, J. E., Jr. (2006). Comparative functional ecology of tropical ferns. Unpublished Ph.D. Thesis, University of Florida, Gainesville, FL.

Watkins, J. E., Jr., Cardelus, C., Colwell, R. K., and Moran, R. C. (2006a). Species richness and distribution of ferns along an elevational gradient in Costa Rica. *American Journal of Botany*, **93**, 73–83.

Watkins, J. E., Jr., Kawahara, A. Y., Leicth, S. A., Auld, L. R., Bicksler, A. J., and Kaiser, K. (2006b). Fern laminar scales protect against photoinhibition from excess light. *America Fern Journal*, **96**, 83–92.

Watkins, J. E., Jr., Mack, M. C., and Mulkey, S. S. (2007a). Gametophyte ecology and demography of epiphytic and terrestrial tropical ferns. *American Journal of Botany*, **94**, 701–708.

Watkins, J. E., Jr., Mack, M. C., Sinclair, T. R., and Mulkey, S. S. (2007b) Ecological and evolutionary consequences of desiccation tolerance in tropical fern gametophytes. *New Phytologist*, **176**, 708–717.

10

Conservation biology

NAOMI N. ARCAND AND TOM A. RANKER

10.1 Introduction

Within the USA, Connecticut was the first state to pass a plant protection law in 1869, and it was for a fern: *Lygodium palmatum* populations were declining due to over-collection for horticultural uses (Yatskievych and Spellenberg, 1993). Over a century later, the same fern was the source of a short article in *The New York Times* on June 16, 1985, which described road relocation negotiations to avoid two patches of the rare *Lygodium palmatum* in Burlington County, New Jersey (Haitch, 1985). The article reads, "The issue was at a stalemate in December. Score one for the ferns. Burlington will move the road 200 ft. east of the originally planned route to bypass the plants . . ." (Haitch, 1985). The beginnings of the American Fern Society in 1893, followed quickly by the publication of the *Fern Bulletin* (for 5¢ each), and later the *American Fern Journal* in 1910, attest to the early importance of ferns and lycophytes to US aficionados (Benedict, 1941).

There are approximately 13 600 named species of ferns and lycophytes globally (Hassler and Swale, 2001; Chapter 14). Because new species continue to be described and because of persistent regional gaps in floristic treatments, the real number of fern and lycophyte species is not yet known. Due to declining abundance or local extirpation, fern and lycophyte species of conservation concern have been identified for certain areas and have become the focus of international conservation efforts. Conservation biology research today is acting in response to environmental degradation across the globe, with increased human impacts to vulnerable ecosystems causing alarming rates of species extinction and habitat loss. Early accounts of conservation threats to fern species in the USA were

Biology and Evolution of Ferns and Lycophytes, ed. Tom A. Ranker and Christopher H. Haufler. Published by Cambridge University Press. © Cambridge University Press 2008.

attributed to habitat destruction and over-collection by fern enthusiasts and commercial horticulturalists (Overacker, 1930). As early as 1912, "a plea for fern conservation" was published in the *American Fern Journal*, lamenting the over-harvesting of local fern species to be shipped to florists (Phelps, 1912). In 1929 the *American Fern Journal* published an editorial concerning the preservation of ferns in the area of Staffordshire, England: donations of rare and native ferns were requested to supplement the new "fern sanctuary," not to the detriment of wild-collected ferns but rather "extra garden specimens" from private holdings (Anonymous, 1929). The New York "state fern law," as it was called, was passed in 1930 as an amendment to the State Conservation Law, which legislated it a misdemeanor to possess whole or part of the hart's tongue fern, *Scolopendrium vulgare* (Overacker, 1930). These early efforts to protect various fern species from local extinction have continued into the present, with conservation biologists rising to the challenge of safeguarding biodiversity and the ecosystem processes that sustain it.

This chapter reviews published research within the context of the multi-dimensional strategies of fern and lycophyte conservation biology. These are exciting times, as research continues to evolve with new approaches to tackle difficult conservation problems. Section 10.2 will describe some of the reasons why ferns and lycophytes are important subjects of conservation efforts. Section 10.3 will expand on the threats to global diversity of ferns and lycophytes. Section 10.4 will summarize published research and conservation implications of the fern and lycophyte life cycle relative to seed plants. Section 10.5 will discuss *ex situ* propagation for conservation. Sections 10.6 and 10.7 will describe geographical regions and ecosystems of concern for ferns and lycophytes, and the specific groups and species of ferns and lycophytes at risk. Section 10.8 will explain the usefulness and growing field of conservation genetics, and Section 10.9 will summarize published research on habitat restoration methods to conserve fern and lycophyte diversity. A concluding section will contain recommendations for future research and action.

10.2 Conservation importance of ferns and lycophytes

Ferns and lycophytes are ancient and diverse lineages with global distribution, yet many taxa are restricted to certain habitats (Given, 2002). The majority of fern and lycophyte species are found in the tropics (Gómez-P., 1985), and often comprise a significant portion of the vascular plant flora in tropical regions. With new fern and lycophyte species still being discovered, as in the case of the newly described *Ctenitis bigarellae* from Paraná, Brazil (Schwartzburd et al., 2007), and their high diversity in the tropical and temperate rainforests, ferns and lycophytes are considered extremely vulnerable to extinction. At the current rate of habitat fragmentation and deforestation of temperate and tropical rainforests, and given the specialized ecological niches that fern and lycophyte species require (e.g., see Chapter 8), it is likely that many species will go extinct before they have been described.

On remote oceanic islands of the tropics and subtropics, ferns and lycophytes comprise a larger proportion of the flora than is typically found in continental regions with similar environmental conditions (Kramer, 1993). Given (1993) stated that, "the relatively large contribution by pteridophytes to the vegetation of many oceanic islands has important consequences for conservation priorities," (p. 295). Indeed, ferns and lycophytes may be considered habitat enhancers for other plant and animal species, in relationships that are currently under investigation. For example, mature forest canopy species of epiphytic ferns provide necessary foraging habitat for endemic avian communities in the Udzungwa Mountains of Tanzania (Fjeldsa, 1999), and sooty owls were more often found in wetter senescent forest associated with the tree ferns *Cyathea australis* and *Dicksonia antarctica* in northeast Victoria, Australia (Loyn *et al.*, 2001). Moran *et al.* (2003) found that tree fern trunks at several sites in Costa Rica supported a greater diversity of fern epiphytic species than did the trunks of neighboring woody angiosperms, with seven of 11 species of epiphytes restricted to tree-fern trunks and none restricted to angiosperm trunks. During winter and spring, the Azores bullfinch *Pyrrhula murina* feeds on fern sporangia and fern fronds (Ramos, 1995), and species of *Drosophila* in Australia use tree fern fronds within the forest interior as important microhabitat sites, due to the cooler temperatures, higher humidity, and a constant thick layer of rotting leaf litter under the interior forest tree ferns (Parsons, 1991). The rusty-winged barbtail (*Premnornis guttuligera*) builds its nest entirely out of the petiole scales of a species of *Cyathea* (Dobbs *et al.*, 2003). Evidence from coprolites suggests that the diet of the extinct, flightless moa-nalo bird (*Thambetochen chauliodous*) of the Hawaiian Islands may have specialized on ferns (James and Burney, 1997).

Humans have long used ferns and lycophytes for medicinal and spiritual purposes. Ethnobotanical studies have documented several species of ferns and lycophytes that are used for treating ailments and illnesses, and in various traditional rituals and ceremonies (e.g., see Manandhar, 1995; Samant *et al.*, 1998; Macía, 2004 and references therein). Macía (2004) recorded the use of 24 species of ferns in two groups of Amerindians from Amazonian Bolivia and Ecuador, mostly for medicinal purposes. Plant geneticists and cytogeneticists have measured mutation rates of fern rhizomes in healthy and polluted ecosystems to detect and screen environmental carcinogenesis caused by toxins in air, water, soil, food, and even manufactured products (Klekowski and Poppel, 1976). For example, a population of *Osmunda regalis* was documented to have a higher incidence of post-zygotic mutational damage in a river heavily polluted with wastes from paper processing compared to populations growing in unpolluted rivers (Klekowski, 1978). Fern bioassays can detect chronic low-dose and episodic high-dose inputs of mutagenic pollutants in aquatic ecosystems (e.g., see Klekowski, 1976, 1978, 1982; Klekowski and Berger, 1976; Klekowski and Levin, 1979; Klekowski and Klekowski, 1982). Ferns have even been discussed in the context of "untapped biodiversity," because the novel extreme environmental adaptations of ferns, including heavy metal tolerance and hyperaccumulation, could be developed as genetic resources to aid the development of stress tolerance in commercial crops (Rathinasabapathi, 2006). One such fern, *Athyrium yokoscense*, was highly tolerant to lead toxicity, and gametophytes accumulated lead and localized it in the cytosol and vacuole of rhizoidal cells (Kamachi *et al.*, 2005). These studies in the detection of environmental toxicity using ferns are especially relevant today as the contamination of soil, water, and air continues with increasing industrialization. They could be useful tools for conservationists and public health officials working to detect and mitigate environmental pollutants.

Ferns and lycophytes are also useful indicators of habitat quality, species diversity, land use types, and disturbance regimes. Because ecological requirements vary across species, their occurrence in certain areas make evident predictable habitat characteristics, such as open disturbed habitat, young secondary forest, or primary forest (Poulsen and Tuomisto, 1996; Tuomisto and Poulsen, 1996; Arens and Baracaldo, 1998; Myster and Sarmiento, 1998; Muller, 2000; Beukema and van Noordwijk, 2004; Banaticla and Buot, 2005). Also, certain fern species are important colonizers in early succession and gaps (Turner *et al.*, 1996; Arens and Baracaldo, 1998; Bey, 2003).

10.3 Threats to global fern and lycophyte diversity

10.3.1 *Habitat fragmentation, degradation, and destruction*

The most commonly identified cause of current global species extinction continues to be the destruction and alteration of natural habitat, and in particular, tropical deforestation. Gómez-P. (1985) states, "it is apparent that whatever conservation strategy is devised, it must be fast and tropically-oriented," (p. 432). Many fern and lycophyte species have high habitat fidelity, occurring only in limited microsites or epiphytically on certain host species (Given, 1993; see Chapter 8). Logging generally reduces forest stand structure and vertical heterogeneity (Lindenmayer *et al.*, 2000; Padmawathe *et al.*, 2004), though effects of logging vary with harvesting intensity. Because vertical distribution is the major gradient in establishment and development of epiphytic ferns (Gardette, 1996), the impact of commercial timber harvesting is particularly alarming in that the largest, tallest trees are often all removed. In the lowland moist forests of India, logging altered the microclimate and substrate features needed to support epiphytic fern abundance and species composition, and species richness was correlated with canopy cover (Padmawathe et al., 2004). Tree ferns (*Cyathea australis* and *Dicksonia antarctica*) were less abundant in young logged forests as compared to old growth forest in Victoria, Australia (Lindenmayer et al., 2000), and these two species exhibited increased mortality after clear-cut logging, with survival continuing to decrease over 6 years post-logging (Ough and Murphy, 2004). These trends of decline are alarming, as tree ferns are generally slow growing and have low recruitment rates, but remain an important component of native forests (Ough and Murphy, 2004). Clear-cut harvesting in Australia alters natural forest regeneration, such that re-sprouting shrubs, tree ferns, and most terrestrial understory species are more abundant in wildfire-regenerating forests than in clear-cut areas in the same stage of development (Ough, 2001).

Similarly, tree plantations in New Zealand disrupt forest structure and diversity (Ogden et al., 1997). However, increased plantation stand age was related to increased species richness and forest composition of woody shrubs, tree ferns, and terrestrial understory ferns, with tree ferns reaching higher densities and species richness in older stands (Ogden et al., 1997). Forest recovery after agricultural use also has long-term effects on forest structure: Dominican Republic forest life-form diversity declined in a 40-year-old secondary forest compared with old-growth forest, with old-growth forest containing fewer introduced species and more arborescent ferns, palms, epiphytic bromeliads, orchids, bryophytes, terrestrial understory ferns, and other herbaceous plants (Martin et al., 2004). These structurally diverse plant communities provide important habitat for ferns and lycophytes and are often biodiversity "hot-spots." The temporal scale of post-agriculture forest recovery depends on the historic intensity of cultivation and proximity to intact native forest, and understanding the dynamics of these land-use impacts as they vary over space, time, and plant communities can aid forest restoration efforts.

Across northern hemisphere agrarian landscapes, native fern species are more often found in riparian and forested areas, and invasive or weedy species are found nearer to farm fields and pastures (Boutin et al., 2003; Heartsill-Scalley and Aide, 2003). Fern species diversity in the patchwork fragmentation of forest islands in agricultural and pastoral landscapes of The Netherlands is greatest in large, old forest areas (Bremer and Smit, 1999). These findings demonstrate the importance of large areas of intact plant communities with minimal human disturbance for conservation of ferns and lycophytes and the ecosystems upon which they depend. However, according to Woolley and Kirkpatrick (1999), even larger and older forest fragments occurring within a severely altered landscape need to be properly managed in order to maintain rare and threatened ferns and lycophytes, with both area-specific and species-specific management approaches.

10.3.2 Commercial collection

In 1913 a note in the *American Fern Journal* entitled "fern protection needed," called attention to the collection industry's early impacts on native fern species (Rugg, 1913). The *American Fern Journal* in 1914 published a report from a newspaper clipping of a lucrative commercial enterprise for fern pickers, where the florist industry paid out a total of $30 000 for the months of September, October, and part of November (Druery et al., 1914). An update on the commercial value and volumes of ferns harvested in Vermont listed *Dryopteris goldiana*, *D. intermedia*, and *Polystichum acrostichoides* as targeted species to supply the florist industry (Winslow and Benedict, 1916). In 1919, a writer for the *American Fern Journal* asked whether an independent observer might make more accurate observations on the real effects of commercial gathering on certain desirable fern species, due to the high volumes of fern fronds that were reported in commercial enterprises and the reliance on reports by the pickers themselves, who anecdotally observed no decline in plants that were harvested for fronds across multiple seasons (Burnham, 1919). In 1922, the *American Fern Journal* ran another article that raised concerns about threatened ferns due to commercial collection, and included the text of Vermont state law that limited the amounts of listed rare ferns that could be collected and imposed a $10 fine per plant for collectors who did not heed the restrictions (Benedict, 1922).

More recently, commercial plant collecting on the southeastern Ivory Coast of Africa was responsible for reduced abundance of the epiphytic fern *Platycerium stemaria*, with note of the commercial value associated with nursery-grown or wild-collected epiphytic ferns across the globe (Porembski and Biedinger, 2001). Tree ferns are also important in the horticultural trade for landscaping, while their trunks are valued for orchid-growing medium and substrate. All species of the tree-fern family Cyatheaceae, all American species of *Dicksonia* (Dicksoniaceae), and *Cibotium barometz* (Cibotiaceae) (Zhang et al., 2002) are listed in Appendix II of the Convention of International Trade of Endangered Species (CITES) for trade monitoring. In 1997, export trade of *C. barometz* from China was prohibited due to the large quantities of individuals harvested and observations of reduced wild populations (Zhang et al., 2002). In Australia, *Dicksonia antarctica* was investigated for sustainable production to supply commercial demand for landscaping, because wild-harvested tree ferns were viewed as unsustainable by the public (Vulcz et al., 2002). Production of *D. antarctica* was also explored in

Tasmania under eucalypt plantations, despite the species being listed on CITES Appendix II (Unwin and Hunt, 1996). Additionally, commercial wild-harvesting of fern fronds can exceed the rate of new frond production, and fern recovery after frond harvesting was found to be very slow for the evergreen fern *Rumohra adiantiformis* from the southern Cape forests of South Africa (Milton and Moll, 1988).

10.3.3 Pathogens, predators, and invasive species

Although many fern taxa are considered resistant to fungal pathogens, the rust fungi (Uredinales) appear to be the most common pathogens of ferns (Helfer, 2006). The three most common fern genera observed to be affected include *Pteridium*, *Athyrium*, and *Dryopteris* (Helfer, 2006). A much more widely observed threat to ferns and lycophytes, especially on oceanic islands, appears to be predation by feral pigs. The major cause of death for individuals of *Botrychium australe* in New Zealand was pig rooting (Kelly, 1994), and the starchy caudex of endemic tree ferns in Hawaii (*Cibotium* spp.) are a major food source for feral pigs (Diong, 1982).

Island flora and fauna are often particularly vulnerable to negative impacts of introduced or invasive species, as has been shown in the Galápagos Islands (Schofield, 1989). Ferns that are not native to an area, especially in island ecosystems with a highly endemic fern flora, can also displace and out-compete native species. There are a total of 32 naturalized alien ferns in Hawaii, of which two are new species established since 1996 (Wilson, 2002). As with other non-native biota, alien ferns and lycophytes are often introduced intentionally for horticultural and landscaping purposes, subsequently escaping cultivation and becoming naturalized. This has been the case in the Hawaiian Islands for the introduced Australian tree fern, *Sphaeropteris cooperi*, which is fast growing and considered extremely hazardous to Hawaiian ecosystems due to its likely displacement of the endemic, slower-growing *Cibotium* tree ferns, and it has spread from Oahu to Maui, Kauai, and Hawaii (Palmer, 2003).

10.3.4 Climate change

Responses of ferns and lycophytes to global climate change are difficult to predict and have been little studied. Because spores provide ferns and lycophytes with superior long-distance dispersal capabilities, these groups may have an advantage over seed-bearing plants in adapting to climate changes (Given, 1993). However, increased habitat fragmentation, physiological stress, and habitat disturbance may cause local or global extinctions of fern and lycophyte species that are unable to survive the changing environmental conditions (Given, 1993). Kelsall *et al.* (2004) compared population census data to climate data for

Phyllitis scolopendrium var. *americana*, limited to 17 known colonies in New York. They found climate was not responsible for much fluctuation in population size, but the authors concluded that this fern species is limited in distribution by the limited occurrence of habitat that can buffer it from climatic fluctuations (Kelsall *et al.*, 2004). With the thinning of the ozone layer, vegetation is exposed to increased ultraviolet-B solar radiation, but there has been little experimental research on the effects of this increased exposure within ferns and lycophytes (Björn, 2007). The aquatic fern *Azolla microphylla* exhibited reduced growth and reduction in its cyanobacterial nitrogen fixation when exposed to ultraviolet-B radiation (Jayakumar *et al.*, 2002), and also curling of the fronds and chlorosis (Jayakumar *et al.*, 1999). However, the lycophyte *Lycopodium annotinum* did not show growth inhibition when exposed to ultraviolet-B radiation that replicated conditions of 15% ozone thinning (Björn, unpublished results). Additional research on fern and lycophyte responses to climate change is much needed.

10.4 Life cycle challenges for conservation

It is important to keep in mind the various life cycle stages of ferns and lycophytes when discussing their protection, conservation, and restoration. Research on spore production, dispersal, viability, and soil spore banks has demonstrated important differences compared with similar research on seed plants. As each fertile frond often produces millions of spores (Peck *et al.*, 1990; see Chapter 9), spores can often be found aerially suspended at high altitudes. For instance, a fern sporangium was trapped by airplane as high as 2400 meters above the Hawai'ian Islands (Gressitt *et al.*, 1961). The closed forest is less easily penetrated by wind, however, and often spores are dropped within the vicinity of the parent plant (Conant, 1978).

Research and knowledge concerning the storage banking of spores has recently advanced, and identifying ideal conditions of storage to maximize spore viability is an important component of *ex situ* fern and lycophyte conservation (see Chapter 11). Studies have identified variation by taxon for ideal spore storage conditions with temperature, moisture, and storage duration (Aragón and Pangua, 2004; Chá-Chá *et al.*, 2005). Additional studies have determined the effects of cryopreservation techniques on spore viability (Pence, 2002) and gametophyte/sporophyte development (Ballesteros *et al.*, 2006) with promising results.

Protecting soil spore bank resources and promoting *in situ* germination has been suggested for the conservation of rare fern taxa (Dyer and Lindsay, 1996), and there is evidence for persistent soil spore banks (Sheffield, 1996; LaDeau and Ellison, 1999; Ranal, 2003; Ramírez-Trejo *et al.*, 2004). However, the longevity of spores in the soil spore bank varies by taxon and habitat (Ranal, 2003). The

Conservation biology 265

viability of tree bark spore banks has also been studied, and interestingly was not observed to be affected by dry seasonality (Ranal, 2004).

The gametophytic stage of ferns and lycophytes is particularly challenging to conservation biologists because of the small size and obscurity of gametophytes (see Chapter 9). Knowledge is growing of gametophyte development and survival *in situ* (see Sheffield, 1996; Bernabe *et al.*, 1999; Ranal, 2004), but further research on *in situ* gametophyte ecology and sporeling establishment is still urgently needed (Dyer and Lindsay, 1996; see Chapters 2 and 9). Many laboratory studies have been conducted on spore germination and gametophyte development, and mechanisms have been identified which prevent self-fertilization (see Chapters 5 and 9), and promote continuous recruitment of gametophytes by intermittent spore germination (Gomes *et al.*, 2006).

Fern and lycophyte sporophytes have been the most prevalently studied, with a focus on sporophyte development and growth rate, species distribution and diversity, and phylogenetic relationships. Ecological studies of ferns and lycophytes and the roles they play in plant communities have been historically rare, but are increasingly needed to inform restoration efforts and plan conservation reserves (see Chapter 8).

10.5   *Ex situ* propagation of ferns and lycophytes

Propagation research for rare or threatened ferns and lycophytes is essential, and *ex situ* collections of sporophytes and spores provide a genetic safety net helping to prevent species extinction. *Ex situ* horticultural propagation of endangered fern and lycophyte species may also reduce harvesting pressures on wild populations if they are made commercially available (Gibby and Dyer, 2002). Propagation techniques specific to certain threatened or endangered taxa have been developed that ensure the most efficient and successful results, including propagation from rhizome cuttings and stipules (Wardlaw, 2002a, 2002b; Zenkteler, 2002; Chiou *et al.*, 2006). For instance, the endangered epiphytic fern *Drynaria quercifolia* was propagated from spore to gametophyte to sporophyte within 4 months, and whole plants were produced from field-grown rhizome explants within 3 months (Hegde and D'Souza, 1996). See Chapter 11 for a thorough review and discussion of *ex situ* conservation of ferns and lycophytes.

10.6   Regional and ecosystem-level conservation

Tropical ferns and lycophytes are especially threatened in areas of high endemism throughout the world (Gómez-P., 1985). To preserve the maximum biodiversity, conservationists must concentrate efforts on biodiversity hotspots and areas of high endemism. Generally, sites of high productivity support higher species diversity (Lehmann et al., 2002), and oceanic islands are also acknowledged as important endemism centers (Given, 1993). For example, 70% of the Hawaiian Island fern and lycophyte flora is endemic (Gagne, 1988). Evidence suggests that the fern flora of east Malesia dispersed from New Guinea and in isolation evolved into several new species (Kato, 1993).

Predictive factors for areas of higher fern and lycophyte endemism and biodiversity generally include terrain and climate (Pausas and Sáez, 2000), and more particularly, regional mean annual temperature and water availability (Lehmann et al., 2002), soil fertility and distance from nearest geological refugium (Lwanga et al., 1998), soil composition and elevation (Duncan et al., 2001; Banaticla and Buot, 2005), and water availability and landscape heterogeneity (Bickford and Laffan, 2006). A pattern emerges between the distribution of endemism and elevation in tropical mountainous areas, where the highest endemism has been observed at mid-elevations in Costa Rica (Watkins et al., 2006), though endemic species at highest elevations are often more common than widespread species (Kluge and Kessler, 2006). Similarly in Peru, fern and lycophyte diversity and endemism are found to concentrate at higher elevation, especially in humid montane forests (León and Young, 1996b).

Smith (2005) has produced a thorough summary of the published regional fern and lycophyte floras and annotated checklists, or lack thereof, which is generally summarized here. Most tropical and subtropical countries/regions are lacking in complete modern fern and lycophyte floras, except Mexico, Mesoamerica, parts of the Antilles, New Zealand, Hawaii, and the Mariana Islands (Smith, 2005). Meanwhile, treatments of fern and lycophyte flora are in progress for Bolivia, East Africa, Venezuela, and China (Smith, 2005). Areas with annotated checklists only include tropical East Africa, Malaysia, and Mount Kinabalu, and areas of partially written floras include Ecuador and Malesia (Smith, 2005). A project to produce an annotated checklist of Sulawesi ferns and lycophytes of Indonesia was also reported underway to encourage further interest in study of the poorly documented fern flora there (Camus and Pryor, 1996). Hotspots with inadequate or outdated floras include Colombia, Brazil, Madagascar, New Guinea, and the Himalayas (Smith, 2005). These regions are noted as important areas of fern and lycophyte diversity, and it is likely that the regional gaps are large and numerous for modern, complete fern and lycophyte floras. Assessments of fern and lycophyte diversity and rarity have also been published for the Askot Wildlife Sanctuary in West Himalaya, India (Samant et al., 1998), the Orongorongo Valley Field Station in New Zealand (Fitzgerald and Gibb, 2001), the Pitcairn Islands (Kingston and Waldren, 2002), and the Prespa National Park of Greece (Pavlides, 1997), among others. These reports, though not conclusive for generalization across entire countries or regions, nonetheless provide valuable insights as to locating potential areas and taxa of high conservation priority.

Previous studies of threatened and rare ferns and lycophytes in various regions are not encouraging, as many species fall within these categories and several are reported to be near extinction. For instance, ferns and lycophytes from Trinidad and Tobago were assigned a risk index rating, and approximately half of the ferns there are deemed possibly at risk (Baksh-Comeau, 1996). Federal protection within the USA (i.e., those species listed by the US Fish and Wildlife Service as threatened or endangered) includes only one third of ferns and fern allies that are currently at risk (Stein et al., 2000). Of the 22 rare species of ferns and lycophytes assessed from the islands of São Tomé e Príncipe (Gulf of Guinea), IUCN Red List Categories were assigned as eight species critically endangered, three endangered, and eleven vulnerable (Figueiredo and Gascoigne, 2001). Fifty percent of the ferns and lycophytes of the Pitcairn Islands have also been assigned to an IUCN threat category, with invasive species, habitat degradation, reduced population and distribution, and over-harvesting listed as the most prevalent threats (Kingston and Waldren, 2002). Similarly, rare ferns in Russia are threatened due to their dependence on restricted habitats, which are additionally stressed by human impacts (Guryeva, 2002). Fern conservation in south tropical Africa faces other obstacles to conservation initiatives that include disrupted economies, civil war, and increased human population pressure on crucial fern habitat (Burrows and Golding, 2002). And in Brazil, habitat degradation, commercial extraction, and lack of population data are considered the largest threats to fern and lycophyte species (Windisch, 2002).

However, an encouraging side-effect of human impact might be found in that the dispersal of ferns in Germany seems to be facilitated by transportation corridors. A study found ferns at more than 90% of railway stations examined, including species on the Red List for Germany (Wittig, 2002). Another encouraging side-effect of human impact was measured by Jamir and Pandey (2003); they found high species richness, high endemism, and high presence of other rare and primitive species of ferns and lycophytes in sacred forest groves of the Jainita Hills in northeast India. Tribal communities have designated these protected areas as part of their religious tradition, and the natural biodiversity preserved there "since time immemorial" (Jamir and Pandey, 2003, p. 1497).

Forest destruction and shifting cultivation are considered the most serious threat to fern and lycophyte habitat and diversity in the Philippines (Amoroso et al., 1996). A diversity assessment found 275 species of ferns and lycophytes at Mount Kitanglad, 249 species at Mount Apulang, and 183 species at Marilog Forest (Amoroso et al., 1996). Of these, a status determination revealed one endangered species, 45 rare, seven depleted, 89 endemic, and 81 economically important (Amoroso *et al.*, 1996). An updated account of Philippines fern and lycophyte diversity calls for more botanical explorations in the Philippines and other tropical regions, as results of field explorations included the rediscovery of the genus *Cyrtomium*, and the rediscovery of rare endemics *Aglaomorpha cornucopia*, *Antrophyum williamsii*, *Ctenitis humilis*, and *Dennstaedtia macgregori* (Barcelona, 2005). The fern and lycophyte floras of tropical regions must be further surveyed, described, and assessed for conservation priority.

10.7 Conservation of fern and lycophyte taxa

The IUCN Red List 2004 categorized approximately 67% of the evaluated species of ferns and allies as threatened (140 out of 210 evaluated species) (Baillie *et al.*, 2004; see www.redlist.org). Although the species evaluated by IUCN only represent 1.6% of the total described species (13 025), they are representative of a large geographical distribution across the world (Baillie *et al.*, 2004). Nevertheless, ferns and lycophytes as a highly diverse group are substantially underrepresented on the IUCN Red List, and it is critical that many more species be evaluated in the near future (Baillie *et al.*, 2004).

Isoëtes has been noted as rare and threatened across several regions and countries. Only two natural populations of *Isoëtes sinensis* are known to exist and both are threatened in mainland China, where populations have decreased by 50% within the last 4 years, most probably due to destruction of wetland habitat (Wang *et al.*, 2005). *Isoëtes georgiana* is now known from 12 locations on the Georgia coastal plain rather than being limited to one location as previously reported; yet this species is still considered rare (Brunton and Britton, 1996). The Andean highlands of Peru were found to contain eight species of *Isoëtes*, of which four are endemic, and these highland areas have been impacted by mining pollution and drainage, agriculture, and grazing (León and Young, 1996a). Musselman (2002) reports that the single species of *Isoëtes* in Syria (*I. olympica*) is reduced to 100 individuals and threatened with extinction by regional irrigated agriculture. All four species of *Isoëtes* in China, (*I. hypsophila, I. sinensis, I. yunguiensis*, and *I. taiwanensis*) are critically endangered and at risk of extinction due to habitat loss and degradation, water pollution, and competitive exclusion by invasive species (Liu *et al.*, 2005). Within the continental USA, *Isoëtes*, *Botrychium*, and *Selaginella* are the top three genera with the highest number of threatened ferns and lycophytes (Grund and Parks, 2002).

Additional ferns that are sparsely distributed and depend upon restricted habitat, such as the newly described New Zealand fern *Asplenium cimmeriorum* of cave entrances, are of high conservation priority due to their vulnerability to over-collection and disturbance (Brownsey and deLange, 1997). *Hymenophyllum tunbrigense*, discovered within the last 10 years in a small area in the western part of the Vosges Mountains, occurs only in *Abies alba* forests on moist sandstone rocks less than 2.5 m high, and requires dense forest canopy for continuous shade and protection from storms (Muller et al., 2006). *Mankyua chejuense* is recommended to be listed as critically endangered, as it is limited to five extant subpopulations and 1300 individuals in the Republic of Korea, and depends on basaltic rock microhabitat (Kim, 2004). And in the Hawaiian Islands, all six species of the endemic fern *Diellia* are considered of conservation priority due to their restricted distribution (five species are single-island endemics) and small population sizes (Aguraiuja et al., 2004). Ferns and lycophytes with restricted distribution and reduced population sizes are of highest risk of extinction, especially within degraded or vulnerable landscapes.

10.8 Genetics in fern and lycophyte conservation strategies

Phylogenetic relationships among ferns and lycophytes and evolutionary biogeographical theory are advancing with the use of molecular systematics and population genetics (see Chapter 4). These phylogenetic studies are important in helping us to understand evolutionary relationships and dispersal patterns among fern and lycophyte taxa (see Chapters 12–16). To explain the evolutionary biogeography of fern floras, research needs to include species-specific biogeographies and analyses of historical large-scale environmental change such as climatic and geological events, which can then be tested and substantiated (or disproved) with molecular methods (e.g., Kato, 1993). Differentiating between dispersal and vicariance in historical patterns of fern biogeography is difficult and needs further study (Wolf et al., 2001).

The erosion of genetic diversity – whether through reduced abundance, loss of unique populations, or extinction of entire species – is the main evil that fern and lycophyte conservation geneticists strive to prevent. Analyses of species-specific genetic diversity aim to identify levels and patterns of genetic variability across time and space (Ranker et al., 1996; Rumsey et al., 1999; Machon et al., 2001), to assess population-level genetic diversity to inform conservation goals and management priorities (Ranker, 1994; Eastwood et al., 2002; Kingston et al., 2004; Su et al., 2004, 2005; Wang et al., 2004; Chen et al., 2005; Kang et al., 2005), and to determine whether morphologically variant populations should be classified as subspecies or distinct species (Perrie et al., 2003). To prevent inbreeding depression in fragmented landscapes and reduced populations, it is suggested to treat each as separate parameters in analyses to identify genotype-specific phenomena from population parameters (Ouborg et al., 2006). The key to these approaches is their integration of genetic and ecological data for a more accurate understanding of conservation needs in the present.

10.9 Protection and restoration

10.9.1 *Planning protected areas*

The maintenance of primary forest has been identified as a conservation priority for ferns and lycophytes, as it is fast disappearing across the globe (Arens and Baracaldo, 1998; Beukema and van Noordwijk, 2004). The World Wildlife Fund (WWF) and World Conservation Union (IUCN) initiated a collaborative effort to identify global Centres of Plant Diversity (WWF and IUCN, 1994–1997). Protected areas must serve many local and global needs, and these include protecting biodiversity, ensuring sustainable use, maintaining local economies and sustenance, protecting water resources, preventing soil erosion, preserving scenic value, and providing novel research opportunities (Phillips, 2002). Fundamentally, it is crucial for conservationists and scientists working towards protecting threatened areas in tropical regions to work with local residents in the planning and decision-making processes, and to be especially considerate of local resource needs (Hamilton, 2002).

The minimum conservation area required for ferns and lycophytes is generally larger than that required for woody plant species (Murakami *et al.*, 2005). Therefore, the scale, reserve coverage, and accuracy in species representation are necessary factors to consider in planning and locating protected areas (Araujo, 2004). For example, Kessler (2001) conducted a study of ferns and lycophytes in a central humid rainforest of Bolivia along an elevational gradient. He concluded that conservation measures should focus on zonal forests in large reserves at low and mid elevations, and smaller reserves at high elevations. In Brazil, fern and lycophyte species richness was negatively affected by fragmentation at the edges of forest remnants, but small forest fragments were also considered important for conservation purposes, as the interior of fragments had similar fern and lycophyte species richness and diversity rates to large forest fragments (Paciencia and Prado, 2005). A study of the fragmented forest landscape in the urban area of Kyoto, Japan found fern and lycophyte diversity and forest patch area to be related, with increased diversity dependent upon larger patch size and similarly with reduced distance from intact mountain forest (Murakami *et al.*, 2005). However, Murakami *et al.* (2005) also showed that microhabitat diversity was an important predictor of fern and lycophyte diversity in fragments, and therefore concluded that protected areas should include both sufficient size and variable microhabitats. Indeed, patterns of biodiversity in the parks of Iberia were analyzed for species richness and range size, and priority areas for fern and lycophyte conservation indicated a dispersed, smaller reserve system would be most effective for including the greatest number of species (Saiz et al., 1996).

Remote sensing techniques for the identification of significant areas of fern subcanopy/understory of forested areas can be useful in locating and targeting fern and lycophyte conservation areas, and promising hyperspectral and hyperspatial imaging technology has recently been developed that may additionally be an effective tool for monitoring changes in land-use and ecosystem processes (Chambers et al., 2007). It is worth noting that surveying biodiversity by identifying to the family level has been found to be an efficient and accurate time-saving substitute for identifying each species present (Williams and Gaston, 1994). Another effective strategy that has been tested for identifying important sites for endangered ferns is analysis by indicator species in global plots-by-species classification, and then looking for conspicuous species in these areas (Vázquez and Norman, 1995). However, tree species richness has not been found to be a good surrogate for fern diversity (Galeano et al., 1998; Williams-Linera et al., 2005).

10.9.2 Restoration

The role of natural disturbances at multiple scales has been important for fern and lycophyte regeneration processes and in maintenance of fern and lycophyte diversity (Arens and Baracaldo, 1998; Page, 2002). Ferns can also dominate degraded sites and impose barriers for tree regeneration by competing for soil moisture, nutrients, and light (George and Bazzaz, 1999; Ashton et al., 2001; Slocum et al., 2004). Slocum et al. (2004) found that clearing fern thickets caused rapid recruitment of woody species and then limited regeneration of fern thickets. On the other hand, in the Andean cloud forests of Colombia, tree ferns (*Lophosoria*, *Cyathea caracasana*, and *Dicksonia*) are important colonizers of open areas and their use in forest restoration is suggested (Arens and Baracaldo, 1998). These disturbance and successional processes should be considered when planning and managing conservation reserves.

The removal of invasive species and weed control may be of high priority for certain fern and lycophyte species, especially on remote islands where endemics have evolved in the absence of competition and predation. *Isoëtes setacea* is considered a keystone species for priority habitat in the European Union, and experimental shrub clearing and increased light levels led to an increase in the number of *I. setacea* individuals (Rhazi et al., 2004). Management and restoration of habitat assessed for the endangered fern *Marsilea villosa* in Hawaii on the island of Oahu initially concluded that weed removal was unnecessary due to this fern's ability to reestablish itself during periodic flooding regardless of weeding efforts, with major threats of extinction from damage by off-road vehicles and a long succession of unusually dry years (Wester, 1994). However, the population health of *Marsilea villosa* as later assessed on Oahu has shown a dramatic decline over 8 years. Other causes of the population decline include a decrease in dominant trees and the invasion of alien grasses, which in turn may reduce the frequency and intensity of flooding (Wester et al., 2006) and increase the risk of fire. These observations demonstrate the cumulative feedback cycle of habitat degradation, where human impacts may be exacerbated by climatic events such as extended periods of drought, and aggravated by biologically adapted invasive species able to seize the upper hand in these windows of opportunity. Limiting the number of introductions of those non-native species with a high potential of becoming invasive in island ecosystems is a conservation priority (Gagne, 1988). As of 2007, Hawaii has failed to achieve political support for implementing stricter regulations for agricultural and biological imports.

In the lowlands of Jambi, Sumatra, ferns normally found in primary forest were also found in agroforest patches, but were lacking in plantation lands where undergrowth was removed or suppressed with herbicides (Beukema and van Noordwijk, 2004). The agroforest patches contained multiple secondary crops, including vegetables, rice, and fruit trees, and eventually secondary forest vegetation regenerated naturally after several years (Beukema and van Noordwijk, 2004). Therefore preserving rainforest biodiversity in commercial forestry may be possible with less intensive production, increased crop diversity, and minimized harvesting impacts to surrounding vegetation. Indeed, Lindenmayer *et al.* (2000) suggest that logging effects should resemble natural disturbance regimes and promote structural complexity in maintaining fern and lycophyte communities. Finding new ways to incorporate sustainable practices in agricultural and commercial harvesting is highly relevant to the future of the world's forests, especially in tropical regions.

10.10 Future directions in fern and lycophyte conservation

Most ferns and lycophytes are dependent on wet tropical montane habitats and are threatened with extinction before scientists have described them all (Page, 1985). The request for more taxonomic, distributional, and abundance documentation of ferns and lycophytes is sustained here, especially for underexplored regions and species of concern. Conservation of ferns and lycophytes *in situ* is possibly the most important issue today with increasing rates of human resource consumption and subsequent deforestation: ". . . it is an inescapable fact that survival for most species is inextricably dependent on conservation of forests worldwide," (Page, 1985, p. 441). Conservation of ferns in the tropics is additionally of high priority due to the concentration of fern and lycophyte diversity in countries with limited funding for research, limited protection for threatened areas, and limited means to enforce existing protections (Burrows and Golding, 2002; Hamilton, 2002).

Public awareness is crucial in conservation efforts (Given, 2002), especially in developed nations where consumer demand is often driving exploitative resource extraction. Demand for rare forest products, including endangered species, often drives global luxury markets into a positive feedback loop, where rarity becomes more valuable and thus reinforces suppliers to seek out the last remaining populations. The collection industry is causing a decline in certain fern and lycophyte taxa, notably tree ferns and other species valuable for the landscaping and florist industries. Consumer product awareness and seller accountability are crucial in today's market economy.

Adopting an ecosystem approach to species conservation, while also considering regional approaches to *in situ* and *ex situ* methods of protection and restoration, are considered the way forward in fern and lycophyte conservation biology (Jermy and Ranker, 2002). In addition to continuing to conduct ecological, demographic, taxonomic, and genetic studies of ferns and lycophytes, we encourage biologists, land managers, and conservationists to expand our knowledge of the functional and structural roles performed by ferns and lycophytes in the Earth's biomes.

References

Aguraiuja, R., Moora, M., and Zobel, M. (2004). Population stage structure of Hawaiian endemic fern taxa of *Diellia*. *Canadian Journal of Botany*, **82**, 1438–1445.

Amoroso, V. B., Acma, F. M., and Pava, H. P. (1996). Diversity, status and ecology of pteridophytes in three forests in Mindanao, Philippines. In *Pteridology in Perspective*, ed. J. M. Camus, M. Gibby, and R. J. Johns. Kew: Royal Botanic Gardens, pp. 53–60.

Anonymous. (1929). Conservation of native plants in England. *American Fern Journal*, **19**, 31.

Aragón, C. F. and Pangua, E. (2004). Spore viability under different storage conditions in four rupicolous *Asplenium* L. taxa. *American Fern Journal*, **94**, 28–38.

Araujo, M. B. (2004). Matching species with reserves – uncertainties from using data at different resolutions. *Biological Conservation*, **118**, 533–538.

Arens, N. C. and Baracaldo, P. S. (1998). Distribution of tree ferns (Cyatheaceae) across the successional mosaic in an Andean cloud forest, Narino, Colombia. *American Fern Journal*, **88**, 60–71.

Ashton, M. S., Gunatilleke, C. V. S., Singhakumara, B. M. P., and Gunatilleke, I. A. U. N. (2001). Restoration pathways for rain forest in southwest Sri Lanka: a review of concepts and models. *Forest Ecology and Management*, **154**, 409–430.

Baillie, J. E. M., Hilton-Taylor, C., and Stuart, S. N. (eds.). (2004). *2004 IUCN Red List of Threatened Species. A Global Species Assessment*. Gland, Switzerland and Cambridge, UK: IUCN.

Baksh-Comeau, Y. S. (1996). Risk index rating of threatened ferns in Trinidad and Tobago. In *Pteridology in Perspective*, ed. J. M. Camus, M. Gibby, and R. J. Johns. Kew: Royal Botanic Gardens, pp. 139–151.

Ballesteros, D., Estrelles, E., and Ibars, A. M. (2006). Responses of pteridophyte spores to ultrafreezing temperatures for long-term conservation in germplasm banks. *The Fern Gazette*, **17**, 293–302.

Banaticla, M. C. N. and Buot, I. E. J. (2005). Altitudinal zonation of pteridophytes on Mt. Banahaw de Lucban, Luzon Island, Philippines. *Plant Ecology*, **180**, 135–151.

Barcelona, J. F. (2005). Noteworthy fern discoveries in the Philippines at the turn of the 21st century. *The Fern Gazette*, **17**, 139–146.

Benedict, R. C. (1922). Game laws for ferns and wild flowers. *American Fern Journal*, **12**, 33–45.

Benedict, R. C. (1941). The American Fern Journal through thirty years. *American Fern Journal*, **31**, 41–48.

Bernabe, N., Williams-Linera, G., and Palacios-Rios, M. (1999). Tree ferns in the interior and at the edge of a Mexican cloud forest remnant: spore germination and sporophyte survival and establishment. *Biotropica*, **31**, 83–88.

Beukema, H. and van Noordwijk, M. (2004). Terrestrial pteridophytes as indicators of a forest-like environment in rubber production systems in the lowlands of Jambi, Sumatra. *Agriculture, Ecosystems and Environment*, **104**, 63–73.

Bey, A. (2003). Evapoclimatonomy modeling of four restoration stages following Krakatau's 1883 destruction. *Ecological Modelling*, **169**, 327–337.

Bickford, S. A. and Laffan, S. W. (2006). Multi-extent analysis of the relationship between pteridophyte species richness and climate. *Global Ecology and Biogeography*, **15**, 588–601.

Björn, L. O. (2007). Stratospheric ozone, ultraviolet radiation, and cryptogams. *Biological Conservation*, **135**, 326–333.

Boutin, C., Jobin, B., and Belanger, L. (2003). Importance of riparian habitats to flora conservation in farming landscapes of southern Quebec, Canada. *Agriculture, Ecosystems and Environment*, **94**, 73–87.

Bremer, P. and Smit, A. (1999). Colonization of polder woodland plantations with particular reference to the ferns. *The Fern Gazette*, **15**, 289–308.

Brownsey, P. J. and deLange, P. J. (1997). *Asplenium cimmeriorum*, a new fern species from New Zealand. *New Zealand Journal of Botany*, **35**, 283–292.

Brunton, D. F. and Britton, D. M. (1996). The status, distribution, and identification of Georgia quillwort (*Isoetes georgiana*; Isoetaceae). *American Fern Journal*, **86**, 105–113.

Burnham, S. H. (1919). Commercial fern gathering. *American Fern Journal*, **9**, 88–93.

Burrows, J. and Golding, J. (2002). Fern conservation in south tropical Africa. *The Fern Gazette*, **16**, 313–318.

Camus, J. M. and Pryor, K. V. (1996). The pteridophyte flora of Sulawesi, Indonesia. In *Pteridology in Perspective*, ed. J. M. Camus, M. Gibby, and R. J. Johns. Kew: Royal Botanic Gardens, pp. 169–170.

Chá-Chá, R., Fernandes, F., and Romano, A. (2005). In vitro spore germination of *Polystichum drepanum*, a threatened fern from Madeira Island. *Journal of Horticultural Science and Biotechnology*, **80**, 741–745.

Chambers, J. Q., Asner, G. P., Morton, D. C., Anderson, L. O., Saatchi, S. S., Espirito-Santo, F. D. B., Palace, M., and Souza C. Jr. (2007). Regional ecosystem structure and function: ecological insights from remote sensing of tropical forests. *Trends in Ecology and Evolution*, **22**, 414–423.

Chen, J. M., Liu, X., Wang, J. Y., Robert, G. W., and Wang, Q. F. (2005). Genetic variation within the endangered quillwort *Isoetes hypsophila* (Isoetaceae) in China as evidenced by ISSR analysis. *Aquatic Botany*, **82**, 89–98.

Chiou, W. L., Huang, Y. M., and Chen, C. M. (2006). Conservation of two endangered ferns, *Archangiopteris somai* and *A. itoi* (Marattiaceae Pteridophyta), by propagation from stipules. *The Fern Gazette*, **17**, 271–278.

Conant, D. S. (1978). A radioisotope technique to measure spore dispersal of the tree fern *Cyathea arborea* Sm. *Pollen et Spores*, **20**, 583–593.

Diong, C. H. (1982). Population biology and management of the feral pig (*Sus scrofa* L.) in Kipahulu Valley, Maui. Unpublished Ph.D. Thesis, University of Hawaii, Hawaii, HI.

Dobbs, R. C., Greeney, H. F., and Martin, P. R. (2003). The nest, nesting behavior, and foraging ecology of the rusty-winged barbtail (*Premnornis guttuligera*). *Wilson Bulletin*, **115**, 367–373.

Druery, C. T., Winslow, E. J., and Benedict, R. C. (1914). $30,000 paid fern pickers. *American Fern Journal*, **4**, 28–29.

Duncan, R. P., Webster, R. J., and Jensen, C. A. (2001). Declining plant species richness in the tussock grasslands of Canterbury and Otago, South Island, New Zealand. *New Zealand Journal of Ecology*, **25**, 35–47.

Dyer, A. F. and Lindsay, S. (1996). Soil spore banks – a new resource for conservation. In *Pteridology in Perspective*, ed. J. M. Camus, M. Gibby, and R. J. Johns. Kew: Royal Botanic Gardens, pp. 153–160.

Eastwood, A., Vogel, J. C., Gibby, M., and Cronk, Q. C. B. (2002). Relationships and genetic diversity of endemic *Elaphoglossum* species from St. Helena: implications for conservation. *The Fern Gazette*, **16**, 411–412.

Figueiredo, E. and Gascoigne, A. (2001). Conservation of pteridophytes in São Tomé e Príncipe (Gulf of Guinea). *Biodiversity and Conservation*, **10**, 45–68.

Fitzgerald, B. M. and Gibb, J. A. (2001). Introduced mammals in a New Zealand forest: long-term research in the Orongorongo Valley. *Biological Conservation*, **99**, 97–108.

Fjeldsa, J. (1999). The impact of human forest disturbance on the endemic avifauna of the Udzungwa Mountains, Tanzania. *Bird Conservation International*, **9**, 47–62.

Gagne, W. C. (1988). Conservation priorities in Hawaiian natural systems: increased public awareness and conservation action are required. *BioScience*, **38**, 264–271.

Galeano, G., Suárez, S., and Balslev, H. (1998). Vascular plant species count in a wet forest in the Chocó area on the Pacific coast of Colombia. *Biodiversity and Conservation*, **7**, 1563–1575.

Gardette, E. (1996). Microhabitats of epiphytic fern communities in large lowland rain forest plots in Sumatra. In *Pteridology in Perspective*, ed. J. M. Camus, M. Gibby, and R. J. Johns. Kew: Royal Botanic Gardens, pp. 655–658.

George, L. O. and Bazzaz, F. A. (1999). The fern understory as an ecological filter: growth and survival of canopy-tree seedlings. *Ecology*, **80**, 846–856.

Gibby, M. and Dyer, A. F. (2002). *Ex situ* conservation of globally-endangered pteridophyte species. *The Fern Gazette*, **16**, 369–370.

Given, D. R. (1993). Changing aspects of endemism and endangerment in pteridophyta. *Journal of Biogeography*, **20**, 293–302.

Given, D. R. (2002). Needs, methods and means. *The Fern Gazette*, **16**, 269–277.

Gomes, G. S., Randi, A. M., Puchalski, A., Santos, D. D. S., and Reis, M. S. D. (2006). Variability in the germination of spores among and within natural populations of the endangered tree fern *Dicksonia sellowiana* Hook. (Xaxim). *Brazilian Archives of Biology and Technology*, **49**, 1–10.

Gómez-P., L. D. (1985). Conservation of pteridophytes. *Proceedings of the Royal Society of Edinburgh Section B, Biological Sciences*, **86**, 431–433.

Gressitt, J. L., Sedlacek, J., Wise, K. A. J., and Yoshimoto, C. M. (1961). A high speed airplane trap for airborne organisms. *Pacific Insects*, **3**, 549–555.

Grund, S. and Parks, J. C. (2002). Conservation status of pteridophytes in the 49 continental United States: a preliminary report. *The Fern Gazette*, **16**, 290–294.

Guryeva, I. I. (2002). Rare fern species of Russia and reasons for their rarity. *The Fern Gazette*, **16**, 319–323.

Haitch, R. (1985). Follow-up on the news: protecting ferns. *The New York Times*, June 16, 1985.

Hamilton, A. C. (2002). Is fern conservation in the tropics possible? *The Fern Gazette*, **16**, 413–416.

Hassler, M. and Swale, B. (2001). *Checklist of World Ferns*. Published by the authors online: http://homepages.caverock.net.nz/~bj/fern/list.htm.

Heartsill-Scalley, T. and Aide, T. M. (2003). Riparian vegetation and stream condition in a tropical agriculture-secondary forest mosaic. *Ecological Applications*, **13**, 225–234.

Hegde, S. and D'Souza, L. (1996). *In vitro* propagation of *Drynaria quercifolia*, an endangered fern. In *Pteridology in Perspective*, ed. J. M. Camus, M. Gibby, and R. J. Johns. Kew: Royal Botanic Gardens, p. 171.

Helfer, S. (2006). Micro-fungal pteridophyte pathogens. *The Fern Gazette*, **17**, 259–261.

James, H. F. and Burney, D. A. (1997). The diet and ecology of Hawaii's extinct flightless waterfowl: evidence from coprolites. *Biological Journal of the Linnean Society*, **62**, 279–297.

Jamir, S. A. and Pandey, H. N. (2003). Vascular plant diversity in the sacred groves of Jainita Hills in northeast India. *Biodiversity and Conservation*, **12**, 1497–1510.

Jayakumar, M., Eyini, M., Selvinthangadurai, P., Lingakumar, K., Premkumar, A., and Kulandaivelu, G. (1999). Changes in pigment composition and photosynthetic activity of aquatic fern (*Azolla Microphylla* Kaulf.) exposed to low doses of UV-C (254 nm) radiation. *Photosynthetica*, **37**, 33–38.

Jayakumar, M., Eyini, M., Lingakumar, K., and Kulandaivelu, G. (2002). Effects of enhanced ultraviolet-B (280–320 nm) radiation on growth and photosynthetic activities in aquatic fern *Azolla Microphylla* Kaulf. *Photosynthetica*, **40**, 85–89.

Jermy, A. C. and Ranker, T. A. (2002). Epilogue – the way forward. *The Fern Gazette*, **16**, 417–424.

Kamachi, H., Komori, I., Tamura, H., Sawa, Y., Karahara, I., Honma, Y., Wada, N., Kawabata, T., Matsuda, K., Ikeno, S., Noguchi, M., and Inoue, H. (2005). Lead tolerance and accumulation in the gametophytes of the fern *Athyrium yokoscense*. *Journal of Plant Restoration*, **118**, 137–145.

Kang, M., Ye, Q., and Huang, H. (2005). Genetic consequence of restricted habitat and population decline in endangered *Isoetes sinensis* (Isoetaceae). *Annals of Botany*, **96**, 1265–1274.

Kato, M. (1993). Biogeography of ferns: dispersal and vicariance. *Journal of Biogeography*, **20**, 265–274.

Kelly, D. (1994). Demography and conservation of *Botrychium australe*, a peculiar, sparse mycorrhizal fern. *New Zealand Journal of Botany*, **32**, 393–400.

Kelsall, N., Hazard, C., and Leopold, D. J. (2004). Influence of climate factors on demographic changes in the New York populations of the federally-listed *Phyllitis scolopendrium* (L.) Newm. var. *americana*. *Journal of the Torrey Botanical Society*, **131**, 161–168.

Kessler, M. (2001). Patterns of diversity and range size of selected plant groups along an elevational transect in the Bolivian Andes. *Biodiversity and Conservation*, **10**, 1897–1921.

Kim, C. H. (2004). Conservation status of the endemic fern *Mankyua chejuense* (Ophioglossaceae) on Cheju Island, Republic of Korea. *Oryx*, **38**, 217–219.

Kingston, N. and Waldren, S. (2002). A conservation assessment of the pteridophyte flora of the Pitcairn Islands, South Central Pacific Ocean. *The Fern Gazette*, **16**, 404–410.

Kingston, N., Waldren, S., and Smyth, N. (2004). Conservation genetics and ecology of *Angiopteris chauliodonta* Copel. (Marattiaceae), a critically endangered fern from Pitcairn Island, South Central Pacific Ocean. *Biological Conservation*, **117**, 309–319.

Klekowski, E. J., Jr. (1976). Mutational load in a fern population growing in a polluted environment. *American Journal of Botany*, **63**, 1024–1030.

Klekowski, E. J., Jr. (1978). Screening aquatic ecosystems for mutagens with fern bioassays. *Environmental Health Perspectives*, **27**, 99–102.

Klekowski, E. J., Jr. (1982). Using components of the native flora to screen environments for mutagenic pollutants. In *Environmental Mutagnesis, Carcinogenesis, and Plant Biology*, ed. E. J. Klekowski, Jr. New York: Praeger Scientific, pp. 91–114.

Klekowski, E. J., Jr. and Berger, B. B. (1976). Chromosome mutations in a fern population growing in a polluted environment: a bioassay for mutagens in aquatic environments. *American Journal of Botany*, **63**, 239–246.

Klekowski, E. J., Jr. and Klekowski, E. (1982). Mutation in ferns growing in an environment contaminated with polychlorinated biphenyls. *American Journal of Botany*, **69**, 721–727.

Klekowski, E. J., Jr. and Levin, D. E. (1979). Mutagens in a river heavily polluted with paper recycling wastes: results of field and laboratory mutagen assays. *Environmental Mutagenesis*, **1**, 209–219.

Klekowski, E. J., Jr. and Poppel, D. M. (1976). Ferns: potential in-situ bioassay systems for aquatic-borne mutagens. *American Fern Journal*, **66**, 75–79.

Kluge, J. and Kessler, M. (2006). Fern endemism and its correlates: contribution from an elevational transect in Costa Rica. *Diversity and Distributions*, **12**, 535–545.

Kramer, K. U. (1993). Distribution patterns in major pteridophyte taxa relative to those of angiosperms. *Journal of Biogeography*, **20**, 287–291.

LaDeau, S. L. and Ellison, A. M. (1999). Seed bank composition of a northeastern US tussock swamp. *Wetlands*, **19**, 255–261.

Lehmann, A., Leathwick, J. R., and Overton, J. M. (2002). Assessing New Zealand fern diversity from spatial predictions of species assemblages. *Biodiversity and Conservation*, **11**, 2217–2238.

León, B. and Young, K. R. (1996a). Aquatic plants of Peru: diversity, distribution and conservation. *Biodiversity and Conservation*, **5**, 1169–1190.

León, B. and Young, K. R. (1996b). Distribution of pteridophyte diversity and endemism in Peru. In *Pteridology in Perspective*, ed. J. M. Camus, M. Gibby, and R. J. Johns. Kew: Royal Botanic Gardens, pp. 77–91.

Lindenmayer, D. B., Cunningham, R. B., Donnelly, C. F., and Franklin, J. F. (2000). Structural features of old-growth Australian montane ash forests. *Forest Ecology and Management*, **134**, 189–204.

Liu, X., Wang, J.-Y., and Wang, Q.-F. (2005). Current status and conservation strategies for *Isoetes* in China: a case study for the conservation of threatened aquatic plants. *Oryx*, **39**, 335–338.

Loyn, R. H., McNabb, E. G., Volodina, L., and Willig, R. (2001). Modelling landscape distributions of large forest owls as applied to managing forests in north-east Victoria, Australia. *Biological Conservation*, **97**, 361–376.

Lwanga, J. S., Balmford, A., and Badaza, R. (1998). Assessing fern diversity: relative species richness and its environmental correlates in Uganda. *Biodiversity and Conservation*, **7**, 1387–1398.

Machon, N., Guillon, J.-M., Dobigny, G., Le Cadre, S., and Moret, J. (2001). Genetic variation in the horsetail *Equisetum variegatum* Schleich., an endangered species in the Parisian region. *Biodiversity and Conservation*, **10**, 1543–1554.

Macía, M. J. (2004). A comparison of useful pteridophytes between two Amerindian groups of Amazonian Bolivia and Ecuador. *American Fern Journal*, **94**, 39–46.

Manandhar, N. P. (1995). A survey of medicinal plants of Jajarkot district, Nepal. *Journal of Ethnopharmacology*, **48**, 1–6.

Martin, P. H., Sherman, R. E., and Fahey, T. J. (2004). Forty years of tropical forest recovery from agriculture: structure and floristics of secondary and old-growth riparian forests in the Dominican Republic. *Biotropica*, **36**, 297–317.

Milton, S. J. and Moll, E. J. (1988). Effects of harvesting on frond production of *Rumohra adiantiformis* (Pteridophyta: Aspidiaceae) in South Africa. *Journal of Applied Ecology*, **25**, 725–743.

Moran, R. C., Klimas, S., and Carlsen, M. (2003). Low-trunk epiphytic ferns on tree ferns versus angiosperms in Costa Rica. *Biotropica*, **35**, 48–56.

Muller, S. (2000). Assessing occurrence and habitat of *Ophioglossum vulgatum* L. and other Ophioglossaceae in European forests. Significance for nature conservation. *Biodiversity and Conservation*, **9**, 673–681.

Muller, S., Jerôme, C., and Mahevas, T. (2006). Habitat assessment, phytosociology and conservation of the Tunbridge Filmy-fern *Hymenophyllum tunbrigense* (L.) Sm. in its isolated locations in the Vosges Mountains. *Biodiversity and Conservation*, **15**, 1027–1041.

Murakami, K., Maenaka, H., and Morimoto, Y. (2005). Factors influencing species diversity of ferns and fern allies in fragmented forest patches in the Kyoto city area. *Landscape and Urban Planning*, **70**, 221–229.

Musselman, L. J. (2002). The only quillwort (*Isoetes olympica* A. Braun) in Syria is threatened with extinction. *The Fern Gazette*, **16**, 324–329.

Myster, R. W. and Sarmiento, F. O. (1998). Seed inputs to microsite patch recovery on two tropandean landslides in Ecuador. *Restoration Ecology*, **6**, 35–43.

Ogden, J., Braggins, J., Stretton, K., and Anderson, S. (1997). Plant species richness under *Pinus radiata* stands on the central North Island volcanic plateau, New Zealand. *New Zealand Journal of Ecology*, **21**, 17–29.

Ouborg, N. J., Vergeer, P., and Mix, C. (2006). The rough edges of the conservation genetics paradigm for plants. *Journal of Ecology*, **94**, 1233–1248.

Ough, K. (2001). Regeneration of wet forest flora a decade after clear-felling or wildfire – is there a difference? *Australian Journal of Botany*, **49**, 645–664.

Ough, K. and Murphy, A. (2004). Decline in tree-fern abundance after clearfell harvesting. *Forest Ecology and Management*, **199**, 153–163.

Overacker, M. L. (1930). A New York state fern law. *American Fern Journal*, **20**, 115–117.

Paciencia, M. L. B. and Prado, J. (2005). Effects of forest fragmentation on pteridophyte diversity in a tropical rain forest in Brazil. *Plant Ecology*, **180**, 87–104.

Padmawathe, R., Qureshi, Q., and Rawat, G. S. (2004). Effects of selective logging on vascular epiphyte diversity in a moist lowland forest of Eastern Himalaya, India. *Biological Conservation*, **119**, 81–92.

Page, C. N. (1985). Epilogue – pteridophyte biology: the biology of the amphibians of the plant world. *Proceedings of the Royal Society of Edinburgh Section B, Biological Sciences*, **86**, 439–442.

Page, C. N. (2002). The role of natural disturbance regimes in pteridophyte conservation management. *The Fern Gazette*, **16**, 284–289.

Palmer, D. (2003). *Hawai'i's Ferns and Fern Allies*. Honolulu, HI: University of Hawai'i Press.

Parsons, P. A. (1991). Biodiversity conservation under global climatic change: the insect *Drosophila* as a biological indicator? *Global Ecology and Biogeography Letters*, **1**, 77–83.

Pausas, J. G. and Sáez, L. (2000). Pteridophyte richness in the NE Iberian Peninsula: biogeographic patterns. *Plant Ecology*, **148**, 195–205.

Pavlides, G. (1997). The flora of Prespa National Park with emphasis on species of conservation interest. *Hydrobiologica*, **351**, 35–40.

Peck, J. H., Peck, C. J., and Farrar, D. R. (1990). Comparative life history studies and the distribution of pteridophyte populations. *American Fern Journal*, **80**, 126–142.

Pence, V. C. (2002). Cryopreservation and *in vitro* methods for *ex situ* conservation of pteridophytes. *The Fern Gazette*, **16**, 362–368.

Perrie, L. R., Brownsey, P. J., Lockhart, P. J., and Large, M. F. (2003). Morphological and genetic diversity in the New Zealand fern *Polystichum vestitum* (Dryopteridaceae), with special reference to the Chatham Islands. *New Zealand Journal of Botany*, **41**, 581–602.

Phelps, O. P. (1912). A plea for fern protection. *American Fern Journal*, **2**, 22–23.

Phillips, A. (2002). Protected areas, and IUCN's World Commission on Protected Areas (WCPA) – how can they help in the conservation of ferns? *The Fern Gazette*, **16**, 278–283.

Porembski, S. and Biedinger, N. (2001). Epiphytic ferns for sale: influence of commercial plant collection on the frequency of *Platycerium stemaria* (Polypodiaceae) in coconut plantations on the southeastern Ivory Coast. *Plant Biology*, **3**, 72–76.

Poulsen, A. D. and Tuomisto, H. (1996). Small-scale to continental distribution patterns of neotropical pteridophytes: the role of edaphic preferences. In *Pteridology in Perspective*, ed. J. M. Camus, M. Gibby, and R. J. Johns. Kew: Royal Botanic Gardens, pp. 551–561.

Ramírez-Trejo, M. d. R., Pérez-García, B., and Orozco-Segovia, A. (2004). Analysis of fern spore banks from the soil of three vegetation types in the central region of Mexico. *American Journal of Botany*, **91**, 682–688.

Ramos, J. A. (1995). The diet of the Azores bullfinch *Pyrrhula murina* and floristic variation within its range. *Biological Conservation*, **71**, 237–249.

Ranal, M. A. (2003). Soil spore bank of ferns in a gallery forest of the ecological station of Panga, Uberlândia, MG, Brazil. *American Fern Journal*, **93**, 97–115.

Ranal, M. A. (2004). Bark spore bank of ferns in a gallery forest of the ecological station of Panga, Uberlândia, MG, Brazil. *American Fern Journal*, **94**, 57–69.

Ranker, T. A. (1994). Evolution of high genetic variability in the rare Hawaiian fern *Adenophorus periens* and implications for conservation management. *Biological Conservation*, **70**, 19–24.

Ranker, T. A., Gemmill, C. E. C., Trapp, P. G., Hambleton, A., and Ha, K. (1996). Population genetics and reproductive biology of lava-flow colonizing species of Hawaiian *Sadleria* (Blechnaceae). In *Pteridology in Perspective*, ed. J. M. Camus, M. Gibby, and R. J. Johns. Kew: Royal Botanic Gardens, pp. 581–598.

Rathinasabapathi, B. (2006). Ferns represent an untapped biodiversity for improving crops for environmental stress tolerance. *New Phytologist*, **172**, 385–390.

Rhazi, M., Grillas, P., Charpentier, A., and Medail, F. (2004). Experimental management of Mediterranean temporary pools for conservation of the rare quillwort *Isoetes setacea*. *Biological Conservation*, **118**, 675–684.

Rugg, H. G. (1913). Notes and news. *American Fern Journal*, **3**, 92–96.

Rumsey, F. J., Vogel, J. C., Russell, S. J., Barrett, J. A., and Gibby, M. (1999). Population structure and conservation biology of the endangered fern *Trichomanes speciosum* Willd. (Hymenophyllaceae) at its northern distributional limit. *Biological Journal of the Linnean Society*, **66**, 333–344.

Saiz, J. C. M., Parga, I. C., Humphries, C. J., and Williams, P. H. (1996). Strengthening the national and natural park system of Iberia to conserve pteridophytes. In *Pteridology in Perspective*, ed. J. M. Camus, M. Gibby, and R. J. Johns. Kew: Royal Botanic Gardens, pp. 101–123.

Samant, S. S., Dhar, U., and Rawal, R. S. (1998). Biodiversity status of a protected area in West Himalaya: Askot Wildlife Sanctuary. *International Journal of Sustainable Development and World Ecology*, **5**, 194–203.

Schofield, E. K. (1989). Effects of introduced plants and animals on island vegetation: examples from the Galápagos Archipelago. *Conservation Biology*, **3**, 227–238.

Schwartzburd, P. B., Labiak, P. H., and Salino, A. (2007). A new species of *Ctenitis* (Dryopteridaceae) from southern Brazil. *Brittonia*, **59**, 29–32.

Sheffield, E. (1996). From pteridophyte spore to sporophyte in the natural environment. In *Pteridology in Perspective*, ed. J. M. Camus, M. Gibby, and R. J. Johns. Kew: Royal Botanic Gardens, pp. 541–549.

Slocum, M. G., Aide, T. M., Zimmerman, J. K., and Navarro, L. (2004). Natural regeneration of subtropical montane forest after clearing fern thickets in the Dominican Republic. *Journal of Tropical Ecology*, **20**, 483–486.

Smith, A. R. (2005). Floristics in the 21st century: balancing user-needs and phylogenetic information. *The Fern Gazette*, **17**, 105–137.

Stein, B., Adams, J., Master, L., Morse, L., and Hammerson, G. (2000). A remarkable array: species diversity in the United States. In *Precious Heritage: The Status of Biodiversity in the United States*, ed. B. Stein, L. Kutner, and J. Adams. New York: Oxford University Press, pp. 55–92.

Su, Y. J., Wang, T., Zheng, B., Jiang, Y., Chen, G., and Gu, H. (2004). Population genetic structure and phylogeographical pattern of a relict tree fern, *Alsophila spinulosa* (Cyatheaceae), inferred from cpDNA *atp*B-*rbc*L intergenic spacers. *Theoretical and Applied Genetics*, **109**, 1459–1467.

Su, Y. J., Wang, T., Zheng, B., Jiang, Y., Chen, G. P., Ouyang, P. Y., and Sun, Y. F. (2005). Genetic differentiation of relictual populations of *Alsophila spinulosa* in southern China inferred from cpDNA trnL-F noncoding sequences. *Molecular Phylogenetics and Evolution*, **34**, 323–333.

Tuomisto, H. and Poulsen, A. D. (1996). Influence of edaphic specialization on pteridophyte distribution in neotropical rain forests. *Journal of Biogeography*, **23**, 283–293.

Turner, I. M., Wong, Y. K., Chew, P. T., and Ibrahim, A. B. (1996). Rapid assessment of tropical rain forest successional status using aerial photographs. *Biological Conservation*, **77**, 177–183.

Unwin, G. L. and Hunt, M. A. (1996). Conservation and management of soft tree fern *Dicksonia antarctica* in relation to commercial forestry and horticulture. In

Pteridology in Perspective, ed. J. M. Camus, M. Gibby, and R. J. Johns. Kew: Royal Botanic Gardens, pp. 125–137.

Vázquez, J. A. and Norman, G. (1995). Identification of site-types important for rare ferns in an area of deciduous woodland in northwest Spain. *Vegetatio*, **116**, 133–146.

Vulcz, R., Vulcz, L., Greer, L. D., and Lawrence, G. (2002). Sustainable production of tree ferns in South-East Australia. *The Fern Gazette*, **16** 388–392.

Wang, T., Su, Y. J., Li, X. Y., Zheng, B., Chen, G. P., and Zeng, Q. L. (2004). Genetic structure and variation in the relict populations of *Alsophila spinulosa* from southern China based on RAPD markers and cpDNA *atp*B-*rbc*L sequence data. *Hereditas*, **140**, 8–17.

Wang, J.-Y., Gituru, R. W., and Wang, Q.-F. (2005). Ecology and conservation of the endangered quillwort *Isoetes sinensis* in China. *Journal of Natural History*, **39**, 4069–4079.

Wardlaw, A. C. (2002a). Conservation of tree ferns *ex situ*. *The Fern Gazette*, **16**, 393–397.

Wardlaw, A. C. (2002b). Horticultural approaches to the conservation of British ferns. *The Fern Gazette*, **16**, 356–361.

Watkins, J. E., Jr., Cardelus, C., Colwell, R. K., and Moran, R. C. (2006). Species richness and distribution of ferns along an elevational gradient in Costa Rica. *American Journal of Botany*, **93**, 73–83.

Wester, L. (1994). Weed management and the habitat protection of rare species: a case study of the endemic Hawaiian fern *Marsilea villosa*. *Biological Conservation*, **68**, 1–9.

Wester, L., Delay, J., Hoang, L., Iida, B., Kalodimos, N., and Wong, T. (2006). Population dynamics of *Marsilea villosa* (Marsileaceae) on Oahu, Hawaii. *Pacific Science*, **60**, 385–403.

Williams, P. H. and Gaston, K. J. (1994). Measuring more of biodiversity – can higher-taxon richness predict wholesale species richness. *Biological Conservation*, **67**, 211–217.

Williams-Linera, G., Palacios-Rios, M., and Hernandez-Gomez, R. (2005). Fern richness, tree species surrogacy, and fragment complementarity in a Mexican tropical montane cloud forest. *Biodiversity and Conservation*, **14**, 119–133.

Wilson, K. A. (2002). Continued pteridophyte invasion of Hawaii. *American Fern Journal*, **92**, 179–183.

Windisch, P. G. (2002). Fern conservation in Brazil. *The Fern Gazette*, **16**, 295–300.

Winslow, E. J. and Benedict, R. C. (1916). Notes and news: the fern-picking industry. *American Fern Journal*, **6**, 18–21.

Wittig, R. (2002). Ferns in a new role as a frequent constituent of railway flora in Central Europe. *Flora*, **197**, 341–350.

Wolf, P. G., Schneider, H., and Ranker, T. A. (2001). Geographic distributions of homosporous ferns: does dispersal obscure evidence of vicariance? *Journal of Biogeography*, **28**, 263–270.

Woolley, A. and Kirkpatrick, J. B. (1999). Factors related to condition and rare and threatened species occurrence in lowland, humid basalt remnants in northern Tasmania. *Biological Conservation*, **87**, 131–142.

WWF–IUCN (1994–1997). *Centres of Plant Diversity: A Guide and Strategy for Their Conservation*. Cambridge: IUCN Publications Unit.

Yatskievych, G. and Spellenberg, R. W. (1993). Plant conservation in the flora of North America region. In *Flora of North America North of Mexico*, Vol. 1, ed. Flora of North America Editorial Committee. New York: Oxford University Press, Chapter 10.

Zenkteler, E. K. (2002). *Ex situ* breeding and reintroduction of *Osmundia regalis* L. in Poland. *The Fern Gazette*, **16**, 371–376.

Zhang, X.-C., Jia, J.-S., and Zhang, G. M. (2002). Survey and evaluation of the natural resources of *Cibotium barometz* (L.) J. Smith in China, with reference to the implementation of the CITES convention. *The Fern Gazette*, **16**, 383–392.

11

Ex situ conservation of ferns and lycophytes – approaches and techniques

VALERIE C. PENCE

11.1 Introduction

Although generally less conspicuous than seed plants, ferns and lycophytes have been the object of human interest since at least the time of the ancient Greeks. As with other plant and animal species, habitat loss is the major threat, but harvesting for medicine, food, or as ornamental plants has also depleted some species in the wild. Of the approximately 13 000 named species, approximately 800 are of conservation concern (Walter and Gillett, 1998). However, some areas of the world are still poorly explored for ferns and lycophytes, and although the exact number of endangered species is uncertain, many species are in habitats that are under pressure. While conserving species *in situ* is the ideal, it cannot always be ensured. As a result, *ex situ* conservation methods can play a complementary role in ensuring the survival of these species, by providing a back-up of genetic diversity for the populations in the wild.

Growing ferns and lycophytes *ex situ* as horticultural or botanically interesting specimens has a long history, but other *ex situ* conservation methods, such as spore banking and cryopreservation are more recent. Whereas spore banking follows protocols similar to those of seed banking, the independent alternation of generations in ferns and lycophytes provides opportunities for cryopreservation that are not available in seed plants. Previous reviews have described some of these possibilities (Page *et al.*, 1992; Pattison *et al.*, 1992; Pence, 2002). This chapter will attempt to provide an updated description of the methods available for *ex situ* conservation of ferns and lycophytes and how they can be used as part of an integrated approach to conservation.

Biology and Evolution of Ferns and Lycophytes, ed. Tom A. Ranker and Christopher H. Haufler. Published by Cambridge University Press. © Cambridge University Press 2008.

11.2 Methods for *ex situ* conservation

11.2.1 Ex situ *cultivation*

Plants have been grown *ex situ*, or away from their natural habitats, in gardens and displays for hundreds of years. Yet, for most of that time, transport of exotic plants was a limiting factor, and plants that could be grown from long-lived seeds or roots had the best chance of being cultivated *ex situ*. In the case of ferns, however, studies of the mode of reproduction and the elusive "fernseed" did not begin until the 1790s and stages in the alternation of generations were not fully elucidated until the mid-nineteenth century.

Transport of whole plants for growth *ex situ* was revolutionized by the protected environment of the Wardian case. It was in pursuit of fern cultivation in the 1830s that Dr. Nathaniel Ward developed the case that bears his name, observing that his ferns fared better in the polluted air of London when they were protected in closed containers (Ward, 1852). The Wardian case gave further momentum to fern collecting in the midst of an era in the mid-nineteenth century known as "Pteridomania" or the Victorian Fern Craze. As an understanding of fern propagation was just beginning, most ferns were wild collected, and in some cases, as in the case of what is now perhaps Britain's rarest fern, *Woodsia ilvensis*, species were severely depleted from the wild, threatening the plants with extirpation (Lusby *et al.*, 2002).

The Fern Craze, however, gave impetus to the display of ferns, and notable collections were assembled at several botanical gardens and displayed in conservatories, gardens, stumperies, and rooteries. These were to provide the basis for collections that aided research and maintained *ex situ* fern germplasm. Present-day botanical gardens continue to play a unique role in *ex situ* fern and lycophyte conservation, and have the capacity to maintain collections, to propagate rare species, and to preserve germplasm worldwide (e.g., Theuerkauf, 1993; Goel, 2002), and some house particularly rare, old, or unique specimens. There are also some personal collections of ferns of high quality that can provide valuable information on the conditions required for growing these plants *ex situ* (Wardlaw, 2002). Although *ex situ* cultivation of ferns is limited by space and resources in the number of genotypes that can be maintained, it can provide protection for individuals of particularly rare species, as well as serve as a resource for educating both researchers and the general public on the growth and development of these species.

11.2.2 Ex situ *banking of spores*

In contrast to the *ex situ* cultivation of whole plants, the banking of spores can preserve thousands of genotypes in much less space than is needed to grow one plant. In many ways similar to seed banking, traditional spore banking is based on the capacity of spores to remain viable through extreme desiccation. Many spores undergo drying in nature when they are dispersed and can survive in that state for from several days to many years, depending on the species. This has formed the basis for the short-term spore banking practiced by the spore exchanges of many fern societies. Organizations such as the American Fern Society and the British Pteridological Society, among others, hold spores that are generally maintained at 4 °C for several years, for distribution to members. Over the past half-century, these societies have played an important role in making genetic material available to a wider audience, thereby securing its growth *ex situ* in horticultural collections.

Long-term spore banking for germplasm conservation, using approaches analogous to seed banking, is a more recent phenomenon, but should be effective in efficiently preserving endangered fern and lycophyte genetic diversity *ex situ*. Seed banks are classified as active or base collections, depending on the activity of use and distribution of the material. An active seed bank is similar to the spore banks of fern societies, in that materials are held for a short time and readily distributed. A base collection is one in which the goal is to maintain the samples for many years or decades, in secure storage, as a resource for conservation efforts if wild populations are threatened or lost. Whereas there are a number of active spore banks around the world, there are fewer base collections. Two that fall into this category are the Germplasm Bank of the Botanical Garden of Valencia University, ICBiBE (Spain) and the Pteridophyte Bank at the Cincinnati Zoo and Botanical Garden (USA). Both are exploring cryopreservation methods for long-term storage of fern and lycophyte spores and tissues and conducting long-term experiments evaluating liquid nitrogen storage in comparison with other methods. The European Plant Conservation Strategy also sets the target of developing protocols for and establishing a spore bank for pteridophytes in Europe (Planta Europa, 2007).

Spore storage is limited by the decline in viability in stored spores over time. The rate of this decline varies with the species, but is slower in brown (non-chlorophyllous) spores and more rapid in green (chlorophyllous) spores. Lloyd and Klekowski (1970) reviewed reports of fern longevity and concluded that green spores had a mean germination time of 1.5 days and a mean viability of 48 days, while brown spores had a mean germination time of 9.5 days and a mean viability of 1045 days. There is significant variation within these groups (Dyer, 1979), with green spores of *Equisetum* losing viability within about 2 weeks (Lebkuecher, 1997), while brown spores of some species can survive for years or decades, as reported for spores of *Pellaea* and sporocarps of *Marsilea* (Allsopp, 1952; Windham *et al.*, 1986). When viability decreases, however, this is reflected not only as a decrease in the percentage germination, but also in the rate at which germination progresses. The normal development of gametophytes and the production of sporophytes from spores that do germinate are also affected (Smith and Robinson, 1975; Beri and Bir, 1993; Chá-Chá et al., 2003).

Because spore longevity is species dependent and ranges from several days to many years, a number of studies have been directed at examining and developing methods for extending viability in stored spores. With seeds, it has been shown that lowering the moisture content and temperature increase longevity (Walters, 2004). Similar methods have been applied to spores of ferns and lycophytes, many of which appear to tolerate desiccation and cold or frozen storage in a manner similar to desiccation tolerant ("orthodox") seeds. Brown spores generally have the longest natural longevity, and dry brown spores of the tree fern, *Cyathea delgadii*, maintained 19% viability after 2 years at $-12\,°C$, while at $25\,°C$, viability was reduced to 8% within 3 months and was lost entirely after 1 year of storage (Simabukuro et al., 1998). At $3\,°C$, this species maintained viability for at least 9 months (Randi and Felippe, 1988).

Green spores also appear to be tolerant of desiccation, but they are generally much more short lived than brown spores, and extending their viability has been of particular interest. Cold storage ($4-5\,°C$) has provided high survival rates after 2 years for dry spores of *Matteuccia struthiopteris*, stored in glass vials capped with parafilm, and of *Onoclea sensibilis*, stored in polyethylene bags (Miller and Miller, 1961; Gantt and Arnott, 1965). The short-lived spores of *Equisetum telmateia* maintained 74% viability for 2 years when stored under glycerine at $-10\,°C$ (Jones and Hook, 1970). In an attempt to simplify *Equisetum* spore storage, Whittier (1996) stored ripe cones of *E. hyemale* at $-70\,°C$ for more than a year and maintained 37% viability of the enclosed spores, although there was some decline in the ability of these spores to form gametophytes. Thus, lowering the temperature of storage of dried green spores appears to hold promise in extending viability well beyond that at ambient temperatures. However, because viability of green spores begins to decline soon after maturity, it is important that protocols directed at extending longevity to green spores be applied before the decline in viability begins.

Although lowering storage temperatures appears to increase longevity in both seeds and spores, the question of whether liquid nitrogen (LN) ($-196\,°C$) will extend viability of spores longer than storage at $-20\,°C$ is still unanswered. Many spores, however, appear to tolerate exposure to LN. No differences were detected in a long-term storage experiment with three brown-spored species (*Pteris* sp., *Cyrtomium falcatum*, and *Polystichum tsus-sinense*) stored at $4\,°C$, $-20\,°C$ and in LN after 75 months of storage, although this experiment is continuing (Pence, 2000a). Spores from several species from different ecological habitats have also survived cryostorage (Ballesteros et al., 2006), and spores of the endangered *Cyathea spinulosa* have shown over 90% survival after exposure to LN and slow thawing (Agrawal et al., 1993). Dried green spores of *Onoclea sensibilis* and *Osmunda regalis* tolerate LN exposure, and spores of *O. regalis* have survived up to 18 months of storage in LN, using the encapsulation dehydration method (Pence, 2000a). However, repeated freeze–thaw cycles, even with dry spore samples, reduce viability (Ballesteros et al., 2004).

Whereas the brown and green spores that have been examined appear tolerant of drying, in some cases, wet storage of spores may be more effective in accomplishing short-term storage goals, compared with dry storage at non-freezing temperatures. Stokey (1951) demonstrated that a portion of a sample of short-lived green spores of *Osmunda regalis* could survive more than 3.5 years when kept moist at 2–6 °C. Green spores of *Lygodium heterodoxum* maintained 40% viability for 1 year when imbibed and kept in darkness at 23–25 °C, whereas dry spores at the same temperature retained less than 1% viability (Pérez-García et al., 1994). Spores of four species of *Asplenium* were compared for survival for up to 12 months through wet and dry storage at −20 °C, 5 °C, and 20 °C. Three of the species showed survival through wet or dry storage at 5 °C and 20 °C, although there was somewhat higher survival with dry storage. In contrast, *A. ruta-muraria* showed more survival through wet storage (Aragón and Pangua, 2004). In similar experiments, spores of five endangered species were stored wet and dry at these same three temperatures for 12 months. Results were species dependent and apparently related to the ecological adaptations of the species (Quintanilla et al., 2002). Survival declined in hygrophilous species after 6 or 12 months dry storage at −20 °C, 5 °C, or 20 °C. *Equisetum arvense* spores stored in a nutrient solution at 5 °C maintained only 7% viability after 3 months of storage (Castle, 1953). In contrast, spores of *Osmunda*, that are short lived when stored dry, showed survival in soil for at least 1 year (Dyer and Lindsay, 1996). The benefits of wet storage may vary with the species and will only be effective at room temperature for species that require light for germination. Wet storage at cold, but not freezing, temperatures can be advantageous in reducing bacterial and fungal contamination and in reducing dark germination (Quintanilla et al., 2002). Studies of wet spore storage *ex situ* may provide insight into the natural history of spores in the soil and their longevity *in situ*. The potential of soil spore banks for use as conservation tools has also been raised (Dyer, 1994; Dyer and Lindsay, 1996). The growth of plants regenerated from soil samples *ex situ* can provide a source of material for increasing genetic diversity or reintroducing species that may have been extirpated from the site of the soil sample.

Wet storage at subfreezing temperatures has generally been lethal, as removal of freezable water is critical for survival through freezing. Spores representing two populations of *Cryptogramma crispa* lost much of their viability when spores were sown on nutrient medium and then exposed to −18 °C, although spores from a third population retained most of their viability (Pangua et al., 1999). Damage may have occurred if the spores imbibed water from the medium before they were frozen. Similarly, the loss of viability of *Equisetum* spores frozen while still in ripe cones (Whittier, 1996) may have been a result of incomplete drying and freezing lethality for those spores with higher moisture contents. In contrast, Hill (1971) reported a high percentage of germination of brown spores after freezing in a liquid medium, but there are several reports in which spores stored wet at −20 °C have not survived (Quintanilla et al., 2002; Aragón and Pangua, 2004).

Although there is empirical evidence that spore longevity can be increased through various protocols, an understanding of the physiology of the loss of spore viability is just beginning. Loss of viability of *Equisetum hyemale* spores equilibrated with 2% relative humidity at 25 °C was primarily because of the loss of the ability to recover water oxidation and photosystem II-core function (Lebkuecher, 1997), whereas partial loss of viability of *Pteris vittata* spores over 100 days of storage was correlated with a decrease in sugar, amino acid, and protein content of the spores (Beri and Bir, 1993). As with seed storage, the longevity of fern spores and the optimal conditions of storage will be closely related to the moisture content of the spores. There has been much research on determining optimal storage moisture levels for seeds (Walters, 2004), but similar work with fern spores is just beginning. Recent results with five species of ferns with brown spores suggest that the affinity of these fern spores for water is low and that they dry quickly but rehydrate more slowly. Physicochemical studies suggest that the viability of these spores will be extended by freeze-storage, but that the optimum water content for freezing is lower and more narrow than that for seeds (Ballesteros and Walters, 2007).

A further consideration for *ex situ* spore storage is the physiological effects of the germination protocols that are used to evaluate viability. Many spore germination experiments are done using surface sterilized spores, but there is evidence that the stress of surface sterilization may interact with other factors. In *Platycerium bifurcatum*, 14 month old spores germinated at the same rate as 2 month old spores, unless they were surface sterilized with NaOCl, in which case germination was reduced in the older spores (Camloh, 1999). NaOCl appears to remove trace elements, such as Ca^{2+} and K^+, which may be needed for germination in some spores (Miller and Wagner, 1987).

As a better understanding of spore physiology is acquired, storage protocols, as well as germination protocols, can be refined and possibly predicted for new species. Long-term germplasm storage protocols for spores of ferns and lycophytes have been tested on a relatively small number of species, and broader studies, including both green and brown spore species from a variety of ecological habitats are needed in order to refine protocols and increase their effective applicability within this group of taxa.

11.3 *In vitro* cultures and collections

In addition to spores, gametophytes and sporophytes offer opportunities for *ex situ* conservation as *in vitro* cultures. Spores can be surface sterilized and sown *in vitro* to initiate gametophyte and subsequently sporophyte cultures. Alternatively, gametophyte and sporophyte tissues can be surface sterilized to initiate cultures directly.

A benefit of maintaining ferns and lycophytes *in vitro* is that they can be isolated from contamination from spores of other species, something that is often difficult to do in greenhouse collections. *In vitro* culture also requires less space than the culture of whole plants and can be kept free from disease and pathogens. Many ferns and a few lycophyte species have been grown *in vitro* (Capellades Queralt *et al.*, 1991; Fernández and Revilla, 2003; Pence, 2003) and these methods appear to be adaptable to a wide variety of species. When applied to endangered taxa, *in vitro* propagation can provide plants for research, display, education, and reintroduction or augmentation projects, *in situ*. Tissue culture propagation, however, can only provide clonal material from each individual source. Thus, the genetic diversity of tissue culture lines will reflect the diversity of the spores or tissues used to initiate those lines. As a result, multiple genotypes used to initiate cultures will increase the value of the collection for conservation. Since *in vitro* cultures can also undergo genetic changes, particularly in long-term cultures, checking for somaclonal variation visually or through biochemical means is also prudent (Karp, 1994).

In vitro methods have been used to propagate sporophytes of several endangered ferns, including *Polystichum drepanum* (Chá-Chá *et al.*, 2003), *Todea barbara* (Oliphant, 1989), *Cibotium schiedei* (Fay, 1992), and *Asplenium heteroresiliens*, among others. Surface sterilized spores of the latter were germinated *in vitro* and the resulting gametophytes produced sporophytes that can be propagated indefinitely *in vitro* (Pence, unpublished data). Rooted plants have been transferred from culture to soil and acclimatized to ambient conditions, producing plants for display and outplanting.

In vitro cultures of gametophytes or sporophytes can also be grown as *ex situ* collections of rare germplasm. Such collections can preserve more genetic diversity in a smaller space than horticultural collections, and the plants can be kept free of pests and pathogens, with less labor than required for traditional greenhouse collections. *In vitro* collections can accumulate genetic or epigenetic changes over time, but slowing growth by incubating the cultures at lower temperatures can decrease this risk (Hvoslef-Eide, 1990).

Growing ferns in sterile soil in enclosed sterile boxes is a method that combines some of the techniques of *in vitro* culture with horticultural methods. When only a few spores of a rare species are available, surface sterilization presents risks of losing some of the spores during the process and of damaging or stressing the spores (Camloh, 1999). Under these circumstances, spores can be sown in sterile containers of moist sterile soil for germination, and this has proven effective for producing sporophytes of several rare species from south Florida. This method can be useful when *in vitro* methods are not required for cloning or cryopreservation, or it can be used as a method for producing sporophytes that can then be sterilized and used to initiate *in vitro* cultures. A further consideration in germinating small amounts of spores from rare species *ex situ* is the possibility of contamination of the spore sample with spores of other species. When germination rates are low, it is important to confirm the identity of the resulting sporophytes, as they may be originating from a small number of contaminating spores in the source sample.

11.4   *Ex situ* cryostorage of gametophytes

When grown *in vitro*, fern gametophyte tissue provides another subject for cryopreservation and germplasm storage. Fern gametophytes are highly regenerative, and can regrow from fragments or areas of tissue that survive freezing. Some fern and lycophyte gametophytes appear to possess natural desiccation tolerance (Mottier, 1914; Page, 1979; Ong and Ng, 1998), and such tissues might be adaptable to cryopreservation in the dry state, but most are not naturally tolerant of drying. However, a 1 week preculture on 10 µM abscisic acid (ABA), the natural plant stress hormone, improved tolerance to rapid drying and subsequent exposure to LN in *in vitro* grown gametophytes of six species of ferns (Pence, 2000b). Although whole gametophytes did not survive, survival was measured as the number of gametophyte pieces with fragments that regenerated growth.

Better survival of gametophyte tissue is obtained, however, when the encapsulation dehydration (Fabre and Dereuddre, 1990) procedure is used in preparation for storage in LN. This technique has been applied to the rare filmy fern, *Hymenophyllum tunbrigense* (Wilkinson, 2002), as well as to six other species of ferns (Pence, 2000b). In this procedure, fragments of gametophytic tissues are encapsulated in beads of alginate gel, exposed overnight to a solution with a high concentration of sucrose, dried to <20% moisture and then rapidly frozen in LN. Upon rewarming, the beads are placed onto a recovery medium where the tissues rehydrate and resume growth. Gametophyte tissue stored in this manner has been recovered after 3.5 years (Pence 2000b) and further samples remain in LN to determine whether even longer storage times are possible.

Cryopreservation of *in vitro* grown gametophytes offers another option, in addition to spores, for long-term storage of fern and lycophyte germplasm *ex situ*. Particularly if only a small number of spores is available, they can be germinated to generate gametophytes, which can be readily multiplied *in vitro*. The clonal gametophytes can then be cryopreserved using the encapsulation dehydration procedures to provide more material for storage than was available from the original spores. Another strategy would be to propagate gametophytes vegetatively *in vitro*, cryopreserving a portion of each genetic line as gametophytes, while using the remaining tissue to produce sporophytes.

11.5 *Ex situ* cryostorage of sporophytes

Cryopreservation methods are also applicable to sporophytic tissues of ferns and lycophytes, particularly to the shoot tips of *in vitro* grown shoots. Shoot tips of many species of seed plants have been successfully cryopreserved (Engelmann and Tagaki, 2000; Towill and Bajaj, 2002), and it is one method for preserving clonal tissues, tissues from species that do not produce seeds or spores, or tissues from species with seeds that cannot survive the drying required in preparation for freezing.

Some fern cultures are more difficult to adapt to shoot tip freezing, since the growing points are basal, and they are more difficult to isolate than the apical shoot tips that are characteristic of many seed plants *in vitro*. However, in preliminary experiments, successful recovery of isolated basal shoot buds from *in vitro* cultures of *Adiantum tenerum* cryopreserved in this laboratory using the encapsulation dehydration procedure provided evidence that these tissues hold potential as subjects for freezing. More easily isolated are apical shoot tips of species such as *Selaginella uncinata*, which have survived cryostorage for at least 1.5 years (Pence, 2001). Species such as *S. uncinata* may possess some natural adaptations to desiccation, as do some other ferns and lycophytes (Proctor and Pence, 2002), and these may adapt more readily than other species to the dehydrating stress involved in preparing tissues for cryopreservation. *Selaginella sylvestris*, adapted to the rainforests of the Neotropics, did not survive the procedures used for *S. uncinata* (unpublished results). However, several methods are available for cryopreservation of shoot tips, including encapsulation vitrification (Tannoury et al., 1991), vitrification (Sakai et al., 1990), and slow freezing (Withers, 1985) in addition to encapsulation dehydration (Fabre and Dereuddre,

1990), and when one method is not effective, another one may be, depending on the species. These methods may be especially useful in preserving *in vitro* lines of endangered ferns and lycophytes that are generated for propagation. Storage of such lines can reduce the space and costs involved in maintaining the cultures long-term.

11.6 *In vitro* collecting for *ex situ* conservation

Although spore germination is the most efficient method for propagation, there are times when spores are not readily available. In those instances, either the entire plant can be removed and grown *ex situ* or pieces of tissue can be removed in order to initiate tissue cultures of the plant for propagation *ex situ*. The latter process is known as "*in vitro* collecting," or IVC, and takes the methods of plant tissue culture to the field. Small pieces of tissue (shoot tips, young leaves, or other young tissues) are removed from the plant, surface sterilized and transferred to containers of sterile medium that are brought into the field (Pence et al., 2002). By using antimicrobial agents in the medium, contamination of the cultures by endogenous bacteria and fungi is kept to a minimum. Although this technique has been used with a number of species, it is best adapted to species with apical shoot tips, so that young shoot tips and immature leaves that are elevated from the ground can be removed. Many ferns, however, have their growth meristems near the ground, which makes them difficult to find, destructive to remove, and difficult to sterilize. There are several reports of regeneration of ferns *in vitro* from leaf tissues (e.g., Camloh et al., 1999; Calderon-Saenz, 2000; Ambrozic-Dolinsek et al., 2002), and these could form the basis for procedures for *in vitro* collecting from fern fronds.

In my laboratory, I have used IVC successfully with two lycophytes, *Selaginella sylvestris*, which has apical shoot tips, and *Isoëtes engelmannii*, with basal meristems. Using a fungicide and antibiotics in the medium, *in vitro* cultures of the tropical *S. sylvestris* were established from shoot tips collected in Trinidad. In the case of *I. engelmannii*, pieces of wild plants were isolated with a vertical cut through the basal corm-like structure. The leaves on the isolated pieces were cut back and the tissue surface sterilized in 70% ethanol, dipped in PPM and placed into vials of medium in the field. The remaining portion of the plant was dug up and transported to the greenhouse, so that its survival after excision could be monitored. *In vitro* cultures were established from the isolated sections of eight plants, and the remaining sections of the plants growing in soil appeared to be undamaged and continued to grow. Work with the common *I. engelmannii* has been used to develop methods for culturing rare species, such as *I. louisianensis* (Figure 11.1).

294 Valerie C. Pence

Figure 11.1 In vitro cultures of the endangered *Isoëtes louisianensis*.

IVC is a very flexible technique that has been adapted for use with a number of species (Pence et al., 2002). When spores are not available or are not viable, IVC can provide a method for efficiently collecting germplasm from many genotypes of rare species for growth *ex situ* while leaving the plants intact and relatively undisturbed *in situ*.

11.7   Current status and future perspectives

With the increasing need for *ex situ* conservation and the development of a number of methods for *ex situ* propagation and germplasm preservation for seed plants, it is natural that these approaches are now beginning to be applied to ferns and lycophytes. It is becoming evident that spores are not entirely equivalent to seeds in their storage physiology (Ballesteros and Walters, 2007), but the similarities that exist have helped direct the development of spore banks that mirror *ex situ* seed banks. The additional opportunities for cryostoring gametophytic and sporophytic tissues and for propagating both gametophytes and sporophytes *in vitro* provide researchers with a variety of tools for *ex situ* conservation that can complement traditional *ex situ* cultivation.

With several methods available for *ex situ* conservation of ferns and lycophytes, the question arises: is there one optimal method for all species? It is likely that there is not. Rather, a number of factors should be considered in developing an *ex situ* conservation strategy for a particular species. These would include the conservation needs of the species and the number of genotypes available. Is there a need for increasing population numbers? Can this be done through spore collecting and germination, or are there few or no spores available? If not, what other propagules are available? If spores are available for *ex situ* storage, is short- or long-term storage (or both) needed? Is wet or dry storage most appropriate? Does the species exhibit desiccation tolerance in its vegetative tissues? By answering these types of questions, a strategy can be formulated and the appropriate techniques applied.

As with seed plants, these techniques were first developed with non-endangered species, but there are now a number of examples where they have been applied to rare and endangered species, such as *Cyathea spinulosa* (Agrawal et al., 1993); *Polystichum drepanum* (Madeira Island) (Chá-Chá et al., 2003); *Polystichum aleuticum* (Alaska) (Holloway and Boyd, 1993). As more germplasm banks are established for ferns and lycophytes, more information will be gathered on the storage physiology of rare species.

Even as these techniques are transferred to deal with the conservation of ferns and lycophytes, there is still much less known about the response of these species to propagation and preservation technologies, compared with seed plants. Much more needs to be learned about the physiology of spores, gametophytes, and sporophytes, and how their ecological adaptations relate to the abilities of these tissues to be stored. There is also a need for more systematic comparative studies of germplasm storage methods using a variety of species adapted to different ecological habitats, in order to understand how broadly applicable the methods are and how much variation there is between species.

There are also far fewer systematic programs for banking fern and lycophyte germplasm compared with seed plants, for which there are numerous seed banks and several massive seed banking initiatives. Ferns and lycophytes represent fewer species than seed-bearing taxa and their spores require less storage space than most seeds. Thus, their addition to existing *ex situ* conservation programs should not pose an unreasonable burden. This could be facilitated by the greater dissemination of information on germplasm banking methods to practitioners who could implement these methods within the broader framework of their programs.

There may also be ways of developing new approaches and new alliances to further develop *ex situ* conservation programs for ferns and lycophytes. Spores from herbarium specimens have proven to be viable in some cases (Windham et al., 1986), but curatorial practices, such as fumigation, that are needed in many herbaria, cause this resource to be of varying reliability. Because spores are easily transported and most are easily adapted to cryostorage, a system whereby spores from fresh herbarium specimens could be collected and routed to spore banks for storage could provide a simple and efficient method for obtaining genetic diversity for spore or tissue banks. It would require appropriate permits and agreements and compliance with the Convention on Biological Diversity, but it could provide a resource that could be easily managed and maintained for the future.

Finally, linking *ex situ* conservation with *in situ* conservation is the ultimate goal (see Chapter 10). The role of *ex situ* conservation is to support the survival of endangered species *in situ* by providing a back-up of preserved germplasm, as well as propagation methods for augmenting small populations and for reintroduction projects. *In situ* soil spore banks may provide a novel resource for spores that can be germinated and grown *ex situ*, providing materials for reintroduction or cryostorage (Lindsay and Dyer, 1990), and studies suggest that it may be possible to recover species or genotypes that are no longer growing as sporophytes in a region.

The need for conserving ferns and lycophytes is as pressing as for seed plants, because habitats that contain both are being lost. The conservation needs of individual species will vary but they can be addressed by looking for opportunities – opportunities in the variety of *ex situ* conservation methods available as well as in the various life stages of ferns and lycophytes. Opportunities must also be sought to integrate these species into already existing *ex situ* conservation programs or to start new initiatives. The technologies are in hand and are being refined. They only need to be implemented and applied to rare and endangered ferns and lycophytes, in order to preserve these ancient plants for the future.

References

Agrawal, D. C., Pawar, S. S., and Mascarenhas, A. F. (1993). Cryopreservation of spores of *Cyathea spinulosa* Wall. ex. Hook. f.: an endangered tree fern. *Journal of Plant Physiology*, **142**, 124–126.

Allsopp, A. (1952). Longevity of *Marsilea* sporocarps. *Nature*, **169**, 79–80.

Ambrozic-Dolinsek, J., Camloh, M., Bohanec, B. and Zel, J. (2002). Apospory in leaf culture of staghorn fern (*Platycerium bifurcatum*). *Plant Cell Reports*, **20**, 791–796.

Aragón, C. F. and Pangua, E. (2004). Spore viability under different storage conditions in four rupicolous *Asplenium* L. taxa. *American Fern Journal*, **94**, 28–38.

Ballesteros, D. and Walters, C. (2007). Water properties in fern spores: sorption characteristics relating to water affinity, glassy states, and storage ability. *Journal of Experimental Botany*, **58**, 1185–1196.

Ballesteros, D., Ibars, A. M., and Estrelles, E. (2004). New data about pteridophytic spore conservation in germplasm banks. *Planta Europa IVth Conference, Valencia, Spain.* www.nerium.net/plantaeuropa/Download/Procedings/Ballesteros_daniel.pdf.

Ballesteros, D., Estrelles, E., and Ibars, A. M. (2006). Responses of pteridophyte spores to ultrafreezing temperatures for long-term conservation in germplasm banks. *The Fern Gazette*, **17**, 293–302.

Beri, A. and Bir, S. S. (1993). Germination of stored spores of *Pteris vittata* L. *American Fern Journal*, **83**, 73–78.

Calderon-Saenz, E. (2000). Production of adventitious buds on the leaves in *Dicksonia sellowiana*. *American Fern Journal*, **90**, 105–108.

Camloh, M. (1999). Spore age and sterilization affects germination and early gametophyte development of *Platycerium bifurcatum*. *American Fern Journal*, **89**, 124–132.

Camloh, M., Vilhar, B., Zel, J., and Ravnikar, M. (1999). Jasmonic acid stimulates development of rhizoids and shoots in fern leaf culture. *Journal of Plant Physiology*, **155**, 798–801.

Capellades Queralt, M., Beruto, M., Vanderschaeghe, A., and DeBergh, P. C. (1991). Ornamentals. In *Micropropagation Technology and Application*, ed. P. C. Debergh and R. H. Zimmerman. Dordrecht: Kluwer, pp. 215–229.

Castle, H. (1953). Notes on the development of the gametophytes of *Equisetum arvense* in sterile media. *Botanical Gazette*, **114**, 323–328.

Chá-Chá, R., Fernandes, F., and Romano, A. (2003). Preservation of *Polystichum drepanum* (Sw.) C. Presl. an endemic pteridophyte of Madeira Island. *Revista de Biologia*, **21**, 7–16.

Dyer, A. F. (1979). The culture of fern gametophytes for experimental investigation. In *The Experimental Biology of Ferns*, ed. A. F. Dyer. London: Academic Press, pp. 253–305.

Dyer, A. F. (1994). Natural soil spore banks. Can they be used to retrieve lost ferns? *Biodiversity and Conservation*, **3**, 160–175.

Dyer, A. F. and Lindsay, S. (1996). Soil spore banks – a new resource for conservation. In *Pteridology in Perspective*, ed. J. M. Camus, M. Gibby, and R. J. Johns. Kew: Royal Botanic Gardens, pp. 153–160.

Engelmann, F. and Takagi, H. (2000). *Cryopreservation of Tropical Plant Germplasm*. Rome: International Plant Genetic Resources Institute.

Fabre, J. and Dereuddre, J. (1990). Encapsulation-dehydration: a new approach to cryopreservation of *Solanum* shoot tips. *Cryoletters*, **11**, 413–426.

Fay, M. F. (1992). Conservation of rare and endangered plants using *in vitro* methods. *In vitro Cellular and Developmental Biology – Plant*, **28**, 1–4.

Fernández, H. and Revilla, M. A. (2003). *In vitro* culture of ornamental ferns. *Plant Cell, Tissue and Organ Culture*, **73**, 1–13.

Gantt, E. and Arnott, H. J. (1965). Spore germination and development of the young gametophyte of the ostrich fern (*Matteuccia struthiopteris*). *American Journal of Botany*, **52**, 82–94.

Goel, A. K. (2002). Botanic gardens: potential source for conservation of plant diversity. *Journal of Economic and Taxonomic Botany*, **26**, 67–74.

Hill, R. H. (1971). Comparative habitat requirements for spore germination and prothallial growth of three ferns of southeastern Michigan. *American Fern Journal*, **61**, 171–182.

Holloway, P. S. and Boyd, D. J. (1993). Spore viability and germination of the endangered Aleutian shield-fern, *Polystichum aleuticum. Hortscience*, **28**, 182A.

Hvoslef-Eide, A. K. (1990). In vitro storage as influenced by growth conditions prior to storage. In *Integration of In vitro Techniques in Ornamental Plant Breeding*, ed. F. De Fong. Wageningen: Eucarpia Symposium, pp. 124–131.

Jones, L. E. and Hook, P. W. (1970). Growth and development in microculture of gametophytes from stored spores of *Equisetum. American Journal of Botany*, **57**, 430–435.

Karp, A. (1994). Origins, causes and uses of variation in plant tissue cultures. In *Plant Cell and Tissue Culture*, ed. I. K. Vasil and T. A. Thorpe. Dordrecht: Kluwer, pp. 139–151.

Lebkuecher, J. G. (1997). Desiccation-time limits of photosynthetic recovery in *Equisetum hyemale* (Equisetaceae) spores. *American Journal of Botany*, **84**, 792–797.

Lindsay, S. and Dyer, A. F. (1990). Fern spore banks: implications for gametophyte establishment. In *Taxonoma, Biogeografa y Conservacion de Pteridofitos*, ed. J. Rita and J. Palma. Mallorca: Societat d'Historia Natural de les Illes Balears IME, pp. 243–253.

Lloyd, R. M. and Klekowski, E. J., Jr. (1970). Spore germination and viability in pteridophyta: evolutionary significance of chlorophyllous spores. *Biotropica*, **2**, 129–137.

Lusby, P., Lindsay, S., and Dyer, A. F. (2002). Principles, practice and problems of conserving the rare British fern *Woodsia ilvensis* (L.) R. Br. In *Fern Flora Worldwide: Threats and Responses*, ed. A. F. Dyer, E. Sheffield, and A. C. Wardlaw. *The Fern Gazette*, **16**, 350–355.

Miller, J. H. and Miller, P. M. (1961). The effect of different light conditions and sucrose on the growth and development of the gametophyte of the fern *Onoclea sensibilis. American Journal of Botany*, **48**, 154–158.

Miller, J. H. and Wagner, P. M. (1987). Co-requirement for calcium and potassium in the germination of spores of the fern *Onoclea sensibilis. American Journal of Botany*, **74**, 1585–1589.

Mottier, D. M. (1914). Resistance of certain fern prothalli to extreme desiccation. *Science*, **39**, 295.

Oliphant, J. L. (1989). *In vitro* cultivation of *Todea babarea* – from spore to sporophyte. *International Plant Propagators Society Proceedings*, **38**, 324–325.

Ong, B.-L. and Ng, M.-L. (1998). Regeneration of drought-stressed gametophytes of the epiphytic fern, *Pyrrosia pilosellodes* (L.) Price. *Plant Cell Reports*, **18**, 225–228.

Page, C. N. (1979). Experimental aspects of fern ecology. In *The Experimental Biology of Ferns*, ed. A. F. Dyer. London: Academic Press, pp. 552–589.

Page, C. N., Dyer, A. F., Lindsay, S., Mann, D. F., Ide, J. M., Jermy, A. C., and Paul, A. M. (1992). Conservation of pteridophytes: the ex situ approach. In *Fern Horticulture: Past, Present and Future Perspectives*, ed. J. M. Ide, A. C. Jermy, and A. M. Paul. Andover: Intercept, pp. 269–278.

Pangua, E., García-Álvarez, L., and Pajarón S. (1999). Studies on *Cryptogramma crispa* spore germination. *American Fern Journal*, **89**, 159–170.

Pattison, G. A., Ide, J. M., Jermy, A. C., and Paul, A. M. (1992). National plant collections: ex situ conservation. In *Fern Horticulture: Past, Present and Future Perspectives*, ed. J. M. Ide, A. C. Jermy, and A. M. Paul. Andover: Intercept, pp. 305–308.

Pence, V. C. (2000a). Survival of chlorophyllous and nonchlorophyllous fern spores through exposure to liquid nitrogen. *American Fern Journal*, **90**, 119–126.

Pence, V. C. (2000b). Cryopreservation of *in vitro* grown fern gametophytes. *American Fern Journal*, **90**, 16–23.

Pence, V. C. (2001). Cryopreservation of shoot tips of *Selaginella uncinata*. *American Fern Journal*, **91**, 37–40.

Pence, V. C. (2002). Cryopreservation and *in vitro* methods for *ex situ* conservation of pteridophytes. *The Fern Gazette*, **16**, 362–368.

Pence, V. C. (2003). *Ex situ* conservation methods for bryophytes and pteridophytes. In *Strategies for Survival*, ed. E. Guerrant, K. Havens, and M. Maunder. Covelo, CA: Island Press, pp. 206–227.

Pence, V. C., Sandoval, J., Villalobos, V., and Engelmann, F. (2002). *In Vitro Collecting Techniques for Germplasm Conservation*. IPGRI Technical Bulletin No. 7. Rome: International Plant Genetic Resources Institute.

Pérez-García, B., Orozco-Segovia, A. and Riba, R. (1994). The effects of white fluorescent light, far-red light, darkness, and moisture on spore germination of *Lygodium heterodoxum* (Schizaeaceae). *American Journal of Botany*, **81**, 1367–1369.

Planta Europa (2007). Targets of the EPCS. www.plantaeuropa.org/pe-EPCS-targets.htm.

Proctor, M. C. F. and Pence, V. C. (2002). Vegetative tissues: bryophytes, vascular "resurrection plants" and vegetative propagules. In *Desiccation and Plant Survival*, ed. M. Black and H. Pritchard. Oxford: CAB International, pp. 207–237.

Quintanilla, L. D., Amigo, J., Pangua, E., and Pajaron, S. (2002). Effect of storage method on spore viability in five globally threatened fern species. *Annals of Botany*, **90**, 461–467.

Randi, A. M. and Felippe, G. M. (1988). Efeito do armazenamento de esporos, da aplicacão de DCMU e da pré-embebicão em PEG na germinacão de *Cyathea delgadii*. *Ciência e Cultura* **40**, 484–489.

Sakai, A., Kobayashi, S., and Oiyama, I. (1990). Cryopreservation of nucellar cells of navel orange (*Citrus sinensis* Osb. var *brasiliensis* Tanaka) by vitrification. *Plant Cell Reports*, **9**, 30–33.

Simabukuro, E. A., Dyer, A. F., and Felippe, G. M. (1998). The effect of sterilization and storage conditions on the viability of the spores of *Cyathea delgadii*. *American Fern Journal*, **88**, 72–80.

Smith, D. L. and Robinson, P. M. (1975). The effects of spore age on germination and gametophyte development in *Polypodium vulgare* L. *New Phytologist*, **74**, 101–108.

Stokey, A. G. (1951). Duration of viability of spores of the Osmundaceae. *American Fern Journal*, **41**, 111–115.

Tannoury, M., Ralambosoa, J., Kaminski, M., and Dereuddre, J. (1991). Cryopreservation by vitrification of coated shoot-tips of carnation (*Dianthus caryophyllus* L.) cultured *in vitro*. *Comptes Rendus de l'Academie des Sciences, Paris, Serie III*, **313**, 633–638.

Theuerkauf, W. D. (1993). South Indian pteridophytes – ex situ conservation. *Indian Fern Journal*, **10**, 219–225.

Towill, L. E. and Bajaj, Y. P. S. (2002). *Crypreservation of Plant Germplasm. II. Biotechnology in Agriculture and Forestry, 50*. Berlin: Springer-Verlag.

Walter, K. S. and Gillett, H. J. (1998). *1997 IUCN Red List of Threatened Plants*. Gland, Switzerland: IUCN The World Conservation Union.

Walters, C. (2004). Principles for preserving germplasm in gene banks. In *Ex Situ Plant Conservation: Supporting Species Survival in the Wild*, ed. E. O. Guerrant Jr., K. Havens, and M. Maunder. Covelo, CA: Island Press, pp. 113–138.

Ward, N. B. (1852). *On the Growth of Plants in Closely Glazed Cases*, 2nd edn. London: John Van Voorst.

Wardlaw, A. C. (2002). Horticultural approaches to the conservation of ferns. In *Fern Flora Worldwide: Threats and Responses*, ed. A. F. Dyer, E. Sheffield, and A. C. Wardlaw. *The Fern Gazette*, **16**, 356–361.

Whittier, D. P. (1996). Extending the viability of *Equisetum hymenale* spores. *American Fern Journal*, **86**, 114–118.

Wilkinson, T. (2002). *In vitro* techniques for the conservation of *Hymenophyllum tunbrigense* (L.) Sm. In *Fern Flora Worldwide: Threats and Responses*, ed. A. F. Dyer, E. Sheffield, and A. C. Wardlaw. *The Fern Gazette*, **16**, 458.

Windham, M. D., Wolf, P. G., and Ranker, T. A. (1986). Factors affecting prolonged stored spore viability in herbarium collections of three species of *Pellaea*. *American Fern Journal*, **76**, 141–148.

Withers, L. A. (1985). Cryopreservation of cultured plant cells and protoplasts. In *Cryopreservation of Plant Cells and Organs*, ed. K. K. Kartha. Boca Raton, FL: CRC Press, pp. 243–267.

PART IV SYSTEMATICS AND EVOLUTIONARY BIOLOGY

12

Species and speciation

CHRISTOPHER H. HAUFLER

12.1 Introduction

Two of the most basic elements of evolutionary biology, species and speciation, are also among the most enigmatic and consistently debated. Systematists seem to have a love/hate relationship with both of these topics, and have devoted literally thousands of pages over the past century and a half to exploring what species are and how they originate. In this chapter, general aspects and contemporary perspectives on species and speciation in ferns and lycophytes will be discussed and interpreted.

12.1.1 Are species real or imagined?

When studying biodiversity, a fundamental question that emerges is, "Do species exist?" Why is the variety of life on earth subdivided into a set of discontinuous and distinct groups rather than existing as a seamless series of intergrading populations? Although this appears to be a central question for biologists to answer, prominent authorities consider it to be "one of the most intriguing unsolved problems of evolutionary biology" (Coyne and Orr, 2004). How do the clearly observable distinctions between the groups we label species arise, and what maintains separate ancestor–descendant lineages through time and space? According to some scientists (including Charles Darwin), species may be arbitrary human constructs erected for our convenience (see also Raven, 1976; Mishler and Donoghue, 1982). On the other hand, we can all detect and give names to non-overlapping distinctions among natural populations of organisms. Supporting this view of clearly discernable boundaries among natural groups,

Biology and Evolution of Ferns and Lycophytes, ed. Tom A. Ranker and Christopher H. Haufler. Published by Cambridge University Press. © Cambridge University Press 2008.

in studies comparing scientific classifications to the names that native peoples apply to animals and plants, a remarkably close correspondence was discovered between the numbers of species and their boundaries as recognized by both approaches (Berlin *et al.*, 1966; Diamond, 1966). Further, as scientists explore the characteristics and origins of species, evidence continues to accumulate demonstrating that some clusters of populations represent coherent, cohesive, interactive biological units that deserve recognition, and that these units are isolated by a remarkable variety of mechanisms from other such clusters (Sites and Marshall, 2003). Thus, it is widely accepted that species actually do constitute real components of the biosphere (Rieseberg and Burke, 2001).

12.2 Species concepts and definitions

If we accept that species are fundamental components of biodiversity, what is the best way to characterize their basic features? What makes species different from other clusters of organisms such as varieties or populations? Beginning with Darwin (1859), scientists considered species from the perspective of history and formulated hypotheses about how species originate. These are two separate but linked components of a successful and accurate framing of the nature of species. Thus, in discussing what species are, I separate "concepts" from "definitions." A concept can be defined as "an abstract idea" whereas a definition is "a statement of the exact meaning of a word" and by separating these two similar but different perspectives on characterizing species, it may be possible to explore broad, fundamental elements of species as well as providing precision in applying evidence to studies of species and speciation.

If a concept is to capture important components of what species are, it should include historical as well as contemporary elements. Species are lineages of ancestor–descendant individuals that (1) originate, (2) develop to occupy unique morphological, geographical, and ecological space, (3) persist as a genetically cohesive set of populations, and ultimately (4) decline and become extinct. An inclusive species concept, therefore, should recognize and be broad enough to include these stages and dimensions (Mayden, 1997; De Queiroz, 1998). The concept that fits these best is the "evolutionary species" concept. As articulated by Wiley (1978; Wiley and Mayden, 2000), and adapted by him from Simpson (1961), "a species is a single lineage of ancestor descendant populations which maintains its identity from other such lineages and has its own evolutionary tendencies and historical fate." This concept accommodates both sexual and asexual lineages, recognizes the importance of history, and includes the many aspects of contemporary isolation from other species by referring to "maintaining its identity" from other species. If we knew enough about the organisms on earth to apply this grand concept to all of them, there would be little left for evolutionary biologists to accomplish. However, to delineate species accurately using this concept is a daunting task. We simply do not have the evidence necessary for all species to show a progression of ancestor–descendant organisms that constitute lineages. The fossil record provides benchmarks, but the many gaps in it must be inferred from a combination of comparative morphology, ecology, and biogeography of contemporary species as well as DNA sequencing studies. Even when applying increasingly sophisticated algorithms, the mass of information necessary to develop robust hypotheses of relationships is available for only a small percentage of species.

Building up to the mass of evidence necessary to characterize "evolutionary species," a series of "species definitions" can be considered as provisional hypotheses. "Morphological" species (also called "taxonomic" species) are those based on a consideration of observable structural differences between clusters of individuals. When explorers encounter new variants during a floristic survey, their initial descriptions of new species are based on morphological evidence. Careful diagnoses indicate how the new discovery may be differentiated from its closest relative based on a novel feature or combination of features. Shared physical characteristics are used to demonstrate common ancestry between the new discovery and presumably related taxa. Such morphological evidence constitutes a first hypothesis and opens the door to further analysis and experimentation. Most fern and lycophyte species conform to morphological species expectations and can be distinguished based on suites of unique traits.

In 1950, Irene Manton published a book that launched a revolution in the study of ferns and lycophytes. She and her students initiated the era of experimental investigations of species boundaries by incorporating studies of chromosome number, meiotic behavior, and artificial hybridization. These studies focused attention on cryptic but fundamental aspects of species limits and demonstrated that some polymorphic "species" were actually species complexes harboring diploid species, allopolyploid derivative species, and hybrid backcrosses. Based on morphological evidence alone, the combination of interactions among the entities in such complexes resulted in an apparently continuous intergradation of morphological features without the discontinuities necessary to circumscribe discrete units. However, when the allopolyploids and hybrid backcrosses are removed, the diploids emerge as reasonably well demarcated independent lineages. As reviewed by Lovis (1977), after Manton's innovative introduction, scientists worldwide began incorporating chromosomal techniques in revisionary studies of ferns and lycophytes. The following are just some of the important contributions that built on Manton's insightful foundation (see Yatskievych and Moran, 1989 for a more inclusive list): *Asplenium*

(Wagner, 1954); *Athyrium* (Schneller, 1979); *Cystopteris* (Blasdell, 1963); *Dryopteris* (Walker, 1955, 1961), *Gymnocarpium* (Sarvella, 1978, 1980); *Isoëtes* (Hickey, 1984; Taylor *et al.*, 1985); *Lycopodium* (Wilce, 1965; Bruce, 1975); *Pellaea* (Tryon and Britton, 1958); *Polystichum* (Wagner, 1979). The perspectives gained through this work brought studies of species biology among ferns and lycophytes to the "biological" species definition era of understanding. Ernst Mayr developed this definition (1942, 1969) as a way to capture the important elements that provide meaningful criteria and an experimental basis for studying species limits. Centered on reproductive biology, Mayr's definition considers species as "groups of interbreeding natural populations that are reproductively isolated from other such groups" (Mayr, 1969). Considering species boundaries to be controlled by processes that take place within and between populations, and involving active participation by individual members of the species, the pros and cons of Mayr's definition have been debated ever since he proposed it. Although widely embraced by zoologists (e.g., Coyne and Orr, 2004), botanists have had more difficulty considering this as a universal definition of species (e.g., Whittemore, 1993). Further, because the biological species definition focuses on contemporary interactions between extant populations, it is difficult to apply when considering species as lineages with histories. It is also not possible to apply this definition to clonal species or asexual species. Nonetheless, by raising the importance of reproductive behavior and by considering species as interactive, cohesive sets of populations, and when combined with Manton's emphasis on genetic aspects of fern and lycophyte species, a link was forged between defining species and a mechanism of speciation. Species are perpetuated as integrated units when their individuals maintain gene flow by sexual reproductive interaction; speciation can occur when gene flow is interrupted.

Building on the foundation laid by Manton and others, fern and lycophyte species were studied in the 1980s and 1990s using enzyme electrophoresis to analyze isozyme variants (Haufler, 1987, 1996, 2002). By developing unique molecular profiles for each lineage, these studies revealed details that were unavailable through earlier approaches. Even in some of the most convoluted species complexes, isozymes were able to clarify the boundaries of species. Isozymes also helped to look within species and provide details of population structure and breeding behavior (see Chapter 4) and, as discussed below, yield insights for understanding aspects of speciation.

By linking the approaches available, it had become possible to delineate species boundaries quite accurately, to separate diploid and polyploid lineages, and, in many cases, to define the details of reticulate patterns of species origins and the pedigrees of polyploid species (reviewed in Barrington *et al.*, 1989; Werth, 1989). Still lacking was clarity concerning the history of each species and their trajectories of ancestor–descendant relationships. Although the databases discussed above merge to build reliable hypotheses for delimiting contemporary species, none of the available lines of evidence yields a firm historical perspective. It is difficult to extract a clear phylogenetic signal from morphology because ecological influences can result in convergent similarities; changes in chromosome complement are quite dynamic and difficult to interpret; isozyme variants cannot be tracked accurately over long periods of time. Developing through the 1990s and continuing into the twenty-first century, the application of DNA sequencing techniques has allowed researchers to chart the genealogy of species, to coordinate fossil benchmarks with molecular clocks, and to formulate evolutionary species definitions for some ferns and lycophytes. We now have species that are firmly rooted in a solid phylogenetic foundation and we are beginning to have an accurate appreciation for the influence that geography, ecology, and breeding system dynamics have on species. By combining fossil and molecular data, we now know that some fern species are remarkably old (e.g., *Osmunda* species date to the Triassic, about 225 mybp (Phipps *et al.*, 1998)) and others are surprisingly young (e.g., epiphytic species in *Polypodium* diversified along with and in response to angiosperms (Schneider *et al.*, 2004)). Such data truly do yield species that conform to evolutionary species definitions. They are well circumscribed and embedded in an extensively documented phylogenetic framework. In many cases we are able to assert confidently that our species hypotheses delineate the boundaries of contemporary species and place the lineage into historical context accurately. As we develop this level of confidence with more and more groups, we will build classifications that reflect phylogenies and that can be used to predict trends and mechanisms of biogeography and speciation.

12.2.1 Asexual species and cryptic species

Ferns and lycophytes present two special problems to those studying species: (1) it is estimated that about 10% of fern and lycophyte species do not reproduce sexually (Walker, 1984), and therefore asexual species deserve special consideration, and (2) when contemporary molecular approaches are applied to widely distributed and (often) morphologically polymorphic species, a series of embedded cryptic species is discovered. Gastony and Windham (1989) reviewed the treatment of agamosporous (asexual) species and since that time, ongoing studies are discovering more examples of groups that are clearly isolated sets of self-perpetuating populations, but are not reproducing sexually (e.g., Windham and Yatskievych, 2003). Gastony and Windham (1989) propose a "genetic" species concept that acknowledges the clearly distinct agamosporous lineages. In many cases, agamosporous groups are also cryptic in that they are triploid autopolyploid derivatives of otherwise "normal" diploid species (e.g., *Notholaena grayii* and

Pellaea andromedifolia discussed in Gastony and Windham, 1989). In cases where the agamosporous lineage cannot be distinguished morphologically from the diploid progenitor, Gastony and Windham (1989) recommend the use of "variety" to identify the agamosporous units. As we learn more about these distinctive evolutionary units, our classification of them can become more sophisticated and responsive to their special characteristics.

One of the more significant advances in fern and lycophyte species biology has been the recognition of cryptic species within formerly circumscribed species. Paris *et al.* (1989) describe cryptic species as (1) poorly differentiated morphologically, (2) distinct evolutionary lineages that are reproductively isolated, and (3) historically misinterpreted as members of a single species. Prior to the incorporation of molecular data from isozymes and DNA analyses, cryptic species in ferns and lycophytes were difficult to detect because of their (1) simple construction, (2) relatively small number of structural characters available for identification, and (3) considerable plasticity in lineage-defining features (Paris *et al.*, 1989). Through the application of molecular methods cryptic species in many fern and lycophyte lineages have been revealed, e.g., *Adiantum pedatum* (Paris and Windham 1988), *Athyrium oblitescens* (Kurihara *et al.*, 1996), the *Asplenium nidus* complex (Yatabe *et al.*, 2001), *Botrychium* subg. *Botrychium* (Hauk and Haufler, 1999), *Ceratopteris thalictroides* (Masuyama *et al.*, 2002; Masuyama and Watano 2005), *Cystopteris* (Haufler and Windham, 1991), *Gymnocarpium dryopteris* (Pryer and Haufler, 1993), *Isoëtes* (Hickey *et al.*, 1989; Taylor and Hickey, 1992), and the *Polypodium vulgare* complex (Haufler and Windham, 1991; Haufler *et al.*, 1995a). In each of these studies, genetically distinct and geographically restricted species were clearly delineated using molecular approaches. By concentrating on these genetically defined groups, it was possible to discover consistent yet often quite subtle morphological features that correlated with the molecular data. It is likely that these are not isolated examples and that investigation of other groups using molecular methods will reveal additional cryptic species masquerading as polymorphisms in currently recognized species.

12.2.2 An example of progress in discovering species boundaries

Studies of the genus *Polypodium* illustrate the progressive improvement in defining species that accurately delimit cohesive and interactive sets of populations and reveal their evolutionary history. Originally named by Linnaeus, the temperate *Polypodium vulgare* "complex" contained two species, *P. vulgare* in the Old World and *P. virginianum* in the New World. When Irene Manton (1950) began counting chromosomes of representative variants, she discovered that there were diploid, tetraploid, and hexaploid sexual species as well as triploid and pentaploid backcross hybrids. Manton's student Molly Shivas (1961) built on

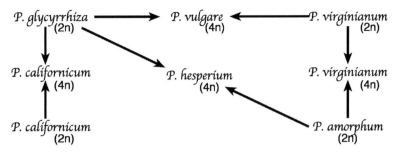

Figure 12.1 Hypothesis of reticulate evolution in the *Polypodium vulgare* complex based on Shivas (1961), Lloyd and Lang (1964), and Lang (1971).

Manton's initial cytotaxonomic survey and made controlled crosses among the various elements of the complex to discover genetic affinities and make predictions about the origins of the polyploid species. Shivas was able to develop solid and straightforward hypotheses for the allopolyploid origin of the European hexaploid, *P. interjectum*, through hybridization and genome doubling involving two primarily European taxa, tetraploid *P. vulgare* and diploid *P. australe*, followed by genome doubling. Analysis of meiotic chromosome behavior in pentaploid backcross hybrids showed pairing consistent with the proposed parentage. Shivas' hypothesis for the origin of tetraploid *P. vulgare* (not found in the Americas) was daring as it proposed an original cross involving diploids found *only* in North America (*P. glycyrrhiza* and *P. virginianum*). Shivas, therefore, clarified some aspects of the *P. vulgare* complex and raised testable hypotheses for other elements. In North America, Lloyd and Lang (1964) and Lang (1971) developed a complex reticulate hypothesis to account for the native diploid and tetraploid *Polypodium* species, and implicating *P. glycyrrhiza* as the second diploid involved with the origin of *P. vulgare* (see Figure 12.1). Lang added another diploid lineage (*P. amorphum*) and an allotetraploid species (*P. hesperium*). Based on the work of Shivas, Lloyd, and Lang, what had been originally two species swelled to eight, some with multiple cytotypes. But at this point, the combination of evidence from morphology, biogeography, and chromosome behavior reached its limit of resolution.

In the 1980s, isozyme analyses were employed to bring more clarity to this species complex. These new data demonstrated that what had been considered a single broadly distributed species (*P. virginianum* ranging from eastern North America, across Canada, and into northern Asia and Siberia) was actually two diploid lineages plus (through allopolyploidy) allotetraploid derivatives. Thus, *P. appalachianum* was circumscribed and limited to eastern North America (Haufler and Windham, 1991), and the distribution of *P. sibiricum*, originally named from a few specimens in Siberia, was broadened to include northern Asia,

310   Christopher H. Haufler

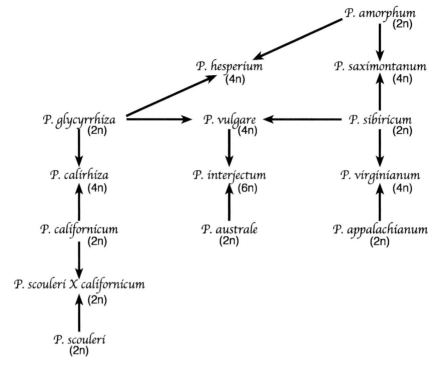

Figure 12.2 Hypothesis of reticular evolution in the *Polypodium vulgare* complex based on Haufler et al. (1995a).

Alaska, and much of Canada. Identifying these cryptic species demonstrated the significance of variation in a special type of soral indument (sporangiasters) that had been described much earlier (Martens, 1943) but not accorded significance in characterizing a lineage. Isozyme evidence also showed that the diploid *P. sibiricum* was responsible for the origins of several allotetraploid species, probably owing to the compression of ranges during glacial incursions. Allopolyploidy involving *P. sibiricum* resulted in forming tetraploid species *P. virginianum*, *P. saximontanum*, and *P. vulgare*. Thus, a species group that was originally recognized as two species swelled to 12 as more and more sophisticated analyses were performed (Figure 12.2). With the allopolyploids and hybrids removed, relationships among the diploid species were explored using DNA sequencing (Haufler and Ranker, 1995; Haufler et al., 1995b), and "evolutionary" species having traceable lineages and origins were hypothesized.

This is not an isolated example; similar techniques were applied to other groups, and greater accuracy of species boundaries and histories were obtained: e.g., *Asplenium* (Werth et al., 1985; Ranker et al., 1994), *Dryopteris* (Werth et al., 1991), *Athyrium* (Schneller, 1979; Kelloff et al., 2002), *Notholaena/Astrolepis/*

Argyrochosma/Pellaea (Gastony, 1983; Windham, 1987; Benham and Windham, 1992); *Gymnocarpium* (Pryer and Haufler, 1993); *Cystopteris* (Haufler et al., 1990); *Polystichum* (Barrington, 1990); *Adiantum* (Paris and Windham, 1988). In these examples, diploid species ranges were often dramatically revised, new species were recognized, and the origins of allopolyploid species were clarified.

12.2.3 *Species summary*

As in most groups, before considering ancestries of species, before developing hypotheses about species origins, and before exploring evolutionary mechanisms, it is vital to discover accurate boundaries of extant representatives (Sites and Marshall, 2003). If working with poorly circumscribed evolutionary units, it is impossible to reconstruct their evolutionary history, especially when species boundaries are drawn too broadly and when the history of lineages includes episodes of hybridization and allopolyploidy. Perhaps because all of these complications are such prominent elements of the biology of fern and lycophyte species, it is especially important to delimit species accurately. It is becoming clear that the assumptions made about broad species boundaries in ferns and lycophytes should not be followed. Far from being more simple and straightforward than species of seed plants, ferns and lycophytes are constrained by most of the same forces acting on other groups, and, especially given the demonstrated complexities of hybridization and allopolyploidy, fern and lycophyte species should be viewed as dynamic, interactive, and often problematic. Continued emphasis at the species level will only enhance studies of the forces that result in the origin and diversification of ferns and lycophytes.

12.3 Speciation

The origin of species has consistently fascinated and perplexed scientists. Originally considered to be divine creations, Wallace (1858) and Darwin (1859) provided the first scientifically testable hypotheses for mechanisms that could result in new lineages. Since that time, considerable attention has been paid to exploring how natural forces can launch populations of organisms carrying novel combinations of features that become established as distinct from their relatives (Orr and Smith, 1998; Coyne and Orr, 2004). Considerable contemporary research on speciation is focused at the level of genes and on discovering what is involved in isolating unique genetic combinations that are perpetuated as cohesive ancestor–descendant populations. By considering the sequence of bases in the DNA of organisms, it is possible to formulate hypotheses about the ancestors of current species and their closest contemporary relatives, and to consider what genes or gene combinations might have contributed to the isolation of new groups.

Current speciation research is built on a foundation of information and hypotheses concerning the modes and mechanisms of species origins. In most discussions, speciation is a two-stage process. The first stage involves some type of isolating event that separates a prospective new lineage from its progenitor. Central questions then focus on determining as precisely as possible the processes that result in such isolation, such as separation by distance, by habitat modification, or by genetic changes that prevent interbreeding (Hey, 2006). It sometimes seems as though there are nearly as many potential mechanisms as there are new species to consider. Following isolation is usually the development of distinctive features, both molecularly and morphologically. Because the discovery of new species usually requires the availability and recognition of distinctive characteristics, by the time scientists describe a new lineage, the actual point of origin from its ancestor is long past. The challenge, therefore, is to develop hypotheses that (1) incorporate observable evidence and natural situations, (2) are based on experimental data, and (3) explain patterns of diversification.

12.3.1 Types of speciation

Previous reviews of fern and lycophyte speciation (Haufler, 1996, 1997) have proposed three different speciation modes (descriptions of the different pathways resulting in speciation) for ferns and lycophytes: **primary** (divergence of a single lineage, usually at the diploid level, to initiate a new lineage), **secondary** (initiation of a new lineage resulting from hybridization and/or polyploidy of existing lineages), and **tertiary** (speciation through changes in polyploid species, e.g., via reciprocal gene silencing). Each of these major categories includes a variety of mechanisms that result in new species. A discussion of each and examples from fern and lycophyte studies follows.

12.3.2 Primary speciation through allopatry

Spatial isolation is the defining factor in allopatric speciation. Simply separating two previously integrated sets of populations may allow changes to accumulate and differentiation to occur. There are different ways to achieve such separation. First, as species modify their ranges (through either expansion or contraction), it is reasonable to expect especially the peripheral populations to become isolated from the main part of the distribution. If such isolation persists, a peripheral population may accumulate genetic differences and thus may diverge and become a separate lineage (e.g., Ranker *et al.*, 1994; Moran and Smith, 2001). Second, because of a vicariant event (usually of geophysical origin, such as continental drift or mountain building, occurring over a long period of time), a once continuous population can become fragmented and the different components isolated from each other (e.g., Paris and Windham, 1988; Kato, 1993). Either of these occurrences can ultimately result in allopatric speciation, or divergence because of isolation by distance. Among ferns and lycophytes, comparing the contemporary distributions of extant species and identifying "centers of species diversity" can be used to support the importance of allopatry in speciation (Barrington, 1993). The limitation in nearly all of these cases of allopatric speciation is that little can be determined concerning the actual mechanism of change within lineages. The studies demonstrate that change has occurred (each lineage has unique features), that closely related lineages have distinct and separate geographic ranges, and that the origin of the change is linked most likely to isolation by distance. We are left to infer that spatial isolation provides opportunities for divergence to occur, and are encouraged to develop compatible scenarios to explain what may result in observable differentiation.

Three case studies provide examples of different groups and different approaches to understanding allopatric speciation. The first (Tryon, 1971) used morphological variation drawn from herbarium specimens to track changes across the ranges of three members of the *Selaginella rupestris* complex. Tryon showed evidence in *S. underwoodii* for a morphological cline in which more variation was detected in the southern range of the species (Texas and New Mexico) and less variation to the north (Wyoming). From collections of *S. densa* ranging across much of western North America, Tryon circumscribed three distinct varieties (currently considered species in Flora North America, 1993). In *S. rupestris*, found across much of eastern North America, Tryon described a series of races showing variation in the distribution of microsporangia in the strobili. For each of these examples, Tryon related the variation to migration of populations in response to climatic changes associated with periods of glaciation, stating that evolutionary migration, "can lead to strongly differentiated peripheral populations that may be recognized as taxa" (p. 99). This descriptive approach posed hypotheses based on progressive changes in morphological features and provided valuable perspectives that may be tested using modern molecular methods.

An example from the ferns that has carried this approach to understanding patterns and processes of allopatric speciation one step further is the temperate *Polypodium sibiricum* group, which contains three related diploid species that all inhabit rock outcrops in forests: *P. sibiricum*, with a circumboreal distribution, extending across Canada and into Japan, Korea, and China; *P. appalachianum*, occurring in the mountains of eastern North America; and *P. amorphum*, found on cliff faces in western North American gorges (Haufler et al., 1995a, 2000). Although differing little in gross leaf morphology and joined by the distinctive morphological synapomorphy of sporangiasters, these three species have an average interspecific genetic identity developed from isozymic comparisons of only 0.460. In the *Selaginella* example, Tryon (1971) proposed that glaciation compacted southern populations which then rebounded post glaciation to establish more northern populations. In the *P. sibiricum* complex, it is hypothesized that periodic glaciation pushed the circumboreal *P. sibiricum* populations south and, when this northern species regained its original distribution following the retreat of the glaciers, southern populations persisted, evolved distinctive traits, and ultimately erected post-zygotic barriers to interbreeding. Combining biogeographic, morphological, and isozymic evidence, this *Polypodium* hypothesis developed a classic allopatric speciation model and suggested further that interspecific distinctions may have been reinforced through contact mediated by subsequent ice ages (Haufler *et al.*, 2000).

Members of the *Athyrium filix-femina* complex provide a good example of temperate speciation in progress. This complex is highly variable and globally distributed and comprises at least six regional taxa for which taxonomic rank is contentious. Two varieties occur in eastern North America: *A. filix-femina* var. *angustum* (Northern Lady Fern) and var. *asplenioides* (Southern Lady Fern), differentiated by various features including degree of frond tapering, rhizome orientation, and spore morphology and color (Butters, 1917; Kelloff *et al.*, 2002). There are substantial differences in allele frequencies between the two varieties, whereas populations within the varieties have similar allele frequencies and do not vary with respect to geographic location (Kelloff *et al.*, 2002). In a more detailed study of the pattern of genetic variation within and between varieties, Sciarretta *et al.* (2005) showed that (1) the spatial distribution of alleles at most loci appears to be even across the populations, but exceptions indicate that the populations are moderately genetically structured; (2) within the region studied, *asplenioides* populations show high levels of genetic similarity, although somewhat differentiated from populations of piedmont/coastal-plain *asplenioides* to the east, and more substantially from *angustum* populations to the north. Thus, the distribution of genetic variation varies in a complex way across the range of *Athyrium filix-femina* in eastern North America. It is uncertain when the divergence between *angustum* and *asplenioides* initially occurred, but likely it was during one of the glacial maxima when a number of eastern North American species became geographically subdivided in remote refugia (Parks *et al.*, 1994). Although the pronounced divergence between the two *Athyrium* taxa in numerous morphological features and allozyme frequencies (Butters, 1917; Kelloff *et al.*, 2002) indicates separation over an extended time period, the formation of fertile hybrids between *angustum* and *asplenioides* (Schneller, 1989; Kelloff *et al.*, 2002) suggests the potential for these diverged taxon gene pools to merge over time. Significant introgression may have taken place between the two taxa, resulting in the lingering elevated frequencies of characteristically *angustum* alleles observed throughout the southern Appalachians. Eastward, on the piedmont and coastal plain, there are no high-elevation refugia for *angustum*, thus frequencies for the *angustum* alleles are much lower. Genetic and biogeographic data coordinate in building well-supported hypotheses about the origin of diversity and the development of lineages that are distinct in much of their range, but whose boundaries are blurred. Further research on this dynamic complex may provide insights into the pattern and process of allopatric primary speciation in ferns and lycophytes.

12.3.3 Primary speciation through adaptation and/or changing ecological preferences

Whereas an allopatric model is consistent with primary speciation of ferns in temperate regions, evidence suggests that, especially in tropical forests, speciation is linked with responses to ecology. With the greatest diversity on Earth occurring in the tropics (see Chapter 14), species are densely packed and may be distributed in response to geological and ecological variation. The edaphic complexity of lowland tropical forests yields patterns of species variation among ferns that correlate with soil features (Tuomisto *et al.*, 2002) and these features may promote segregation of species edaphically at the landscape scale within a uniform-looking forest (Tuomisto *et al.*, 1998, 2003). Montane tropical forests are stratified both elevationally (Gentry, 1988; Lieberman *et al.*, 1996) and within tree canopies (Terborgh, 1985; Richardson *et al.*, 2000) and fern species composition changes along elevational gradients with the greatest diversity occurring at middle elevations (Kluge and Kessler, 2006; Watkins *et al.*, 2006).

In contrast to the pattern of low genetic identities among cryptic or nearly cryptic congeneric fern species in temperate regions (Haufler, 1987; Soltis and Soltis, 1989), isozymic analyses of species in Hawaii (Ranker, 1992a,1992b, 1994; Ranker *et al.*, 1996) and tropical America (Hooper and Haufler, 1997) showed that some morphologically distinct species have very high genetic identities. Combined with an analysis of DNA sequences among members of a putatively monophyletic cluster of *Pleopeltis* species, it was possible to demonstrate ecological differentiation among these species, and to suggest that changes in ecological preferences promoted species diversification (Haufler *et al.*, 2000). Speciation in this set of *Pleopeltis* species may have occurred as follows. (1) An ancestral species expands its range, colonizing new areas. Disturbance may play a role in generating opportunities for colonization by providing open habitats for spore germination and gametophyte development. Alternatively, gametophytes of tropical species may have growth requirements and abilities that could narrow or widen the possible sites for successful establishment and growth.

(2) Range expansion may result in two outcomes that are significant in speciation. The new peripheral populations can (a) become separated by distance from the main distribution of the species or (b) encounter novel habitats that provide selective pressure for change to occur. (3) In either of these cases, the peripheral population could become genetically differentiated from the main range of the ancestral species. It seems likely that peripheral populations are changing relatively rapidly through selective adaptation to new environmental pressures. (4) These changes are reflected in a modified morphology for the residents of the peripheral populations and the changes may take place quickly. (5) Because the peripheral populations are adapted to the new ecology, they would develop pre-zygotic isolation from their ancestor populations, i.e., residents of the two ecologically differentiated populations may not tolerate the conditions of their neighbors even though they may be within spore dispersal range. Speciation following this scenario would be rapid relative to that described for temperate species (Figure 12.3).

These case studies of ferns and lycophytes provide evidence of considerable variation in the modes of primary speciation. Further, they illustrate the complexities involved in developing and testing hypotheses for the mechanisms that result in the origin of species. There is fertile ground for more studies in both temperate and tropical habitats, and for consideration of the many different patterns and processes by which primary divergence can lead to the diversification of fern and lycophyte species.

12.3.4 Introduction to secondary speciation

When species originate through genome-level changes involving primary species, the term "secondary" speciation has been applied (Grant, 1981). Although debate continues on defining terms and especially in determining types of polyploids (Soltis and Rieseberg, 1986), a "taxonomic" approach to defining the different kinds of secondary speciation is applied here. Thus, secondary speciation can occur through genome multiplication within a lineage (autopolyploidy), through hybridization between lineages that brings together two distinct genomes but does not involve genome doubling (allohomoploidy), or through hybridization between lineages followed by genome doubling in the (usually) sterile hybrid (allopolyploidy). Although substantial secondary speciation has been documented in the ferns and among members of the Lycopodiaceae and Isoëtaceae, and despite the discovery of hybrids (e.g., Tryon, 1955; Somers and Buck, 1975) and dysploidy (Marcon et al., 2005) in the Selaginellaceae, and hybrids and triploidy (Bennert et al., 2005) in the Equisetaceae, secondary speciation appears absent from these latter two families.

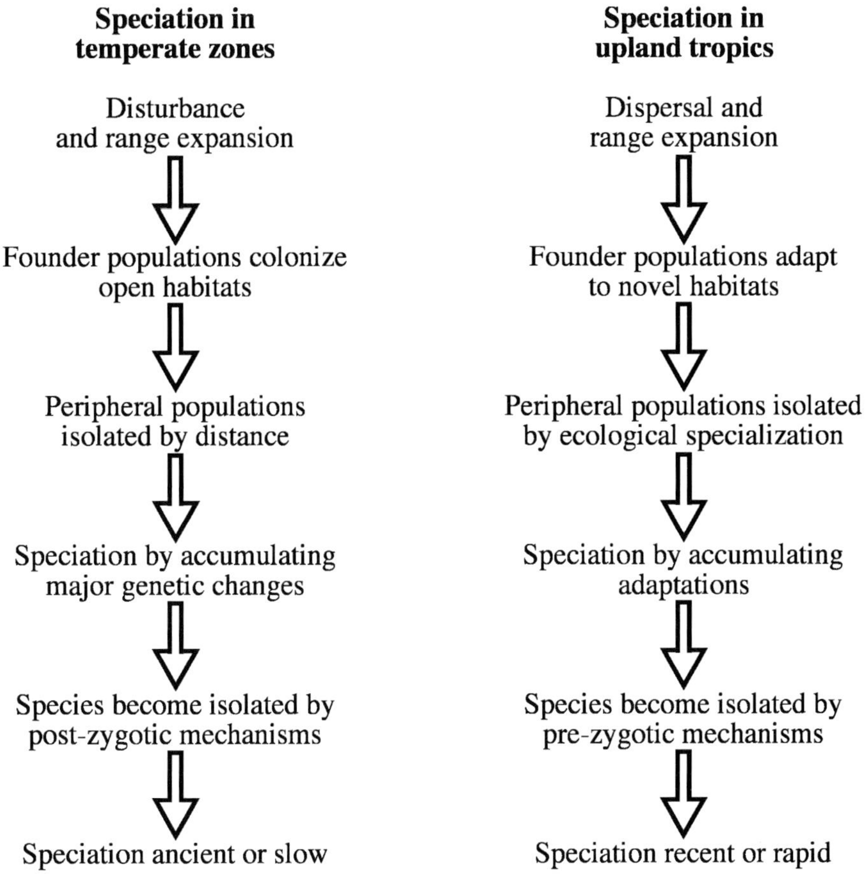

Speciation in temperate zones	Speciation in upland tropics
Disturbance and range expansion	Dispersal and range expansion
⇓	⇓
Founder populations colonize open habitats	Founder populations adapt to novel habitats
⇓	⇓
Peripheral populations isolated by distance	Peripheral populations isolated by ecological specialization
⇓	⇓
Speciation by accumulating major genetic changes	Speciation by accumulating adaptations
⇓	⇓
Species become isolated by post-zygotic mechanisms	Species become isolated by pre-zygotic mechanisms
⇓	⇓
Speciation ancient or slow	Speciation recent or rapid

Figure 12.3 Contrasting modes of speciation in temperate and upland tropical habitats. (Adapted from Haufler et al., 2000.)

12.3.5 Secondary speciation through autopolyploidy

As reviewed by Gastony (1986), speciation by chromosome doubling within a species has been largely overlooked as a significant mechanism. Soltis et al. (2003) reviewed numerous examples of autopolyploid species origins in angiosperms, but noted that it is likely that autopolyploidy is not as common as allopolyploidy. In some fern groups, however, especially when accompanied by apomixis, autopolyploidy may be a frequent, rapid, and important speciation mechanism (Windham and Yatskievych, 2003). The most clearly and completely documented examples are triploid apomicts derived from diploid progenitors (Gastony, 1983), but some studies have also showed that mechanisms are operating that could generate fertile tetraploids through autopolyploid processes (Gastony, 1986; Haufler et al., 1985; Rabe and Haufler, 1992).

12.3.6 Secondary speciation through hybridization

More is understood about speciation that is initiated following interspecific hybridization. This mode of species origin is both more easily deduced from morphological comparisons and easier to catch in its early stages than that resulting from primary divergence or autopolyploidy. As noted elsewhere (Haufler, 2002), ferns and lycophytes lack the sophisticated breeding systems that have evolved in the seed plants and evidence of barriers that prevent their eggs and sperms from uniting (except between genera; Schneller, 1981) is lacking. Thus, in most cases when fern and lycophyte species co-occur, hybrids are expected. The vast majority of these hybrids are sterile (often quite vigorous and long lived, but sterile nonetheless). However, there are cases in which either the hybrids are fertile (allohomoploidy) or when the rare event of chromosome doubling in progeny of the hybrid plants (they still produce sori and sporangia, and most are abortive) yields an allopolyploid individual. Based on the difference between these two outcomes, two modes of hybrid speciation have been described. The more difficult to recognize and characterize is allohomoploidy, and it may also be less frequent.

12.3.7 Secondary speciation through allohomoploidy

The first demonstration of apparent allohomoploid speciation was in *Pteris* (Walker, 1958, 1962). By analyzing chromosome behavior and morphology, Walker showed that diploid species may be ecologically isolated, but not always reproductively isolated. Under some circumstances, hybrid swarms developed and apomictic species were frequent. Unfortunately, Walker's perceptive and seminal hypotheses have not been tested by the application of molecular methods such as isozymes or DNA. Nonetheless, Walker's work demonstrates the great evolutionary complexity that can be revealed through biosystematic studies of ferns, and current generations should consider testing his hypotheses.

More recent studies of a second instance of allohomoploid hybridization involve species of *Polystichum* in western North America. Using chloroplast DNA and isozyme characters, Mullenniex *et al.* (1998) demonstrated that when two morphologically distinct species of *Polystichum* (*P. imbricans* and *P. munitum*) co-occur, hybrid swarms can form. Typically, these swarms occur in partially shaded ecotonal habitats, and although there is not yet evidence of any stabilization of these interspecific but fully fertile hybrids, this could be the first stage of a process that may result in a new, ecologically distinct lineage.

The most extensively documented case of allohomoploidy involves tree fern species on Caribbean islands (Conant and Cooper-Driver, 1980). Tree ferns are distinctive among the ferns by lacking polyploidy and by containing numerous instances of interspecific hybrids that are fertile (Conant, 1990). In the allohomoploid model of speciation, the process is initiated through primary speciation that isolates species rapidly based on different ecological specificities. Isozyme studies have demonstrated that although these species are morphologically distinct, they have very high congeneric genetic identities (Barrington and Conant, 1989) and they are interfertile. Without disturbance, they might remain ecologically isolated from each other and eventually accumulate more genetic differentiation. Conant and Cooper-Driver (1980) used a combination of morphological and biochemical data to demonstrate, however, that the barriers to species isolation had broken down and fertile hybrids had formed. These hybrids were sufficiently morphologically distinct from their progenitors that they had been accorded species status and, at least in some circumstances, were ecologically isolated from their progenitors. Given that this is an island system, the most provocative component of this analysis is the possibility that the hybrid-derived species could migrate to a neighboring island that could contain only one or neither of the parental species.

In several groups of lycophytes, allohomoploid complexes have been reported. As summarized by Wagner (1992), there are apparently stabilized hybrids with normal meiosis and producing abundant, apparently viable spores in the genera *Diphasiastrum*, *Lycopodiella*, and *Lycopodium*. The *Diphasiastrum* complex includes three species that all cross with a fourth species, resulting in three fertile hybrids. Backcrosses have not been discovered and the stabilization of large clusters of hybrid shoots has been attributed to the considerable cloning ability of the species. Hybrids with the same chromosome complement as the parents, and producing apparently viable spores, have also been reported between two pairs of species in *Lycopodiella* and between two tropical species of *Lycopodium* (Øllgaard, 1987). Thus, even among species that form subterranean gametophytes, hybridization occurs and can result in fertile progeny. Given the demonstration that random mating occurs in some populations of Lycopodiaceae species (Soltis and Soltis, 1988), discoveries of frequent hybridization should not be surprising. However, the indication that these hybrids have normal meioses and produce apparently viable spores suggests that post-zygotic, genetic barriers between species are lacking and that speciation via ecological specialization may be occurring.

Allohomoploidy, therefore, is a mechanism that emphasizes the role of ecology in promoting and maintaining diversity despite only minor genetic modification. If this mechanism is prominent, accurate characterization of species and speciation in some regions will be especially challenging. At the same time, demonstration that such a mechanism is operating provides important clues to the special complexity of some systems. In developing strategies for exploring species and speciation, the possibility of allohomoploid models must be entertained.

12.3.8 Secondary speciation through allopolyploidy

Much more widely recognized is the process of allopolyploidy. In this case, the progenitors of hybrid-derived species are not interfertile and form sterile offspring when they cross. Manton's (1950) pioneering chromosomal approach to species and speciation in the ferns and lycophytes focused considerable attention on allopolyploid species. Devoting whole chapters to species complexes such as *Dryopteris filix-mas* and *Polypodium vulgare*, Manton raised the consciousness of botanists concerning the importance of understanding how allopolyploid speciation contributes to diversity. More recently, Otto and Whitton (2000) estimated that polyploidization may be involved in about 7% of speciation events in ferns (as opposed to only 5% in angiosperms), and Haufler (1987, 2002) suggested that the basic qualities of seed-free plants contribute to the prominence of allopolyploid complexes in pteridophytes. There appear to be few barriers to the initiation of interspecific zygotes, and field studies have demonstrated a high frequency of vigorous but sterile hybrids in some complexes (e.g., Carlson, 1979). The persistence of these perennial herbs and the huge number of aborted sporangia that they contain provide the opportunity for initiating polyploid offspring through meiotic (non-disjunction and chromosome doubling) "mistakes." Because of redundant copies of genes, polyploid gametophytes are more tolerant of intragametophytic selfing than their diploid progenitors (e.g., Masuyama, 1979) and thus allopolyploid sporophytes may be initiated relatively frequently in ferns and lycophytes (Figure 12.4). However, even though it is an accepted and frequent mode of speciation, there are certainly many open questions about so-called species complexes that involve allopolyploidy.

Results from molecular studies of such well-studied fern groups as *Cystopteris* (Haufler et al., 1990), *Dryopteris* (Werth, 1991), *Gymnocarpium* (Pryer and Haufler, 1993), *Polypodium* (Haufler et al., 1995a), and *Polystichum* (Soltis et al., 1990) as well as the lycophyte *Isoëtes* (Hickey et al., 1989; Taylor and Hickey, 1992) demonstrate how dynamic and reticulate these species complexes are. The *Polypodium vulgare* complex (Figures 12.1 and 12.2) discussed earlier demonstrates the necessity of identifying allopolyploid origins so that hypotheses concerning species and speciation events can be accurately proposed and tested. In the species complexes listed above, testing hypotheses of allopolyploid origins has led to significant revisions of species boundaries at the diploid (primary speciation) level. Surveys of chromosome numbers in tropical habitats (e.g., Jermy and Walker, 1985) have demonstrated that polyploidy is frequent and we should be alert to the possibility of allopolyploid complexes as biosystematic and molecular studies expand

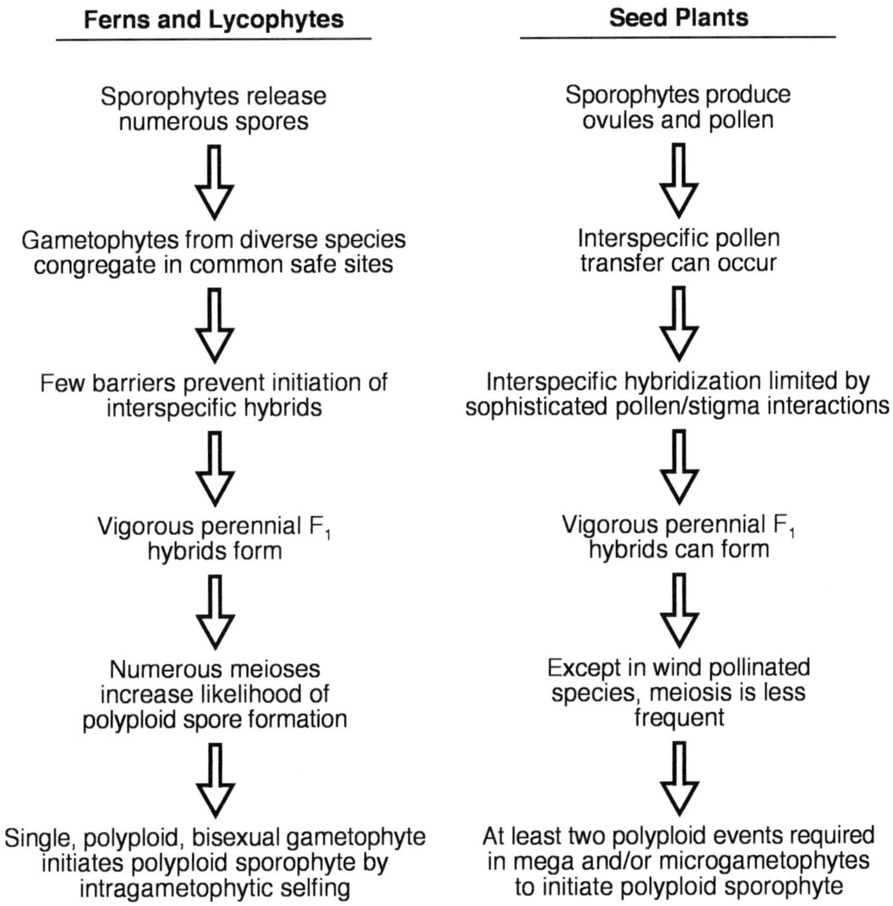

Ferns and Lycophytes

Sporophytes release
numerous spores

⇓

Gametophytes from diverse species
congregate in common safe sites

⇓

Few barriers prevent initiation of
interspecific hybrids

⇓

Vigorous perennial F_1
hybrids form

⇓

Numerous meioses
increase likelihood of
polyploid spore formation

⇓

Single, polyploid, bisexual gametophyte
initiates polyploid sporophyte by
intragametophytic selfing

Seed Plants

Sporophytes produce
ovules and pollen

⇓

Interspecific pollen
transfer can occur

⇓

Interspecific hybridization limited by
sophisticated pollen/stigma interactions

⇓

Vigorous perennial F_1
hybrids can form

⇓

Except in wind pollinated
species, meiosis is less
frequent

⇓

At least two polyploid events required
in mega and/or microgametophytes
to initiate polyploid sporophyte

Figure 12.4 Comparison of the likelihood of polyploid origins in ferns and lycophytes versus seed plants. (Adapted from Haufler, 2002.)

into the tropics. In addition, the application of appropriate DNA techniques to resolve species boundaries and test hypotheses of recurring allopolyploid origins should increase the accuracy of our conclusions (e.g., Vogel et al., 1996; Trewick et al., 2002; Adjie, et al., 2007).

12.3.9 Tertiary speciation

One of the most dramatic statements in a recent review of polyploidy by Soltis et al. (2003) is "A major discovery of the past decade is the extent and rapidity of genome reorganization in polyploids" (p. 183). It is now becoming clear that change can occur quickly in polyploid lineages, and can result in what have been called "tertiary" species (Haufler, 1989). Given that (1) homosporous ferns and lycophytes contain a higher percentage of polyploid species than other plant groups (Otto and Whitton, 2000), (2) the nuclear genome in these lineages appears unique in having diploid genetic expression despite large basal chromosome numbers, and (3) the many questions that remain unanswered about the fate of polyploid species (see Chapter 7), it is unfortunate that there is little research exploring the mechanisms that resulted in this combination of features. When Werth and Windham (1991) proposed that reciprocal gene silencing within polyploid lineages could result in genetically (but probably cryptically) isolated populations, polyploids were generally considered to be static entities whose bloated, polygenic systems prevented rapid evolutionary change. It seems clear that if we are to develop a synthesis of the evolutionary modes and mechanisms of ferns and lycophytes, we must consider the fate of polyploids, and the role that they play in generating biodiversity.

12.4 Summary and future prospects

Contemporary analyses are providing a progressive illumination of species boundaries, and an improved perspective on how ferns and lycophytes evolve. More widespread application of molecular methods will continue the trend of discovering cryptic species, especially within those currently considered to be polymorphic and widespread. It is critical, however, that we carry this technology into tropical regions, as it is clear that the bulk of fern and lycophyte species occur there. The power of these tools has been well demonstrated in the many studies completed on temperate species complexes, and we need to build on this foundation.

As presented in Chapter 4, changes in our understanding of fern and lycophyte breeding systems (from primarily inbreeding to an emphasis on outcrossing) have promoted a shift from considering ferns and lycophytes (with their easily wind-dispersed propagules) as unlimited in dispersal capacity to a combination of dispersal and vicariance-driven mechanisms of biogeographical change (Wolf et al., 2001; Haufler, 2007). As a consequence, we have a better perspective on the shift from a relatively small number of polymorphic, widely dispersed species to many, more narrowly distributed, often cryptic species. These population-level discoveries extend to speciation and a consideration of a range of mechanisms including changes in genetic composition, geographic barriers, ecology, and ploidal level. As presented in Chapter 9, it is critical that we learn more about the ecology of the gametophyte generation as these colonizers are likely the agents that are selecting and defining habitat specificities of many groups, especially in species-rich tropical regions.

Studies of speciation in the ferns and lycophytes remain based primarily on circumstantial evidence. Our definition of monophyletic groups continues to improve (see Chapters 13, 15, and 16), and, therefore, so does the accuracy of comparing levels of diversity within and among lineages. With accurate assessment of the amount of change that is associated with the origins of species, we can consider more confidently the events and/or mechanisms that have caused that change. As it has now been demonstrated that ferns and lycophytes have breeding systems that are directly comparable to those of seed plants and as sophisticated molecular tools provide even more fine-grained interpretations of the genetic variation in populations, it will be possible to draw on the many models and hypotheses that have been proposed in work with other groups of plants. Even now it is possible to develop hypotheses that explain population-level mechanisms for maintaining and modifying genetic variation (see Chapter 4), providing a much improved foundation for understanding speciation mechanisms. Just as with other plant groups, combining mechanisms operating in populations with regional and global patterns of biogeography and vicariance will generate robust hypotheses to explain and model speciation in ferns and lycophytes.

At the same time as the basis for establishing a solid understanding of primary speciation is developed, current interest in and understanding of polyploidy as a powerful force in generating opportunities for radical reorganization of genomes and rapid change in lineages should extend to the ferns and lycophytes. Especially considering the unique combination in the homosporous groups of high chromosome numbers and genetic diploidy, ferns and lycophytes should yield important perspectives on this emerging and exciting element of the genetics of organisms. Clearly, polyploidy is an important agent of genetic change and without exploring the fern and lycophyte genomes, a full understanding of the history of evolutionary progress will not be obtained (see Chapter 7).

Tertiary speciation through polyploid genetic revolutions or through passive silencing of reciprocal loci expands the realm of possibilities for explaining the origins of biodiversity. Especially among widespread polyploid species, considering the patterns of variation within and among populations will yield a superior picture of (1) multiple origins of polyploids, (2) how polyploids change as they migrate, and (3) whether polyploids differ from diploids in how and how fast they become modified over time and distance.

The time is ripe for building on the advances we have made in proposing accurate hypotheses about the phylogenetic trees of relationships among fern and lycophyte lineages (see Chapters 15 and 16). These studies are generating a clear picture of the patterns of change. We should use this phylogenetic foundation to propose and test hypotheses about the processes that resulted in the patterns. These studies also provide robust hypotheses about major lineages, but often the circumscription of the elements of these major lineages requires additional investigation. Especially in tropical regions, where it appears that changes may be rapid and tied to the complex ecology of such areas, our understanding of evolutionary mechanisms is limited at best. Targeting efforts at these regions is critical, and should yield valuable insights and perspectives.

References

Adjie, B., Masuyama, S., Ishikawa, H., and Watano, Y. (2007). Independent origins of tetraploid cryptic species in the fern *Ceratopteris thalictroides*. *Journal of Plant Research*, **120**, 129–138.

Barrington, D. S. (1990). Hybridization and allopolyploidy in Central American *Polystichum*: cytological and isozyme documentation. *Annals of the Missouri Botanical Garden*, **77**, 297–305.

Barrington, D. S. (1993). Ecological and historical factors in fern biogeography. *Journal of Biogeography*, **20**, 275–279.

Barrington, D. S., and Conant, D. S. (1989). Breeding system, genetic distance, and hybridization in *Alsophila*. *American Journal of Botany*, **76**, Supplement, 201 (abstract).

Barrington, D. S., Haufler, C. H., and Werth, C. R. (1989). Hybridization, reticulation and species concepts in the ferns. *American Fern Journal*, **79**, 55–64.

Benham, D. M. and Windham, M. D. (1992). Generic affinities of the star-scaled cloak ferns. *American Fern Journal*, **82**, 47–58.

Bennert, W., Lubiensky, M., Körner, S., and Steinberg, M. (2005). Triploidy in *Equisetum* subgenus *Hippochaete* (Equisetaceae, Pteridophyta). *Annals of Botany*, **95**, 807–815.

Berlin, B., Breedlove, D. E., and Raven, P. R. (1966). Folk taxonomies and biological classification. *Science*, **154**, 273–275.

Blasdell, R. F. (1963). A monographic study of the fern genus *Cystopteris*. *Memoirs of the Torrey Botanical Club*, **21**, 1–102.

Bruce, J. G. (1975). Systematics and morphology of subgenus *Lepidotis* of the genus *Lycopodium* (Lycopodiaceae). Unpublished Ph.D. Thesis, University of Michigan, Ann Arbor, MI.

Butters, F. K. (1917). Taxonomic and geographic studies in North American ferns. I. The genus *Athyrium* and the North American ferns allied to *Athyrium filix-femina*. *Rhodora*, **19**, 169–207.

Carlson, T. J. (1979). The comparative ecology and frequencies of interspecific hybridization of Michigan woodferns. *Michigan Botanist*, **18**, 47–56.

Conant, D. S. (1990). Observations on the reproductive biology of *Alsophila* species and hybrids (Cyatheaceae). *Annals of the Missouri Botanical Garden*, **77**, 290–296.

Conant, D. S. and Cooper-Driver, G. (1980). Autogamous allohomoploidy in *Alsophila* and *Nephelea* (Cyatheaceae): a new hypothesis for speciation in homoploid homosporous ferns. *American Journal of Botany*, **67**, 1269–1288.

Coyne, J. A. and Orr, H. A. (2004). *Speciation*. Sunderland, MA: Sinauer.

Darwin, C. (1859). *On the Origin of Species by Means of Natural Selection or the Preservation of Favoured Races in the Struggle for Life*. London: J. Murray.

De Queiroz, K. (1998). The general lineage concept of species, species criteria, and the process of speciation. In *Endless Forms: Species and Speciation*, ed. D. J. Howard and S. H. Berlocher. Oxford: Oxford University Press, pp. 57–75.

Diamond, J. M. (1966). Zoological classification system of a primitive people. *Science*, **151**, 1102–1104.

Flora of North America Editorial Committee (eds.) (1993). *Flora of North America North of Mexico*, Vol. 2. New York: Oxford University Press.

Gastony, G. J. (1983). The *Pellaea glabella* complex: electrophoretic evidence for the derivations of the agamosporous taxa and a revised taxonomy. *American Fern Journal*, **78**, 44–67.

Gastony, G. J. (1986). Electrophoretic evidence for the origin of fern species by unreduced spores. *American Journal of Botany*, **73**, 1563–1569.

Gastony, G. J. and Windham, M. D. (1989). Species concepts in pteridophytes: the treatment and definition of agamosporous species. *American Fern Journal*, **79**, 65–77.

Gentry, A. H. (1988). Changes in plant community diversity and floristic composition on environmental and geographical gradients. *Annals of the Missouri Botanical Garden*, **75**, 1–34.

Grant, V. (1981). *Plant Speciation*. New York: Columbia University Press.

Haufler, C. H. (1987). Electrophoresis is modifying our concepts of evolution in homosporous pteridophytes. *American Journal of Botany*, **74**, 953–966.

Haufler, C. H. (1989). Toward a synthesis of evolutionary modes and mechanisms in homosporous pteridophytes. *Biochemical Systematics and Ecology*, **17**, 109–115.

Haufler, C. H. (1996). Species concepts and speciation in pteridophytes. In *Pteridology in Perspective*, ed. J. M. Camus, M. Gibby, and R. J. Johns. Kew: Royal Botanic Gardens, pp. 291–305.

Haufler, C. H. (1997). Modes and mechanisms of speciation in pteridophytes. In *Evolution and Diversification of Land Plants*, ed. K. Iwatsuki and P. H. Raven. Tokyo: Springer-Verlag, pp. 291–308.

Haufler, C. H. (2002). Homospory 2002: an odyssey of progress in pteridophyte genetics and evolutionary biology. *Bioscience*, **52**, 1081–1093.

Haufler, C. H. (2007). Genetics, phylogenetics, and biogeography: considering how shifting paradigms and continents influence fern diversity. *Brittonia*, **59**, 108–114.

Haufler, C. H. and Ranker, T. A. (1995). *rbcL* sequences provide phylogenetic insights among sister species of the fern genus *Polypodium*. *American Fern Journal*, **85**, 359–372.

Haufler, C. H. and Windham, M. D. (1991). New species of North American *Cystopteris* and *Polypodium*, with comments on their reticulate relationships. *American Fern Journal*, **81**, 6–22.

Haufler, C. H., Windham, M. D., Britton, D. M., and Robinson, S. J. (1985). Triploidy and its evolutionary significance in *Cystopteris protrusa*. *Canadian Journal of Botany*, **63**, 1855–1863.

Haufler, C. H., Windham, M. D., and Ranker, T. A. (1990). Biosystematic analysis of the *Cystopteris tennesseensis* complex. *Annals of the Missouri Botanical Garden*, **77**, 314–329.

Haufler, C. H., Windham, M. D., and Rabe, E. W. (1995a). Reticulate evolution in the *Polypodium vulgare* complex. *Systemetic Botany*, **20**, 89–109.

Haufler, C. H., Soltis, D. E., and Soltis, P. S. (1995b). Phylogeny of the *Polypodium vulgare* complex: insights from chloroplast DNA restriction site data. *Systematic Botany*, **20**, 110–119.

Haufler, C. H., Hooper, E. A., and Therrien, J. P. (2000). Modes and mechanisms of speciation in pteridophytes: implications of contrasting patterns in ferns representing temperate and tropical habitats. *Plant Species Biology*, **15**, 223–236.

Hauk, W. D. and Haufler, C. H. (1999). Isozyme variability among cryptic species of *Botrychium* subgenus *Botrychium* (Ophioglossaceae). *American Journal of Botany*, **86**, 614–633.

Hey, J. (2006). Recent advances in assessing gene flow between diverging populations and species. *Current Opinion in Genetics and Development*, **16**, 592–596

Hickey, R. J. (1984). Chromosome numbers in neotropical *Isöetes*. *American Fern Journal*, **74**, 9–13.

Hickey, R. J., Taylor, W. C., and Luebke, N. T. (1989). The species concept in Pteridophyta with special reference to *Isöetes*. *American Fern Journal*, **79**, 78–89.

Hooper, E. A. and Haufler, C. H. (1997). Genetic diversity and breeding system in a group of neotropical epiphytic ferns (*Pleopeltis*; Polypodiaceae). *American Journal of Botany*, **84**, 1664–1674.

Jermy, A. C. and Walker, T. G. (1985). Cytotaxonomic studies of the ferns of Trinidad. *Bulletin of the British Museum (Natural History), Botany*, **13**, 133–276.

Kato, M. (1993). Biogeography of ferns: dispersal and vicariance. *Journal of Biogeography*, **20**, 265–274.

Kelloff, C., Skog, J., Adamkewicz, L., and Werth, C. R. (2002). Differentiation of two taxa of eastern North American *Athyrium*: evidence from allozymes and spores. *American Fern Journal*, **92**, 185–213.

Kluge, J. and Kessler, M. (2006). Fern endemism and its correlates: contribution from an elevational transect in Costa Rica. *Diversity and Distributions*, **12**, 535–545.

Kurihara T., Watano, Y., Takamiya, M., and Shimizu, T. (1996). Electrophoretic and cytological evidence for genetic heterogeneity and hybrid origin of *Athyrium oblitescens*. *Journal of Plant Research*, **109**, 29–36.

Lang, F. A. (1971). The *Polypodium vulgare* complex in the Pacific Northwest. *Madroño*, **21**, 235–254.

Lieberman, D., Lieberman, M., Peralta, R., and Hartshorn, G. S. (1996). Tropical forest structure and composition on a large-scale altitudinal gradient in Costa Rica. *The Journal of Ecology*, **84**, 137–152.

Lloyd, R. M. and Lang, F. A. (1964). The *Polypodium vulgare* complex in North America. *British Fern Gazette*, **9**, 168–177.

Lovis, J. D. (1977). Evolutionary patterns and processes in ferns. In *Advances in Botanical Research*, Vol. 4, ed. R. D. Preston and H. W. Woolhouse. London: Academic Press, pp. 229–415.

Manton, I. (1950). *Problems of Cytology and Evolution in the Pteridophyta*. Cambridge: Cambridge University Press.

Marcon, A. B., Barros, I. C. L., and Guerra, M. (2005). Variation in chromosome numbers, CMA Bands and 45S rDNA sites in species of *Selaginella* (Pteridophyta). *Annals of Botany*, **95**, 271–276.

Martens, P. (1943). Les organes gladuleux de *Polypodium virginianum* (*P. vulgare* var. *virginianum*). I. Valeur systématique et répartition géographique. *Bulletin du Jardin Botanique de L'État, Bruxelles*, **17**, 1–14.

Masuyama, S. (1979). Reproductive biology of the fern *Phegopteris decursive-pinnata*. I. The dissimilar mating systems of diploids and tetraploids. *Botanical Magazine (Tokyo)*, **92**, 275–289.

Masuyama, S. and Watano, Y. (2005). Cryptic species in the fern *Ceratopteris thalictroides* (L.) Brongn. (Parkeriaceae). II. Cytological characteristics of three cryptic species. *Acta Phytotaxonomica et Geobotanica*, **56**, 231–240.

Masuyama S., Yatabe, Y., Murakami, N., and Watano, Y. (2002). Cryptic species in the fern *Ceratopteris thalictroides* (L.) Brongn. (Parkeriaceae). I. Molecular analyses and crossing tests. *Journal of Plant Research*, **115**, 87–97.

Mayden, R. L. (1997). A hierarchy of species concepts: the denouement in the saga of the species problem. In *Species: The Units of Biodiversity*, ed. M. F. Claridge, A. H. Dawah, and M. R. Wilson. London: Chapman and Hall, pp. 381–424.

Mayr, E. (1942). *Systematics and the Origin of Species*. New York: Columbia University Press.

Mayr, E. (1969). The biological meaning of species. *Biological Journal of the Linnean Society*, **1**, 311–320.

Mishler, B. D. and Donoghue, M. J. (1982). Species concepts: a case for pluralism. *Systematic Zoology*, **31**, 491–503.

Moran, R. C. and Smith, A. R. (2001). Phytogeographic relationships between neotropical and African-Madagascan pteridophytes. *Brittonia*, **53**, 304–351.

Mullenniex, A., Hardig, T. M., and Mesler, M. R. (1998). Molecular confirmation of hybrid swarms in the fern genus *Polystichum* (Dryopteridaceae). *Systematic Botany*, **23**, 421–426.

Øllgaard, B. (1987). A revised classification of the Lycopodiaceae s. lat. *Opera Botanica*, **92**, 153–178.

Orr, M. R. and Smith, T. B. (1998). Ecology and speciation. *Trends in Ecology and Evolution*, **13**, 502–506.

Otto, S. P. and Whitton, J. (2000). Polyploid incidence and evolution. *Annual Review of Genetics*, **34**, 401–437.

Paris, C. A. and Windham, M. D. (1988). A biosystematic investigation of the *Adiantum pedatum* complex in eastern North America. *Systematic Botany*, **13**, 240–255.

Paris, C. A., Wagner, F. S., and Wagner, W. H. (1989). Cryptic species, species delimitation, and taxonomic practice in the homosporous ferns. *American Fern Journal*, **79**, 46–54.

Parks, C. R., Wendel, J. F., Sewell, M. M., and Qiu, Y.-L. (1994). The significance of allozyme variation and introgression in the *Liriodendron tulipifera* complex (Magnoliaceae). *American Journal of Botany*, **81**, 878–889.

Phipps, C. J., Taylor, T. N., Taylor, E. L., Cuneo, N. R., Boucher, L. D., and Yao, X. (1998). *Osmunda* (Osmundaceae) from the Triassic of Antarctica: an example of evolutionary stasis. *American Journal of Botany*, **85**, 888–895.

Pryer, K. M. and Haufler, C. H. (1993). Isozymic and chromosomal evidence for the allotetraploid origin of *Gymnocarpium dryopteris* (Dryopteridaceae). *Systematic Botany*, **18**, 150–172.

Rabe, E. W. and Haufler, C. H. (1992). Incipient polyploid speciation in the maidenhair fern (*Adiantum pedatum*; Adiantaceae). *American Journal of Botany*, **79**, 701–707.

Ranker, T. A. (1992a). Genetic diversity of endemic Hawaiian epiphytic ferns: implications for conservation. *Selbyana*, **13**, 131–137.

Ranker, T. A. (1992b). Genetic diversity, mating systems, and interpopulational gene flow in neotropical *Hemionitis palmata* L. (Adiantaceae). *Heredity*, **69**, 175–183.

Ranker, T. A. (1994). Evolution of high genetic variability in the rare Hawaiian fern *Adenophorus periens* and implications for conservation management. *Biological Conservation*, **70**, 19–24.

Ranker, T. A., Floyd, S. K., Windham, M. D., and Trapp, P. G. (1994). Historical biogeography of *Asplenium adiantum-nigrum* (Aspleniaceae) in North America and implications for speciation theory in homosporous pteridophytes. *American Journal of Botany*, **81**, 776–781

Ranker, T. A., Gemmill, C. E. C., Trapp, P. G., Hambleton, A., and Ha, K. (1996). Population genetics and reproductive biology of lava-flow colonising species of Hawaiian *Sadleria* (Blechnaceae). In *Pteridology in Perspective*, ed. J. M. Camus, M. Gibby, and R. J. Johns. Kew: Royal Botanic Gardens, pp. 581–598.

Raven, P. H. (1976). Systematics and plant population biology. *Systematic Botany*, **1**, 284–316.

Richardson, B. A., Richardson, M. J., Scatena, F. N., and McDowell, W. H. (2000). Effects of nutrient availability and other elevational changes on bromeliad populations and their invertebrate communities in a humid tropical forest in Puerto Rico. *Journal of Tropical Ecology*, **16**, 167–188.

Rieseberg, L. H. and Burke, J. M. (2001). The biological reality of species, gene flow, selection, and collective evolution. *Taxon*, **50**, 47–67.

Sarvella, J. (1978). A synopsis of the fern genus *Gymnocarpium*. *Annales Botanici Fennici*, **15**, 101–106.

Sarvella, J. (1980). *Gymnocarpium* hybrids from Canada and Alaska. *Annales Botanici Fennici*, **17**, 292–295.

Schneider, H., Schuettpelz, E., Pryer, K. M., Cranfill, R., Magallón, S., and Lupia, R. (2004). Ferns diversified in the shadow of angiosperms. *Nature*, **428**, 553–557.

Schneller, J. J. (1979). Biosystematic investigations on the lady fern (*Athyrium filix-femina*). *Plant Systematics and Evolution*, **132**, 255–277.

Schneller, J. J. (1981), Evidence for intergeneric incompatibility in ferns. *Plant Systematics and Evolution*, **137**, 45–56.

Schneller, J. J. (1989). Remarks on hereditary regulation of spore wall pattern in intra- and interspecific crosses of *Athyrium*. *Botanical Journal of the Linnean Society*, **99**, 115–123.

Sciarretta, K. L., Arbuckle, E. P., Werth, C. R., and Haufler, C. H. (2005). Patterns of genetic variation in southern Appalachian populations of *Athyrium filix-femina* var. *asplenioides* (Dryopteridaceae). *International Journal of Plant Science*, **166**, 761–780.

Shivas, M. G. (1961). Contributions to the cytology and taxonomy of *Polypodium* in Europe and America. I. Cytology. *Botanical Journal of the Linnean Society*, **58**, 13–25.

Simpson, G. G. (1961). *Principles of Animal Taxonomy*. New York: Columbia University Press.

Sites, J. D. and Marshall, J. C. (2003). Delimiting species: a Renaissance issue in systematic biology. *Trends in Ecology and Evolution*, **18**, 462–470.

Soltis, D. E. and Rieseberg, L. H. (1986). Autopolyploidy in *Tolmiea menziezii* (Saxifragaceae): evidence from enzyme electrophoresis. *American Journal of Botany*, **73**, 1171–1174.

Soltis, D. E. and Soltis, P. S. (1988). Estimated rates of intragametophytic selfing in lycopods. *American Journal of Botany*, **75**, 248–256.

Soltis, D. E. and Soltis, P. S. (1989). Polyploidy, breeding systems, and genetic differentiation in homosporous pteridophytes. In *Isozymes in Plant Biology*, ed. D. E. Soltis and P. S. Soltis. Portland, OR: Dioscorides Press, pp. 241–258.

Soltis, P. S., Soltis, D. E. and Wolf, P. G. (1990). Allozymic divergence in North American *Polystichum* (Dryopteridaceae). *Systematic Botany*, **15**, 205–215.

Soltis, D. E., Soltis, P. S., and Tate, J. A. (2003). Advances in the study of polyploidy since *Plant Speciation*. *New Phytologist*, **161**, 173–191.

Somers, P. and Buck, W. R. (1975). *Selaginella ludoviciana*, *S. apoda* and their hybrids in the southeastern United States. *American Fern Journal*, **65**, 76–82.

Taylor, W. C. and Hickey, R. J. (1992). Habitat, evolution, and speciation in *Isöetes*. *Annals of the Missouri Botanical Garden*, **79**, 613–622.

Taylor, W. C., Luebke, N. T., and M. B. Smith. (1985). Speciation and hybridization in North American quillworts. *Proceedings of the Royal Society of Edinburgh*, **86B**, 259–263.

Terborgh, J. (1985). The vertical component of plant species diversity in temperate and tropical forests. *The American Naturalist*, **126**, 760–776.

Trewick, S. A., Morgan-Richards, M., Russell, S. J., Henderson, S., Rumsey, F. J., Pintér, J. A., Barrett, J. A., Gibby, M., and Vogel, J. C. (2002). Polyploidy, phylogeography and Pleistocene refugia of the rockfern *Asplenium ceterach*: evidence from chloroplast DNA. *Molecular Ecology*, **11**, 2003–2012.

Tryon, R. M. (1955). *Selaginella rupestris* and its allies. *Annals of the Missouri Botanical Garden*, **42**, 1–99.

Tryon, R. (1971). The process of evolutionary migration in species of *Selaginella*. *Brittonia*, **23**, 89–100.

Tryon, A. F. and Britton, D. M. (1958). Cytotaxonomic studies on the fern genus *Pellaea*. *Evolution*, **12**, 137–145

Tuomisto, H., Poulsen, A. D. O., and Moran, R. C. (1998). Edaphic distribution of some species of the fern genus *Adiantum* in western Amazonia. *Biotropica*, **30**, 392–399.

Tuomisto, H., Ruokolainen, K., Poulsen, A. D., Moran, R. C., Quintana, C., Cañas, G., and Celi, J. (2002). Distribution and diversity of pteridophytes and Melastomataceae along edaphic gradients in Yasuní National Park, Ecuadorian Amazonia. *Biotropica*, **34**, 516–533.

Tuomisto, H., Ruokolainen, K., Aguilar, M., and Sarmiento, A. (2003). Floristic patterns along a 43-km long transect in an Amazonian rain forest. *Journal of Ecology*, **91**, 743–756.

Vogel, J. C., Russell, S. J, Barrett, J. A., and Gibby, M. (1996). A non-coding region of chloroplast DNA as a tool to investigate reticulate evolution in European *Asplenium*. In *Pteridology in Perspective*, ed. J. M. Camus, M. Gibby, and R. J. Johns. Kew: Royal Botanic Gardens, pp. 313–327.

Wagner, W. H., Jr. (1954). Reticulate evolution in the Appalachian aspleniums. *Evolution*, **8**, 103–118.

Wagner, D. H. (1979). Systematics of *Polystichum* in western North America, north of Mexico. *Pteridologia*, **1**, 1–64.

Wagner, F. S. (1992). Cytological problems in *Lycopodium* sens. lat. *Annals of the Missouri Botanical Garden*, **79**, 718–729.

Walker, S. (1955). Cytogenetic studies in the *Dryopteris spinulosa* complex I. *Watsonia*, **3**, 193–209.

Walker, T. G. (1958). Hybridization in some species of *Pteris* L. *Evolution*, **12**, 82–92.

Walker, S. (1961). Cytogenetic studies in the *Dryopteris spinulosa* complex. II. *American Journal of Botany*, **48**, 607–614.

Walker, T. G. (1962). Cytology and evolution in the fern genus *Pteris* L. *Evolution*, **16**, 27–43.

Walker, T. G. (1984). Chromosomes and evolution in pteridophytes. In *Chromosomes in Evolution of Eukaryotic Groups*, Vol. 2, ed. A. K. Sharma and A. Sharma. Boca Raton, FL: CRC Press.

Wallace, A. R. (1858). On the tendency of varieties to depart indefinitely from the original type. *Journal of the Proceedings of the Linnean Society (Zoology)*, **3**, 53–62.

Watkins, J. E., Cardelus, C., Colwell, R. K., and Moran, R. C. (2006). Species richness and distribution of ferns along an elevational gradient in Costa Rica. *American Journal of Botany*, **93**, 73–83.

Werth, C. R. (1989). The use of isozyme data for inferring ancestry of polyploid species of pteridophytes. *Biochemical Systematics and Ecology*, **17**, 117–130.

Werth, C. R. (1991). Isozyme studies on the *Dryopteris "spinulosa"* complex, I: the origin of the log fern *Dryopteris celsa*. *Systematic Botany*, **10**, 184–192.

Werth, C. R. and Windham, M. D. (1991). A model for divergent, allopatric speciation of polyploid pteridophytes resulting from silencing of duplicate gene expression. *American Naturalist*, **137**, 515–526.

Werth, C. R., Guttman, S. I. and Eshbaugh, W. H. (1985). Electrophoretic evidence of reticulate evolution in the Appalachian *Asplenium* complex. *Systematic Botany*, **16**, 446–461.

Whittemore, A. T. (1993). Species concepts: a reply to Mayr. *Taxon*, **42**, 573–583.

Wilce, J. H. (1965). Section *Complanata* of the genus *Lycopodium*. *Nova Hedwigia*, **19**, 1–233.

Wiley, E. O. (1978). The evolutionary species concept reconsidered. *Systematic Zoology*, **27**, 17–26.

Wiley, E. O. and Mayden, R. (2000). The evolutionary species concept. In *Species Concepts and Phylogenetic Theory: A Debate*, ed. Q. D. Wheeler and R. Meier. New York: Columbia University Press, pp. 70–89.

Windham, M. D. (1987). *Argyrochosma*, a new genus of cheilanthoid ferns. *American Fern Journal*, **77**, 37–41.

Windham, M. D. and Yatskievych, G. (2003). Chromosome studies of cheilanthoid ferns (Pteridaceae: Cheilanthoideae) from the western United States and Mexico. *American Journal of Botany*, **90**, 1788–1800.

Wolf, P. G., Schneider, H., and Ranker, T. A. (2001). Geographic distributions of homosporous ferns: does dispersal obscure evidence of vicariance? *Journal of Biogeography*, **28**, 263–270.

Yatabe Y., Masuyama, S., Darnaedi, D., and Murakami, N. (2001). Molecular systematics of the *Asplenium nidus* complex from Mt. Halimun National Park, Indonesia: evidence for reproductive isolation among three sympatric *rbc*L sequence types. *American Journal of Botany*, **88**, 1517–1522.

Yatskievych, G. and Moran, R. C. (1989). Primary divergence and species concepts in ferns. *American Fern Journal*, **79**, 36–45.

13

Phylogeny and evolution of ferns: a paleontological perspective

GAR W. ROTHWELL AND RUTH A. STOCKEY

13.1 Introduction

Ferns traditionally have been identified as megaphyllous plants that reproduce by sporangia borne on leaves (i.e., fronds; Bower 1923; Kaplan and Groff, 1995). Although not all species have the entire set of characters, ferns typically are recognized by sporophytes that display unipolar growth (Rothwell, 1995), are dominated by fronds (Kaplan and Groff, 1995), and are devoid of secondary growth. Many have highly branched fronds, mesarch xylem maturation, and a rhizome stele that is dissected by leaf gaps. Plants that display various combinations of these features occur in the fossil record from the Middle Devonian (i.e., 390 million years ago) to the Recent, but there has been considerable systematic turnover with several prominent clades replacing one another through geological time (Rothwell, 1999). In practice, botanists have traditionally recognized as "ferns" those species that are left over after all other euphyllophytes have been removed to clades with clearly identifiable synapomorphies (Rothwell, 1999).

Up to the present, attempts to define ferns within a phylogenetic framework have met with only limited success (Figure 13.1) due to a combination of (1) limited information about many extinct ferns and fern-like plants, and (2) the restricted taxon sampling available for phylogenetic analyses that include only living species. The paleontological record of ferns is incomplete, and thus far has been sampled for only a small fraction of the available fossils (Stockey and Rothwell, 2006). This leaves us with an often confusing picture of inadequately known extinct species that can be difficult to comprehend and appreciate.

Biology and Evolution of Ferns and Lycophytes, ed. Tom A. Ranker and Christopher H. Haufler. Published by Cambridge University Press. © Cambridge University Press 2008.

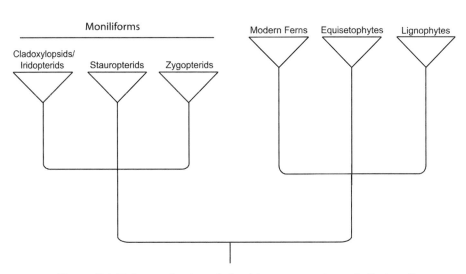

Figure 13.1 Phylogeny showing relationships among major euphyllophyte lineages. Some trimerophyte grade euphyllophytes and the psilotophytes have been omitted for clarity.

Within this context, evolutionary and phylogenetic studies that focus on living species are in danger of excluding or misrepresenting the wealth of available paleontological data.

This chapter is designed primarily for the non-paleontologist with the goals of imparting a general understanding of what is known about the fossil record of ferns, and providing an entrée to the paleobotanical literature for extinct ferns and fern-like plants. The information that follows summarizes the geological history of ferns (Figure 13.2), relates data upon which concepts of the fossil ferns and extinct fern taxa are based, and interprets those data in a phylogenetic context. We include explanations of both the promise and constraints of fossils for improving our understanding of ferns through time, and an analysis of important information that extinct species can provide to help achieve the goal of ultimately resolving the overall pattern of euphyllophyte phylogeny. We hope this information will help infuse the generation of new hypotheses with known paleontological data and enhance phylogenetic studies that include fossils. Our coverage of the fossil record is both eclectic and spotty, as the chapter has been figured primarily from specimens housed in the University of Alberta Paleobotanical Collections. More comprehensive treatments of specific groups are noted in the text and included in the literature cited.

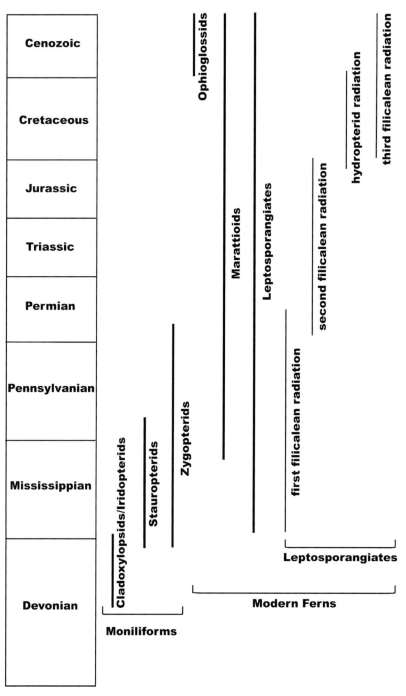

Figure 13.2 Stratigraphic distribution of ferns s.l. Extinct groups, including taxa used as exemplars for moniliforms by Kenrick and Crane (1997), stauropterids, and zygopterids are on the left. Among clades with modern representatives (on the right), the fossil record for leptosporangiate ferns is separated into three filicalean radiations and the hydropterid radiation (far right).

13.2 Nature of the fossil record

Because of an open indeterminate pattern of growth and relatively simple structure, there has been a great deal of parallel and convergent evolution of plant organs (Stewart and Rothwell, 1993; Taylor and Taylor, 1993; Boyce and Knoll, 2002). Plants routinely shed organs during life and commonly disarticulate after death. Therefore, a majority of the paleobotanical record consists of isolated organs and fragments of shoot systems that may be hard to classify with precision. When fossils consist of plant parts, and/or modes of preservation that cannot be recognized as representing a particular species, genus, or even family of plants, they are considered to be "morphotaxa" (Greuter et al., 2000). Whereas morphotaxa can provide solid evidence for the evolution of structural features and hint at the occurrence of clades in specific geological strata, they are of only limited utility for phylogenetic analysis. By contrast, when large numbers of fossils from a small number of sources are studied carefully (e.g., Holmes, 1977, 1981) or when fossil sporophytes are discovered intact, species of extinct plants can be characterized in great detail, and their structure and biology understood well. In such cases concepts of extinct plant species have the potential to generate a more complete understanding of the extinct organisms than has been developed for the vast majority of living plant species (e.g., Taylor et al., 2005). Morphotaxa of plant fossils yield information for interpreting the evolution of plant organs through transformational series of structure (Stewart, 1964), and species of extinct plants (commonly understood as the result of plant reconstructions) constitute direct evidence of extinct organisms at particular points in time. Therefore, species of extinct plants contribute solid data for formulating phylogenetic hypotheses through cladistic analyses and for testing systematic hypotheses developed from other sources of data.

13.3 Systematic relationships among ferns, fern-like plants, and other euphyllophytes

Competing hypotheses regarding the evolutionary history of vascular plants have been published over the past ten years. In 1997, Kenrick and Crane developed a comprehensive phylogenetic treatment of ancient vascular plants and their more recent sister groups (i.e., polysporangiophytes) and included three Devonian genera that traditionally have been linked to modern ferns (i.e., *Pseudosporochnus* and *Rhacophyton*) and equisetophytes (i.e., *Ibyka*). Results of those phylogenetic analyses resolved a clade consisting of (*Pseudosporochnus* + *Rhacophyton*) + *Ibyka* as the sister group to lignophytes (i.e., Fig. 4.31 of Kenrick and Crane, 1997). Hypothesizing sister group relationships between the fossil (i.e., *Pseudosporochnus, Rhacophyton*, and *Ibyka*) and modern "fern-like" euphyllophytes, Kenrick and Crane (1997) recognized the fossil and living members as a clade that they named infradivision "Moniliformopsis" (Kenrick and Crane, 1997). According to the Kenrick and Crane hypothesis, Moniliformopsis includes cladoxylopsid and iridopterid "ferns" (e.g., *Pseudosporochnus*), zygopterid "ferns" (e.g., *Rhacophyton*), the three clades of ferns with living representatives (i.e., ophioglossids, marattioids, and leptosporangiates), as well as living psilotophytes (i.e., *Psilotum* and *Tmesipteris*) and equisetophytes (to which *Ibyka* sometimes is assigned).

A subsequent cladistic analysis of the hypothesized relationships using a combination of living and extinct taxa (Rothwell, 1999) produced trees that resolved the moniliform taxa as a non-monophyletic assemblage. In this analysis, the various "moniliform" taxa were removed to several more distantly related groups, some of which are more closely related to lignophytes than to other moniliform clades (Rothwell, 1999). More specifically, cladoxylopsids and zygopterids were resolved as the sister group to equisetophytes, which formed the sister group to lignophytes (i.e., [[cladoxylopsids + zygopterids] + equisetophytes] + lignophytes). That clade formed the sister group to a clade that consisted of the three major groups of living ferns (i.e., [ophioglossids + [marattioids + leptosporangiates]]), with stauropterid "ferns" and psilotophytes attaching to the stem of the tree at two of the more basal nodes (i.e., Fig. 2 of Rothwell, 1999). Rothwell (1999) did not consider any members of the Iridopteridales well enough understood at that time to be included in the analysis. Thus, Rothwell's results falsify the hypothesis that *Pseudosporochnus*, *Rhacophyton*, and *Ibyka* form a clade with living ferns, equisetophytes, and psilotophytes, thus restricting Moniliformopsis to the Devonian taxa included in the Kenrick and Crane analyses (1997) plus related fossil taxa (Fig. 13.1).

Pryer *et al.* (2001) analyzed relationships among living euphyllophytes using a combination of nucleotide sequences and morphological characters. Their results supported the Kenrick and Crane hypothesis and resolved living ferns, equisetophytes, and psilotophytes as a clade that forms the sister group to living seed plants (i.e., Fig. 1 of Pryer *et al.*, 2001). Rothwell and Nixon (2006) provided alternative analyses of the competing hypotheses of euphyllophyte relationships and demonstrated that neither of the recent hypotheses (i.e., Rothwell, 1999; Pryer *et al.*, 2001) is robust. The analyses of Rothwell and Nixon (2006) also indicated that none of the living euphyllophyte taxa are closely related to the Devonian plants upon which the Moniliformopsis concept is based. These results indicate that moniliforms are a grade or clade of extinct euphyllophytes that includes neither living ferns nor equisetophytes. A condensed tree depicting this synthesis is presented in Figure 13.1. Moniliforms are attached to the stem of the tree at a node below a polytomy that includes modern ferns, equisetophytes, and lignophytes.

Given the competing hypotheses regarding relationships among euphyllophyte clades, we present the major groups of extinct and living ferns and fern-like plants in this chapter roughly in the order that they appear in the fossil record. Structural features, stratigraphic ranges, and known relationships are reviewed for each group, and representative plant concepts are presented. Fossil ferns and fern-like taxa are organized as (1) cladoxylopsids and iridopterids, (2) stauropterids, (3) zygopterids, (4) ophioglossids, (5) marattioids, and (6) leptosporangiates. The leptosporangiate clade is further subdivided by families that represent the first, second, and third evolutionary radiations of homosporous filicaleans, and the heterosporous Hydropteridales. Stratigraphic ranges for the various "fern" groups are summarized in Figure 13.2.

13.4 Moniliforms – the most ancient fern-like plants

Moniliform "ferns" consist primarily of assemblages of compressed and/or anatomically preserved plant fragments that occur in deposits ranging from the Middle Devonian to the Permian (Figure 13.2). Representatives with the simplest structure have features that intergrade with trimerophyte-grade euphyllophytes (e.g., Hilton, 1999), whereas, those with somewhat more complex structure intergrade with each other and with plants that represent lignophytes and equisetophytes (e.g., Berry and Wang, 2006). As a result, moniliform fossils sometimes are allied with modern ferns and at other times are regarded as ancestral (or sister groups) to equisetophytes (Stein 1982), but hypothesized sister group relationships with modern euphyllophytes have not been confirmed for any of the Devonian moniliforms.

Up to the present, only a few moniliform species have been reconstructed to produce whole plant concepts. To complicate the identification and classification of moniliform plants further, a large percentage of species are known only from small segments of the branching systems. Some generic and specific concepts are based on compression/impression fossils that show primarily gross morphology, while others are derived from segments of permineralized axes that show primarily internal anatomy and branching pattern for only one or two branching orders (e.g., Stein and Hueber 1989; Berry and Wang, 2006). Only a few are known from a variety of modes of preservation that represent all of the organs of the sporophyte.

In our current treatment, moniliforms are organized as cladoxylopsids/iridopterids, stauropterids, and zygopterids (Figure 13.1), but many of the most ancient moniliform fossils are hard to classify with precision (see Berry and Wang, 2006, for an illustration of the difficulties). Plants display relatively simple gross morphologies. Branching patterns typically vary from level to level within a shoot system, and clear stem/leaf organography is absent from all but the zygopterids. Likewise, in all but the zygopterids, branching systems are made up of terete axes that lack laminar distal segments (i.e., pinnules) and have fertile regions of varying complexities that produce terminal sporangia. In some species the sporangia are erect, while in others they are variously reflexed and/or recurved, and many show considerable variation in this feature. Some species are homosporous and others are heterosporous. Where known, spores are typically trilete.

13.4.1 Cladoxylopsids and iridopterids

Plants range up to moderate sized trees (Berry and Fairon-Demaret, 2002; Stein et al., 2007), and often have a trunk that is anchored by adventitious roots basally, and that forms a crown of branches distally (Figure 13.3a). Individual branches display one of several distinctive branching patterns (Skog and Banks, 1973; Stein, 1982), many being three dimensional, while others appear more-or-less planar (Figures 13.3a–c). Distal units fork more-or-less equally and are often considered to be the equivalent of pinnules, even though there is no laminar tissue interconnecting the terete terminal segments (Figure 13.3b). In contrast to modern ferns, the branching systems of cladoxylopsids and iridopterids do not display clear stem/leaf organography. Fertile species are typically assigned to one of several genera that consist of branching systems with terminal units that are more-or-less similar to the vegetative systems, except that each axis terminates as a sporangium (cf., Figures. 13.2 and 13.3). In some genera the sporangia are erect (Figure 13.3c) whereas in others they are reflexed or even recurved (Stewart and Rothwell, 1993; Taylor and Taylor, 1993). Those with recurved sporangia are more likely to be allied to modern equisetophytes, while those with erect sporangia tend to be related to modern ferns.

Large anatomically preserved axes often have a stele formed by a complex system of anastomosing xylem segments (Soria et al., 2001; Figure 13.3e) that divide radially to produce traces to branches. In somewhat smaller axes the xylem segments tend to be connected centrally, forming radiating ribs of a protostele (Figure 13.3d). Xylem of the smallest axes consists of a trilobed, bilobed, or terete bundle. The largest axes typically have an orthotropic branching pattern, whereas successively smaller axes often branch in a four-ranked (i.e., quadriseriate) and then a two-ranked pattern. Protoxylem occurs near the periphery of the stele, and in some species there may be additional strands toward the center. Xylem maturation is mesarch, and often there is a hollow channel or parenchyma bundle associated with each protoxylem strand (Figure 13.3d). In

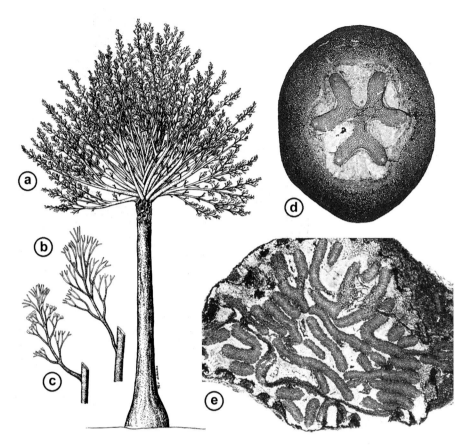

Figure 13.3 Devonian and Lower Carboniferous ferns (Moniliforms). (a)–(c) *Pseudosporochnus nodosus*. (a) Reconstruction of plant (from Berry and Fairon-Demaret, 2002). (b), (c) Reconstructions of fertile and vegetative branches respectively, from Stewart and Rothwell (1993). (d) Cross-section of *Arachnoxylon kopfii* showing protostele of interconnected xylem ribs with mesarch protoxylem near tips of ribs (from Stein, 1981). (e) Cross-section of *Pseudosporochnus hueberi* showing partly interconnected xylem ribs with mesarch protoxylem strands near the periphery and also toward the interior of stele (from Stein and Hueber, 1989)

species that have been related to equisetophytes, the hollow is often compared to the carinal canals of *Equisetum*, but usually it is difficult to determine whether the hollow represents the living condition of the plant, or whether it has resulted from incomplete tissue preservation (Scheckler, 1974). In species where the parenchyma strand increases in size at levels of branching, and accompanies the xylem into a branch, these structures are called "peripheral loops." Some axes show radial rows of tracheids toward the margins of the xylem segments, but these rows more likely represent radially aligned metaxylem tracheids than secondary xylem that has been produced by a vascular cambium (Esau, 1943).

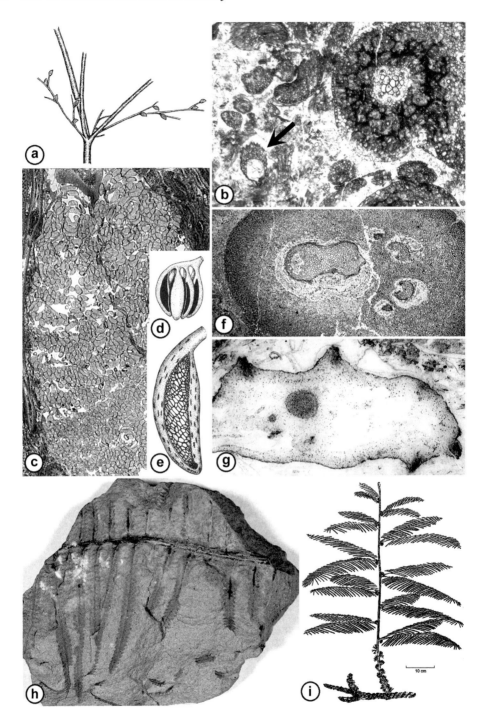

Pseudosporochnus nodosus is one of the most completely known of the cladoxylopsids, and can be used to exemplify such plants. *Pseudosporochnus* formed a small tree that was anchored by adventitious roots, and that formed a crown of highly dissected branches sometimes compared to fronds (Leclercq and Banks, 1962; Berry and Fairon-Demaret, 2002). Branching is generally three dimensional, becoming somewhat planar distally (Figure 13.3a), with individual branches either terminating as terete tips (Figure 13.3b) or sporangia (Figure 13.3c).

13.4.2 Stauropterids

Stauropterid moniliforms are apparently small plants that consist of highly branched systems of terete axes. Rooting systems have not been found attached to the branching systems, so the maximum size of stauropterid sporophytes has not been determined for certain. As with cladoxylopsids and iridopterids, there is no evidence for the evolution of distinct stem/leaf organography among stauropterid ferns. Branching typically is alternate, with pairs of smaller laterals diverging from the side of a larger axis forming a quadriseriate system (Figure 13.4a; Surange, 1952). Two small branching systems, termed

---

Figure 13.4 Additional representatives of extinct fern clades (Moniliforms), assignable to Stauropteridales and Zygopteridales. (a), (b) Stauropteridales. (a) Reconstruction of stauropterid fern *Gillespiea randolphensis* showing three-dimensional pattern of branching, and terminal sporangia attached to small branches, or aphlebiae, that occur at branching levels (from Erwin and Rothwell, 1989). (b) Section views of *Stauropteris burntislandica* Bertrand, including relatively large axis with exarch/marginally mesarch protostele, smaller axes, and megasporangium (at arrow), from the Lower Carboniferous (Visean) of Scotland. (c)–(i) Zygopteridales. (c)–(e) *Biscalitheca musata* sporangial structures. (c) Large aggregation of sporangia from the Upper Pennsylvanian of Ohio. (d) Reconstruction of sporangial cluster. (e) Reconstruction of sporangium in longitudinal section showing one of two large annuli that characterize this genus (d and e from Stewart and Rothwell, 1993). (f) Cross-section through a rachis of Lower Carboniferous *Metaclepsydropsis duplex*, showing divergence of C-shaped traces to quadriseriately arranged primary pinnae (at right) and smaller traces to aphlebiae, from the Visean of Scotland. (g) Cross-section through a rhizome of *Zygopteris illinoiensis* from the Upper Pennsylvanian of Illinois, showing protostele with radially aligned tracheids of putative secondary origin, and the bases of aphlebiae (dark protruding areas) at the periphery of the cortex. (h) Compression specimen assignable to the Paleozoic fern-like foliage morphogenus *Alloiopteris*. This specimen is equivalent to a primary pinna of a *Zygopteris illinoiensis/Biscalitheca musata* type plant. (i) Reconstruction of *Zygopteris illinoiensis* plant from Phillips and Galtier (2005) as amplified from Dennis (1974).

aphlebiae, occur at each level of branching (Fig. 13.4a). Internally, stauropterid axes display a protostele that may be lobed (Fig. 13.4b) or consist of a small number of adjacent xylem stands. Xylem maturation ranges from marginally mesarch to exarch, and a parenchyma strand accompanies the protoxylem in some species.

Some stauropterids are heterosporous plants with the number of megaspores per sporangium being as few as two (i.e., *Stauropteris burntislandica*; Figure 13.4b, at arrow) or even one (i.e., *Gillespiea randolphensis*; Figure 13.4a). Where known, all of the sporangia of other stauropterid species contain a relatively large number of radial, trilete spores that fall within a single range of size variation. The latter species is considered to have a homosporous life cycle. Sporangia are terminal on either the terete branches or tips of the branching aphlebiae (Figure 13.4a).

13.4.3 Zygopterids

Zygopterid ferns extend from the Upper Devonian to the Permian (Figure 13.2), and recently have been comprehensively reviewed by Phillips and Galtier (2005). Among the moniliform groups, zygopterids show the greatest similarity to living ferns. However, zygopterids also have a distinct suite of synapomorphies that distinguishes them from all modern fern clades (Rothwell, 1999). Plants are homosporous and display clear stem/leaf organography. Some, such as the Pennsylvanian age genus *Zygopteris*, produce elongated horizontal rhizomes with three-dimensional fronds (Figure 13.4i), contributing to the ground cover beneath the canopy of Carboniferous and Early Permian wetland communities (Dennis, 1974; Phillips and Galtier, 2005). Other zygopterids, such as the Late Devonian *Rhacophyton*, may have short stems, producing fronds with indeterminate growth and that root adventitiously when they touch the ground. Plants of the latter type are thought to have formed dense thickets in Upper Devonian peat-forming wetlands of the paleotropics (Scheckler, 1986).

Zygopterid fronds conform to a complex and characteristic pattern of three- and two-dimensional branching (Figure 13.4i; Phillips and Galtier, 2005) not known to occur in modern ferns. They also have laminar pinnules of characteristic sizes and shapes (Figure 13.4h). The rachis branches alternately to produce paired primary pinnae in a quadriseriate pattern (Figure 13.4i). More distal branching is planar (Figures 13.4h, 13.4i), and the elongated pinnules show pointed alternate lobes (Figure 13.4h). In addition to the large fronds, zygopterids are characterized by the production of small, consistently arranged, branching aphlebiae on both the rhizome and fronds (Figure 13.4i). Sporangia are either terminal on highly branched frond segments that form large dense clusters (Figure 13.4c; e.g., *Rhacophyton* and *Biscalitheca*; Millay and Rothwell, 1983) or else occur superficially in radial sori on the abaxial surface of fertile pinnules (i.e., *Corynepteris*; Phillips and Galtier, 2005).

In contrast to modern ferns, zygopterids show evidence of secondary growth in the form of radially aligned tracheids and interspersed rays at the periphery of the rhizome stele (Figure 13.4g; Dennis, 1974). Small traces diverge from the stele and traverse the parenchymatous cortex in a consistent helical arrangement before entering the base of the aphlebiae (i.e., dark protrusions at the periphery of the rhizome in Figure 13.4g; Dennis, 1974). The rachis trace is large and dumbbell (Figure 13.4f) or clepsydroid in shape, with a prominent peripheral loop at each end (Figure 13.4f, at left). A pair of primary pinna traces (larger traces on the right in Figure 13.4f) diverge alternately from each end of the rachis trace producing the quadriseriate frond (Figure 13.4i). Small aphlebiae traces flank the primary pinna traces at the level of branching (Figure 13.4f). More distal branching of the frond is planar, but at right angles to the plane of branching for the rachis (Figure 13.4i).

Sporangia are large and thick walled with a large number of trilete spores (Figures 13.4c–e), and are apparently the products of eusporangial development (Rothwell, 1999). Devonian and Lower Carboniferous zygopterid sporangia (e.g., *Rhacophyton* and *Musatea*; Galtier, 1968) apparently dehisce along a longitudinal slit, but in the Upper Carboniferous species, dehiscence results from the activity of a prominent annulus. In *Biscalitheca* there is a pair of longitudinally elongated annuli, one on each side of the sporangium (Figures 13.4d, e). In *Corynepteris* the two segments of annulus are connected near the tip of the sporangium, forming a single U-shaped dehiscence apparatus (Phillips and Galtier, 2005).

13.5 Ophioglossid ferns

Ophioglossales is one of two clades of eusporangiate ferns with living representatives that appear to be more closely related to each other and to leptosporangiate ferns than to the other clades of living pteridophytic euphyllophytes (i.e., psilotophytes and equisetophytes). However, as stressed above, there are competing hypotheses for the latter relationships. Ophioglossids have a poor fossil record, consisting only of isolated spores from Jurassic and Cretaceous deposits of Russia, and of a single species of megafossils assignable to the genus *Botrychium* s.l. from basal-most Paleogene deposits of western Canada (Rothwell and Stockey, 1989). *Botrychium wightonii* is represented by pinnate pinnatifid vegetative frond segments (i.e., trophophores; Figure 13.5a) and bipinnately dissected fertile frond segments (i.e., sporophores; Figure 13.5b) with large eusporangia that are borne abaxially and submarginally on extremely narrow pinnule laminae. Spores are radial, subglobose, and trilete. In all of these features, *B. wightonii*

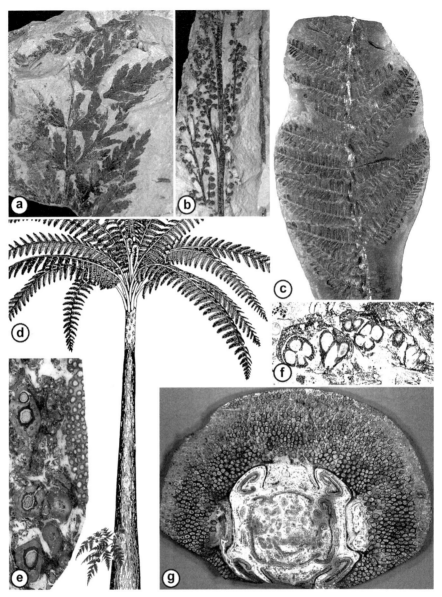

Figure 13.5 Extinct representatives of modern eusporangiate fern clades Ophioglossales (a, b) and Marattiales (c–g). (a) Vegetative portion of a frond (trophophore) of *Botrychium wightonii* from the Paleocene of Alberta, ×2. (b) Fertile portion of a frond (sporophore) of *Botrychium wightonii* from the Paleocene of Alberta, ×1. (c) Segment of *Pecopteris* sp. from the Middle Pennsylvanian of Illinois, ×0.5. (d) Reconstruction of *Psaronius* plant with epiphytic *Botryopteris tridentata* growing on the root mantle of the trunk, amplified from Morgan (1959) and Rothwell (1991). (e) Cross-section of stems, frond bases and branches of the vine *Ankyropteris brongniartii* growing on the root mantle of *Psaronius* from the Middle Pennsylvanian of Illinois, ×1. (f) Cross-section of a *Scolecopteris* pinnule with several abaxial synangia from the Upper Pennsylvanian of Illinois, ×9. (g) Cross-section of a *Psaronius brasiliensis* stem with surrounding root mantle from the Permian of Brazil, ×0.6.

compares favorably with the larger sized species of modern *Botrychium* s.l., such as *B. virginianum*. Because of the extremely limited fossil record for Ophioglossales, paleontology offers little evidence to augment systematic data from living species, other than to document that *Botrychium* s.l. had evolved essentially modern species by the beginning of the Paleogene.

13.6 Marattioid ferns

In contrast to ophioglossid ferns, the Marattiales have a rich and extensive fossil record that extends from at least the Lower Pennsylvanian onward (Stewart and Rothwell, 1993; Taylor and Taylor, 1993). Marattioid fossils are particularly abundant in the Pennsylvanian and Permian paleotropical wetlands of Euramerica (Morgan, 1959; Mickle, 1984; Millay, 1997; Rößler, 2000), but are well known from other regions as well, including Antarctic Gondwana (Delevoryas et al., 1992). Fossils diminish in abundance and occurrence through the Triassic and Jurassic, becoming essentially absent from Cretaceous and more recent deposits (Hu et al., 2006). The reason for this reduction in marattioid fossils through time is not fully understood, but may be correlated with a shift of habitat from wetlands to more arid exposures through time and with the progressive decrease in frost-free climates during the Tertiary and Quaternary.

Highly detailed plant concepts have been developed for several species of the Pennsylvanian Permian age *Psaronius* (Millay, 1997). Species of *Psaronius* are basically similar to living marattioid plants except that they typically have a much more elongated trunk (Figure 13.5d), and distinctive morphologies of the synangia (Figure 13.5f). The stem is upright and surrounded by a thick mantle of prop roots (Figure 13.5g) that increase in thickness toward the base of the plant (Figure 13.5d). As with living marattioid ferns, roots have a polyarch actinostele with long narrow ribs of xylem, and an aerenchymatous cortex (Ehret and Phillips, 1977). The oldest species known from anatomy displays a solenostele (DiMichele and Phillips, 1977), but in most species the stem stele increases in size and complexity distally, forming a polycyclic dictyostele (Figure 13.5g) from which frond traces diverge in helical, opposite-decussate, and/or whorled arrangements. In several species, the frond arrangement changes in a characteristic pattern from level to level (Morgan, 1959; Mickle, 1984). Prominent sclerenchyma bundles and sheaths occur in the ground tissues of all organs.

Fronds unroll from croziers (Figure 13.5d), showing several orders of pinnate dissection, and are up to several meters long (Stidd, 1971). Pinnules are broadly attached in an alternate fashion and have a strong midvein (Figure 13.5c) from which laterals diverge at a wide angle. Fertile pinnules bear a row of radial synangia (Figure 13.5f) on each side of the midvein. Sporangia dehisce via a longitudinal slit located on the inward facing sporangial wall, and typically display a pointed apex (Figure 13.5f). As with living species, spores are both trilete and monolete with a prominent perispore (Millay, 1997).

Species of *Psaronius* became the canopy dominants in many habitats of the equatorial tropics during the Late Pennsylvanian and basal Permian (Phillips et al., 1985). These ferns also provide some of the first conclusive fossil evidence for the evolution of complex plant–plant interrelationships during the late Paleozoic. Because of the unique architecture and development of tree fern trunks, vines and epiphytes become buried in the expanding root mantle of *Psaronius*, where they are preserved in the position of growth (Rothwell, 1991, 1996; Rößler, 2000). Examples of interactions with species of *Psaronius* have been documented for both filicalean ferns (Figures 13.5d, e) and for the equisetophyte *Sphenophyllum* (Mickle, 1984). A detailed explanation of such interactions is presented by Rößler (2000).

13.7 Patterns of diversification among leptosporangiate ferns

There is strong phylogenetic evidence that leptosporangiates form a monophyletic group (Rothwell, 1999; Pryer et al., 2001). Among fern clades with living representatives, homosporous leptosporangiates (i.e., filicaleans) have the longest and most informative fossil record. The leptosporangiate clade is also the most species rich of all non-angiospermous vascular plants (Rothwell, 1996).

Data for extinct leptosporangiate ferns are available from several types of fossil preservation. Many fossils consist of compressed frond material. When fertile, such fronds typically are informative of family and generic relationships (e.g., see Figure 13.7a, s). However, the majority of compressed fossil fern fronds are vegetative, and those are much more difficult to relate to a given family with certainty (e.g., Figure 13.7f). Other fern fossils are permineralized, often providing anatomical evidence of affinities even when fertile structures are not available (e.g., see Figures 13.6q, 13.7h–i, k–m, o). While internal anatomy can be diagnostic for families, genera, and even species of filicalean ferns (e.g., Figure 13.6q), anatomical features have been surveyed for only a tiny fraction of living species. Therefore, the taxonomic and systematic utility of anatomical features is limited by a paucity of comparative information about living species. The best currently available general reference for inferring fern relationships from internal anatomy is Ogura (1972).

13.7.1 Three evolutionary radiations of homosporous leptosporangiate ferns

Paleobotanical studies reveal a pattern of leptosporangiate evolution in which there have been three major pulses of evolutionary radiation (Rothwell, 1987). Diversity among the Paleozoic fossils is summarized by Stewart and

Rothwell (1993), whereas the Mesozoic and Cenozoic records of ferns recently have been compiled by Skog (2001) and Collinson (2001) respectively.

Pulses of filicalean evolution were first recognized by Lovis (1977), who concluded correctly that the species richness of filicaleans increased dramatically, rather than diminishing as the result of flowering plant diversification in the Cretaceous (Rothwell, 1996). Lovis (1977) hypothesized that Cretaceous filicalean diversification resulted from opportunistic evolution within the rapidly increasing complexity of angiosperm dominated plant communities. Two earlier pulses of filicalean evolution were subsequently recognized and were also correlated with opportunistic evolution (Rothwell, 1987). The first occurred during the Mississippian and Pennsylvanian in conjunction with increasing complexity of seed plant dominated plant communities, and the second extended from the Permian through the Jurassic. Whereas families of the second radiation are represented in the modern flora as the most basal representatives of living leptosporangiates (A. Smith *et al.*, 2006), all of the first radiation became extinct before the end of the Paleozoic (Figure 13.2).

The first radiation

The oldest evidence for leptosporangiates consists of small annulate filicalean sporangia with trilete spores that occur in Mississippian (i.e., "Lower Carboniferous" in European terminology) deposits of France (Galtier, 1981). Throughout the Mississippian, Pennsylvanian, and Lower Permian, filicaleans diversified into several families that all became extinct before the Late Permian (Figure 13.2; Rothwell, 1999). This Paleozoic diversification of homosporous filicaleans represents the first of the three leptosporangiate radiations (Figure 13.2; Rothwell, 1987). Paleozoic filicaleans displayed well-developed stem/leaf organography, but the occurrence of epiphyllous buds on many species led to confusion about this issue for many years. Whereas most earlier workers interpreted such leaf-borne stems as evidence for incomplete evolution of stem/leaf organography, we now realize that such structures represent specialized mechanisms for opportunistic growth (Figure 13.6f), as they do in living filicaleans (Tomescu *et al.*, 2006).

Extinct families of the Paleozoic radiation that have been characterized from whole plants include the Botryopteridaceae (e.g., Figures 13.6a–c, f–k; Rothwell, 1991; Rothwell and Good, 2000), Kaplanopteridaceae (Figures 13.6l–p; Tomescu *et al.*, 2006), Psalixochlaenaceae (Holmes, 1977, 1981), Sermayaceae (Eggert and Delevoryas, 1967), and Tedeleaceae (Figures 13.5e, 13.6d, e; Eggert and Taylor, 1966). Species of *Botryopteris* produce short rhizomes from which pinnately dissected fronds diverge to produce a crown (Figures 13.6c, f). Stipe traces of *Botryopteris* conform to a distinctive omega-shape when viewed in cross-section

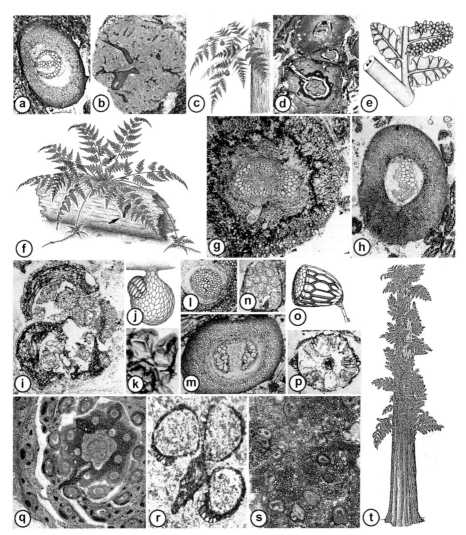

Figure 13.6 Ancient representatives of basal filicalean fern families Botryopteridaceae (a–c, f–k), Tedeleaceae (d, e), Kaplanopteridaceae (l–p), Osmundaceae (q, r), and Tempskyaceae (s, t). (a)–(c) Middle and Upper Pennsylvanian *Botryopteris forensis* from Illinois and Ohio. (a) Stipe, ×3. (b) Massive aggregation of sporangia, ×1. (c) Reconstruction of the plant as an epiphyte (from Rothwell, 1991). (d) Middle Pennsylvanian *Ankyropteris brongniartii* stem with diverging frond base and epiphyllous ("axillary") branch, from Illinois, ×2. (e) *Tedelea glabra*, reconstruction of fertile frond segment with abaxial sporangia (from Eggert and Taylor, 1966). (f)–(k) Middle Pennsylvanian *Botryopteris tridentata*. (f) Reconstruction of a plant producing plantlets (from Rothwell and Good, 2000). (g) Cross-section of solenostelic rhizome, ×9. (h) Cross-section of stipe, ×3. (i) Cross-section of enrolled fertile pinnule with attached sporangia, ×10. (j) Reconstruction of sporangium (from Rothwell and Good, 2000). (k) Trilete spores, ×350. (l)–(p) Upper Pennsylvanian

(Figures 13.6a, h). *Botryopteris forensis* is a trunk epiphyte of the tree fern *Psaronius* (Figures 13.5d and 13.6c; Rothwell, 1991) that flourished from the Middle Pennsylvanian to the basal Permian. The *B. forensis* plant is protostelic and produces epiphyllous branches to establish new shoots (Figure 13.6c). This species bears annulate sporangia on terminal segments of highly branched pinnae that form large tightly compacted fertile units (Figures 13.6b, c).

The Middle Pennsylvanian *Botryopteris tridentata* plant shows the oldest evidence of a solenostele in the fossil record (Figure 13.6g; Rothwell and Good, 2000). Individuals produce annulate sporangia similar to those of modern Osmundaceae (Figure 13.6j) on the abaxial surface on enrolled pinnules (Figure 13.6i) that are aggregated on fertile pinnae (Figure 13.6f, at arrow). Spores are radial and trilete (Figure 13.6k). Some fronds of *B. tridentata* are specialized for vegetative propagation forming the foliar equivalent of a stolon. Such fronds grew unbranched until they touched the ground, and then developed an epiphyllous shoot that produced a new ramet (Figure 13.6f).

The family Tedeleaceae consists of protostelic ferns with elongated rhizomes that produce a branch in the axil of each frond (Figure 13.6d; Mickle, 1980). Isolated vegetative organs of Pennsylvanian age Tedeleaceae are assigned to the morphogenus *Ankyropteris*. Until fertile structures (Figure 13.6e) were discovered and the new filicalean family Tedeleaceae was described (Eggert and Taylor, 1966), *Ankyropteris* was classified with the zygopterid ferns. *Tedelea glabra* is one of the species of fossil fern plants (Mickle, 1980) that grew as a vine on trunks of *Psaronius* (Figure 13.5e).

The Kaplanopteridaceae (Tomescu et al., 2006) is the most recently described family of the first filicalean radiation. This family is typified by *Kaplanopteris clavata* (Figures 13.6l–p). Frond segments and epiphyllous shoots of *K. clavata* were known as the morphospecies *Anachoropteris clavata* for many years before distinctive fertile structures were discovered (Rothwell, 1987), and the plant was reconstructed (Tomescu et al., 2006). The *Kaplanopteris clavata* plant has short protostelic stems (Figure 13.6l) that produce fronds with four orders of alternate,

Figure 13.6 (cont.) *Kaplanopteris clavata* from Ohio. (l) Cross-section of protostelic rhizome, ×8. (m) Cross-section of stipe, ×10. (n) Longitudinal section of indusiate sorus, ×55. (o) Reconstruction of a sporangium (from Tomescu et al., 2006). (p) Cross-section of indusiate sorus with enclosed sporangia, ×90. (q) Upper Cretaceous *Osmunda cinnamomea* rhizome in cross-section, from Alberta, ×3. (r) Lower Cretaceous *Osmunda vancouverensis* sporangia attached to a narrow frond segment, from Vancouver Island, British Columbia, ×35. (s) Cretaceous false stem of *Tempskya* sp. with several solenostelic rhizomes and stipe bases embedded in the root mantle, from Idaho, ×1. (t) Reconstruction of *Tempskya* plant (from Andrews and Kern, 1947).

pinnate branching. Sporangia occur in indusiate sori (Figures 13.6n, p), and are characterized by a long narrow stalk and capsule with a horizontal multiseriate annulus (Figure 13.6o). Among first-radiation filicaleans, *K. clavata* has two distinctive characters that previously were thought to be restricted to subsequent radiations. These are gradate maturation of sporangia within the sorus, and the presence of a soral indusium. Vegetatively, *Kaplanopteris* displays the most complex syndrome of vegetative reiteration known among both extinct and living ferns. Two types of reiterative units characterize the fronds. Latent croziers on otherwise mature fronds impart the capacity for indeterminate growth like that of the living schizaeaceous fern, *Lygodium japonicum* (Trivett and Rothwell, 1988). Also borne on the fronds are epiphyllous shoots that grow as ramets when the frond touches the substrate (Tomescu *et al.*, 2006). Repeated discovery of *Kaplanopteris* frond material among vegetative and fertile frond parts of *Psaronius* suggest that *K. clavata* was a facultative liana.

The second radiation

The second radiation of filicaleans produced all the basal families of living leptosporangiates by the Jurassic (Figure 13.2). Using the recent taxonomy of A. Smith *et al.*, (2006) families of the second radiation include Osmundaceae, Gleicheniaceae, Hymenophyllaceae, Dipteridaceae, Matoniaceae, Schizaeaceae s.l. (the Schizaeales of A. Smith *et al.*, 2006), and Cyatheaceae s.l (the Cyatheales of A. Smith *et al.*, 2006). This second radiation also resulted in evolution of the heterosporous leptosporangiate clade Hydropteridales (Rothwell and Stockey, 1994). Modern members of the second filicalean radiation show a great deal of structural diversity, but little species richness as compared to ferns of the third filicalean radiation.

The Osmundaceae has the richest fossil record of all filicalean families (Tidwell and Ash, 1994). In addition to compression fossils of fronds that extend from the Permian to the Recent, there is a particularly rich record of osmundaceous anatomical evolution in the form of permineralized rhizomes with attached stipe bases and roots (Tidwell and Ash, 1994; Cantrill, 1997; Stockey and Smith, 2000). Whereas many of the compressed fronds and trunks from Permian and Mesozoic deposits are recognized as morphotaxa, others are assignable to modern genera, revealing both a much higher level of evolutionary stasis and greater species longevity for ferns than is commonly recognized (Rothwell, 1996). For example, the genus *Osmunda* s.l. is represented in the Triassic of Antarctica (Phipps *et al.*, 1998), and the living species *O. cinnamomea* has a fossil record extending back through the Tertiary to 75 million year old sediments of the Cretaceous (Figure 13.6q; Serbet and Rothwell, 1999). Still other fossils reveal that the osmundaceous sporangial and spore characters have changed little during the Tertiary and Quaternary (Figure 13.6r; Vavrek et al., 2006).

Filmy ferns are relatively uncommon as fossils, but there is good evidence for the family as far back as the Late Triassic. Compressed remains from North America described as *Hopetedia praetermissa* reveal that modern appearing species of the Hymenophyllaceae evolved by the early Mesozoic (Axsmith et al., 2001). Gleichenioid ferns (i.e., Gleicheniales *sensu* A. Smith et al., 2006) may have evolved as early as the Pennsylvanian and are well documented from the Permian (Yao and Taylor, 1988) onward. Fossils include both compressed fronds such as fertile *Matonidium* sp. from the Cretaceous of Utah, (Figure 13.7j) and anatomically preserved protostelic rhizomes (Figure 13.7i) and branching rachides. The recently described *Gleichenia appianensis* (Figure 13.7i) extends the fossil record of Gleicheniaceae to the Tertiary of North America.

Fossils of the Dipteridaceae range from Triassic to Cretaceous (Tidwell and Ash, 1994; Cantrill, 1995). The recent discovery of anatomically preserved specimens of *Hausmannia* in Lower Cretaceous deposits (Figures 13.7a–e) reveals that this fossil genus is clearly distinct from the two living genera *Dipteris* and *Cheiropleuria* (Stockey et al., 2006b). Specimens show characteristic anastomosing venation of several branching orders (Figures 13.7a, c), with scattered sporangia located in abaxial depressions between the veins (Figures 13.7b, c). Sporangia have a short stalk and vertical annulus, and produce radial, trilete spores (Figures 13.7d, e).

The family Schizaeaceae *s.l.* is represented by both extinct genera and extinct species of genera with living representatives (Collinson, 2001; Skog, 2001). Included are both compressed foliage (e.g., Figure 13.7f) and anatomically preserved remains of vegetative and fertile structures (Figure 13.7g) for several of the living genera. At one time the family was thought to extend from the Mississippian onward, but those Paleozoic fossils (i.e., *Senftenbergia*) are now recognized as representing the Tedeleaceae of the first filicalean radiation (Jennings and Eggert, 1977). Beginning in the Jurassic, extinct genera such as *Klukia* and *Ruffordia* enter the fossil record, with diversity increasing through the Cretaceous and Paleogene (Yoshida et al., 1997; Trivett et al., 2006). The genus *Lygodium* diversified throughout the Tertiary (Manchester and Zavada, 1987; Collinson, 2001; Skog, 2001), and a species of *Anemia* has recently been described from the Lower Cretaceous (Hernandez-Castillo et al., 2006). *Schizaea* may not have evolved until the Oligocene (Collinson, 2001).

Tree fern fossils are relatively abundant in Jurassic and more recent deposits. These include vegetative and fertile frond segments (Collinson, 2001; Skog, 2001) as well as several species of anatomically preserved stems with attached stipe bases (Figure 13.7m), dispersed frond segments, and sori (Figure 13.7n) that are

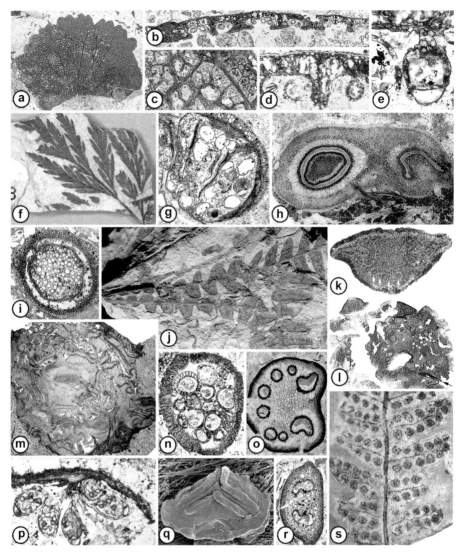

Figure 13.7 Cretaceous and Tertiary representatives of modern filicalean families. (a)–(e) *Hausmannia morinii* (Dipteridaceae), Lower Cretaceous of Vancouver Island, Canada. (a) Tip of blade with several orders of closed venation and paired marginal teeth, ×0.8. (b) Cross-section of blade with sporangia in abaxial cavities, ×15. (c) Paradermal section of blade showing several orders of anastomosing veins, ×10. (d) Cross-section of blade showing histology and sporangia, ×60. (e) Longitudinal section of attached sporangium with enclosed spores, ×150. (f) *Anemia fremontii* foliage, Upper Cretaceous of Utah, ×0.8. (g) Fertile pinnule of Eocene *Paralygodium vancouverensis* (Schizaeaceae) with enclosed sporangia from Vancouver Island, British Columbia, ×20. (h) Cross-section of rhizome with diverging stipe trace of Eocene *Dennstaedtiopsis aerenchymata* (Dennstaedtiaceae), British Columbia, ×6. (i) Cross-section of Eocene protostelic rhizome of *Gleichenia appianensis* (Gleicheniaceae),

assignable to both living and extinct genera (Stockey and Rothwell, 2004). The genus *Cyathea s.s.* can be traced to the Lower Cretaceous on the basis of anatomically preserved frond segments with attached indusiate sori of annulate sporangia that produce trilete spores (Figure 13.7n; Smith *et al.*, 2003).

Polypodiaceous ferns of the second radiation (i.e., Polypodiales of A. Smith *et al.*, 2006) are represented by the Dennstaedtiaceae *s.l.* in Upper Cretaceous and Tertiary deposits (Serbet and Rothwell, 2003). Many of these fossils show characteristic anatomical features of the solenostelic rhizomes and stipe bases (Figure 13.7h). Even older solenostelic fern rhizomes occur in Triassic deposits of Antarctica, but the latter have not been definitely assigned to known fern lineages (Millay and Taylor, 1990).

One additional group that probably represents the second radiation of filicaleans is the extinct family Tempskyaceae (Figures 13.6s, t). Cretaceous species of *Tempskya* consist of false stems made up of highly branched systems of solenostelic rhizomes and diverging stipe traces that are embedded in a dense matrix of roots (Figure 13.6s). This produces a plant with an elongated trunk from which fronds extend in several directions (Figure 13.8t; Andrews and Kern, 1947). Little is known for certain about the foliage and sporangial features of *Tempskya* plants, but associational evidence suggests that fossils assigned to *Anemia fremontii* (Figure 13.7f; also suspected of affinities to the Schizaeaceae *s.l.*) could belong to *Tempskya*.

The third radiation

The third radiation of filicaleans has produced the greatest species diversity among homosporous leptosporangiates. The oldest evidence for the third

Figure 13.7 (*cont.*) Vancouver Island, British Columbia, ×25. (j) *Matonidium* sp. with sori, Cretaceous, Utah, ×1.5. (k) (l) Upper Cretaceous *Midlandia nishidae* (Blechnaceae) from Alberta. (k) Cross-section of stipe, ×18. (l) Cross-section of dictyostelic rhizome with diverging stipe bases, ×5. (m) Cross-section of stem of *Rickwoodopteris hirsuta* (dicksoniaceous grade of Cyatheaceae *s.l.*), Cretaceous of Vancouver Island, British Columbia, ×1. (n) Lower Cretaceous indusiate sorus showing sections of sporangia with spores of *Cyathea cranhamii* (Cyatheaceae *s.s.*), Vancouver Island, British Columbia, ×35. (o) Eocene stipe of *Trawetsia princetonensis* (Blechnaceae) in cross-section, ×5. (p), (q) Lower Cretaceous *Pterisorus radiata* (Pteridaceae?), Vancouver Island, British Columbia. (p) Sorus of sporangia attached to surface of pinnule, ×50. (q) Proximal face of spore, ×1000. (r) Cross-section of Middle Eocene stipe of *Makotopteris princetonensis* (Athyriaceae), British Columbia, ×20. (s) Paleocene fertile pinna of *Speirseopteris orbiculata* (Thelypteridaceae), Vancouver Island, British Columbia, ×4.

radiation is found in Lower Cretaceous deposits of North America (Rothwell and Stockey, 2006) and China (Deng, 2002). The fossil record is incomplete with respect to the entire spectrum of families recognized from living species (A. Smith et al., 2006), but many of the most common families are represented by the Paleogene. These include species of the Pteridaceae as well as the two terminal clades of the Smith et al. tree (Figure 2 of A. Smith et al., 2006), which they refer to as "Eupolypods I" (i.e., Dryopteridaceae, Lomariopsidaceae, Tectariaceae, Oleandraceae, Davalliaceae, and Polypodiaceae) and "Eupolypods II" (i.e., Aspleniaceae, Woodsiaceae, Thelypteridaceae, Blechnaceae, and Onocleaceae).

The oldest probable evidence for the third radiation of leptosporangiate ferns occurs in 135 million year old deposits of western Canada (Rothwell and Stockey, 2006). *Pterisorus radiata* (Figures 13.7p, q) consists of radial sori of stalked sporangia with a near-vertical annulus (Figure 13.7p). Spores are radial and trilete, with a prominent equatorial ridge, deltoid distal structure, and ridges that flank the trilete suture (Figure 13.7q). In all of these features *Pterisorus* conforms closely to living genera of the Pteridaceae, particularly the genus *Pterozonium*. Only the sporangial annulus of *Pterisorus* (which is not interrupted by the stalk) and the quadriseriate stalk of *Pterisorus* are not found in modern Pteridaceae. The quadriseriate sporangial stalk of *Pterisorus* falls between the wider stalks (i.e., 5–6 cells) of most tree ferns and the narrower stalks (up to 3 cells) of living Pteridaceae. An annulus that is not interrupted by the stalk compares favorably with sporangia of the tree ferns. While *Pterisorus* appears to represent the third radiation of filicaleans, the genus is probably intermediate between Pteridaceae and a potential tree fern sister group.

A large number of compressed fern remains from the Lower Cretaceous of China represent the third filicalean radiation (Deng, 2002). Sporangia of these ferns are characterized by a narrow stalk (where known) and a vertical annulus that is not interrupted by the stalk. The genus *Adiantopteris* (Adiantaceae) has trilete spores, while the rest of the early Cretaceous Chinese ferns produce monolete spores and are assigned to the Pteridaceae, and Dryopteridaceae (*sensu* Kramer and Green, 1990). This unexpected diversity of highly derived Filicales representing the Pteridaceae as well as both the Eupolypod I and Eupolypod II clades of A. Smith et al. (2006) suggests that the third radiation of homosporous leptosporangiates was initiated in northern China by the beginning of the Cretaceous.

Characteristic Eupolypod I (of A. Smith et al., 2006) fossils include the fully reconstructed, anatomically preserved sporophyte of *Makotopteris princetonensis* (Dryopteridaceae) from the Middle Eocene of western Canada (Figure 13.7r; Stockey et al., 1999). The *Makotopteris* plant is characterized by pinnately compound fronds produced by a dictyostelic horizontal rhizome. Fronds display two hippocampiform bundles in the stipe, and sori of annulate sporangia with a narrow stalk and vertical annulus that arise from a swollen receptacle. Spores are monolete with a smooth exine and loose spiny perispore.

The Eupolypod II clade includes compressed fertile fronds that represent the oldest evidence for Thelypteridaceae in Paleocene deposits (Stockey et al., 2006a). This fern, *Speirseopteris orbiculata*, is characterized by bipinnate fronds with fertile primary pinnae (Figure 13.7s). Sori are round, consisting of sporangia with a vertical annulus and monolete, bean-shaped spores.

Several morphotaxa of anatomically preserved Blechnaceae consist of dictyostelic rhizomes that produce helically arranged stipes with a C-shaped arch of traces, the most adaxial of which comprise of a pair of hippocampiform bundles. These include the Upper Cretaceous *Midlandia nishidae* (Figures 13.7k, l), which is the oldest evidence for Blechnaceae (Serbet and Rothwell, 2006). *Trawetsia princetonensis* reveals that blechnoid ferns with pinnate pinnatifid fronds and frond vascularization that undergoes the same gradational series of frond vasculature as modern species were present by the Middle Eocene (S. Smith et al., 2006). One intriguing athyrioid species of the Blechnaceae shows a sympodial architecture of the cauline vasculature that is reminiscent of seed plants (Karafit et al., 2006). The Eocene morphotaxon *Dickwhitea allenbyensis* confirms earlier reports that filicaleans may have evolved sympodial cauline vasculature in parallel with lignophytes (White and Weidlich, 1995), thus emphasizing that the structural diversity of Filicales is greater than commonly recognized.

Two particularly interesting fern fossils are of importance because they represent ancient specimens of species that have living representatives. Virtually entire sporophytes of the eastern chain fern *Woodwardia virginica* are preserved in 16 million year old deposits of Washington State in northwestern USA (Figures 13.8i–k, Pigg and Rothwell, 2001). Specimens show the characteristic rhizome and frond morphology, rhizome and stipe anatomy (Figure 13.8j), deciduous primary pinnae, pinnule morphology and venation (Figure 13.8i), and soral and sporangial characters (Figure 13.8k) of the living plants.

Even more ancient fossils of the sensitive fern *Onoclea sensibilis*; Figure 13.8e–h) are preserved in 57 million year old deposits of western Canada (Rothwell and Stockey, 1991). Although somewhat younger than the Upper Cretaceous fossils of *Onoclea* sp. that are the oldest evidence for the family Onocleaceae, these virtually entire sporophytes represent plants that were buried *in situ* showing features of the horizontal rhizome, vegetative fronds (Figure 13.8e), fertile fronds with enrolled fertile pinnules (Figure 13.8f), sporangia with a vertical annulus (Figure 13.8g) and monolete spores (Figure 13.8h) with spiny perispore.

The 16 million year old (*W. virginica*), 57 million year old (*O. sensibilis*), and 75 million year old (*O. cinnamomea*) fossil specimens of living species belong to

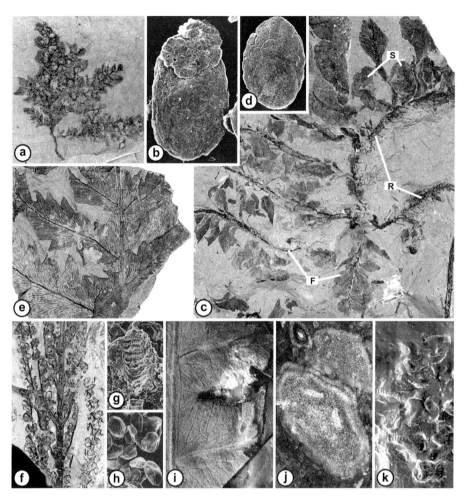

Figure 13.8 Cretaceous and Tertiary representatives of Hydropteridales and fossilized representatives of living homosporous leptosporangiate fern species. (a), (b) Paleocene *Azolla stanleyi* (Salviniaceae, Hydropteridales), Alberta. (a) Floating fertile shoot, ×3. (b) Megaspore apparatus with attached microsporangiate massulae, ×110. (c), (d) Upper Cretaceous *Hydropteris pinnata*, Alberta. (c) Branching rhizome with attached roots (at R) diverging fronds (at F), and sporocarps (at S), ×1.3. (d) Megaspore apparatus with numerous distal floats, × 70. (e)–(h) Paleocene (i.e., 57 million years old) *Onoclea sensibilis* (Dryopteridaceae), Alberta. (e) Vegetative frond segment, ×1. (f) Fertile frond segment, ×0.6. (g) Sporangium with vertical annulus, ×100. (h) Monolete spores without perispore, ×210. (i)–(k) Miocene (i.e., 15 million years old) *Woodwardia virginica* (Blechnaceae), Idaho. (i) Lobed pinnule with characteristic "chair fern" venation pattern, ×6. (j) Cross-section of rhizome and stipe base, ×8. (k) Sorus of annulate sporangia with long narrow stalks, ×44.

the Blechnaceae, Onocleaceae, and Osmundaceae respectively. Obviously, species longevity among a wide array of leptosporangiate ferns is far greater than for flowering plants, suggesting a dramatically different mode and tempo of evolution for some homosporous plants than for seed plants (Rothwell, 1996).

13.7.2 Hydropteridales, the heterosporous leptosporangiate clade

The evolution of heterospory has occurred numerous times throughout the geological history of plants (Rothwell, 1996). Nevertheless, the systematic diversity and species richness of free sporing heterosporous plants is extremely low, with most heterosporous species restricted to wetland and aquatic habitats. Among living leptosporangiates there are two families, the amphibious Marsileaceae and the floating aquatic Salviniaceae (*sensu* A. Smith *et al.*, 2006), that traditionally have been regarded as representing parallel evolution of heterospory (e.g., Gifford and Foster, 1989). However, more recent phylogenetic analyses using morphological characters of living and extinct species (Rothwell and Stockey, 1994) and nucleotide sequence characters of the chloroplast gene *rbcL* (Hasebe *et al.*, 1995) have independently resolved these families as monophyletic. This agreement of results from the analysis of different subsets of the data has dramatically strengthened the hypothesis of a single origin of heterospory among leptosporangiates, emphasizing that concordance of results from different methodologies produces the strongest hypotheses of systematics and phylogeny.

Paleobotanical evidence for monophylesis of heterosporous leptosporangiates consists of the Upper Cretaceous *Hydropteris pinnata* (Figures 13.8c, d; Rothwell and Stockey, 1994). *Hydropteris* combines the amphibious general plant morphology of *Marsilea, Regnellidium*, and *Pilularia* (Marsileaceae) with the complex megaspore complexes (Figure 13.8d) and microsporangiate massulae of *Salvinia* and *Azolla* (Salviniaceae). This species also displays pinnate fronds like those of homosporous filicaleans (Figure 13.8c) and typifies the order Hydropteridales for nomenclatural purposes. Megaspore complexes of *Hydropteris* compare favorably with those of Salviniaceae, and if found dispersed would be assigned to the morphotaxon *Parazolla*.

Several other whole plant concepts have been developed for extinct species of hydropteridalean fern plants, but all except *H. pinnata* either conform to, or are extremely similar to, genera with living species (but see *Regnellites*, below). A representative example of the latter is the Paleocene species *Azolla stanleyi* (Hoffman and Stockey, 1994), which has gross morphology of the sporophyte (Figure 13.8a), microsporangiate massulae (Figure 13.8b, at top), and megasporangiate complexes (Figure 13.8b, at bottom), all of which fall within the ranges of variation that characterize living species of the genus.

The oldest evidence for Hydropteridales consists of isolated megaspore complexes (Lupia *et al.*, 2000) in late Jurassic deposits, and nearly whole marsileaceous plants from Upper Jurassic/Lower Cretaceous deposits of Japan. The structures of those marsileaceous vegetative sporophytes and sporocarps are intermediate between living species of *Marsilea* and *Regnellidium* (Yamada and Kato, 2002). The occurrence of both the megaspore complexes and of *Regnellites nagashimae* in pre-Cretaceous deposits is stratigraphically consistent with the evolution of hydropterid ferns as part of the second leptosporangiate radiation (Figure 13.2). It is also concordant with the interpretation that species of Marsileaceae display ancestral characters of growth form, sporophyte morphology, megaspore complex structure, and reproductive biology comparable to Salviniaceae (Rothwell and Stockey, 1994).

Several types of morphotaxa reveal that the diversity of Hydropteridales was greater in the Cretaceous than the Recent. Compression specimens of vegetative sporophytes that are similar to living genera of Marsileaceae, but that have distinctive pinnule morphologies and venation patterns, recently have been assigned to the morphogenus *Marsileaceaephyllum* (Nagalingum, 2007). Although whole plant concepts are not yet developed for the vegetative sporophytes represented by these fossils, their distinctive pinnule characters and associated megaspores that differ from those of *Marsilea* imply that *Marsileaceaephyllum* represents an extinct lineage of Hydropteridales.

Megaspore complexes and microsporangiate massulae are the most common evidence for the diversity of hydropteridalean ferns through time (e.g., Lupia *et al.*, 2000). Whereas a few morphospecies based on megaspore complexes and microsporangiate massulae have been found attached to fossil sporophytes of known structure, most have been associated with either Marsileaceae or Salviniaceae by similarities to living species. In the past, some authors have attempted to infer the pattern of phylogeny for hydropteridaleans from the patterns of structural variation in megaspore complexes. However, among even the small number of examples of fossils for which both sporophyte morphology and megaspore complex structure are known, we already have discovered that the two patterns are not concordant (Rothwell and Stockey, 1994). Therefore, while fossil megaspore complexes are a reliable indicator for hydropteridalean fern diversity through time, in isolation they cannot be used to infer patterns of phylogeny.

13.8 Historical context, popular practices, and the upward outlook

This chapter has been organized as an overview of the phylogeny and evolution of ferns inferred from a combination of the fossil record and living species. Many of the phylogenetic relationships and evolutionary conclusions presented here are concordant with those found elsewhere in the current literature (including other chapters of this book). Others differ significantly, reflecting the healthy dissonance that accompanies scientific interchange during the testing of competing hypotheses. These differences in interpretation derive both from a differential application of alternative subsets of the data, and from fundamentally different approaches to reaching an understanding of plant phylogeny.

In the past few years, the practice of formulating phylogenetic hypotheses from living taxa only has become increasingly popular. When extinct taxa do appear on trees that result from that methodology, they are routinely added in a *post hoc* fashion (e.g., Fig. 17–13 of Raven *et al.*, 2005) such that inaccuracies resulting from the original inadequate taxon sampling (i.e., living taxa only) are perpetuated. As applied to ferns, this approach has led to the popular perception that phylogenetic relationships among ferns and other euphyllophytes are well resolved (e.g., Moran, 2004; A. Smith *et al.*, 2006). But are they? If anything, hypotheses of phylogeny that either include or exclude data from the fossil record (and the evolutionary predictions that flow from them) appear to be increasingly discordant (cf., Friedman *et al.*, 2004, and Crane *et al.*, 2004 with Schneider *et al.*, 2002).

While it may seem natural for those who work primarily or exclusively with extant organisms to develop phylogenetic concepts from the analysis of living species only, omitting paleontological information from systematic data sets all too often yields dramatically different patterns than analyses that do include paleontological and other types of data (Friedman *et al.*, 2004; Bateman *et al.*, 2006; Rothwell and Nixon, 2006). In the broader context, the study of only living plants represents a "downward looking" approach for reconstructing phylogeny and evolutionary patterns (Stewart, 1964; Bateman *et al.*, 2006). Can one realistically expect to discern accurately the ancestry and/or overall patterns of diversity from the tiny fraction of species that happen to be alive at any particular time?

An alternative approach, routinely practiced by both paleontologists and those who utilize data from a combination of living and fossil plants, employs knowledge of "ancestral forms" to help understand the living. Indeed, virtually all of the major advances in resolving relationships among major groups of vascular plants (i.e., deep internal nodes of the polysporangiophyte tree) to date have originated from studies that utilize morphological characters of fossil plants in the formulation of phylogenetic hypotheses (e.g., Rothwell and Nixon, 2006). Whereas this methodology frequently provides lower levels of resolution and lower support values for shallower branches of phylogenetic trees than do some other approaches, it has been considered the most sound approach currently available for resolving the deep internal nodes of the tree (Mishler, 2000; Bateman et al., 2006; Rothwell and Nixon, 2006). Bower (1935) advocated this "upward outlook" over 70 years ago, and it subsequently has guided several generations of evolutionary plant biologists (Stewart, 1964).

Characterizations of the ferns and fern-like plants presented in this chapter summarize the fossil record and the paleontological data upon which a still incomplete understanding of euphyllophyte phylogeny and evolution currently is based (i.e., Figure 13.1). While incompletely resolved phylogenies are inherently less satisfying than well resolved trees, the pattern presented in Figure 13.1 of this chapter utilizes the fossil record to understand more clearly what we know and what we have yet to discover. In this regard, we hope that this presentation will help inform a broader audience about the nature, promise, and realistic limitations of the fossil record of ferns, provide greater accessibility to the paleopteridological literature, and encourage a wider range of systematists to adopt an upward outlook by including paleontological data in phylogenetic analyses.

References

Andrews, H. N. and Kern, E. M. (1947). The Idaho tempskyas and associated fossil plants. *Annals of the Missouri Botanical Garden*, **34**, 119–186.

Axsmith, B. J., Krings, M., and Taylor, T. N. (2001). A filmy fern from the Upper Triassic of North Carolina (USA). *American Journal of Botany*, **88**, 1558–1567.

Bateman, R. M., Hilton, J., and Rudall, P. J. (2006). Morphological and molecular phylogenetic context of the angiosperms: contrasting the "top-down" and "bottom-up" approaches used to infer the likely characteristics of the first flowers. *Journal of Experimental Botany*, **57**, 3471–3503.

Berry, C. M. and Fairon-Demaret, M. (2002). The architecture of *Pseudosporochnus nodosus* Leclerq and Banks: a Middle Devonian cladoxylopsid from Belgium. *International Journal of Plant Sciences*, **163**, 699–713.

Berry, C. M. and Wang, Y. (2006). *Eocladoxylon* (*Protopteridium*) *minimum* (Halle) Koidzume from the Middle Devonian of Yunnan, China: an early *Rhacophyton*-like plant. *International Journal of Plant Sciences*, **167**, 551–556.

Bower, F. O. (1923). *The Ferns*, Vol. 1, *Analytical Examination of the Criteria of Comparison*. Cambridge: Cambridge University Press.

Bower, F. O. (1935). *Primitive Land Plants*. London: Macmillan.

Boyce, C. K. and Knoll, A. H. (2002). Evolution of developmental potential and the multiple independent origins of leaves in Paleozoic vascular plants. *Paleobiology*, **28**: 70–100.

Cantrill, D. J. (1995). The occurrence of the fern *Hausmannia* Dunker (Dipteridaceae) in the Cretaceous of Alexander Island, Antarctica. *Alcheringa*, **19**: 243–254.

Cantrill, D. J. (1997). The pteridophyte *Ashicaulis livingstonensis* (Osmundaceae) from the Upper Cretaceous of Williams Point, Livingston Island, Antarctica. *New Zealand Journal of Geology and Geophysics*, **40**, 315–323.

Collinson, M. E. (2001). Cainozoic ferns and their distribution. *Brittonia*, **53**, 173–235.

Crane, P. R., Herendeen, P., and Friis, E. M. (2004). Fossils and plant phylogeny. *American Journal of Botany*, **91**, 1683–1699.

Delevoryas, T., Taylor, T. N., and Taylor, E. L. (1992). A marattialean fern from the Triassic of Antarctica. *Review of Palaeobotany and Palynology*, **74**, 101–107.

Deng, S. (2002). Ecology of the early Cretaceous ferns of northeast China. *Review of Palaeobotany and Palynology*, **119**, 93–112.

Dennis, R. L. (1974). Studies of Paleozoic ferns: *Zygopteris* from the Middle and Late Pennsylvanian of the United States. *Palaeontographica*, **148B**, 95–136.

DiMichele, W. A. and Phillips, T. L. (1977). Monocyclic *Psaronius* from the Lower Pennsylvanian of the Illinois Basin. *Canadian Journal of Botany*, **55**, 2514–2524.

Eggert, D. A. and Delevoryas, T. (1967). Studies of Paleozoic ferns: *Sermaya*, gen. nov. and its bearing on filicalean evolution in the Paleozoic. *Palaeontographica*, **120B**, 169–180.

Eggert, D. A. and Taylor, T. N. (1966). Studies of Paleozoic ferns: on the genus: *Tedelea* gen. nov. *Palaeontographica*, **118B**, 52–73.

Ehret, E. L. and Phillips, T. L. (1977). *Psaronius* root systems – morphology and development. *Palaeontographica*, **161B**, 147–164.

Erwin, D. M. and Rothwell, G. W. (1989). *Gillespiea randolphensis* gen. et sp. nov. (Stauropteridales), from the Upper Devonian of West Virginia. *Canadian Journal of Botany*, **67**, 3063–3077.

Esau, K. (1943). Origin and development of primary vascular tissues in seed plants. *The Botanical Review*, **9**, 125–206.

Friedman, W. E., Moore, R. C., and Purugganan, M. D. (2004). The evolution of plant development. *American Journal of Botany*, **91**, 1726–1741.

Galtier, J. (1968). Un nouveau type de fructification filicinéenne du Carbonifère Inférieur. *Comptes Rendues de l'Académie des Sciences de Paris*, **266D**, 1004–1007.

Galtier, J. (1981). Structures foliaires de fougères et Pteridospermales du Carbonifère Inférieur et leur signification évolutive. *Palaeontographica*, **180B**, 1–38.

Gifford, E. M. and Foster, A. S. (1989). *Morphology and Evolution of Vascular Plants*, 3rd edn. New York: W. H. Freeman.

Greuter, W., McNeill, J., Barrie, F. R., Burdet, H. M., Demoulin, V., Filgueiras, T. S., Nicolson, D. H., Silva, P. C., Skog, J. E., Trehane, P., Turland, N. J., and Hawksworth, D. L. (ed.) (2000). International code of botanical nomenclature: Saint Louis code: adopted by the Sixteenth International Botanical Congress, St Louis, Missouri, July–August 1999. Königstein: Koeltz Scientific Books.

Hasebe, M. T., Wolf, P. G., Pryer, K. M., Ueda, K, Ito, M., Sano, R., Gastony, G. J., Yokoyama, J., Manhart, J. R., Murakami, N., Crane, E. H., Haufler, C. H., and Hauk, W. D. (1995). A global analysis of fern phylogeny based on *rbcL* nucleotide sequences. *American Fern Journal*, **85**, 134–181.

Hernandez-Castillo, G. R., Stockey, R. A., and Rothwell, G. W. (2006). *Anemia quatsinoensis* sp. nov. (Schizaeaceae), a permineralized fern from the Lower Cretaceous of Vancouver Island. *International Journal of Plant Sciences*, **167**, 665–674.

Hilton, J. (1999). A Late Devonian plant assemblage from the Avon Gorge, west England: taxonomic, phylogenetic and stratigraphic implications. *Botanical Journal of the Linnean Society*, **129**, 1–54.

Hoffman, G. L. and Stockey, R. A. (1994). Sporophytes, megaspores and massulae of *Azolla stanleyi* from the Paleocene Joffre Bridge locality, Alberta. *Canadian Journal of Botany*, **72**, 301–308.

Holmes, J. (1977). The Carboniferous fern *Psalixochlaena cylindrica* as found in Westphalian A coal balls from England. Part I. Structure and development of the cauline system. *Palaeontographica*, **164B**, 33–75.

Holmes, J. (1981). The Carboniferous fern *Psalixochlaena cylindrica* as found in Westphalian A coal balls from England. Part II. Structure and development of the cauline system. *Palaeontographica*, **176B**, 147–173.

Hu, S., Dilcher, D. L., Schneider, H., and D. M. Jarsen. (2006). Eusporangiate ferns from the Dakota Formation, Minnesota, U.S.A. *International Journal of Plant Sciences*, **167**, 579–589.

Jennings, J. R. and Eggert, D. A. (1977). Preliminary report on permineralized *Senftenbergia* from the Chester Series of Illinois. *Review of Palaeobotany and Palynology*, **24**, 221–225.

Kaplan, D. R. and Groff, P. A. (1995). Developmental themes in vascular plants: functional and evolutionary significance. In *Experimental and Molecular Approaches to Plant Biosystematics*, ed. P. C. Hoch and A. G. Stephenson. St. Louis, MO: Missouri Botanical Garden, pp. 111–146.

Karafit, S. J., Rothwell, G. W., Stockey, R. A., and Nishida, H. (2006). Evidence for sympodial vascular architecture in a filicalean fern rhizome: *Dickwhitea allenbyensis* gen. et sp. nov. (Athyriaceae). *International Journal of Plant Sciences*, **167**, 721–727.

Kenrick, P. and Crane, P. R. (1997). *The Origin and Early Diversification of Land Plants*. Washington, DC: Smithsonian Institution Press.

Kramer, K. U. and Green, P. S. (1990). *The Families and Genera of Vascular Plants*, Vol. I, Pteridophytes and Gymnosperms. Berlin: Springer-Verlag.

Leclercq, S. and Banks, H. P. (1962). *Pseudosporochnus nodosus* sp. nov., a Middle Devonian plant with cladoxylalean affinities. *Palaeontographica*, **110B**, 1–34.

Lovis, D. J. (1977). Evolutionary patterns and processes in ferns. In *Advances in Botanical Research*, ed. R. D. Preston and H. W. Woolhouse. London: Academic Press, pp. 229–440.

Lupia, R., Schneider, H. Moeser, G. M., Pryer, K. M., and Crane, P. R. (2000). Marsileaceae sporocarps and spores from the Late Cretaceous of Georgia, U.S.A. *International Journal of Plant Sciences*, **161**, 976–988.

Manchester, S. R. and Zavada, M. S. (1987). *Lygodium* foliage with intact sporophores from the Eocene of Wyoming. *Botanical Gazette*, **148**, 392–399.

Mickle, J. E. (1980). *Ankyropteris* from the Pennsylvanian of eastern Kentucky. *Botanical Gazette*, **141**, 230–243.

Mickle, J. E. (1984). Taxonomy of specimens of the Pennsylvanian age marattialean fern *Psaronius* from Ohio and Illinois. *Illinois State Museum Scientific Paper*, **19**, 1–64.

Millay, M. A. (1997). A review of permineralized Euramerican Carboniferous tree ferns. *Review of Palaeobotany and Palynology*, **95**, 191–209.

Millay, M. A. and Rothwell, G. W. (1983). Fertile pinnae of *Biscalitheca* (Zygopteridales) from the Upper Pennsylvanian of the Appalachian Basin. *Botanical Gazette*, **144**, 589–599.

Millay, M. A. and Taylor, T. N. (1990). New fern stems from the Triassic of Antarctica. *Review of Palaeobotany and Palynology*, **62**, 41–64.

Mishler, B. D. (2000). Deep phylogenetic relationships among "plants" and their implications for classification. *Taxon*, **49**, 133–155.

Moran, R. C. (2004). *A Natural History of Ferns*. Portland, OR: Timber Press.

Morgan, J. (1959). The morphology and anatomy of American species of the genus *Psaronius*. *Illinois Biological Monographs*, **27**, 1–108.

Nagalingum, N. S. (2007). *Marsileaceaephyllum*, a new genus for marsileaceous macrofossils: leaf remains from the Early Cretaceous (Albian) of southern Gondwana. *Plant Systematics and Evolution*, **264**, 41–55.

Ogura, Y. (1972). *Comparative Anatomy of Vegetative Organs of the Pteridophytes*, 2nd edn. Berlin: Borntraeger.

Phillips, T. L. and Galtier, J. (2005). Evolutionary and ecological perspectives of Late Paleozoic ferns. Part I. Zygopteridales. *Review of Palaeobotany and Palynology*, **135**, 165–203.

Phillips, T. L., Peppers, R. A., and DiMichele, W. A. (1985). Stratigraphic and interregional changes in Pennsylvanian coal-swamp vegetation: environmental inferences. *International Journal of Coal Geology*, **5**, 43–109.

Phipps, C. J., Taylor, T. N., Taylor, E. L. Cuneo, N. R., Boucher, L. D., and Xao, X. (1998). *Osmunda* (Osmundaceae) from the Triassic of Antarctica: an example of evolutionary stasis. *American Journal of Botany*, **85**, 888–895.

Pigg, K. B. and Rothwell, G. W. (2001). Anatomically preserved *Woodwardia virginica* (Blechnaceae) and a new filicalean fern from the Middle Miocene Yakima Canyon flora of central Washington, USA. *American Journal of Botany*, **88**, 777–787.

Pryer, K. M., Schneider, H., Smith, A. R., Cranfill, R., Wolf, P. G., Hunt, J. S., and Sipes, S. D. (2001). Horsetails and ferns are a monophyletic group and the closest living relatives to seed plants. *Nature*, **409**, 618–622.

Raven, P. H., Evert, R. F., and Eichhorn, S. E. (2005). *Biology of Plants*, 7th edn. New York: W. H. Freeman.

Rößler, R. (2000). The late Palaeozoic tree fern *Psaronius* – an ecosystem unto itself. *Review of Palaeobotany and Palynology*, **108**, 55–74.

Rothwell, G. W. (1987). Complex Paleozoic Filicales in the evolutionary radiation of ferns. *American Journal of Botany*, **74**, 458–461.

Rothwell, G. W. (1991). *Botryopteris forensis* (Botryopteridaceae), a trunk epiphyte of the tree fern *Psaronius*. *American Journal of Botany*, **78**, 782–788.

Rothwell, G. W. (1995). The fossil history of branching: implications for the phylogeny of land plants. In *Experimental and Molecular Approaches to Plant Biosystematics*, ed. P. C. Hoch and A. G. Stephenson. St. Louis, MO: Missouri Botanical Garden, pp. 71–86.

Rothwell, G. W. (1996). Pteridophytic evolution: an often under appreciated phytological success story. *Review of Palaeobotany and Palynology*, **90**, 209–222.

Rothwell, G. W. (1999). Fossils and in the resolution of land plant phylogeny. *The Botanical Review*, **65**: 188–218.

Rothwell, G. W. and Good, C. W. (2000). Reconstructing the Pennsylvanian-age filicalean fern *Botryopteris tridentata* (Felix) Scott. *International Journal of Plant Sciences*, **161**, 495–507.

Rothwell, G. W. and Nixon, K. C. (2006). How does the inclusion of fossil data change our conclusions about the phylogenetic history of euphyllophytes? *International Journal of Plant Sciences*, **167**, 737–749.

Rothwell, G. W. and Stockey, R. A. (1989). Fossil Ophioglossales in the Paleocene of western North America. *American Journal of Botany*, **76**, 637–644.

Rothwell, G. W. and Stockey, R. A. (1991). *Onoclea sensibilis* in the Paleocene of North America, a dramatic example of structural and ecological stasis. *Review of Palaeobotany and Palynology*, **70**, 113–124.

Rothwell, G. W. and Stockey, R. A. (1994). The role of *Hydropteris pinnata* gen. et sp. nov. in reconstructing the cladistics of heterosporous ferns. *American Journal of Botany*, **81**, 387–394.

Rothwell, G. W. and Stockey, R. A. (2006). Combining the characters of Pteridaceae and tree ferns: *Pterisorus radiata* gen. et sp. nov., a permineralized Lower Cretaceous filicalean with radial sori. *International Journal of Plant Sciences*, **167**, 695–701.

Scheckler, S. E. (1974). Systematic characters of Devonian ferns. *Annals of the Missouri Botanical Garden*, **61**, 462–473.

Scheckler, S. E. (1986). Geology, floristics and paleoecology of Late Devonian coal swamps from Appalachian Laurentia (U.S.A.). *Annales de la Société géologique de Belgique*, **209**, 209–222.

Schneider, H. K. Pryer, K. M., Cranfill, R, Smith, A. R., and Wolf, P. G. (2002). Evolution of vascular plant body plans: a phylogenetic perspective. In *Developmental Genetics and Plant Evolution*, ed. Q. C. B. Cronk, R. M. Bateman, and J. A. Hawkins. New York: Taylor and Francis, pp. 330–364.

Serbet, R. and Rothwell, G. W. (1999). *Osmunda cinnamomea* (Osmundaceae) in the Upper Cretaceous of western North America: additional evidence for exceptional species longevity among filicalean ferns. *International Journal of Plant Sciences*, **160**, 425–433.

Serbet, R. and Rothwell, G. W. (2003). Anatomically preserved ferns from the Late Cretaceous of western North America: Dennstaedtiaceae. *International Journal of Plant Sciences*, **164**, 1041–1051.

Serbet, R. and Rothwell, G. W. (2006). Anatomically preserved ferns from the Late Cretaceous of western North America. II. Blechnaceae/Dryopteridaceae. *International Journal of Plant Sciences*, **167**, 703–709.

Skog, J. E. (2001). The biogeography of Mesozoic leptosporangiate ferns related to extant ferns. *Brittonia*, **53**, 236–269.

Skog, J. E. and Banks, H. P. (1973). *Ibyka amphikoma* gen. et sp. n., a new protoarticulate precursor from the late Middle Devonian of New York State. *American Journal of Botany*, **60**, 366–380.

Smith, A. R., Pryer, K. M., Schuettpelz, E., Korall, P., Schneider, H., and Wolf, P. G. (2006). A classification for extant ferns. *Taxon*, **55**, 705–731.

Smith, S. Y., Rothwell, G. W., and Stockey, R. A. (2003). *Cyathea cranhamii* sp. nov., anatomically preserved tree fern sori from the Lower Cretaceous of Vancouver Island, British Columbia. *American Journal of Botany*, **90**, 755–760.

Smith, S. Y., Stockey, R. A., and Rothwell, G. W. (2006). *Trawetsia princetonensis* gen. et sp. nov. (Blechnaceae): a permineralized fern from the Middle Eocene Princeton Chert. *International Journal of Plant Sciences*, **167**, 711–719.

Soria, A., Meyer-Berthaud, B., and Scheckler, S. E. (2001). Reconstructing the architecture and growth habit of *Pietzschia levis* sp. nov. (Cladoxylopsida) from the Late Devonian of southeastern Morocco. *International Journal of Plant Sciences*, **162**, 911–926.

Stein, W. E. (1981). Reinvestigation of *Arachnoxylon kopfii* from the Middle Devonian of New York State, USA. *Palaeontographica*, **177B**, 90–117.

Stein, W. E. (1982). *Iridopteris eriensis* from the Middle Devonian of North America, with systematics of apparently related taxa. *Botanical Gazette*, **143**, 401–416.

Stein, W. E. and Hueber, F. M. (1989). The anatomy of *Pseudosporochnus*: *P. hueberi* from the Devonian of New York. *Review of Palaeobotany and Palynology*, **60**, 311–359.

Stein, W. E., Mannolini, F., Hernick, V. L., Landing, E., and Berry, C. M. (2007). Giant cladoxylopsid trees resolve the enigma of the Earth's earliest forest stumps at Gilboa. *Nature*, **446**, 904–907.

Stewart, W. N. (1964). An upward outlook in plant morphology. *Phytomorphology*, **14**, 120–134.

Stewart, W. N. and Rothwell, G. W. (1993). *Paleobotany and the Evolution of Plants*, 2nd edn. Cambridge: Cambridge University Press.

Stidd, B. M. (1971). Morphology and anatomy of the frond of *Psaronius*. *Palaeontographica*, **134B**, 87–123.

Stockey, R. A. and Rothwell, G. W. (2004). Cretaceous tree ferns of western North America: *Rickwoodopteris hirsuta* gen. et sp. nov. (Cyatheaceae s.l.). *Review of Palaeobotany and Palynology*, **132**, 103–114.

Stockey, R. A. and Rothwell, G. W. (2006). Introduction: evolution of modern ferns. *International Journal of Plant Sciences*, **167**, 613–614.

Stockey, R. A. and Smith, S. Y. (2000). A new species of *Millerocaulis* (Osmundaceae) from the Lower Cretaceous of California. *International Journal of Plant Sciences*, **161**, 159–166.

Stockey, R. A., Nishida, H., and Rothwell, G. W. (1999). Permineralized ferns from the Middle Eocene Princeton Chert. I. *Makotopteris princetonensis* gen et sp. nov. (Athyriaceae). *International Journal of Plant Sciences*, **160**, 1047–1055.

Stockey, R. A., Lantz, T. C., and Rothwell, G. W. (2006a). *Speirseopteris orbiculata* gen. et sp. nov. (Thelypteridaceae), a derived fossil filicalean from the Paleocene of western North America. *International Journal of Plant Sciences*, **167**, 729–736.

Stockey, R. A., Rothwell, G. W., and Little, S. A. (2006b). Relationships among fossil and living Dipteridaceae: anatomically preserved *Hausmannia* from the Lower Cretaceous of Vancouver Island. *International Journal of Plant Sciences*, **167**, 649–663.

Surange, K. R. (1952). The morphology of *Stauropteris burntislandica* P. Bertrand and its megasporangium *Bensonites fusiformis* R. Scott. *Philosophical Transactions of the Royal Society of London*, **237B**, 73–91.

Taylor, T. N. and Taylor, E. L. (1993). *The Biology and Evolution of Fossil Plants*. Englewood Cliffs, NJ: Prentice-Hall.

Taylor, T. N., Kerp, H., and Hass, H. (2005). Life history biology of early land plants: deciphering the gametophyte phase. *Proceedings of the National Academy of Sciences of the United States of America*, **102**, 5892–5897.

Tidwell, W. D. and Ash, S. R. (1994). A review of selected Triassic to early Cretaceous ferns. *Journal of Plant Research*, **107**, 417–442.

Tomescu, A. M. F., Rothwell, G. W., and Trivett, M. L. (2006). Kaplanopteridaceae fam. nov., additional diversity in the inital radiation of filicalean ferns. *International Journal of Plant Sciences*, **167**, 615–630.

Trivett, M. L. and Rothwell, G. W. (1988). Modeling the growth architecture of fossil plants: a Paleozoic filicalean fern. *Evolutionary Trends in Plants*, **2**, 25–29.

Trivett, M. L., Stockey, R. A., Rothwell, G. W., and Beard, G. (2006). *Paralygodium vancouverensis* sp. nov. (Schizaeaceae), additional evidence for filicalean diversity in the Paleogene of North America. *International Journal of Plant Sciences*, **167**, 675–681.

Vavrek, M. J., Stockey, R. A., and Rothwell, G. W. (2006). *Osmunda vancouverensis* sp. nov. (Osmundaceae), permineralized fertile frond segments from the Lower Cretaceous of British Columbia, Canada. *International Journal of Plant Sciences*, **167**, 631–637.

White, R. A. and Weidlich, W. H. (1995). Organization of the vascular system in stems of *Diplazium* and *Blechnum* (Filicales). *American Journal of Botany*, **82**, 982–991.

Yao, Z. and Taylor, T. N. (1988). On a new gleicheniaceous fern from the Permian of South China. *Review of Palaeobotany and Palynology*, **54**, 121–134.

Yamada, T. and Kato, M. (2002). *Regnellites nagashimae*, gen. et sp. nov., the oldest macrofossil of Marsileaceae, from the Upper Jurassic and Lower Cretaceous of western Japan. *International Journal of Plant Sciences*, **163**, 715–723.

Yoshida, A., Nishida, H., and Nishida, M. (1997). Permineralized schizaeaceous fertile pinnules from the Upper Cretaceous of Hokkaido, Japan II. *Paralygodium yezoense* gen. et sp. nov. *Research Institute for Evolutionary Biology Scientific Reports*, **9**, 1–10.

14

Diversity, biogeography, and floristics

ROBBIN C. MORAN

14.1 Introduction

The biogeography of ferns and lycophytes can be studied from several points of view and with various methods. It might, for instance, examine the distribution of species on a single tree (Krömer and Kessler, 2006; Schuettpelz and Trapnell, 2006), or the frequency and abundance of species over large regions (Ruokolainen *et al.*, 1997; Lwanga *et al.*, 1998; Tuomisto *et al.*, 2003; Jones *et al.*, 2005; Tuomisto and Ruokolainen, 2005), or the relationships of species on different continents (e.g., Moran and Smith, 2001; Parris, 2001). Methods can be as varied as producing lists of plants growing on different soil types (e.g., Young and León, 1989; van der Werff, 1992), calculating the percentage of floristic similarity between different regions (Dzwonko and Kornás, 1978, 1994; Pichi Sermolli, 1979), or analyzing the phylogeny of a clade in relation to its geography and geological history (e.g., Geiger and Ranker, 2005; Hoot *et al.*, 2006). These and other approaches have contributed to what is now an overwhelming amount of literature on the subject. To limit the subject for this chapter, three themes have been chosen: diversity, long-distance dispersal, and vicariance. After discussing these, a summary of the current state of floristics is given because biogeography is ultimately based on that subject.

14.2 Historical review

The earliest works on fern and lycophyte biogeography were mostly tabular summaries of the percentages and/or occurrences of species in different

Biology and Evolution of Ferns and Lycophytes, ed. Tom A. Ranker and Christopher H. Haufler. Published by Cambridge University Press. © Cambridge University Press 2008.

regions of the world (D'Urville, 1835; Baker, 1868; Lyell, 1879). These were enlarged upon by Christ (1910) and Winkler (1938), who described floras region by region. The work of Christ (1910) was influential for its comprehensiveness and because it compared the distributions of ferns to those of flowering plants. Christ dispelled the then prevailing notion that ferns (and other spore-bearing plants) had unlimited dispersal ability and thus tended to show less definite distribution patterns compared to angiosperms. He demonstrated that ferns and flowering plants could show similar distribution patterns such as the famous disjunction between eastern North America and eastern Asia. Christ's work, however, suffered from the poor taxonomic circumscriptions of his day. Many of the genera that he and other pteridologists used for analysis were paraphyletic or polyphyletic, thus hindering valid biogeographical conclusions. (For a scholarly review of Christ's life and contributions to pteridology, see Gómez, 1978.)

After Christ and Winkler, a noted contribution to fern biogeography came from Copeland (1939). Intrigued by the apparent centering of the Southern Hemisphere fern and lycophyte taxa on Antarctica, he attempted a "reconstruction of the history of ferns by the interpretation of present geographic distribution" and summarized the information available on widely discontinuous far-southern fern distributions. He concluded that most fern genera were of Antarctic origin. This might seem tenable if Gondwana were substituted for Antarctica, but Copeland's assessment of the distribution data was generally dogmatic, and his paper has been downplayed or ignored by most recent pteridologists (e.g., Parris, 2001).

Since Copeland's work, scores of journal articles (many of which are cited below) have treated fern and lycophyte biogeography. Within the past 15 years, three symposia proceedings dedicated to fern and lycophyte biogeography have been published (Barrington and Kato, 1993 and references therein; Moran, 2001, 2007). These works contain key papers and reflect current interest in the topic.

14.3 Diversity

How many species of ferns and lycophytes are there in the world? If the number of species estimated to occur in various continents and regions (Figure 14.1) is summed, the total is about 17 000. This total is an overestimate because many species occur in more than one region. Assuming 20% of the species in each region also occur in other regions, then the number of species worldwide would be reduced by that same percentage; that is, the total would be about 136 00. This agrees with a previous estimate of 12 000 to 15 000 (Roos, 1996). Because the number of lycophyte species worldwide is about 13 600 (Kramer and Green, 1990), the number of fern species would be about 12 240.

Diversity, biogeography, and floristics 369

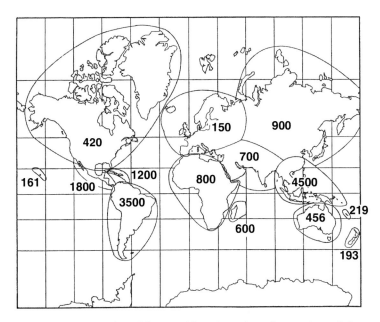

Figure 14.1 The number of ferns and lycophytes in various regions of the world.

Worldwide, fern and lycophyte species show a dominant pattern called the latitudinal diversity gradient. In both hemispheres, as one goes from the pole toward the equator, the number of species per unit area increases. For instance, in eastern Asia, 42 species of ferns and lycophytes are found on the Kamchatka Peninsula of Russia, 140 on Hokkaido Island and 430 on Honshu Island of Japan, 560 in Taiwan, 960 in the Philippines, and about 1200 in Borneo (Figure 14.2). In the Americas, the pattern is the same: 30 species of ferns and lycophytes in Greenland, 98 in New England, 113 in Florida, 652 in Guatemala, and 1250 in Ecuador. Thus, in both hemispheres going from high latitudes to the equator, the number of species increases over 30 times (Moran, 2004).

This latitudinal diversity gradient has exceptions, mostly in small genera. Some lack tropical species, such as the northern temperate *Phegopteris* (three species; Holttum, 1969) and *Gymnocarpium* (eight species; Pryer, 1993), and the southern temperate *Synammia* (four species; Schneider et al., 2006). Other genera have tropical species but show a reverse latitudinal gradient, being more diverse in the temperate zones, such as *Huperzia*, sensu stricto (i.e., the gemmiferous *Selago* group), with one species in tropical America (Mexico) and seven in Canada and the USA. *Equisetum* has three species in South America and 11 in Canada and the USA; and *Dryopteris* has three species in South America and 14 in Canada and the USA. Despite these exceptions, the latitudinal diversity gradient holds well among genera of ferns and lycophytes.

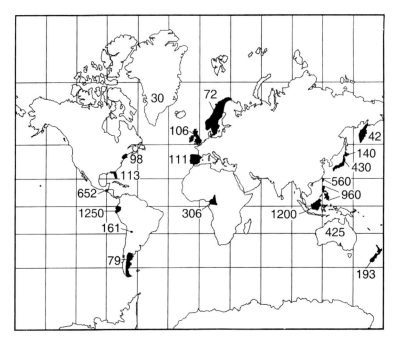

Figure 14.2 The latitudinal diversity gradient as shown by the numbers of ferns and lycophytes in various parts of the world.

The high species richness of the tropics gives rise to some striking contrasts with temperate regions. Costa Rica, a country slightly smaller than the state of West Virginia, harbors about 1165 species of ferns and lycophytes – nearly three times as many as in Canada and the continental USA (Wagner and Smith, 1993). On the Caribbean side of Costa Rica lies the La Selva Biological Station. Within its borders are 15 square kilometers (9.5 square miles) of rainforest harboring 150 species of ferns and lycophytes, or roughly the same number as in the entire northeastern USA (Grayum and Churchill, 1987). Similarly, a total sampling area of 79 hectares in western Amazonia (Colombia, Ecuador, and Peru) contains 323 species of ferns and lycophytes (Tuomisto and Ruokolainen, 2005); a plot of 1 hectare in Amazonian Ecuador contains 50 species (Poulsen and Nielsen, 1995); and a single tree in a wet premontane forest in Costa Rica contains 45 species (Schuettpelz and Trapnell, 2006). These numbers are remarkably high compared to larger regions in the temperate zones, such as nearly 40 species for Armenia (Gabrielian and Greuter, 1984), 57 for Luxembourg (Krippel, 2005), 100 for the New England states (Tryon and Moran, 1997), and 125 for the Mediterranean Region (Pichi Sermolli, 1991).

Within the tropics, fern and lycophyte diversity is not distributed uniformly. The mountains hold far more species than the lowlands (Tryon, 1986; Moran,

1995). For instance, the Andes have about 2500 species of ferns and lycophytes (my estimate), whereas the relatively flat Amazonian lowlands of central Brazil include only about 235 species (Tryon and Conant, 1975). In the Old World tropics, New Guinea harbors about 2000 species whereas Borneo has about 1000. This difference is thought to be generated primarily by New Guinea's greater number of high mountains and the more extensive habitat they provide (Parris, 1985).

Yet even within the mountains, species richness is not distributed uniformly. It increases from the lowlands (<800 m) to middle elevations (800–2000 m) and then diminishes toward the tree line. Represented graphically, this distribution pattern is hump shaped, with the hump occurring in the middle elevations. This pattern has been documented in the New and Old Worlds (Lellinger, 1985; Jacobsen and Jacobsen, 1989; Parris et al., 1992; Parris, 1997; Kessler, 1999, 2000, 2001a, 2001b, 2002; Moran et al., 2003; Watkins et al., 2006). It holds for terrestrial and epiphytic species (Parris et al., 1992; Kessler, 2001a; Kessler et al., 2001; Kluge and Kessler, 2006; Watkins et al., 2006). The middle-elevation bulge is not surprising in ferns and lycophytes because it also occurs in most groups of plants and animals (Rahbek, 2005).

Endemism in ferns and lycophytes tracks the hump-shaped pattern of species richness, being highest at middle elevations in the mountains (Tryon, 1972; Tryon and Gastony, 1975; Kluge and Kessler, 2006). The endemism of ferns and lycophytes in central Amazonian Brazil is only 1.4% (four species; Tryon and Conant, 1975), whereas in the mountains of Costa Rica and Panama, a much smaller area by comparison, it is 15% (205 species; statistic compiled from Moran and Riba, 1995). Endemism is also higher in the Andes than in Amazonia. A sample of 190 Andean species treated in monographs had nearly 40% endemism (Tryon, 1972).

Why do mountains harbor more species and endemics than lowlands? The explanation usually given is that mountains provide greater environmental heterogeneity, such as different slopes, exposures, parent rocks, soils, microclimates, and elevations (e.g., Moran, 1995). All these factors would allow more species to inhabit mountains. But why should mountains be most diverse at middle elevations? One explanation is the mid-domain effect (Colwell and Hurtt, 1994; Colwell and Lees, 2000; Grytnes, 2003; Colwell et al., 2004). This claims that high species richness at middle elevations is a mathematical artifact. Species will be most numerous in the middle of *any* gradient bounded by definite limits, or "hard boundaries," which represent absolute barriers to the dispersal or survival of species. For elevation gradients, the hard boundaries would be ocean water at sea level and the air above a mountaintop – places where ferns and lycophytes physically cannot occur. Within such a bounded gradient (domain), the ranges of most species will tend to overlap, by chance alone, toward the middle. The result is a mid-domain peak of species richness. This theory does not deny the importance of ecological and historical factors; it claims that, in aggregate, the distribution of many species along a gradient conforms to a stochastic model. The theory is controversial (Hawkins and Diniz-Filho, 2002; Ferrer-Castan and Vetaas, 2005), but evidence supporting it has been found in the distribution of ferns along a bounded elevation gradient from 30 to 2960 m in Costa Rica (Watkins et al., 2006).

One aspect of mountains often not emphasized is that they can act as barriers to migration. Fern and lycophyte biogeographers often stress, and rightly so (see below), that ferns and lycophytes have dust-like spores readily capable of dispersing long distances. As a result, the species occur almost everywhere they are potentially capable of growing (Tryon, 1986; Barrington, 1993). The mountains, however, can act as formidable barriers, and this can be seen in the Andes of Colombia and Ecuador. Here the eastern and western slopes harbor many of the same species (Balslev, 1988), suggesting that similar forest habitat exists on both sides. Yet nearly 25% of the fern and lycophyte species on the western side of the Andes (including the adjacent Pacific lowlands of the Chocó region) are restricted to that side; that is, they do not occur on the eastern side (Moran, 1996). For these species, the mountains have acted as a barrier to migration (Moran, 1995). Many species of the western Andean slopes and Pacific lowlands (the Chocó region) extend northward into Panama and Costa Rica (Lellinger, 1975, 1985). The resulting floristic similarity is why the ferns and lycophytes of the Chocó region have been treated floristically with those of Costa Rica and Panama (Lellinger, 1989).

One assertion about tropical fern diversity has been called the American Paradox (Kramer, 1990). This claims that although the number of species in the Neotropics and Paleotropics is comparable, the number of genera, subgenera, and sections is much higher in the eastern Paleotropics. This suggests there has been greater morphological divergence in the Paleotropics. The claim, however, can have no firm basis without some objective, quantitative means of comparing fern morphology. It seems more likely that the difference reflects an artifact of taxonomy, with more genera, subgenera, and sections having been named in the Paleotropics, instead of any actual greater amount of morphological diversity.

14.4 Long-distance dispersal

Ferns have lightweight, dust-like spores generally 30–70 μm long (Tryon and Lugardon, 1991). Given these characteristics, they might be expected to be picked up by winds and dispersed readily over long distances. What evidence, if any, supports this expectation?

First, atmospheric sampling has found fern spores at high altitudes, including in the jet stream (Erdtman, 1937; Polunin, 1951; Aylor and Ferrandino, 1985; Puentha, 1991; Caulton et al., 2000). In fact, at 2400 m over the Hawaiian Islands, an entire fern sporangium was found (Gressitt et al., 1961). Once airborne, fern spores can tolerate the low temperatures and high UV light in the upper atmosphere, as has been shown experimentally (Gradstein and Van Zanten, 2001). Thus, a spore picked up and transported by winds could conceivably survive an upper-atmospheric journey.

Strong evidence for the great vagility of ferns and lycophytes is seen in the floras of oceanic islands (Tryon, 1970; Smith, 1972; Carlquist, 1980; Tryon, 1986; Moran, 2004; Geiger et al., 2007). Being both volcanic and of relatively recent origin from the sea floor (generally within the past 10 MY), most oceanic islands have never been connected to continents. The species occurring on them must have arrived by long-distance dispersal or, for endemic species, had progenitors that once did. A striking feature of oceanic island floras is their high percentage of ferns and lycophytes, generally 16–60% (Table 14.1). In contrast, wet tropical forests on continents harbor only 5–10% (Table 14.2). Why the difference? Because of their greater buoyancy in the air, fern spores are more likely to be transported long distances than are the heavier propagules (i.e., seeds and fruits) of seed plants. Thus, ferns and lycophytes are more likely to migrate to islands, and over time more species accumulate.

A more accurate way of comparing the higher vagility of ferns and lycophytes is to examine the number of original colonizing species (i.e., exclude the number of endemics that evolved on the islands). This has been done for the Hawaiian Islands where ferns and lycophytes compose 17% of the total vascular plant flora (Table 14.1). Between 270 and 280 original colonizing species of angiosperms migrated to the archipelago, in contrast to 140 ferns and lycophytes (Wagner et al., 1990; Wagner, 1995). In other words, nearly 50% of the original colonists were ferns and lycophytes, which is much higher than their percentage of the total flora in the archipelago and those of the mainland source areas (Table 14.2). This is strong evidence of their greater vagility. Yet even this 50% is low. It would be higher if the number of times a given species migrated to the archipelago could be taken into account. It has been shown, for instance, that *Asplenium adiantum-nigrum* migrated to the Hawaiian Islands at least three times and perhaps as many as 17 (Ranker et al., 1994).

Another comparison from the Hawaiian Islands shows the greater vagility of ferns. Angiosperms have 89% endemism within the archipelago, whereas ferns have only 71%. For single-island endemics, the percentages differ dramatically:

Table 14.1 *Comparison of ferns and lycophytes in the total vascular plant flora of various oceanic islands*

Island or archipelago	Number of indigenous fern and lycophyte species (percent endemic)		Number of indigenous seed plant species	Ferns and lycophytes as % of total vascular plant flora	Source
Antipodes	20	(0)	49	29	Godley, 1989
Cocos	80	(22)	127	38	Rojas-Alvarado and Trusty, 2004
Easter	18	(22)	30	60	Christensen and Skottsberg, 1920; Looser, 1958; Zizka, 1991
Fiji	310	(?)	1318	19	Smith, 1996
Galapagos	113	(7)	491	19	Adsersen, 1990
Hawaiian Islands	161	(71)	941	17	Palmer, 2003
Juan Fernández Islands	58	(45)	156	27	Marticorena et al., 1998; Boudrie, 2003
Marion	7	(28)	15	46	Alston and Schelpe, 1957; M. C. Rutherford, personal communication
Marquesas	108	(16)	212	34	Wagner, 1991; Florence and Lorence, 1997
Pitcairn	21	(10)	64	33	Brownlie, 1961; Florence et al., 1995
Pohnpei	124	(6)	264	35	D. Lorence, personal communication
Rodrigues	25	(8)	127	16	Lorence, 1976; Guého, 1988
Tristan da Cuhna	30	(37)	41	42	Wace and Dickson, 1965; Tryon, 1966; Tryon, 1970
Wallis and Futuna Islands	59	(?)	240	25	Badré and Hoff, 1995

80% for angiosperms and 6% for ferns (Ranker et al., 2000). Ferns most likely have fewer single-island endemics because they disperse more easily among the islands.

Although oceanic islands show a high percentage of ferns and lycophytes in their floras, the endemism of these plants is lower than for angiosperms (Table 14.2 of Tryon, 1970; Table 14.2 of Smith, 1972; Wagner, 1995). Why? One idea is that speciation is slower in ferns, resulting in fewer endemics (Smith,

Table 14.2 *Comparison of ferns and lycophytes in the total vascular plant flora of various continental regions*

Region	Number of indigenous fern and lycophyte species	Number of indigenous seed plant species	Ferns and lycophytes as % of total vascular plant flora	Source
Acre, Brazil	214	3762	5	D. Daly, personal communication
Barro Colorado Island, Panama	104	1246	8	Croat, 1978
Central French Guiana	194	1910	9	Mori et al., 1997, 2002
Florida, USA	152	2689	7	R. Wunderlin, personal communication
La Selva Biological Station, Costa Rica	173	1727	10	Grayum and Churchill, 1987
Mesoamerica (southern Mexico through Panama)	1358	16650	8	Moran and Riba, 1995; G. Davidse, personal communication
Missouri, USA	68	1981	3	G. Yatskievych, personal communication
Río Palenque Science Center, Ecuador	75	1037	7	Dodson and Gentry, 1978
Venezuelan Guayana (states of Amazonas and Bolívar)	669	8613	8	Smith, 1995

1972). Speciation might be slower in ferns because they lack many of the isolating mechanisms found in angiosperms, such as differences in flowering time, structural differences of the flowers, incompatibility between the pollen and stigmatic surface, and the potential to switch to different pollinators (see Chapter 12). Another idea is that repeated migration from the source areas maintains gene flow so that speciation of the island populations does not occur (Ranker et al., 1994, 2000).

Long-distance dispersal in ferns and lycophytes is also evident by the many wide range disjunctions exhibited at the species level that seem unlikely to be explained by extirpation of intervening populations (Wagner, 1972). A few examples from temperate species of *Asplenium* illustrate this: *Asplenium platyneuron* is disjunct between eastern North America and South Africa (Wagner et al., 1993); *A. subglandulosum* has three subspecies, one each in Australia and New Zealand, southern South America, and Spain and Morocco (Salvo et al., 1982); and A. dalhousiae is found in the western Himalayan region, and extends disjunctly to northeastern Africa (Yemen) and to western North America (Arizona and northwestern Mexico) (Knobloch and Correll, 1962; Wagner et al., 1993; Mickel and Smith, 2004). Examples of wide disjunctions also come from tropical American and African species. Several species occurring above 1000 m in the Andes also occur in Africa (e.g., *Huperzia saururus, Melpomene flabelliformis, Pleopeltis macrocarpa*). Given the geologically young age of the Andes (less than 15 MY; Kroonenberg et al., 1990), it seems likely that the disjunctions resulted from long-distance dispersal, not continental drift (Moran and Smith, 2001).

The fern and lycophyte flora of high latitudes in the Southern Hemisphere also provides good examples of long-distance dispersal. A total of 103 species occur on two or more of the southern continents, and the probable explanation is long-distance dispersal, not continental drift (Brownsey, 2001; Parris, 2001). Of these species, 22 are completely circum-Antarctic, 39 occur in Australia and New Zealand and Africa–Madagascar but not South America, 29 occur in Africa and South America but not elsewhere, and 13 occur in South America and Australia and New Zealand (Parris, 2001). On islands in the Southern Hemisphere, floristic similarity correlates more strongly with wind direction from source areas than with geographical proximity to continents, supporting the idea that wind is the main dispersal agent for these widespread ferns (Muñoz et al., 2004). This is illustrated by the middle Atlantic island of Tristan da Cunha, which is influenced by the prevailing westerly winds. All but one of its fern and lycophyte species occur in South America or, if endemic, have their closest relatives there (Tryon, 1966).

Further evidence for the ease of long-distance dispersal comes from the rapidity with which some introduced fern species have spread in various parts of the world. In the Neotropics before 1900, *Macrothelypteris torresiana, Thelypteris dentata*, and *T. opulenta* were nearly unknown. Presently they are common and widespread (Strother and Smith, 1971; Leonard, 1972; Taylor and Johnson, 1979; Smith, 1993). In 1966 *Lygodium microphyllum*, native to Asia, was found naturalized in southeastern Florida. By 1993 it had spread to cover 11 200 hectares, and by 1999 it was found over 43 300 hectares (Volin et al., 2004). In the Old World, the neotropical *Pityrogramma calomelanos* has escaped and spread rapidly in Africa, India, Southeast Asia, and Australia where it is considered a weed (Wardlaw, 1962; Verma, 1966; Schelpe, 1975; Bostock, 1998). In Southeast Asia, the Neotropical *Adiantum latifolium* was introduced in the 1970s and has increased rapidly (Piggott, 1988; Chin, 1997). All these species disperse by spores; they have no leaf buds or gametophytic gemmae for vegetative dispersal. Their rapid spread attests to the tremendous potential for increase by spores.

A famous example of the efficacy of long-distance fern dispersal by spores is the re-vegetation of Krakatau, located between Java and Sumatra. Its eruption on 26 August 1883 destroyed most, if not all, of the original vegetation, and ferns were among the earliest plants to re-colonize the island (Turrill, 1935). Three years after the eruption, 11 species of ferns were thriving, including widespread species such as *Pityrogramma calomelanos*, *Blechnum orientale*, *Pteris vittata*, *Nephrolepis exaltata*, and *Pteridium aquilinum*. After 25 years, 15 fern species and one lycophyte (*Lycopodiella cernua*) had established, and after 40 years, two species of *Trichomanes* abounded on the misty summit. After 53 years, a total of 60 species were present. All of these species must have established themselves from air-borne spores settling on the open lava slopes of the island (Doctors van Leeuwen, 1936).

Among epiphytic ferns, long-distance migration may be enhanced in those groups that have long-lived, colonial, gemmiferous gametophytes (Dassler and Farrar, 1997, 2001; see Chapter 9). The evidence for this is supported by comparison of epiphytes with and without such gametophytes in the floras of islands and mainlands. The percentage of epiphytic ferns with long-lived, colonial, gemmiferous gametophytes (grammitids, vittarioids, Hymenophyllaceae) is statistically higher on islands than on mainlands, whereas epiphytic ferns with short-lived, isolated, non-gemmiferous gametophytes (polypods and *Elaphoglossum*) are under-represented (Dassler and Farrar, 2001). Similarly, epiphytes with long-lived, colonial, gemmiferous gametophytes are over-represented among the same-species or sister-species disjunctions between the Neotropics and Africa–Madagascar (Moran and Smith, 2001). What accounts for this more successful colonization after long-distance dispersal? Species with gametophytic gemmae can bridge gaps between clones from isolated spores and thus facilitate sexual reproduction by increasing the opportunity for outbreeding through sperm and antheridiogen transfer. Also, the dispersed gemmae can become a source of new tissue for the formation of antheridia when they land near a mature gametophyte (Dassler and Farrar, 2001). (It should be noted, however, that these estimates have not been corrected for phylogenetic history; such analyses must await the completion of many more phylogenetic studies of island groups.)

Instances of long-distance dispersal are increasingly being inferred from molecular phylogenetic studies. For example, two African species of *Lomariopsis* (*L. guineensis* and *L. palustris*) nest within a neotropical clade of five species. This suggests long-distance dispersal from the Neotropics to Africa for the ancestor of the two African species (Rouhan et al., 2007). Dispersal in this direction has also been inferred for many clades of Polypodiaceae (Janssen et al., 2007). Dispersal in the opposite direction, from Africa to the Neotropics, was inferred for the origin of *Platycerium andinum*, the sole American species in the genus (Kreier and Schneider, 2006).

Estimates of divergence times based on molecular data have helped decide whether dispersal or vicariance took place. For instance, the New Zealand species of *Polystichum* form a clade sister to two species from Lord Howe Island, and these in turn are sister to *Polystichum* in Australia (Perrie et al., 2003). The divergence time between the New Zealand–Lord Howe clade and the Australian clade was estimated to be within the last 20 MY. This post-dates continental drift (New Zealand and Australia had drifted far apart by about 70 MY), suggesting that long-distance dispersal accounts for the disjunct distribution. A similar conclusion was reached for New Zealand Gleicheniaceae (Perrie et al., 2007) and Aspleniaceae (Perrie and Brownsey, 2005a, 2005b). The number of such examples is expected to increase as more phylogenetic studies become available.

14.5 Vicariance

In addition to long-distance dispersal, vicariance explains the present-day distribution patterns of some clades. A well-known example is the disjunction between eastern North America and eastern Asia. These two regions share more genera and species than any other two continental regions on Earth, which is remarkable given the distance between them (Boufford and Spongberg, 1983; Kato and Iwatsuki, 1983; Kato, 1993). Whereas angiosperms show these similarities at the generic level, ferns and lycophytes show them mostly at the species level (Li, 1952; Moran, 2004). For example, Hokkaido, the northernmost island of Japan, has roughly the same number of fern species as are found in northeastern North America: 122 and 116, respectively. Of these species, 47 (40%) are common to both regions (Kato and Iwatsuki, 1983; Iwatsuki, 1995). Examples include *Adiantum pedatum*, *Deparia acrostichoides*, *Onoclea sensibilis*, *Osmunda claytoniana*, and *Thelypteris palustris* var. *pubescens*. Besides sharing some of the same species, sister species occur, with one in eastern Asia and the other in eastern North America. A well-known example – but one that has yet to be tested by incorporation into a cladistic analysis – is *Asplenium ruprechtii* (Asia) and *A. rhizophyllum* (North America) (Tryon, 1969). Revisions of some groups are showing that cryptic sister species (e.g., *Polypodium sibiricum* and *P. appalachianum*; Haufler and Windham, 1991) may be hiding within what have been considered wide-ranging single species in the past.

What were the vicariant events that generated these eastern North American–eastern Asian disjunctions? The answer goes back to the early Tertiary when the Earth's climate was the warmest in the history of plant life (Parrish, 1987). Tropical and temperate forests thrived at high northern latitudes across

North America, Europe, and Asia, which were connected at various times by land bridges (Tiffney, 1985a, 1985b; Tiffney and Manchester, 2001). This circumboreal forest provided nearly continuous habitat for millions of years, resulting in a relatively homogeneous flora, as can be seen in the fossil record (e.g., Rothwell and Stockey, 1991). During the latter half of the Tertiary, the Earth's climate became cooler and more seasonal – a trend that culminated in the Ice Ages – and the ranges of warm-loving species receded southward, severing their connections across Eurasia and North America (Graham, 1993). Ranges were further fragmented in North America by the uplift of the Rocky Mountains during the middle Tertiary, creating a rain shadow that favored grasslands instead of forests in the interior of the continent. By the late Tertiary, large blocks of this former forest remained in eastern Asia and eastern North America, resulting in the floristic similarity of these regions.

Instances of vicariance have been repeatedly inferred from molecular phylogenetic studies. Splits at deep nodes into New World and Old World clades have been found in *Isoëtes* (Hoot et al., 2006), Cyatheaceae (Conant et al., 1995), Hymenophyllaceae (Pryer et al., 2001), *Lomariopsis* (Rouhan et al., 2007), *Polystichum* (Little and Barrington, 2003; Driscoll and Barrington, 2007), thelypteroid ferns (Smith and Cranfill, 2002), *Trichomanes* (Dubuisson et al., 2003), and woodwardioid ferns (Cranfill and Kato, 2003). Some studies have found a separation of deeper branches into tropical and temperate clades, such as in *Asplenium* (Schneider et al., 2004), and Polypodiaceae (Wolf et al., 2001).

One particularly well-documented example of vicariance is tropical *Huperzia*, which is partitioned into Neotropical and Paleotropical clades (Wikström and Kenrick, 1997, 2000). The two clades are estimated to have diverged before the final breakup of Africa and South America that occurred about 95 MY in the Cretaceous (Wikström and Kenrick, 2001). The present-day species richness of these clades resulted from subsequent diversification in the late Cretaceous and Tertiary. For both clades, epiphytism was the ancestral condition. Reversion to the terrestrial condition occurred at least twice in the Andes, most likely within the past 15 MY (Wikström et al., 1999). Overlying this pattern, but not obscuring it, are several trans-Atlantic dispersal events leading to the formation of sister taxa in South America and Africa (Wikström et al., 1999; Moran and Smith, 2001).

14.6 Floristics

Biogeography ultimately depends on knowledge of what the species are and where they occur – that is, the study of floristics. Much remains to be done in this field. Many ferns and lycophytes are still undescribed, as suggested by the number of new species proposed every year. Between 1991 and 1995, about 620 new species, subspecies, and varieties – or about 125 per year – were described worldwide, mostly from the species-rich tropics (statistic compiled from Johns, 1997). In Mesoamerica (southern Mexico through Panama), 138 new ferns and lycophytes were described from 1985 to 1995, or nearly 10% of the region's fern and lycophyte flora (personal observation). Even in the botanically well-known temperate regions, new species are still being found. Between 1985 and 1993, 29 new ferns and lycophytes were described from the USA and Canada (Wagner and Smith, 1993).

More collecting activity is needed to document species ranges, especially in the poorly explored tropics. Just how incompletely known the ranges of some species are is shown by recent reports of new range extensions and country records (e.g., Burrows and Crouch, 1995; Smith *et al.*, 1999; Figueiredo, 2001; Rojas-Alvarado, 2001, 2002; Ranker *et al.*, 2005; Cremers and Boudrie, 2006). The following three instances show how incomplete knowledge of species ranges can be. For over 170 years, *Schizaea pusilla* was known only from coastal northeastern North America, between Delaware and Newfoundland. In 1987 it was reported from central Peru, about 5200 km from the nearest known populations in Delaware – a remarkable disjunction between the temperate zone and the tropics (Stolze, 1987). *Alsophila salvinii*, a large and conspicuous tree fern, was long known from only southern Mexico to Nicaragua (Gastony, 1973), but has now been found in Peru (Smith *et al.*, 2005). Several species long considered endemic to the coastal mountains of southeastern Brazil (Tryon, 1972) have been collected recently farther inland, some as far north as the state of Minas Gerais (Alexandre Salino, personal communication). This discovery lowers the percentage endemism of the fern flora attributed to the coastal mountains. A collection program for ferns and lycophytes is urgently needed given the destruction of tropical habitats.

Over the past 15 years, floristics has been advanced by the publication of floras covering all ferns and lycophytes within certain continents or large regions. Examples include North America north of Mexico (Flora of North America Editorial Committee, 1993), Mexico (Mickel and Smith, 2004), Mesoamerica (Moran and Riba, 1995), the Venezuelan Guyana (Smith, 1995), central French Guiana (Mori *et al.*, 1997), Patagonia (de la Sota *et al.*, 1998), Chile (Rodríguez, 1995), Europe (Tutin *et al.*, 1993; Prelli, 2001), Swaziland (Roux, 2003), Hunan (Yu and Chengsen, 2007), Japan (Iwatsuki *et al.*, 1995), Australia (Orchard, 1998), Tasmania (Garrett, 1996), New Zealand (Brownsey and Smith-Dodsworth, 2000), and the Hawaiian Islands (Palmer, 2003). Of these floras, the most comprehensive is *The Pteridophytes of Mexico* (Mickel and Smith, 2004). It illustrates every species, provides indented dichotomous keys, and gives complete information about nomenclature, synonyms, type specimens, descriptions, habitats, geographic ranges, and discussions. Modern floras are still lacking for some of the most species-rich regions of the earth, such as Colombia, Bolivia, and Brazil. Other diverse regions, such as Ecuador and Malesia, have only partially completed floras.

14.7　Important questions

Fern and lycophyte biogeography still presents important questions. Do ferns speciate less frequently than angiosperms, as suggested by their lower endemism on oceanic islands? Ferns might speciate less rapidly because they have fewer isolating mechanisms or, alternatively, because their greater vagility promotes an infrequent but steady amount of gene flow from source areas, thus hindering speciation (Klekowski, 1972, 1973; Ranker et al., 1994). More experiments and observations are needed to answer this intriguing question (see Chapter 12).

Further research is needed on the influence of morphological and life history traits on geographical range (Peck et al., 1990; Guo et al., 2003). Evidence has already shown that epiphytic ferns with colonial, gemmiferous gametophytes are statistically over-represented on oceanic islands and intercontinental disjunctions (Dassler and Farrar, 1997, 2001; Moran and Smith, 2001). Also, apomicts and polyploids may be more common among source species of oceanic island ferns because of their capacity for single-spore reproduction (e.g., Masuyama and Watano, 1990). Other traits might influence dispersal and establishment, but more investigation is required. For example, is there a correlation between range size and spore traits such as green versus non-green, monolete versus trilete, homosporous versus heterosporous, winged perispores versus non-winged? How does selfing versus outcrossing affect the range and ability to establish after dispersal (Haufler, 2007)? Do tropical and subtropical species tend to have narrower latitudinal ranges compared to temperate species – a pattern called the Rapoport effect (Kessler, 2001b)? Some of these questions have been examined in *Asplenium*. In European *Asplenium*, many species have restricted ranges and, in general, the diploids have smaller ranges than polyploids (Vogel et al., 1999). Several studies have positively correlated long-distance dispersal with characters such as polyploidy, and breeding system for some widely distributed asplenoioid ferns (Crist and Farrar, 1983; Pichi Sermolli, 1990; Ranker et al., 1994; Schneller, 1996; Schneller and Holderegger, 1996; Suter et al., 2000; Trewick et al., 2002).

One biogeographical scenario that has not yet been inferred for ferns is the boreotropics hypothesis. It postulates an exchange of species between North America and Europe during the thermal maxima of the early Tertiary (Eocene) when tropical plants occurred at higher latitudes than they do today (Lavin and Luckow, 1993). Species migrated along land bridges connecting Greenland

and Western Europe (Tiffney, 1985b, 1985a; Tiffney and Manchester, 2001) with their descendents eventually entering tropical Africa. The present-day result is sister taxa, one of which occurs in the American tropics and the other in Africa (e.g., Davis *et al.*, 2002). To test this hypothesis, we need fossils showing that the taxa were present at high northern latitudes during the early Tertiary and a phylogeny showing sister-group relationships between the taxa. Some species of *Asplenium* fit the hypothesis but the data are inconclusive (Schneider *et al.*, 2004). This biogeographical scenario needs to be considered when analyzing phylogenies of temperate ferns.

The geographical distribution of different ploidy levels within species badly needs investigation in the tropics. Although considerable work has been done on this subject for temperate species (e.g., Bruce, 1976; Takamiya and Tanaka, 1982; Takamiya and Kurita, 1983; Takamiya, 1989; Pichi Sermolli, 1990, 1991), many tropical species remain uninvestigated. Once a correlation has been established between different ploidy levels and the lengths of spores or guard cells, cytogeographical studies can be done from spore samples taken from herbarium specimens (Barrington *et al.*, 1986). Such studies can reveal unexpected distributional patterns, such as the occurrence of the different ploidy levels inhabiting separate rock types (e.g., Moran, 1982), different elevations (e.g., Matsumoto, 1982), or in distinct regions of the species ranges (e.g., Murakami and Moran, 1993).

14.8 Future directions

Floristic research will rely increasingly on the worldwide web to disseminate information. Already available on the web are essential taxonomic works such as *Taxonomic Literature-2* and the *International Code of Botanical Nomenclature*. A vast amount of nomenclatural information is available from the Missouri Botanical Garden's TROPICOS, and the International Plant Names Index. Various on-line geographic gazetteers facilitate the location of place names for mapping specimens. Information about and images of type specimens are available on-line, for example for African plants (www.aluka.org/) and for over 92 000 types housed at the New York Botanical Garden. Also, images of many ferns and lycophytes are now publicly available, for example on the Diversity of Life website. It seems likely that all this information will eventually be downloadable, and botanists of the future will go into the field equipped with handheld computers capable of providing electronic keys, descriptions, and images for identification.

Although these web-based resources greatly aid research, most significant for biogeographical studies will be the continued improvement in methods of sequencing DNA and analyzing the data obtained. The resulting phylogenetic trees will provide explicit hypotheses to serve as frameworks for posing and answering interesting questions about biogeography and the origins of diversity (see Chapter 15). Such trees have already stimulated more research in biogeography than ever before, and this trend is expected to continue. Advances in estimating evolutionary divergence times from molecular data should also provide exciting results for further research.

References

Adsersen, H. (1990). Intra-archipelago distribution patterns of vascular plants in Galapagos. *Monographs in Systematic Botany, Missouri Botanical Garden*, **32**, 67–78.

Alston, A. H. G. and Schelpe, E. A. C. L. E. (1957). The Pteridophyta of Marion Island. *Journal of South African Botany*, **23**, 105–109.

Aylor, D. E. and Ferrandino, F. J. (1985). Rebound of pollen and spores during deposition on cylinders by inertial impact. *Atmospheric Environment*, **19**, 803–806.

Badré, F. and Hoff, M. (1995). Les ptéridophytes des Iles Wallis et Futuna (Pacifique Sud): écologie et répartition. *Feddes Repertorium*, **106**, 271–290.

Baker, J. G. (1868). On the geographical distribution of ferns. *Transactions of the Linnean Society*, **26**, 305–352.

Balslev, H. (1988). Distribution patterns of Ecuadorean plant species. *Taxon*, **37**, 567–577.

Barrington, D. S. (1993). Ecological and historical factors in fern biogeography. *Journal of Biogeography*, **20**, 275–280.

Barrington, D. S. and Kato, M. (1993). Changing concepts in the biogeography of pteridophytes: the biogeography symposium at the 1990 Progress in Pteridology Conference – Ann Arbor, Michigan. *Journal of Biogeography*, **20**, 253.

Barrington, D. S., Paris, C. A., and Ranker, T. A. (1986). Systematic inferences from spore and stomate size in the ferns. *American Fern Journal*, **76**, 149–159.

Bostock, P. D. (1998). Pityrogramma. *Flora of Australia*, **48**, 263–264.

Boudrie, M. (2003). Les ptéridophytes de l'île Robinson Crusoe (archipel Juan Fernández, Chili). *Le Journal de Botanique de la Societe de France*, **24**, 32–48.

Boufford, D. B. and Spongberg, S. A. (1983). Eastern Asian–eastern North American phytogeographical relationships – a history from the time of Linnaeus to the twentieth century. *Annals of the Missouri Botanical Garden*, **70**, 423–439.

Brownlie, G. (1961). Studies on Pacific ferns, Part IV, the pteridophyte flora of Pitcairn Island. *Pacific Science*, **25**, 297–300.

Brownsey, P. J. (2001). New Zealand's pteridophyte flora – plants of ancient lineage but recent arrival? *Brittonia*, **53**, 284–303.

Brownsey, P. J. and Smith-Dodsworth, J. C. (2000). *New Zealand Ferns and Allied Plants*. Aukland: David Bateman.

Bruce, J. G. (1976). Comparative studies in the biology of *Lycopodium carolinianum*. *American Fern Journal*, **66**, 125–137.

Burrows, J. E. and Crouch, N. R. (1995). Pteridophyta: new distribution records of South African pteridophytes. *Bothalia*, **25**, 236–238.

Carlquist, S. (1980). *Island Biology*. New York: Columbia University Press.

Caulton, E., Keddie, S., Carmichael, R., and Sales, J. (2000). A ten year study of the incidence of spores of bracken (*Pteridium aquilinum* (L.) Kuhn.) in an urban rooftop airstream in south east Scotland. *Aerobiologia*, **16**, 29–33.

Chin, W. Y. (1997). *Ferns of the Tropics*. Singapore: Times Editions.

Christ, H. (1910). *Die Geographie der Farne*. Jena: Gustav Fischer.

Christensen, C. and Skottsberg, C. (1920). The ferns of Easter Island. In *The Natural History of Juan Fernandez and Easter Island*, ed. C. Skottsberg. Uppsala: Almquist and Wiksells, pp. 47–53.

Colwell, R. K. and Hurtt, G. C. (1994). Nonbiological gradients in species richness and a spurious Rapoport effect. *American Naturalist*, **144**, 570–595.

Colwell, R. K. and Lees, D. C. (2000). The mid-domain effect: geometric constraints on the geography of species richness. *Trends in Ecology and Evolution*, **15**, 70–76.

Colwell, R. K., Rahbek, C., and Gotelli, N. J. (2004). The mid-domain effect and species richness patterns: what have we learned so far? *American Naturalist*, **163**, E1–E23.

Conant, D. S., Raubeson, L. A., Attwood, D. K., and Stein, D. B. (1995). The relationships of Papuasian Cyatheaceae to New World tree ferns. *American Fern Journal*, **85**, 328–340.

Copeland, E. B. (1939). Fern evolution in Antarctica. *Philippine Journal of Science*, **70**, 157–188.

Cranfill, R. and Kato, M. (2003). Phylogenetics, biogeography, and classification of the woodwardioid ferns (Blechnaceae). In *Pteridology in the New Millennium*, ed. S. Chandra and M. Srivastava. Boston, MA: Kluwer, pp. 25–48.

Cremers, G. and Boudrie, M. (2006). Ptéridophytes de Guyane française non récoltées depuis plus d'un siècle ou récemment retrouvées. *Acta Botanica Gallica*, **153**, 3–48.

Crist, K. C. and Farrar, D. R. (1983). Genetic load and long-distance dispersal in *Asplenium platyneuron*. *Canadian Journal of Botany*, **61**, 1809–1814.

Croat, T. B. (1978). *Flora of Barro Colorado Island*. Stanford, CA: Stanford University Press.

Dassler, C. L. and Farrar, D. R. (1997). Significance of form in fern gametophytes: clonal, gemmiferous gametophytes of *Callistopteris baueriana* (Hymenophyllaceae). *International Journal of Plant Sciences*, **158**, 622–639.

Dassler, C. L. and Farrar, D. R. (2001). Significance of gametophyte form in long-distance colonization by tropical, epiphytic ferns. *Brittonia*, **53**, 325–369.

Davis, C. C., Bell, C. D., Mathews, S., and Donoghue, M. J. (2002). Laurasian migration explains Gondwanan disjunctions: evidence from Malpighiaceae. *Proceedings of the National Academy of Science of the United States of America*, **99**, 6833–6837.

de la Sota, E., Ponce, M. M., Morbelli, M. A., and Cassá de Pazos, L. (1998). Pteridophyta. In *Flora Patagonica*, ed. M. N. Correa. Buenos Aires: Instituto Nacional de Tecnología Agropecuaria, pp. 282–369.

Doctors van Leeuwen, W. M. (1936). Krakatau 1883–1933. A. Botany. *Annales du Jardin Botanique de Buitenzorg*, **5**, 46–47.

Dodson, C. H. and Gentry, A. H. (1978). Flora of the Río Palenque Science Center, Los Ríos, Ecuador. *Selbyana*, **4**, 1–628.

Driscoll, H. E. and Barrington, D. S. (2007). Origin of Hawaiian *Polystichum* (Dryopteridaceae) in the context of a world phylogeny. *American Journal of Botany*, **94**, 1413–1424.

Dubuisson, J., Hennequin, S., Douzery, E. J. P., Cranfill R. B., Smith, A. R., and Pryer, K. M. (2003). rbcL phylogeny of the fern genus *Trichomanes* (Hymenophyllaceae), with special reference to neotropical taxa. *International Journal of Plant Sciences*, **164**, 753–761.

D'Urville, J. (1835). Mémoire sur la distribution géographique des fougères à la surface du globe. *Annales des Sciences Naturelles (Paris)*, **6**, 1–51.

Dzwonko, Z. and Kornás, J. (1978). A numerical analysis of the distribution of pteridophytes in Zambia. *Zeszyty Naukowe Uniwersitetu Jagiellonskiego, Prace Botaniczne*, **6**, 39–49.

Dzwonko, Z. and Kornás, J. (1994). Patterns of species richness and distribution of pteridophytes in Rwanda (Central Africa): a numerical approach. *Journal of Biogeography*, **21**, 491–501.

Erdtman, G. (1937). Pollen grains recovered from the atmosphere over the Atlantic. *Acta Horti Gotoburgensis*, **12**, 185–196.

Ferrer-Castan, D. and Vetaas, O. R. (2005). Pteridophyte richness, climate and topography in the Iberian Peninsula, comparing spatial and nonspatial models of richness patterns. *Global Ecology and Biogeography*, **14**, 155–165.

Figueiredo, E. (2001). New findings of pteridophytes from the mountain rainforests of São Tomé and Príncipe. *Fern Gazette*, **16**, 191–193.

Flora of North America Editorial Committee (1993). *Pteridophytes and Gymnosperms*, Vol. 2, *Flora of North America North of Mexico*. New York: Oxford University Press.

Florence, J. and Lorence, D. H. (1997). Introduction to the flora and vegetation of the Marquesas Islands. *Allertonia*, **7**, 226–237.

Florence, J., Waldren, S., and Chepstow-Lusty, A. J. (1995). The flora of the Pitcairn Islands: a review. *Biological Journal of the Linnean Society*, **56**, 79–119.

Gabrielian, E. C. and Greuter, W. (1984). A revised catalogue of the Pteridophyta of the Armenian SSR. *Willdenowia*, **14**, 145–158.

Garrett, M. (1996). *The Ferns of Tasmania, Their Ecology and Distribution*. Hobart: Tasmanian Forest Research Council.

Gastony, G. J. (1973). A revision of the fern genus *Nephelea*. *Contributions from the Gray Herbarium of Harvard University*, **203**, 81–148.

Geiger, J. M. and Ranker, T. A.. (2005). Molecular phylogenetics and historical biogeography of Hawaiian *Dryopteris* (Dryopteridaceae). *Molecular Phylogenetics and Evolution*, **34**, 392–407.

Geiger, J. M. O., Ranker, T. A., Neale, J. M. R., and Klimas, S. T. (2007). Molecular biogeography and origins of the Hawaiian fern flora. *Brittonia*, **59**, 142–158.

Godley, E. J. (1989). The flora of the Antipodes Island. *New Zealand Journal of Botany*, **27**, 531–563.

Gómez, L. D. (1978). Contribuciones a la pteridología costarricense. XI. Hermann Christ, su vida, obra e influencia en la botánica nacional. *Brenesia*, **12–13**, 25–79.

Gradstein, S. R. and Van Zanten, B. O. (2001). High altitude dispersal of spores: an experimental approach. *XVI International Botanical Congress, St. Louis*, Abstract Number 15.14.13.

Graham, A. (1993). History of the vegetation: Cretaceous (Maastrichtian) – Tertiary. In *Flora of North America North of Mexico*, ed. Flora of North America Editorial Committee. New York: Oxford University Press, pp. 57–70.

Grayum, M. H. and Churchill, H. W. (1987). An introduction to the pteridophyte flora of Finca La Selva, Costa Rica. *American Fern Journal*, **77**, 73–89.

Gressitt, J. L., Sedlacek, J., Wise, K. A. J., and Yoshimoto, C. M. (1961). A high speed airplane trap for air-borne organisms. *Pacific Insects*, **3**, 549–555.

Grytnes, J. A. (2003). Ecological interpretations of the mid-domain effect. *Ecology Letters*, **6**, 883–888.

Guého, J. (1988). *La Vegetation de l'Île Maurice*. Stanley, Rose Hill, Ile Maurice: Editions de l'Ocean Indien.

Guo, Q., Kato, M., and Ricklefs, R. E. (2003). Life history, diversity and distribution: a study of Japanese pteridophytes. *Ecography*, **26**, 129–138.

Haufler, C. H. (2007). Genetics, phylogenetics, and biogeography: considering how shifting paradigms and continents influence fern diversity. *Brittonia*, **59**, 108–114.

Haufler, C. H. and Windham, M. D. (1991). New species of North American *Cystopteris* and *Polypodium*, with comments on their reticulate relationships. *American Fern Journal*, **81**, 6–22.

Hawkins, B. A. and Diniz-Filho, J. A. F. (2002). The mid-domain effect cannot explain the diversity gradient of Nearctic birds. *Global Ecology and Biogeography*, **11**, 419–426.

Holttum, R. E. (1969). Studies in the family Thelypteridaceae. The genera *Phegopteris*, *Pseudophegopteris*, and *Macrothelypteris*. *Blumea*, **17**, 5–32.

Hoot, S. B., Taylor, W. C., and Napier, N. S. (2006). Phylogeny and biogeography of *Isoëtes* (Isoëtaceae) based on nuclear and chloroplast DNA sequence data. *Systematic Botany*, **31**, 449–460.

Iwatsuki, K. (1995). Comparison of pteridophyte flora between North America and Japan. In *Vegetation in Eastern North America*, ed. A. Mayawaki, K. Iwatsuki, and M. Grandtner. Tokyo: University of Tokyo Press, pp. 89–97.

Iwatsuki, K., Yamazaki, T., Boufford, D. E., and Ohbah, H. (1995). *Flora of Japan*, Vol. 1, Pteridophyta and Gymnospermae. Tokyo: Kodansha Press.

Jacobsen, W. B. G. and Jacobsen, N. H. G. (1989). Composition of the pteridophyte floras of southern and eastern Africa, with special reference to high-altitude species. *Bulletin Jardin Botanique Belgique*, **59**, 261–317.

Janssen, T., Kreier, H.-P., and Schneider, H. (2007). Origin and diversification of African ferns with special emphasis on Polypodiaceae. *Brittonia*, **59**, 159–181.

Johns, R. J. (1997). *Index Filicum. Supplementum septimum, pro annis 1991–1995*. Kew: Royal Botanic Gardens.

Jones, M. M., Tuomisto, H., Clark, D. B., and Olivas, P. (2005). Effects of mesoscale environmental heterogeneity and dispersal limitation on floristic variation in rain forest ferns. *Journal of Ecology*, **94**, 181–195.

Kato, M. (1993). Biogeography of ferns: dispersal and vicariance. *Journal of Biogeography*, **20**, 265–274.

Kato, M. and Iwatsuki, K. (1983). Phytogeographic relationships of pteridophytes between temperate North America and Japan. *Annals of the Missouri Botanical Garden*, **70**, 724–733.

Kessler, M. (1999). Plant species richness and endemism during natural landslide succession in perhumid montane forest in the Bolivian Andes. *Ecotropica*, **5**, 123–136.

Kessler, M. (2000). Elevational gradients in species richness and endemism of selected plant groups in the central Bolivian Andes. *Plant Ecology*, **149**, 181–193.

Kessler, M. (2001a). Pteridophyte species richness in Andean forests in Bolivia. *Biodiversity and Conservation*, **10**, 1473–1495.

Kessler, M. (2001b). Patterns of diversity and range size of selected plant groups along an elevational transect in the Bolivian Andes. *Biodiversity and Conservation*, **10**, 1897–1921.

Kessler, M. (2002). The elevational gradient of Andean plant endemism: varying influences of taxon-specific traits and topography at different taxonomic levels. *Journal of Biogeography*, **29**, 1159–1166.

Kessler, M., Parris, B. S., and Kessler, E. (2001). A comparison of the tropical montane pteridophyte floras of Mount Kinabalu, Borneo, and Parque Nacional Carrasco, Bolivia. *Journal of Biogeography*, **28**, 611–622.

Klekowski, E. J. (1972). Genetical features of ferns as contrasted to seed plants. *Annals of the Missouri Botanical Garden*, **59**, 138–151.

Klekowski, E. J. (1973). Genetic endemism of Galapagos *Pteridium*. *Botanical Journal of the Linnean Society*, **66**, 181–188.

Kluge, J. and Kessler, M. (2006). Fern endemism and its correlates: contribution from an elevational transect in Costa Rica. *Diversity and Distribution*, **12**, 535–545.

Knobloch, I. W. and Correll, D. S. (1962). *Ferns and Fern Allies of Chihuahua, Mexico*. Renner, TX: Texas Research Foundation.

Kramer, K. U. (1990). The American paradox in the distribution of fern taxa above the rank of species. *Annals of the Missouri Botanical Garden*, **77**, 330–333.

Kramer, K. U. and Green, P. S. (1990). *The Families and Genera of Vascular Plants*, Vol. 1, *Pteridophytes and Gymnosperms*. Berlin: Springer-Verlag.

Kreier, H.-P. and Schneider, H. (2006). Phylogeny and biogeography of the staghorn fern genus *Platycerium* (Polypodiaceae, Polypodiidae). *American Journal of Botany*, **93**, 217–225.

Krippel, Y. (2005). Pteridophyte diversity in Luxembourg. *Fern Gazette*, **17**, 216.

Krömer, T. and Kessler, M. (2006). Filmy ferns (Hymenophyllaceae) as high-canopy epiphytes. *Ecotropica*, **12**, 57–63.

Kroonenberg, S. B., Bakker, J. G. M., and van der Wiel, A. M. (1990). Late Cenozoic uplift and paleogeography of the Colombian Andes: constraints on the development of high-Andean biota. *Geologie en Mijnbouw*, **69**, 279–290.

Lavin, M. and Luckow, M. (1993). Origins and relationships of tropical North America in the context of the boreotropics hypothesis. *American Journal of Botany*, **80**, 1–14.

Lellinger, D. B. (1975). Phytogeographic analysis of Chocó pteridophytes. *Fern Gazette*, **11**, 105–114.

Lellinger, D. B. (1985). The distribution of Panama's pteridophytes. *Monographs in Systematic Botany from the Missouri Botanical Garden*, **10**, 43–47.

Lellinger, D. B. (1989). The ferns and fern-allies of Costa Rica, Panama, and the Chocó (part 1: Psilotaceae through Dicksoniaceae). *Pteridologia*, **2A**, 1–364.

Leonard, S. W. (1972). The distribution of *Thelypteris torresiana* in the southeastern United States. *American Fern Journal*, **62**, 97–99.

Li, H.-L. (1952). Floristic relationships between eastern Asia and eastern North America. *Transactions of the American Philosophical Society*, **42**, 371–429.

Little, D. P. and Barrington, D. S. (2003). Major evolutionary events in the origin and diversification of the fern genus *Polystichum* (Dryopteridaceae). *American Journal of Botany*, **90**, 508–514.

Looser, G. (1958). Los helechos de la Isla de Pascua. *Revista Universitaria, Santiago, Chile*, **43**, 39–64.

Lorence, D. H. (1976). The pteridophytes of Rodrigues Island. *Botanical Journal of the Linnean Society*, **72**, 269–283.

Lwanga, J. S., Balmford, A., and Badaza, R. (1998). Assessing fern diversity: relative species richness and its environmental correlates in Uganda. *Biodiversity and Conservation*, **7**, 1387–1398.

Lyell, K. M. (1879). *A Geographical Handbook of All the Known Ferns with Tables to Show their Distribution*. London: John Murray.

Marticorena, C., Stuessy, T. F., and Baeza, C. M. (1998). Catalogue of the vascular flora of the Robinson Crusoe or Juan Fernández Islands, Chile. *Gayana, Botany*, **55**, 187–211.

Masuyama, S. and Watano, Y. (1990). Trends for inbreeding in polyploid pteridophytes. *Plant Species Biology*, **5**, 13–17.

Matsumoto, S. (1982). Distribution patterns of two reproductive types of *Phegopteris connectilis* in eastern Japan. *Bulletin of the National Science Museum, Series B (Botany)*, **8**, 101–110.

Mickel, J. T. and Smith, A. R. (2004). Pteridophytes of Mexico. *Memoirs of the New York Botanical Garden*, **88**, 1–1055.

Moran, R. C. (1982). The *Asplenium trichomanes* complex in the United States and adjacent Canada. *American Fern Journal*, **72**, 5–11.

Moran, R. C. (1995). The importance of mountains to pteridophytes, with emphasis on neotropical montane forests. In *Biodiversity and Conservation of Neotropical Montane Forests*, ed. S. P. Churchill, H. Balslev, E. Forero, and J. L. Luteyn. New York: New York Botanical Garden, pp. 359–363.

Moran, R. C. (1996). The importance of the Andes as a barrier to migration, as illustrated by the pteridophytes of the Chocó phytogeographic region. In *Pteridology in Perspective*, ed. J. M. Camus, M. Gibby, and R. J. Johns. Kew: Royal Botanic Gardens, p. 75.

Moran, R. C. (2001). Introduction: papers from the Pteridophyte Biogeography Symposium, International Botanical Congress. *Brittonia*, **53**, 171–172.

Moran, R. C. (2004). *A Natural History of Ferns*. Portland, OR: Timber Press.

Moran, R. C. (2007). Introduction to papers from the Fern Biogeography Symposium, International Botanical Congress, 2005. *Brittonia*, **59**, 107.

Moran, R. C. and Riba, R. (eds.) (1995). *Psilotaceae a Salviniaceae*, Vol. 1. In *Flora Mesoamericana*, ed. G. Davidse, M. S. Souza, and S. Knapp. Ciudad Universitaria: Universidad Nacional Autónoma de México.

Moran, R. C. and Smith, A. R. (2001). Phytogeographic relationships between neotropical and African-Madagascan pteridophytes. *Brittonia*, **53**, 304–351.

Moran, R. C., Klimas, S., and Carlsen, M. (2003). Low-trunk epiphytic ferns on tree ferns versus angiosperms in Costa Rica. *Biotropica*, **35**, 48–56.

Mori, S., Cremers, G., Gracie, C. A., de Granville, J.-J., Heald, S. V., Hoff, M., and Mitchell, J. D. (1997). Guide to the vascular plants of Central French Guiana. Part 1. Pteridophytes, gymnosperms, and monocotyledons. *Memoirs of the New York Botanical Garden*, **71**, 1–422.

Mori, S., Cremers, G., Gracie, C. A., de Granville, J.-J., Heald, S. V., Hoff, M., and Mitchell, J. D. (2002). Guide to the vascular plants of Central French Guiana. Part 2. Dicotyledons. *Memoirs of the New York Botanical Garden*, **76**, 1–776.

Muñoz, J., Felicisimo, Á. M., Cabezas, F., Burgaz, A. R., and Martínez, I. (2004). Wind as a long-distance dispersal vehicle in the Southern Hemisphere. *Science*, **304**, 1144–1147.

Murakami, N. and Moran, R. C. (1993). Monograph of the neotropical species of *Asplenium* sect. *Hymenasplenium* (Aspleniaceae). *Annals of the Missouri Botanical Garden*, **80**, 1–38.

Orchard, A. E. (1998). *Flora of Australia*, Vol. 48, *Ferns, Gymnosperms and Allied Groups*. Melbourne: CSIRO Publishing.

Palmer, D. D. (2003). *Hawai'i's Ferns and Fern Allies*. Honolulu, HI: University of Hawaii Press.

Parris, B. S. (1985). Ecological aspects of distribution and speciation in Old World tropical ferns. *Proceedings of the Royal Society of Edinburgh*, **86B**, 341–346.

Parris, B. S. (1997). The ecology and phytogeography of Mt. Kinabalu pteridophytes. *Sandakania*, **9**, 87–88.

Parris, B. S. (2001). Circum-Antarctic continental distribution patterns in pteridophyte species. *Brittonia*, **53**, 270–283.

Parris, B. S., Beaman, R. S., and Beaman, J. H. (1992). *The Plants of Mount Kinabalu. I. Ferns and Fern Allies*. Kew: Royal Botanic Gardens.

Parrish, J. T. (1987). Global palaeogeography and palaeoclimate of the late Cretaceous and early Tertiary. In *The Origins of Angiosperms and Their Biological Consequences*, ed. E. M. Friis, W. G. Chaloner, and P. R. Crane. Cambridge: Cambridge University Press, pp. 51–74.

Peck, J. H., Peck, C. J., and Farrar, D. R. (1990). Influences of life history events on formation of local and distant fern populations. *American Fern Journal*, **80**, 126–142.

Perrie, L. R. and Brownsey, P. J. (2005a). New Zealand *Asplenium* (Aspleniaceae: Pteridophyta) revisited – DNA sequencing and AFLP fingerprinting. *Fern Gazette*, **17**, 235–242.

Perrie, L. R. and Brownsey, P. J. (2005b). Insights into the biogeography and polyploid evolution of New Zealand *Asplenium* from chloroplast DNA sequence data. *American Fern Journal*, **95**, 1–21.

Perrie, L. R., Brownsey, P. J., Lockhart, P. J., Brown, E. A., and Large, M. F. (2003). Biogeography of temperate Australian *Polystichum* ferns as inferred from chloroplast sequence and AFLP. *Journal of Biogeography*, **30**, 1729–1736.

Perrie, L. R., Bayly, M. J., Lehnebach, C. A., and Brownsey, P. J. (2007). Molecular phylogenetics and molecular dating of the New Zealand Gleicheniaceae. *Brittonia*, **59**, 129–141.

Pichi Sermolli, R. E. G. (1979). A survey of the pteridological flora of the Mediterranean region. *Webbia*, **34**, 175–242.

Pichi Sermolli, R. E. G. (1990). Speciazione e distribuzione geografica nelle Pteridophyta. *Anales del Jardin Botanico de Madrid*, **46**, 489–518.

Pichi Sermolli, R. E. G. (1991). Considerazioni sull'affinitá ed origine della flora pteridologica della Regione Mediterranea. *Acta Botanica Malacitana*, **16**, 235–280.

Piggott, A. G. (1988). *Ferns of Malaysia in Colour*. Kuala Lumpur: Tropical Press.

Polunin, N. (1951). Seeking airborne botanical particles about the North Pole. *Svensk Botanisk Tidskrift*, **45**, 320–354.

Poulsen, A. D. and Nielsen, I. H. (1995). How many ferns are there in one hectare of tropical rain forest? *American Fern Journal*, **85**, 29–35.

Prelli, R. (2001). *Les Fougères et Plantes Alliees de France et d'Europe Occidentale*. Paris: Belin.

Pryer, K. M. (1993). *Gymnocarpium*. In *Flora of North America North of Mexico*, Vol. 2, *Pteridophytes and Gymnosperms*, ed. Flora of North America Editorial Committee. New York: Oxford University Press, pp. 258–262.

Pryer, K. M., Smith, A. R., Hunt, J. S., and Dubuisson, J.-Y. (2001). rbcL data reveal two monophyletic groups of filmy ferns (Filicopsida: Hymenophyllaceae). *American Journal of Botany*, **88**, 1118–1130.

Puentha, N. (1991). Studies on atmospheric fern spores at Pithorgarh (northwest Himalaya) with particular reference to distribution of ferns in the Himalayas. *Annual Review of Plant Science*, **13**, 146–161.

Rahbek, C. (2005). The role of spatial scale and the perception of large-scale species-richness patterns. *Ecology Letters*, **8**, 224–239.

Ranker, T. A., Floyd, S. K., and Trapp, P. G. (1994). Multiple colonizations of *Asplenium adiantum-nigrum* onto the Hawaiian archipelago. *Evolution*, **48**, 1364–1370.

Ranker, T. A., Gemmill, C. E. C., and Trapp, P. G. (2000). Microevolutionary patterns and processes of the native Hawaiian colonizing fern *Odontosoria chinensis* (Lindsaeaceae). *Evolution*, **54**, 828–839.

Ranker, T. A., Trapp, P. G., Smith, A. R., Moran, R. C., and Parris, B. S. (2005). New records of lycophytes and ferns from Moorea, French Polynesia. *American Fern Journal*, **95**, 126–127.

Rodríguez, R. (1995). Pteridophyta. In *Flora of Chile*, Vol. 1, *Pteridophyta – Gymnospermae*, ed. C. Marticorena and R. Rodríguez. Chile: Universidad de Concepción, pp. 119–309.

Rojas-Alvarado, A. F. (2001). New species, newly used names and new ranges of tree ferns (Filicales: Cyatheaceae) in the Neotropics. *Revista de Biologa Tropical*, **49**, 453–465.

Rojas-Alvarado, A. F. (2002). New species, new combinations and new distributions in neotropical species of *Elaphoglossum* (Lomariopsidaceae). *Revista de Biologa Tropical*, **50**, 969–1006.

Rojas-Alvarado, A. F., and Trusty, J. (2004). Diversidad pteridofítica de la Isla del Coco, Costa Rica. *Brenesia*, **62**, 1–14.

Roos, M. (1996). Mapping the world's pteridophyte diversity – systematics and floras. In *Pteridology in Perspective*, ed. J. M. Camus, M. Gibby, and R. J. Johns. Kew: Royal Botanic Gardens, pp. 29–42.

Rothwell, G. W. and Stockey, R. A. (1991). *Onoclea sensibilis* in the Paleocene of North America, a dramatic example of structural and ecological stasis. *Review of Palaeobotany and Palynology*, **70**, 113–124.

Rouhan, G., Garrison Hanks, J., McClelland, D., and Moran, R. C. (2007). Preliminary phylogenetic analysis of the fern genus *Lomariopsis* (Lomariopsidaceae). *Brittonia*, **59**, 115–128.

Roux, J. P. (2003). Swaziland ferns and fern allies. *Southern African Botanical Diversity Network Report*, **19**, 1–241.

Ruokolainen, K., Linna, A., and Tuomisto, H. (1997). Use of Melastomataceae and pteridophytes for revealing phytogeographical patterns in Amazonian rain forests. *Journal of Tropical Ecology*, **13**, 243–256.

Salvo, A. E., Prada, C., and Díaz, T. (1982). Revisión del género *Asplenium* L., subgénero *Pleurosorus* (Fée) Salvo, Prada and Díaz. *Candollea*, **37**, 457–484.

Schelpe, E. A. C. L. E. (1975). Observations on the spread of the American fern *Pityrogramma calomelanos*. *Fern Gazette*, **11**, 101–104.

Schneider, H., Russell, S. J., Cox, C. J., Bakker, H. J., Henderson, S., Rumsey, F. J., Barrett, J. A., Gibby, M., and Vogel, J. C. (2004). Chloroplast phylogeny of asplenioid ferns based on *rbcL* and *trnL-F* spacer sequences (Polypodiidae, Aspleniaceae) and its implications for biogeography. *Systematic Botany*, **29**, 260–274.

Schneider, H., Kreier, H.-P., Wilson, R., and Smith, A. R. (2006). The *Synammia* enigma: evidence for a temperate lineage of polygrammoid fens (Polypodiaceae; Polypodiiae) in southern South America. *Systematic Botany*, **31**, 31–41.

Schneller, J. J. (1996). Outbreeding depression in the fern *Asplenium ruta-muraria* L.: evidence from enzyme electrophoresis data, meiotic irregularities and reduced spore viability. *Biological Journal of the Linnean Society*, **59**, 281–295.

Schneller, J. J. and Holderegger, R. (1996). Colonization events and genetic variability within populations of *Asplenium ruta-muraria* L. In *Pteridology in Perspective*, ed. J. M. Camus, M. Gibby, and R. J. Johns. Kew: Royal Botanic Gardens, pp. 571–580.

Schuettpelz, E. and Trapnell, D. W. (2006). Exceptional epiphyte diversity on a single tree in Costa Rica. *Selbyana*, **27**, 65–71.

Smith, A. R. (1972). Comparison of fern and flowering plant distributions with some evolutionary interpretations for ferns. *Biotropica*, **4**, 4–9.

Smith, A. R. (1993). Phytogeographic principles and their use in understanding fern relationships. *Journal of Biogeography*, **20**, 255–264.

Smith, A. R. (1995). Pteridophytes. In *Flora of the Venezuelan Guayana*, ed. J. A. Steyermark, P. E. Berry, and B. K. Holst. St. Louis, MO: Missouri Botanical Garden and Timber Press, pp. 1–334.

Smith, A. C. (1996). *Flora Vitiensis Nova, A New Flora of Fiji (Spermatophytes only); Comprehensive Indices*. Lawai, HI: National Tropical Botanical Garden.

Smith, A. R. and Cranfill, R. (2002). Intrafamilial relationships of the thelypteroid ferns (Thelypteridaceae). *American Fern Journal*, **92**, 131–149.

Smith, A. R., Kessler, M., and Gonzáles, J. (1999). New records of pteridophytes from Bolivia. *American Fern Journal*, **89**, 244–266.

Smith, A. R., León, B., Tuomisto, H., van der Werff, H., Moran, R. C., Lehnert, M., and Kessler, M. (2005). New records of pteridophytes for the flora of Peru. *Sida Contributions to Botany*, **21**, 2321–2342.

Stolze, R. G. (1987). *Schizaea pusilla* discovered in Peru. *American Fern Journal*, **77**, 64–65.

Strother, J. L. and Smith, A. R. (1971). Chorology, collection dates, and taxonomic responsibility. *Taxon*, **19**, 871–874.

Suter, M., Schneller, J. J., and Vogel, J. C. (2000). Investigations into the genetic variation, population structure, and breeding systems of the fern *Asplenium trichomanes* subsp. *quadrivalens*. *International Journal of Plant Sciences*, **161**, 233–244.

Takamiya, M. (1989). Cytological and ecological studies on the speciation of *Lycopodium clavatum* L. in the Japanese Archipelago. *Journal of Science of the Hiroshima University, Series B, Division 2 (Botany)*, **22**, 353–430.

Takamiya, M. and Kurita, S. (1983). Cytotaxonomic studies on Japanese species of the genus *Lycopodium* sensu lato. *Acta Phytotaxonomica et Geobotanica*, **34**, 66–79.

Takamiya, M. and Tanaka, R. (1982). Polyploid cytotypes and their habitat preferences in *Lycopodium clavatum*. *Botanical Magazine (Tokyo)*, **95**, 419–434.

Taylor, W. C. and Johnson, D. M. (1979). *Thelypteris* in Arkansas. *American Fern Journal*, **69**, 26–28.

Tiffney, B. H. (1985a). The Eocene North Atlantic land bridge: its importance in Tertiary and modern phytogeography of the Northern Hemisphere. *Journal of the Arnold Arboretum*, **66**, 243–273.

Tiffney, B. H. (1985b). Perspectives on the origin of the floristic similarity between eastern Asia and eastern North America. *Journal of the Arnold Arboretum*, **66**, 73–94.

Tiffney, B. H. and Manchester, S. R. (2001). The use of geological and paleontological evidence in evaluating plant phylogeographic hypotheses in the northern hemisphere. *International Journal of Plant Sciences*, **162** (supplement), S3–S17.

Trewick, S. A., Morgan-Richards, M., Russell, S. J., Henderson, S., Rumsey, F. J., Pinter, I., Barrett, J. A., Gibby, M., and Vogel, J. C. (2002). Polyploidy, phylogeography and Pleistocene refugia of the rockfern *Asplenium ceterach*: evidence from chloroplast DNA. *Molecular Ecology*, **11**, 2003–2012.

Tryon, A. F. (1966). Origin of the fern flora of Tristan da Cunha. *Fern Gazette*, **9**, 269–276.

Tryon, A. F. and Lugardon, B. (1991). *Spores of the Pteridophyta*. New York: Springer-Verlag.

Tryon, A. F. and Moran, R. C. (1997). *The Ferns and Allied Plants of New England*. Lincoln, MA: Massachusetts Audubon Society.

Tryon, R. M. (1969). Taxonomic problems in the geography of North American ferns. *BioScience*, **19**, 790–795.

Tryon, R. M. (1970). Development and evolution of fern floras of oceanic islands. *Biotropica*, **2**, 76–84.

Tryon, R. M. (1972). Endemic areas and speciation in tropical American ferns. *Biotropica*, **4**, 76–84.

Tryon, R. M. (1986). The biogeography of species, with special reference to ferns. *Botanical Review*, **52**, 118–156.

Tryon, R. M. and Conant, D. S. (1975). The ferns of Brazilian Amazonia. *Acta Amazonica*, **5**, 23–34.

Tryon, R. M. and Gastony G. J. (1975). The biogeography of endemism in the Cyatheaceae. *Fern Gazette*, **11**, 73–79.

Tuomisto, H. and Ruokolainen, K. (2005). Environmental heterogeneity and the diversity of pteridophytes and Melastomataceae in western Amazonia. *Biologiske Skrifter*, **55**, 37–56.

Tuomisto, H., Ruokolainen, K., Aguilar, M., and Sarmiento, A. (2003). Floristic patterns along a 43-km long transect in an Amazonian rain forest. *Journal of Ecology*, **91**, 743–756.

Turrill, W. B. (1935). Krakatau and its problems. *New Phytologist*, **34**, 442–442.

Tutin, T. G., Burges, N. A., Chater, A. O., Edmondson, J. R., Heywood, V. H., Moore, D. M., Valentine, D. H., Walters, S. M., and Webb, D. A. (1993). *Flora Europea*, Vol. 1, 2nd edn. Cambridge: Cambridge University Press.

van der Werff, H. (1992). Substrate preference of Lauraceae and ferns in the Iquitos area, Peru. *Candollea*, **47**, 11–20.

Verma, S. C. (1966). A note on *Pityrogramma calomelanos* in India. *Bulletin of the Botanical Survey of India*, **8**, 99–100.

Vogel, J. C., Rumsey, F. J., Schneller, J. J., Barrett, J. A., and Gibby, M. (1999). Where are the glacial refugia in Europe? Evidence from pteridophytes. *Botanical Journal of the Linnean Society*, **66**, 23–37.

Volin, J. C., Lott, M. S., Muss, J. D., and Owen, D. (2004). Predicting rapid invasion of the Florida Everglades by Old World climbing fern (*Lygodium microphyllum*). *Diversity and Distributions*, **10**, 439–446.

Wace, N. M. and Dickson, J. H. (1965). The biological report of the Royal Society expedition to Tristan da Cunha, 1962. Part II. The terrestrial botany of the Tristan da Cunha Islands. *Philosophical Transactions of the Royal Society of London, Series B*, **249**, 273–360.

Wagner, W. H., Jr. (1972). Disjunctions in homosporous vascular plants. *Annals of the Missouri Botanical Garden*, **59**, 203–217.

Wagner, W. L. (1991). Evolution of waif floras: a composition of the Hawaiian and Marquesan Archipelagoes. In *The Unity of Evolutionary Biology: Proceedings of the Fourth International Congress of Systematic and Evolutionary Biology, University of Maryland, College Park, USA, July 1990, co-hosted by the Smithsonian Institution*, ed. E. C. Dudley. Portland, OR: Dioscorides Press, pp. 267–284.

Wagner, W. H., Jr. (1995). Evolution of Hawaiian ferns and fern allies in relation to their conservation status. *Pacific Science*, **49**, 31–41.

Wagner, W. H., Jr. and Smith, A. R. (1993). Pteridophytes. In *Flora of North America North of Mexico*, Vol. 2, *Pteridophytes and Gymnosperms*, ed. Flora of North America Editorial Committee. New York: Oxford University Press, pp. 247–266.

Wagner, W. L., Herbst, D. R., and Sohmer, S. H. (1990). *Manual of the Flowering Plants of Hawai'i*, Vol. 1. Honolulu, HI: University of Hawai'i Press and Bishop Museum Press (Special Publication 83).

Wagner, W. H., Jr., Moran, R. C., and Werth, C. R. (1993). Asplenium. In *Flora of North America North of Mexico*, Vol. 2, *Pteridophytes and Gymnosperms*, ed. Flora of North America Editorial Committee. New York: Oxford University Press, pp. 228–245.

Wardlaw, C. W. (1962). A note on *Pityrogramma calomelanos* (L.) Link, a fern nuisance in Cameroons plantations. *Journal of Ecology*, **50**, 129–131.

Watkins, J. E., Cardelús, C., Colwell, R. K., and Moran, R. C. (2006). Species richness and distribution of ferns along an elevational gradient in Costa Rica. *American Journal of Botany*, **93**, 73–83.

Wikström, N. and Kenrick, P. (1997). Phylogeny of Lycopodiaceae (Lycopsida) and the relationship of *Phylloglossum drummondii* Kunze based on *rbcL* sequence data. *International Journal of Plant Sciences*, **160**, 862–871.

Wikström, N. and Kenrick, P. (2000). Phylogeny of epiphytic *Huperzia* (Lycopodiaceae): paleotropical and neotropical clades corroborated by rbcL sequences. *Nordic Journal of Botany*, **20**, 165–171.

Wikström, N. and Kenrick, P. (2001). Evolution of Lycopodiaceae (Lycopsida): estimating divergence times from *rbcL* gene sequences by use of nonparametric rate smoothing. *Molecular Phylogenetics and Evolution*, **19**, 177–186.

Wikström, N., Kenrick, P., and Chase, M. (1999). Epiphytism and terrestrialization in tropical *Huperzia* (Lycopodiaceae). *Plant Systematics and Evolution*, **218**, 221–243.

Winkler, H. (1938). Geographie. In *Manual of Pteridology*, ed. F. Verdoorn. The Hague: Martinus Nijhoff, pp. 451–472.

Wolf, P. G., Schneider, H., and Ranker, T. A. (2001). Geographic distributions of homosporous ferns: does dispersal obscure evidence of vicariance? *Journal of Biogeography*, **28**, 263–270.

Young, K. and León, B. (1989). Pteridophyte species diversity in the Central Peruvian Amazon: importance of edaphic specialization. *Brittonia*, **41**, 388–395.

Yu, J. and Chengsen, L. (2007). *Yunnan Ferns of China (Supplement)*. Beijing: Chinese Scientific Book Service.

Zizka, G. (1991). Flowering plants of Easter Island. *Palmarum Hortus Francofortensis: Palmengaeten Wissenschaftliche Berichte*, **3**, 1–108.

15

Fern phylogeny

ERIC SCHUETTPELZ AND KATHLEEN M. PRYER

15.1 Introduction

As a consequence of employing DNA sequence data and phylogenetic approaches, unprecedented progress has been made in recent years toward a full understanding of the fern tree of life. At the broadest level, molecular phylogenetic analyses have helped to elucidate which of the so-called "fern allies" are indeed ferns, and which are only distantly related (Nickrent *et al.*, 2000; Pryer *et al.*, 2001a; Wikström and Pryer, 2005; Qiu *et al.*, 2006). Slightly more focused analyses have revealed the composition of, and relationships among, the major extant fern clades (Hasebe *et al.*, 1995; Wolf, 1997; Pryer *et al.*, 2004b; Schneider *et al.*, 2004c; Schuettpelz *et al.*, 2006; Schuettpelz and Pryer, 2007). A plethora of analyses, at an even finer scale, has uncovered some of the most detailed associations (numerous references cited below). Together, these studies have helped to answer many long-standing questions in fern systematics.

In this chapter, a brief synopsis of vascular plant relationships – as currently understood – is initially provided to place ferns within a broader phylogenetic framework. This is followed by an overview of fern phylogeny, with most attention devoted to the leptosporangiate clade that accounts for the bulk of extant fern diversity. Discussion of finer scale relationships is generally avoided; instead, the reader is directed to the relevant literature, where more detailed information can be found. The phylogeny presented (Figure 15.1) – which serves as a guide – integrates the results of two distinct phylogenetic analyses of fern relationships: one focused on vascular plants but emphasizing ferns (Pryer *et al.*, 2001a), the other focused specifically on overall leptosporangiate

Biology and Evolution of Ferns and Lycophytes, ed. Tom A. Ranker and Christopher H. Haufler. Published by Cambridge University Press. © Cambridge University Press 2008.

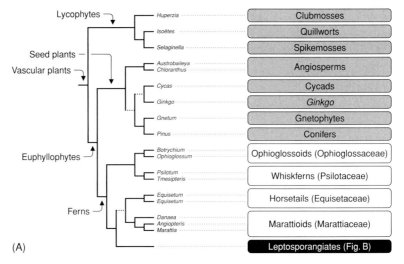

Figure 15.1 (A) Early vascular plant divergences. Phylogeny from maximum likelihood analysis of nuclear 18S and plastid *rbcL*, *atpB*, and *rps4* data (Pryer *et al.*, 2001a; leptosporangiate sampling collapsed here to a single terminal); all branches received maximum likelihood bootstrap support ≥70 except those drawn with dotted lines. Thirteen major vascular plant lineages (four of which are fern families; Smith *et al.*, 2006b) are indicated, as are other clades discussed in the text. (B–H) Leptosporangiate divergences. Phylogeny from maximum likelihood analysis of plastid *rbcL*, *atpB*, and *atpA* data (Schuettpelz and Pryer, 2007); all branches received maximum likelihood bootstrap support ≥70 except those drawn with dotted lines. Families recognized in the most recent fern classification (Smith *et al.*, 2006b; see also Smith *et al.*, Chapter 16) are indicated, as are other clades discussed in the text. (B) Early leptosporangiate divergences. (C) Early core leptosporangiate divergences. (D) Early polypod divergences. (E) Divergences within eupolypods I, part 1. (F) Divergences within eupolypods I, part 2. (G) Divergences within eupolypods II, part 1. (H) Divergences within eupolypods II, part 2.

fern relationships (Schuettpelz and Pryer, 2007). The most recent fern classification (Smith *et al.*, 2006b; Chapter 16) took the results of these studies into consideration; therefore, it is largely consistent with the phylogeny presented here.

15.2 Early vascular plant divergences

The deepest phylogenetic dichotomy among extant vascular plant lineages separates lycophytes from euphyllophytes (Figure 15.1A; Raubeson and Jansen, 1992; Kenrick and Crane, 1997; Doyle, 1998; Nickrent *et al.*, 2000; Renzaglia *et al.*, 2000; Pryer *et al.*, 2001a; 2004a; Wikström and Pryer, 2005;

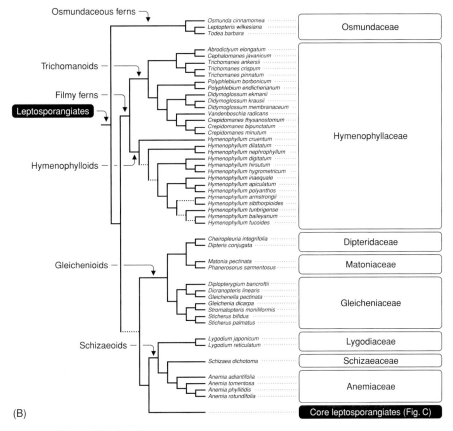

Figure 15.1 (*cont.*)

Qiu *et al.*, 2006). The lycophytes, which are characterized by lycophylls (leaves with an intercalary meristem) and account for less than 1% of vascular plant diversity, comprise three distinct lineages – the homosporous clubmosses, the heterosporous quillworts, and the heterosporous spikemosses. Each of these has been the focus of several phylogenetic studies (Wikström and Kenrick, 1997; Wikström, 2001; Korall and Kenrick, 2002, 2004; Rydin and Wikström, 2002; Hoot *et al.*, 2006).

Within the euphyllophytes, characterized by euphylls (leaves with marginal or apical meristems and an associated leaf gap in the vascular stele), a deep split subsequently separates seed plants from ferns (Figure 15.1A; Kenrick and Crane, 1997; Doyle, 1998; Nickrent *et al.*, 2000; Renzaglia *et al.*, 2000; Pryer *et al.*, 2001a, 2004a, 2004b; Wikström and Pryer, 2005; Qiu *et al.*, 2006; Schuettpelz *et al.*, 2006). Seed plants, united by the presence of seeds and wood, account for some 96% of extant vascular plant diversity. Accordingly, these plants have received considerable phylogenetic attention (for an overview of relationships, see Burleigh

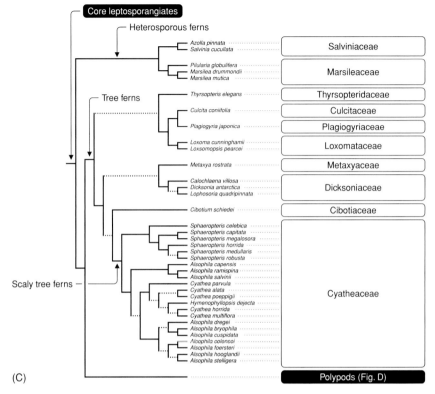

Figure 15.1 (cont.)

and Mathews, 2004; Soltis and Soltis, 2004). Ferns, while accounting for little more than 3% of vascular plant species, display morphological and ecological disparity rivaling that of their equally ancient sister group. A clear morphological synapomorphy for this clade is lacking. However, ferns are routinely resolved as monophyletic in analyses of morphology (e.g., Renzaglia *et al.*, 2000; Rothwell and Nixon, 2006), and are consistently well supported as a natural group in analyses of DNA sequence data (e.g., Nickrent *et al.*, 2000; Renzaglia *et al.*, 2000; Pryer *et al.*, 2001a, 2004b; Wikström and Pryer, 2005; Qiu *et al.*, 2006; Rothwell and Nixon, 2006; Schuettpelz *et al.*, 2006).

Ferns, like lycophytes, are spore bearing and "seed-free." Because of this, members of these two lineages were traditionally lumped as "pteridophytes" or "ferns and fern allies." Although these terms served the botanical community well when there was little resolution near the base of the vascular plant phylogeny, robustly supported hypotheses resolving these deepest divergences are now available. Thus, "ferns" (in a somewhat more inclusive than traditional sense; see Figure 15.1A) and "lycophytes," which specify clade membership, are

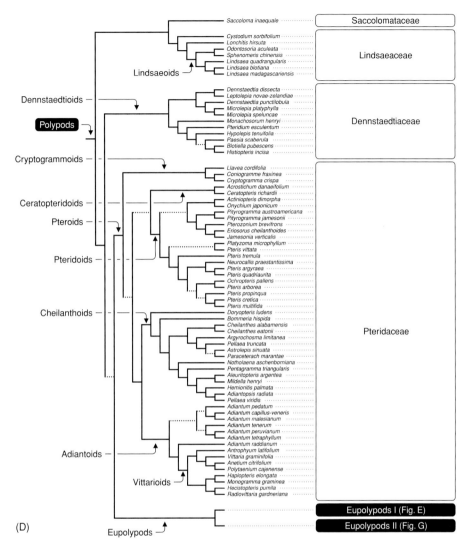

Figure 15.1 (cont.)

preferable to the terms "pteridophytes" and "ferns and fern allies" that unite paraphyletic assemblages of plants.

15.3 Early fern divergences

Within ferns, the first divergence separates a clade including whisk-ferns and ophioglossoids from a clade including horsetails, marattioids, and leptosporangiates (Figure 15.1A; Nickrent et al., 2000; Pryer et al., 2001a, 2004a, 2004b; Wikström and Pryer, 2005; Qiu et al., 2006; Rothwell and Nixon, 2006;

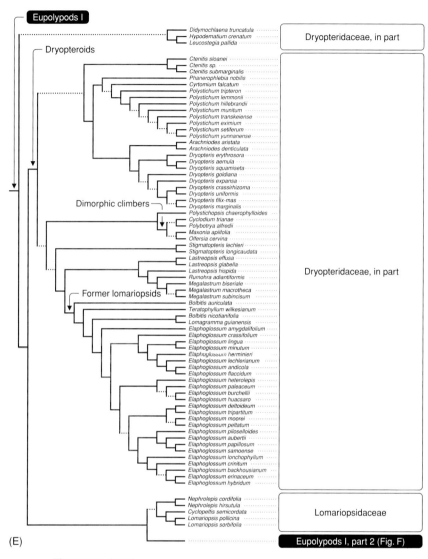

Figure 15.1 (cont.)

Schuettpelz et al., 2006). Whiskferns (Psilotaceae) and ophioglossoids (Ophioglossaceae) are both relatively small lineages, and both exhibit considerable morphological simplification. Some whiskferns, in fact, have such strikingly simplified body plans that they were long thought to be direct descendants of some of the first vascular plants appearing in the fossil record (Parenti, 1980; Bremer, 1985). This has made unique shared characteristics especially difficult to identify, but reduction of the root system may actually constitute such a trait – ophioglossoids have simple unbranched roots that lack root hairs, and

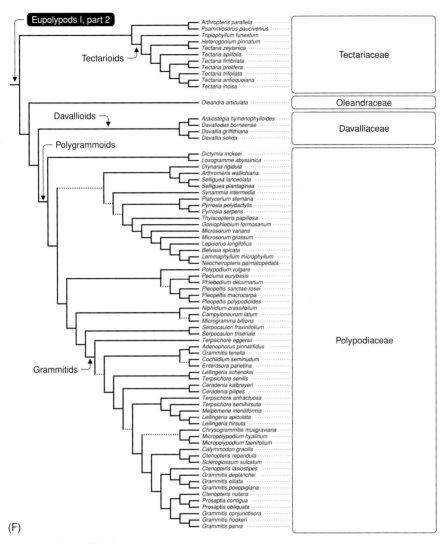

Figure 15.1 (cont.)

whiskferns lack roots altogether (Schneider et al., 2002a). Phylogenetic studies within this clade have focused primarily on the ophioglossoids (Hauk et al., 2003).

Although well supported together as a clade, the relationships among the horsetail, marattioid, and leptosporangiate lineages remain somewhat elusive (Nickrent et al., 2000; Pryer et al., 2001a, 2004b; Wikström and Pryer, 2005; Qiu et al., 2006; Rothwell and Nixon, 2006; Schuettpelz et al., 2006; see Chapter 13). Their divergence from one another in rather rapid succession deep in time – all arose in the Paleozoic – could be responsible. Horsetails today account for

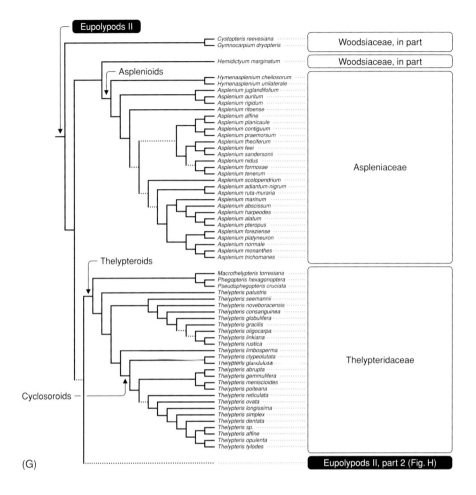

Figure 15.1 (cont.)

just a handful of species (in a single family, Equisetaceae), apparently resulting from a Cenozoic diversification (Des Marais et al., 2003). The marattioids (Marattiaceae) are somewhat more species rich, but their diversity pales in comparison to the leptosporangiates, which account for well over 95% of extant fern diversity. Leptosporangiates compose a well-supported monophyletic group (Pryer et al., 2001a, 2004b; Schneider et al., 2004c; Wikström and Pryer, 2005; Qiu et al., 2006; Schuettpelz et al., 2006; Schuettpelz and Pryer, 2007) characterized by sporangia that develop from a single cell, have mature walls just one cell thick, and possess a distinctive annulus that serves to eject the spores. Specific features of the sporangia – including the shape and position of the annulus – have figured prominently in the classification of these ferns, and are generally consistent with their phylogeny (see below).

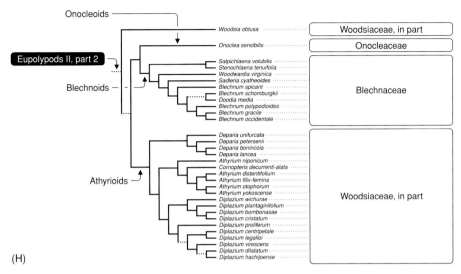

Figure 15.1 (cont.)

15.4 Early leptosporangiate divergences

The osmundaceous ferns are well supported as sister to all other leptosporangiates (Figure 15.1B; Pryer et al., 2001a, 2004b; Schneider et al., 2004c; Wikström and Pryer, 2005; Schuettpelz et al., 2006; Schuettpelz and Pryer, 2007), a position consistent with the fossil record because the oldest leptosporangiate fossils assignable to an extant lineage are members of this clade (Miller, 1971; Tidwell and Ash, 1994; Phipps et al., 1998; Galtier et al., 2001; Rößler and Galtier, 2002). These ferns are placed in a single family, Osmundaceae (Figure 15.1B; Smith et al., 2006b; Chapter 16), and are generally divided into three genera. But although recent studies have confirmed the monophyly of *Leptopteris* and *Todea*, they have found strong support for the paraphyly of *Osmunda*; *Osmunda cinnamomea* is resolved sister to the remaining osmundaceous ferns, advocating the recognition of a fourth genus (*Osmundastrum*; Yatabe et al., 1999; Metzgar et al., in press).

The filmy ferns, composing a single large family (Hymenophyllaceae) and the gleichenioid ferns, with three smaller families (Dipteridaceae, Matoniaceae, and Gleicheniaceae) are both now understood to be monophyletic (Figure 15.1B; Pryer et al., 2004b; Schuettpelz et al., 2006; Schuettpelz and Pryer, 2007). However, the relationships of these two lineages to one another and to the remaining leptosporangiate ferns remain unclear. Within filmy ferns, two clades of roughly equal size are resolved (Figure 15.1B; Pryer et al., 2001b; Schuettpelz and Pryer, 2006) – the hymenophylloid clade with a single genus (*Hymenophyllum*) and the trichomanoid clade with eight genera (*Abrodictyum*, *Callistopteris*, *Cephalomanes*, *Crepidomanes*, *Didymoglossum*, *Polyphlebium*, *Trichomanes*, and *Vandenboschia*; Ebihara et al., 2006). Each of these two large filmy fern clades has been the subject of several focused phylogenetic studies (Ebihara et al., 2002, 2004; Dubuisson et al., 2003; Hennequin et al., 2003, 2006a, 2006b); however, because most of these analyses relied on a single gene, relationships were sometimes unsupported. A three-gene analysis (Schuettpelz and Pryer, 2007) still does not find strong support within the epiphytic genus *Hymenophyllum*, but does find good support for relationships among the trichomanoid genera (Figure 15.1B). Two large trichomanoid subclades emerge, one of which is mostly terrestrial (*Abrodictyum*, *Cephalomanes*, and *Trichomanes*), the other of which is mostly epiphytic (*Crepidomanes*, *Didymoglossum*, *Polyphlebium*, and *Vandenboschia*).

Strong support for the monophyly of the gleichenioid ferns (Figure 15.1B) has only recently been obtained (Schuettpelz et al., 2006; Schuettpelz and Pryer, 2007), although earlier morphological (Jarrett, 1980) and molecular (Hasebe et al., 1995; Pryer et al., 2004b) data suggested such a clade. Within gleichenioids, Dipteridaceae is sister to Matoniaceae and these together are sister to the Gleicheniaceae (Figure 15.1B; Schuettpelz and Pryer, 2007).

The schizaeoid ferns are well supported as sister to the so-called "core leptosporangiates" – a large clade composed of heterosporous, tree, and polypod ferns (Figures 15.1B, C; Pryer et al., 2004b; Schuettpelz and Pryer, 2007). The schizaeoids comprise three morphologically and molecularly distinct clades (Figure 15.1B; Skog et al., 2002; Wikström et al., 2002), recognized as distinct families in the most recent fern classification (Lygodiaceae, Schizaeaceae, and Anemiaceae; Smith et al., 2006b; Chapter 16).

The heterosporous, or water, ferns are also monophyletic and comprise two major clades (Figure 15.1C; Hasebe et al., 1995; Pryer et al., 2004b), treated as families (Smith et al., 2006b). The Salviniaceae consists of two free-floating genera: *Azolla* and *Salvinia*. The Marsileaceae consists of three genera, all of which are rooted in the soil: *Marsilea*, *Pilularia*, and *Regnellidium*. These ferns have been the focus of several recent and ongoing phylogenetic studies that have addressed their relationships in greater detail (Pryer, 1999; Reid et al., 2006; Metzgar et al., 2007; Nagalingum et al., 2007).

The tree ferns are well supported as monophyletic in molecular analyses (Pryer et al., 2001a; 2004b; Wikström and Pryer, 2005; Korall et al., 2006b; Schuettpelz et al., 2006; Schuettpelz and Pryer, 2007), but lack an obvious morphological synapomorphy. Many species do indeed have trunk-like stems, but this character is not ubiquitous throughout the clade. The phylogeny of these ferns was specifically examined by Korall et al. (2006b), and the phylogenetic branching pattern presented here (Figure 15.1C; Schuettpelz and Pryer, 2007) is in agreement with these results. The Culcitaceae, Loxomataceae, Plagiogyriaceae, and Thyrsopteridaceae together form a clade, as do the Cibotiaceae, Cyatheaceae, Dicksoniaceae, and Metaxyaceae. Within the large scaly tree fern clade (Figure 15.1C; note that this clade is equivalent to Cyatheaceae), four primary subclades emerge: *Sphaeropteris*, *Cyathea* (with *Hymenophyllopsis* embedded within it), and two distinct *Alsophila* clades (Korall et al., 2007; but see Conant et al., 1995, 1996).

Although not always thought to form a natural group, the polypod ferns (Figures 15.1D–H), have received solid support in all recent analyses (Pryer et al., 2001a, 2004b; Schneider et al., 2004c; Wikström and Pryer, 2005; Schuettpelz et al., 2006; Schuettpelz and Pryer, 2007). This clade is united by an unequivocal morphological synapomophy – sporangia each with a vertical annulus interrupted by the stalk.

15.5 Early polypod divergences

The much smaller of the two clades arising from the first divergence within the polypods contains the lindsaeoid ferns and a few rather enigmatic fern genera (Figure 15.1D). Two of the smaller genera (*Lonchitis* and *Saccoloma*) were traditionally placed in the Dennstaedtiaceae; the other (*Cystodium*) was traditionally placed in the Dicksoniaceae, a tree fern family. In the most recent and comprehensive analysis (Schuettpelz and Pryer, 2007), these three genera and the lindsaeoids together form a well-supported clade, but one that was not recovered, in its entirety, in earlier analyses (Hasebe et al., 1995; Pryer et al., 2004b; Schneider et al., 2004c; Korall et al., 2006a; Schuettpelz et al., 2006). In the most recent classification (Smith et al., 2006b; Chapter 16), this clade is divided into two families – Saccolomataceae and Lindsaeaceae. The former comprises only the genus *Saccoloma*; the latter includes eight genera (note that *Ormoloma*, *Tapeinidium*, and *Xyropteris* do not appear in Figure 15.1D).

The remaining polypods compose three well-supported clades: the small dennstaedtioid clade, the large pteroid clade, and the hyperdiverse eupolypod fern clade (Figure 15.1D; Schneider et al., 2004c; Schuettpelz and Pryer, 2007). But again, the relationships among these three lineages remain unclear. Within the dennstaedtioids, two approximately equally diverse subclades are resolved; the genus *Dennstaedtia* itself is strongly supported as paraphyletic (Wolf et al., 1994; Wolf, 1995; Schuettpelz and Pryer, 2007).

The pteroid ferns account for roughly 10% of extant fern diversity, placed in a single family (Pteridaceae; Smith et al., 2006b; Chapter 16). Five primary clades are resolved in molecular phylogenetic analyses (Prado et al., 2007; Schuettpelz et al., 2007; Schuettpelz and Pryer, 2007): cryptogrammoids, ceratopteridoids, pteridoids, cheilanthoids, and adiantoids (with the vittarioid ferns apparently embedded within the genus *Adiantum*; Figure 15.1D). Some finer-scale relationships within most of these groups were addressed in earlier studies (Crane et al., 1995; Gastony and Rollo, 1995, 1998; Nakazato and Gastony, 2003; Sánchez-Baracaldo, 2004; Zhang et al., 2005).

Within the eupolypod ferns, two large clades are resolved, dubbed "eupolypods I" and "eupolypods II" (Figure 15.1D; Schneider et al., 2004c; Schuettpelz and Pryer, 2007). This split is well supported by molecular data, but also by a frequently overlooked morphological character, namely the vasculature of the petiole. Eupolypods I have three or more vascular bundles (with the exception of the diminutive grammitid ferns with one, and the genus *Hypodematium* with two); whereas eupolypods II have only two (with the exception of the well-nested blechnoid ferns with three or more).

15.6 Divergences within eupolypods I

Three genera not traditionally thought to be closely related to one another (*Didymochlaena*, *Hypodematium*, and *Leucostegia*) form a small, but poorly supported clade sister to the rest of eupolypods I (Figure 15.1E). In earlier classifications (e.g., Kramer, 1990a; Kramer et al., 1990), *Didymochlaena* was considered to be associated with the dryopteroid ferns (Figure 15.1E), *Hypodematium* with the athyrioid ferns (Figure 15.1H), and *Leucostegia* among the davallioid ferns (Figure 15.1F). Previous studies found these genera to be rather isolated (Hasebe et al., 1995; Schneider et al., 2004c; Tsutsumi and Kato, 2006), but all three were only recently included in the same analysis (Schuettpelz and Pryer, 2007). The finding of good support for the monophyly of the remaining eupolypods I – excluding *Didymochlaena*, *Hypodematium*, and *Leucostegia* – is the first convincing evidence that these three genera should indeed be segregated from the Dryopteridaceae (Smith et al., 2006b), as they render it paraphyletic (Figure 15.1E).

The dryopteroid ferns form a very large and well-supported clade, with most "former lomariopsid" genera nested within it (Figure 15.1E; Schuettpelz and Pryer, 2007). Notably absent from the dryopteroid clade, however, is the genus *Lomariopsis* itself, which is resolved elsewhere in the eupolypods I (Figure 15.1E). This suggests that the distinctive rhizome anatomy (elongated ventral meristele) characteristic of lomariopsid ferns has apparently evolved at least twice (once in the Dryopteridaceae and once in the Lomariopsidaceae). Within dryopteroids, the well-studied genera *Dryopteris* (Geiger and Ranker, 2005; Li and Lu, 2006) and *Polystichum* (Little and Barrington, 2003; Li et al., 2004) compose a large, well-supported clade together with *Phanerophlebia*, *Cyrtomium*, and *Arachniodes* (Figure 15.1E; Schuettpelz and Pryer, 2007). The genus *Polystichopsis* – which is often synonymized under *Arachniodes* (e.g., Kramer et al., 1990) – is, however, not closely related. Rather, it is sister to a clade of "dimorphic climbers" – dryopteroid genera with creeping (to climbing) stems and dimorphic leaves (Figure 15.1E). *Stigmatopteris*, as well as *Ctenitis*, are both rather isolated, but *Megalastrum* (a relatively recent segregate of *Ctenitis*; Holttum, 1986) forms a clade with *Rumohra* and the paraphyletic genus *Lastreopsis*. This clade is in turn sister to the "former lomariopsid" genera (Figure 15.1E), within which *Bolbitis* is resolved as polyphyletic. *Elaphoglossum*, the largest of the "former lomariopsid" genera is clearly monophyletic (Figure 15.1E), and has received considerable phylogenetic attention (Rouhan *et al.*, 2004; Skog *et al.*, 2004).

Nephrolepis, *Cyclopeltis*, and *Lomariopsis* also form a clade within the eupolypods I, (Figure 15.1E; Tsutsumi and Kato, 2006; Schuettpelz and Pryer, 2007). Although this assemblage is generally not well supported in analyses of molecular data, its monophyly is reinforced by a morphological synapomorphy – specifically the presence of articulate pinnae (Smith *et al.*, 2006b). Oleandroid ferns, on the other hand, which were thought to compose a natural group based on morphology (Kramer, 1990b), are resolved in molecular analyses as definitively not monophyletic (Schuettpelz and Pryer, 2007). Two oleandroid genera (*Arthropteris* and *Psammiosorus*) are sister to the tectarioid ferns (Figure 15.1F); they are now included in the Tectariaceae (Smith *et al.*, 2006b). *Oleandra* itself is sister to a large clade of davallioid and polygrammoid ferns (Figure 15.1F), and is now considered to be the sole genus in Oleandraceae (Smith *et al.*, 2006b).

The phylogeny of davallioid and polygrammoid ferns has been extremely well studied in recent years (Schneider *et al.*, 2002b, 2004a, 2004d, 2006a, 2006b; Haufler *et al.*, 2003, Ranker *et al.*, 2003, 2004; Janssen and Schneider, 2005; Tsutsumi and Kato, 2005, 2006; Kreier and Schneider, 2006a, 2006b), and the relationships presented (Figure 15.1F) are generally consistent with those resolved in earlier studies. As previously determined, the grammitid ferns (Grammitidaceae *sensu* Parris, 1990) are nested firmly within the Polypodiaceae *sensu* Hennipman *et al.* (1990). There is now strong support for the newly described genus *Serpocaulon* (Smith *et al.*, 2006a) as sister to the grammitid clade (Figure 15.1F; Schuettpelz and Pryer, 2007).

15.7 Divergences within eupolypods II

Eupolypods II consists of several large well-supported clades with a number of small genera interspersed among them. Among the smaller genera, *Cystopteris* and *Gymnocarpium* are notable in being sister to the rest of eupolypods II; *Hemidictyum* is sister to the large asplenioid clade; and *Woodsia* is sister to the onocleoid, blechnoid, and athyrioid ferns (Figure 15.1G, H; Schuettpelz and

Pryer, 2007). All four of these genera were tentatively placed in the Woodsiaceae (Smith et al., 2006b); however, it seems that this circumscription is paraphyletic, relative to the other families in the eupolypods II, and in need of further study.

The asplenioid ferns (Figure 15.1G) are among the larger eupolypods II clades that have consistently received strong support in molecular analyses (Gastony and Johnson, 2001; Schuettpelz and Pryer, 2007). Earlier studies have clearly demonstrated that nearly all genera previously segregated from *Asplenium* (e.g., *Camptosorus*, *Diellia*, and *Loxoscaphe*) nest well within this large genus (Murakami and Schaal, 1994; Murakami et al., 1999; Pinter et al., 2002; Schneider et al., 2004b, 2005; Perrie and Brownsey, 2005). Thus, in the most recent classification (Smith et al., 2006b; Chapter 16) only two genera were recognized in the Aspleniaceae, *Hymenasplenium* being sister to *Asplenium* (Figure 15.1G).

The thelypteroid ferns are also strongly supported as monophyletic (Figure 15.1G; Smith and Cranfill, 2002; Schuettpelz and Pryer, 2007) and are recognized as a large family (Thelypteridaceae) with five genera (Smith et al., 2006b). The three smaller genera (*Macrothelypteris*, *Phegopteris*, and *Pseudophegopteris*) form a clade sister to the two larger genera (*Cyclosorus* and *Thelypteris*; Figure 15.1G; Smith and Cranfill, 2002; Schuettpelz and Pryer, 2007). *Thelypteris* (*sensu* Smith, 1990) is, however, definitively paraphyletic to the cyclosoroids (Figure 15.1G; note that all species potentially assignable to *Cyclosorus* are presented in Figure 15.1G under *Thelypteris* to circumvent a variety of nomenclatural issues). The thelypteroid clade is in need of considerable phylogenetic study.

The onocleoid ferns, including *Onoclea* (and three other small genera not appearing in Figure 15.1H; Gastony and Ungerer, 1997; Smith et al., 2006b), are sister to a larger blechnoid clade that was the subject of three recent studies (Nakahira, 2000; Cranfill, 2001; Cranfill and Kato, 2003). The results presented here (from Schuettpelz and Pryer, 2007) are in general accord with their more densely sampled analyses: *Blechnum* is definitely not monophyletic (Figure 15.1H) and blechnoid taxonomy thus requires further attention.

The athyrioid ferns, which account for most of the diversity in the paraphyletic Woodsiaceae, are themselves monophyletic (Figure 15.1H; Schuettpelz and Pryer, 2007). However, the largest genera in this clade, *Athyrium* and *Diplazium*, are likely not (see Sano et al., 2000; Wang et al., 2003; not entirely evident in Figure 15.1G).

15.8 Future prospects

DNA sequence data and phylogenetic approaches have revolutionized fern systematics, providing an unparalleled framework within which to explore evolutionary patterns. However, many questions remain at every scale, and the inclusion of many more taxa and additional data will be required to ultimately bring the fern tree of life into focus. Only with a continued effort to improve our understanding of phylogeny will a full appreciation of fern, as well as lycophyte, evolution and diversification be possible.

References

Bremer, K. (1985). Summary of green plant phylogeny and classification. *Cladistics*, **1**, 369–385.

Burleigh, J. G. and Mathews, S. (2004). Phylogenetic signal in nucleotide data from seed plants: implications for resolving the seed plant tree of life. *American Journal of Botany*, **91**, 1599–1613.

Conant, D. S., Raubeson, L. A., Attwood, D. K., and Stein, D. B. (1995). The relationships of Papuasian Cyatheaceae to New World tree ferns. *American Fern Journal*, **85**, 328–340.

Conant, D. S., Raubeson, L. A., Attwood, D. K., Perera, S., Zimmer, E. A., Sweere, J. A., and Stein, D. B. (1996). Phylogenetic and evolutionary implications of combined analysis of DNA and morphology in the Cyatheaceae. In *Pteridology in Perspective*, ed. J. M. Camus, M. Gibby, and R. J. Johns. Kew: Royal Botanic Gardens, pp. 231–248.

Crane, E. H., Farrar, D. R., and Wendel, J. F. (1995). Phylogeny of the Vittariaceae: convergent simplification leads to a polyphyletic *Vittaria*. *American Fern Journal*, **85**, 283–305.

Cranfill, R. B. (2001). Phylogenetic Studies in the Polypodiales (Pteridophyta) with an Emphasis on the Family Blechnaceae. Unpublished Ph.D. Thesis, University of California, Berkeley, CA.

Cranfill, R. and Kato, M. (2003). Phylogenetics, biogeography, and classification of the woodwardioid ferns (Blechnaceae). In *Pteridology in the New Millenium*, ed. S. Chandra and M. Srivastava. Dordrecht: Kluwer, pp. 25–48.

Des Marais, D. L., Smith, A. R., Britton, D. M., and Pryer, K. M. (2003). Phylogenetic relationships and evolution of extant horsetails (*Equisetum*) based on chloroplast DNA sequence data *rbcL* and *trnL-F*. *International Journal of Plant Sciences*, **164**, 737–751.

Doyle, J. A. (1998). Phylogeny of vascular plants. *Annual Review of Ecology and Systematics*, **29**, 567–599.

Dubuisson, J.-Y. Hennequin, S., Douzery, E. J. P., Cranfill, R. B., Smith, A. R., and Pryer, K. M. (2003). *rbcL* phylogeny of the fern genus *Trichomanes* (Hymenophyllaceae), with special reference to neotropical taxa. *International Journal of Plant Sciences*, **164**, 753–761.

Ebihara, A., Iwatsuki, K., Kurita, S., and Ito, M. (2002). Systematic position of *Hymenophyllum rolandi-principis* Rosenst. or a monotypic genus *Rosenstockia* Copel. (Hymenophyllaceae) endemic to New Caledonia. *Acta Phytotaxonomica et Geobotanica*, **53**, 35–49.

Ebihara, A., Hennequin, S., Iwatsuki, K., Bostock, P. D., Matsumoto, S., Jaman, R., Dubuisson, J.-Y., and Ito, M. (2004). Polyphyletic origin of *Microtrichomanes* (Prantl) Copel. (Hymenophyllaceae), with a revision of the species. *Taxon*, **53**, 935–948.

Ebihara, A., Dubuisson, J.-Y., Iwatsuki, K., Hennequin, S., and Ito, M. (2006). A taxonomic revision of Hymenophyllaceae. *Blumea*, **51**, 221–280.

Galtier, J., Wang, S. J., Li, C. S., and Hilton, J. (2001). A new genus of filicalean fern from the Lower Permian of China. *Botanical Journal of the Linnean Society*, **137**, 429–442.

Gastony, G. J. and Johnson, W. P. (2001). Phylogenetic placements of *Loxoscaphe thecifera* (Aspleniaceae) and *Actiniopteris radiata* (Pteridaceae) based on analysis of *rbcL* nucleotide sequences. *American Fern Journal*, **91**, 197–213.

Gastony, G. J. and Rollo, D. R. (1995). Phylogeny and generic circumscriptions of cheilanthoid ferns (Pteridaceae: Cheilanthoideae) inferred from *rbcL* nucleotide sequences. *American Fern Journal*, **85**, 341–360.

Gastony, G. J. and Rollo, D. R. (1998). Cheilanthoid ferns (Pteridaceae: Cheilanthoideae) in the southwestern United States and adjacent Mexico – a molecular phylogenetic reassessment of generic lines. *Aliso*, **17**, 131–144.

Gastony, G. J. and Ungerer, M. C. (1997). Molecular systematics and a revised taxonomy of the onocleoid ferns (Dryopteridaceae: Onocleeae). *American Journal of Botany*, **84**, 840–849.

Geiger, J. M. O. and Ranker, T. A. (2005). Molecular phylogenetics and historical biogeography of Hawaiian *Dryopteris* (Dryopteridaceae). *Molecular Phylogenetics and Evolution*, **34**, 392–407.

Hasebe, M., Wolf, P. G., Pryer, K. M., Ueda, K., Ito, M., Sano, R., Gastony, G. J., Yokoyama, J., Manhart, J. R., Murakami, N., Crane, E. H., Haufler, C. H., and Hauk, W. D. (1995). Fern phylogeny based on *rbcL* nucleotide sequences. *American Fern Journal*, **85**, 134–181.

Haufler, C. H., Grammer, W. A., Hennipman, E., Ranker, T. A., Smith, A. R., and Schneider, H. (2003). Systematics of the ant-fern genus *Lecanopteris* (Polypodiaceae): testing phylogenetic hypotheses with DNA sequences. *Systematic Botany*, **28**, 217–227.

Hauk, W. D., Parks, C. R., and Chase, M. W. (2003). Phylogenetic studies of Ophioglossaceae: evidence from *rbcL* and *trnL-F* plastid DNA sequences and morphology. *Molecular Phylogenetics and Evolution*, **28**, 131–151.

Hennequin, S., Ebihara, A., Ito, M., Iwatsuki, K., and Dubuisson, J.-Y. (2003). Molecular systematics of the fern genus *Hymenophyllum* s.l. (Hymenophyllaceae) based on chloroplastic coding and noncoding regions. *Molecular Phylogenetics and Evolution*, **27**, 283–301.

Hennequin, S., Ebihara, A., Ito, M., Iwatsuki, K., and Dubuisson, J.-Y. (2006a). New insights into the phylogeny of the genus *Hymenophyllum* s.l. (Hymenophyllaceae): revealing the polyphyly of *Mecodium*. *Systematic Botany*, **31**, 271–284.

Hennequin, S., Ebihara, A., Ito, M., Iwatsuki, K., and Dubuisson, J.-Y. (2006b). Phylogenetic systematics and evolution of the genus *Hymenophyllum* (Hymenophyllaceae: Pteridophyta). *Fern Gazette*, **17**, 247–257.

Hennipman, E., Veldhoen, P., and Kramer, K. U. (1990). Polypodiaceae. In *The Families and Genera of Vascular Plants*, Vol. 1, *Pteridophytes and Gymnosperms*, ed. K. U. Kramer and P. S. Green. Berlin: Springer-Verlag, pp. 203–230.

Holttum, R. E. (1986). Studies in the fern-genera allied to *Tectaria* Cav. VI. A conspectus of genera in the Old World regarded as related to *Tectaria*, with descriptions of two genera. *Gardens' Bulletin (Singapore)*, **39**, 153–167.

Hoot, S. B., Taylor, W. C., and Napier, N. S. (2006). Phylogeny and biogeography of *Isoetes* (Isoetaceae) based on nuclear and chloroplast DNA sequence data. *Systematic Botany*, **31**, 449–460.

Janssen, T. and Schneider, H. (2005). Exploring the evolution of humus collecting leaves in drynarioid ferns (Polypodiaceae, Polypodiidae) based on phylogenetic evidence. *Plant Systematics and Evolution*, **252**, 175–197.

Jarrett, F. M. (1980). Studies in the classification of the leptosporangiate ferns: I. The affinities of the Polypodiaceae sensu stricto and the Grammitidaceae. *Kew Bulletin*, **34**, 825–833.

Kenrick, P. and Crane, P. R. (1997). *The Origin and Early Diversification of Land Plants: A Cladistic Study*. Washington, DC: Smithsonian Press.

Korall, P. and Kenrick, P. (2002). Phylogenetic relationships in Selaginellaceae based on *rbcL* sequences. *American Journal of Botany*, **89**, 506–517.

Korall, P. and Kenrick, P. (2004). The phylogenetic history of Selaginellaceae based on DNA sequences from the plastid and nucleus: extreme substitution rates and rate heterogeneity. *Molecular Phylogenetics and Evolution*, **31**, 852–864.

Korall, P., Conant, D. S., Schneider, H., Ueda, K., Nishida, H., and Pryer, K. M. (2006a). On the phylogenetic position of *Cystodium*: it's not a tree fern – it's a polypod! *American Fern Journal*, **96**, 45–53.

Korall, P., Pryer, K. M., Metzgar, J. S., Schneider, H., and Conant, D. S. (2006b). Tree ferns: monophyletic groups and their relationships as revealed by four protein-coding plastid loci. *Molecular Phylogenetics and Evolution*, **39**, 830–845.

Korall, P., Conant, D. S., Metzgar, J. S., Schneider, H., and Pryer, K. M. (2007). A molecular phylogeny of scaly tree ferns (Cyatheaceae). *American Journal of Botany*, **94**, 873–886.

Kramer, K. U. (1990a). Davalliaceae. In *The Families and Genera of Vascular Plants*, Vol. 1, *Pteridophytes and Gymnosperms*, ed. K. U. Kramer and P. S. Green. Berlin: Springer-Verlag, pp. 74–80.

Kramer, K. U. (1990b). Oleandraceae. In *The Families and Genera of Vascular Plants*, Vol. 1, *Pteridophytes and Gymnosperms*, ed. K. U. Kramer and P. S. Green. Berlin: Springer-Verlag, pp. 190–193.

Kramer, K. U., Holttum, R. E., Moran, R. C., and Smith, A. R. (1990). Dryopteridaceae. In *The Families and Genera of Vascular Plants*, Vol. 1, *Pteridophytes and Gymnosperms*, ed. K. U. Kramer and P. S. Green. Berlin: Springer-Verlag, pp. 101–144.

Kreier, H.-P. and Schneider, H. (2006a). Phylogeny and biogeography of the staghorn fern genus *Platycerium* (Polypodiaceae, Polypodiidae). *American Journal of Botany*, **93**, 217–225.

Kreier, H.-P. and Schneider, H. (2006b). Reinstatement of *Loxogramme dictyopteris* for a New Zealand endemic fern known as *Anarthropteris lanceolata* based on phylogenetic evidence. *Australian Systematic Botany*, **19**, 309–314.

Li, C.-X. and Lu, S.-G. (2006). Phylogenetics of Chinese Dryopteris (Dryopteridaceae) based on the chloroplast rps4-trnS sequence data. *Journal of Plant Research*, **119**, 589–598.

Li, C.-X., Lu, S.-G., and Yang, Q. (2004). Asian origin for *Polystichum* (Dryopteridaceae) based on *rbcL* sequences. *Chinese Science Bulletin*, **49**, 1146–1150.

Little, D. P. and Barrington, D. S. (2003). Major evolutionary events in the origin and diversification of the fern genus *Polystichum* (Dryopteridaceae). *American Journal of Botany*, **90**, 508–514.

Metzgar, J. S., Schneider, H., and Pryer, K. M. (2007). Phylogeny and divergence time estimates for the fern genus *Azolla* (Salviniaceae). *International Journal of Plant Sciences*, **168**, 1045–1053.

Metzgar, J. S., Skog, J. E., Zimmer, E. A., and Pryer, K. M. (in press). The paraphyly of *Osmunda* is confirmed by phylogenetic analyses of seven plastid loci. *Systematic Botany*.

Miller, C. N. (1971). Evolution of the fern family Osmundaceae based on anatomical studies. *Contributions from the Museum of Paleontology, University of Michigan*, **28**, 105–169.

Murakami, N. and Schaal, B. A. (1994). Chloroplast DNA variation and the phylogeny of *Asplenium* sect. *Hymenasplenium* (Aspleniaceae) in the New World tropics. *Journal of Plant Research*, **107**, 245–251.

Murakami, N., Nogami, S., Watanabe, M., and Iwatsuki, K. (1999). Phylogeny of Aspleniaceae inferred from *rbcL* nucleotide sequences. *American Fern Journal*, **89**, 232–243.

Nagalingum, N. S., Schneider, H., and Pryer, K. M. (2007). Molecular phylogenetic relationships and morphological evolution in the heterosporous fern genus *Marsilea*. *Systematic Botany*, **32**, 16–25.

Nakahira, Y. (2000). A Molecular Phylogenetic Analysis of the Family Blechnaceae, Using the Chloroplast Gene rbcL. Unpublished M.S. Thesis, Graduate School of Science, University of Tokyo, Tokyo.

Nakazato, T. and Gastony, G. J. (2003). Molecular phylogenetics of *Anogramma* species and related genera (Pteridaceae: Taenitidoideae). *Systematic Botany*, **28**, 490–502.

Nickrent, D. L., Parkinson, C. L., Palmer, J. D., and Duff, R. J. (2000). Multigene phylogeny of land plants with special reference to bryophytes and the earliest land plants. *Molecular Biology and Evolution*, **17**, 1885–1895.

Parenti, L. R. (1980). A phylogenetic analysis of the land plants. *Biological Journal of the Linnean Society*, **13**, 225–242.

Parris, B. S. (1990). Grammitidaceae. In *The Families and Genera of Vascular Plants*, Vol. 1, *Pteridophytes and Gymnosperms*, ed. K. U. Kramer and P. S. Green. Berlin: Springer-Verlag, pp. 153–156.

Perrie, L. R. and Brownsey, P. J. (2005). Insights into the biogeography and polyploid evolution of New Zealand *Asplenium* from chloroplast DNA sequence data. *American Fern Journal*, **95**, 1–21.

Phipps, C. J., Taylor, T. N., Taylor, E. L., Cuneo, N. R., Boucher, L. D., and Yao, X. (1998). *Osmunda* (Osmundaceae) from the Triassic of Antarctica: an example of evolutionary stasis. *American Journal of Botany*, **85**, 888–895.

Pinter, I., Bakker, F., Barrett, J., Cox, C., Gibby, M., Henderson, S., Morgan-Richards, M., Rumsey, F., Russell, S., Trewick, S., Schneider, H., and Vogel, J. (2002). Phylogenetic and biosystematic relationships in four highly disjunct polyploid complexes in the subgenera *Ceterach* and *Phyllitis* in *Asplenium* (Aspleniaceae). *Organisms, Diversity, and Evolution*, **2**, 299–311.

Prado, J., Del Nero Rodrigues, C., Salatino, A., and Salatino, M. L. F. (2007). Phylogenetic relationships among Pteridaceae, including Brazilian species, inferred from *rbcL* sequences. *Taxon*, **56**, 355–368.

Pryer, K. M. (1999). Phylogeny of marsileaceous ferns and relationships of the fossil *Hydropteris pinnata* reconsidered. *International Journal of Plant Sciences*, **160**, 931–954.

Pryer, K. M., Schneider, H., Smith, A. R., Cranfill, R., Wolf, P. G., Hunt, J. S., and Sipes, S. D. (2001a). Horsetails and ferns are a monophyletic group and the closest living relatives to seed plants. *Nature*, **409**, 618–622.

Pryer, K. M., Smith, A. R., Hunt, J. S., and Dubuisson, J.-Y. (2001b). *rbcL* data reveal two monophyletic groups of filmy ferns (Filicopsida: Hymenophyllaceae). *American Journal of Botany*, **88**, 1118–1130.

Pryer, K. M., Schneider, H., and Magallón, S. (2004a). The radiation of vascular plants. In *Assembling the Tree of Life*, ed. J. Cracraft and M. J. Donoghue. New York: Oxford University Press, pp. 138–153.

Pryer, K. M., Schuettpelz, E., Wolf, P. G., Schneider, H., Smith, A. R., and Cranfill, R. (2004b). Phylogeny and evolution of ferns (monilophytes) with a focus on the early leptosporangiate divergences. *American Journal of Botany*, **91**, 1582–1598.

Qiu, Y. L., Li, L. B., Wang, B., Chen, Z. D., Knoop, V., Groth-Malonek, M., Dombrovska, O., Lee, J., Kent, L., Rest, J., Estabrook, G. F., Hendry, T. A., Taylor, D. W., Testa, C. M., Ambros, M., Crandall-Stotler, B., Duff, R. J., Stech, M., Frey, W., Quandt, D., and Davis, C. C. (2006). The deepest divergences in land plants inferred from phylogenomic evidence. *Proceedings of the National Academy of Sciences of the United States of America*, **103**, 15511–15516.

Ranker, T. A., Geiger, J. M. O., Kennedy, S. C., Smith, A. R., Haufler, C. H., and Parris, B. S. (2003). Molecular phylogenetics and evolution of the endemic Hawaiian genus *Adenophorus* (Grammitidaceae). *Molecular Phylogenetics and Evolution*, **26**, 337–347.

Ranker, T. A., Smith, A. R., Parris, B. S., Geiger, J. M. O., Haufler, C. H., Straub, S. C. K., and Schneider, H. (2004). Phylogeny and evolution of grammitid ferns (Grammitidaceae): a case of rampant morphological homoplasy. *Taxon*, **53**, 415–428.

Raubeson, L. A. and Jansen, R. K. (1992). Chloroplast DNA evidence on the ancient evolutionary split in vascular land plants. *Science*, **255**, 1697–1699.

Reid, J. D., Plunkett, G. M., and Peters, G. A. (2006). Phylogenetic relationships in the heterosporous fern genus *Azolla* (Azollaceae) based on DNA sequence data from three noncoding regions. *International Journal of Plant Sciences*, **167**, 529–538.

Renzaglia, K. S., Duff, R. J., Nickrent, D. L., and Garbary, D. J. (2000). Vegetative and reproductive innovations of early land plants: implications for a unified phylogeny. *Philosophical Transactions of the Royal Society B: Biological Sciences*, **355**, 769–793.

Rößler, R. and Galtier, J. (2002). First *Grammatopteris* tree ferns from the Southern Hemisphere – new insights in the evolution of the Osmundaceae from the Permian of Brazil. *Review of Palaeobotany and Palynology*, **121**, 205–230.

Rothwell, G. W. and Nixon, K. C. (2006). How does the inclusion of fossil data change our conclusions about the phylogenetic history of euphyllophytes? *International Journal of Plant Sciences*, **167**, 737–749.

Rouhan, G., Dubuisson, J.-Y., Rakotondrainibe, F., Motley, T. J., Mickel, J. T., Labat, J.-N., and Moran, R. C. (2004). Molecular phylogeny of the fern genus *Elaphoglossum* (Elaphoglossaceae) based on chloroplast non-coding DNA sequences: contributions of species from the Indian Ocean area. *Molecular Phylogenetics and Evolution*, **33**, 745–763.

Rydin, C. and Wikström, N. (2002). Phylogeny of *Isoetes* (Lycopsida): resolving basal relationships using *rbcL* sequences. *Taxon*, **51**, 83–89.

Sánchez-Baracaldo, P. (2004). Phylogenetic relationships of the subfamily Taenitoideae, Pteridaceae. *American Fern Journal*, **94**, 126–142.

Sano, R., Takamiya, M., Ito, M., Kurita, S., and Hasebe, M. (2000). Phylogeny of the lady fern group, tribe Physematieae (Dryopteridaceae), based on chloroplast *rbcL* gene sequences. *Molecular Phylogenetics and Evolution*, **15**, 403–413.

Schneider, H., Pryer, K. M., Cranfill, R., Smith, A. R., and Wolf, P. G. (2002a). Evolution of vascular plant body plans: a phylogenetic perspective. In *Developmental Genetics and Plant Evolution*, ed. Q. Cronk, R. M. Bateman, and J. A. Hawkins. London: Taylor and Francis, pp. 330–363.

Schneider, H., Smith, A. R., Cranfill, R., Haufler, C. H., Ranker, T. A., and Hildebrand, T. (2002b). *Gymnogrammitis dareiformis* is a polygrammoid fern (Polypodiaceae) – resolving an apparent conflict between morphological and molecular data. *Plant Systematics and Evolution*, **234**, 121–136.

Schneider, H., Janssen, T., Hovenkamp, P., Smith, A. R., Cranfill, R., Haufler, C. H., and Ranker, T. A. (2004a). Phylogenetic relationships of the enigmatic Malesian fern *Thylacopteris* (Polypodiaceae, Polypodiidae). *International Journal of Plant Sciences*, **165**, 1077–1087.

Schneider, H., Russell, S. J., Cox, C. J., Bakker, F., Henderson, S., Gibby, M., and Vogel, J. C. (2004b). Chloroplast phylogeny of asplenioid ferns based on *rbcL* and *trnL-F* spacer sequences (Polypodiidae, Aspleniaceae) and its implications for the biogeography. *Systematic Botany*, **29**, 260–274.

Schneider, H., Schuettpelz, E., Pryer, K. M., Cranfill, R., Magallón, S., and Lupia, R. (2004c). Ferns diversified in the shadow of angiosperms. *Nature*, **428**, 553–557.

Schneider, H., Smith, A. R., Cranfill, R., Hildebrand, T. E., Haufler, C. H., and Ranker, T. A. (2004d). Unraveling the phylogeny of polygrammoid ferns (Polypodiaceae and Grammitidaceae): exploring aspects of the diversification of epiphytic plants. *Molecular Phylogenetics and Evolution*, **31**, 1041–1063.

Schneider, H., Ranker, T. A., Russell, S. J., Cranfill, R., Geiger, J. M. O., Aguraiuja, R., Wood, K. R., Grundmann, M., Kloberdanz, K., and Vogel, J. C. (2005). Origin of the endemic fern genus *Diellia* coincides with the renewal of Hawaiian terrestrial life in the Miocene. *Proceedings of the Royal Society B: Biological Sciences*, **272**, 455–460.

Schneider, H., Kreier, H.-P., Perrie, L. R., and Brownsey, P. J. (2006a). The relationships of *Microsorum* (Polypodiaceae) species occurring in New Zealand. *New Zealand Journal of Botany*, **44**, 121–127.

Schneider, H., Kreier, H.-P., Wilson, R., and Smith, A. R. (2006b). The *Synammia* enigma: evidence for a temperate lineage of polygrammoid ferns (Polypodiaceae, Polypodiidae) in southern South America. *Systematic Botany*, **31**, 31–41.

Schuettpelz, E. and Pryer, K. M. (2006). Reconciling extreme branch length differences: decoupling time and rate through the evolutionary history of filmy ferns. *Systematic Biology*, **55**, 485–502.

Schuettpelz, E. and Pryer, K. M. (2007). Fern phylogeny inferred from 400 leptosporangiate species and three plastid genes. *Taxon*, **56**, 1037–1050.

Schuettpelz, E., Korall, P., and Pryer, K. M. (2006). Plastid *atpA* data provide improved support for deep relationships among ferns. *Taxon*, **55**, 897–906.

Schuettpelz, E., Schneider, H., Huiet, L., Windham, M. D., and Pryer, K. M. (2007). A molecular phylogeny of the fern family Pteridaceae: assessing overall relationships and the affinities of previously unsampled genera. *Molecular Phylogenetics and Evolution*, **44**, 1172–1185.

Skog, J. E., Zimmer, E., and Mickel, J. T. (2002). Additional support for two subgenera of *Anemia* (Schizaeaceae) from data for the chloroplast intergenic spacer region *trnL-F* and morphology. *American Fern Journal*, **92**, 119–130.

Skog, J. E., Mickel, J. T., Moran, R. C., Volovsek, M., and Zimmer, E. A. (2004). Molecular studies of representative species in the fern genus *Elaphoglossum* (Dryopteridaceae) based on cpDNA sequences *rbcL*, *trnL-F*, and *rps4-trnS*. *International Journal of Plant Sciences*, **165**, 1063–1075.

Smith, A. R. (1990). Thelypteridaceae. In *The Families and Genera of Vascular Plants*, Vol. 1, *Pteridophytes and Gymnosperms*, ed. K. U. Kramer and P. S. Green. Berlin: Springer-Verlag, pp. 263–272.

Smith, A. R. and Cranfill, R. B. (2002). Intrafamilial relationships of the thelypteroid ferns (Thelypteridaceae). *American Fern Journal*, **92**, 131–149.

Smith, A. R., Kreier, H.-P., Haufler, C. H., Ranker, T. A., and Schneider, H. (2006a). *Serpocaulon* (Polypodiaceae), a new genus segregated from *Polypodium*. *Taxon*, **55**, 919–930.

Smith, A. R., Pryer, K. M., Schuettpelz, E., Korall, P., Schneider, H., and Wolf, P. G. (2006b). A classification for extant ferns. *Taxon*, **55**, 705–731.

Soltis, P. S. and Soltis, D. E. (2004). The origin and diversification of angiosperms. *American Journal of Botany*, **91**, 1614–1626.

Tidwell, W. D. and Ash, S. R. (1994). A review of selected Triassic to Early Cretaceous ferns. *Journal of Plant Research*, **107**, 417–442.

Tsutsumi, C. and Kato, M. (2005). Molecular phylogenetic study on Davalliaceae. *Fern Gazette*, **17**, 147–162.

Tsutsumi, C. and Kato, M. (2006). Evolution of epiphytes in Davalliaceae and related ferns. *Botanical Journal of the Linnean Society*, **151**, 495–510.

Wang, M.-L., Chen, Z.-D., Zhang, X.-C., Lu, S.-G., and Zhao, G.-F. (2003). Phylogeny of the Athyriaceae: evidence from chloroplast *trnL-F* region sequences. *Acta Phytotaxonomica Sinica*, **41**, 416–426.

Wikström, N. (2001). Diversification and relationships of extant homosporous lycopods. *American Fern Journal*, **91**, 150–165.

Wikström, N. and Kenrick, P. (1997). Phylogeny of Lycopodiaceae (Lycopsida) and the relationships of *Phylloglossum drummondii* Kunze based on *rbcL* sequences. *International Journal of Plant Sciences*, **158**, 862–871.

Wikström, N. and Pryer, K. M. (2005). Incongruence between primary sequence data and the distribution of a mitochondrial *atp1* group II intron among ferns and horsetails. *Molecular Phylogenetics and Evolution*, **36**, 484–493.

Wikström, N., Kenrick, P., and Vogel, J. C. (2002). Schizaeaceae: a phylogenetic approach. *Review of Palaeobotany and Palynology*, **119**, 35–50.

Wolf, P. G. (1995). Phylogenetic analyses of *rbcL* and nuclear ribosomal RNA gene sequences in Dennstaedtiaceae. *American Fern Journal*, **85**, 306–327.

Wolf, P. G. (1997). Evaluation of *atpB* nucleotide sequences for phylogenetic studies of ferns and other pteridophytes. *American Journal of Botany*, **84**, 1429–1440.

Wolf, P. G., Soltis, P. S., and Soltis, D. E. (1994). Phylogenetic relationships of dennstaedtioid ferns: evidence from *rbcL* sequences. *Molecular Phylogenetics and Evolution*, **3**, 383–392.

Yatabe, Y., Nishida, H., and Murkami, N. (1999). Phylogeny of Osmundaceae inferred from *rbcL* nucleotide sequences and comparison to the fossil evidence. *Journal of Plant Research*, **112**, 397–404.

Zhang, G., Zhang, X., and Chen, Z. (2005). Phylogeny of cryptogrammoid ferns and related taxa based on *rbcL* sequences. *Nordic Journal of Botany*, **23**, 485–493.

16

Fern classification

ALAN R. SMITH, KATHLEEN M. PRYER, ERIC SCHUETTPELZ,
PETRA KORALL, HARALD SCHNEIDER, AND PAUL G. WOLF

16.1 Introduction and historical summary

Over the past 70 years, many fern classifications, nearly all based on morphology, most explicitly or implicitly phylogenetic, have been proposed. The most complete and commonly used classifications, some intended primarily as herbarium (filing) schemes, are summarized in Table 16.1, and include: Christensen (1938), Copeland (1947), Holttum (1947, 1949), Nayar (1970), Bierhorst (1971), Crabbe *et al.* (1975), Pichi Sermolli (1977), Ching (1978), Tryon and Tryon (1982), Kramer (in Kubitzki, 1990), Hennipman (1996), and Stevenson and Loconte (1996). Other classifications or trees implying relationships, some with a regional focus, include Bower (1926), Ching (1940), Dickason (1946), Wagner (1969), Tagawa and Iwatsuki (1972), Holttum (1973), and Mickel (1974). Tryon (1952) and Pichi Sermolli (1973) reviewed and reproduced many of these and still earlier classifications, and Pichi Sermolli (1970, 1981, 1982, 1986) also summarized information on family names of ferns. Smith (1996) provided a summary and discussion of recent classifications.

With the advent of cladistic methods and molecular sequencing techniques, there has been an increased interest in classifications reflecting evolutionary relationships. Phylogenetic studies robustly support a basal dichotomy within vascular plants, separating the lycophytes (less than 1% of extant vascular plants) from the euphyllophytes (Figure 16.1; Raubeson and Jansen, 1992, Kenrick and Crane, 1997; Pryer *et al.*, 2001a, 2004a, 2004b; Qiu *et al.*, 2006). Living euphyllophytes, in turn, comprise two major clades: spermatophytes (seed plants), which are in excess of 260 000 species (Thorne, 2002; Scotland and Wortley,

Biology and Evolution of Ferns and Lycophytes, ed. Tom A. Ranker and Christopher H. Haufler. Published by Cambridge University Press. © Cambridge University Press 2008.

Table 16.1 *Comparison of fern classifications relative to Smith et al., 2006b*

+, indicates a circumscription the same as, or essentially the same as, the one adopted by Smith et al. (2006b); −, indicates a family with different circumscription than in Smith et al. (2006b); 0, family not recognized, i.e., included in another family, either implicitly or explicitly; ?, no mention of family (family not treated), or circumscription unclear. The last row gives percentages of classifications, for a given family, with the same circumscription as Smith et al. (2006b) and excluding those classifications for which there is doubt, or no mention (?) of circumscription. Oph, Ophioglossaceae; Psi, Psilotaceae; Equ, Equisetaceae; Mar, Marattiaceae; Osm, Osmundaceae; Hym, Hymenophyllaceae; Gle, Gleicheniaceae; Dip, Dipteridaceae; Mat, Matoniaceae; Lyg, Lygodiaceae; Ane, Anemiaceae; Sch, Schizaeaceae; Mrs, Marsileaceae; Sal, Salviniaceae; Thy, Thyrsopteridaceae; Lox, Loxomataceae; Cul, Culcitaceae; Pla, Plagiogyriaceae; Cib, Cibotiaceae; Cya, Cyatheaceae; Dic, Dicksoniaceae; Met, Metaxyaceae; Lin, Lindsaeaceae; Sac, Saccolomataceae; Den, Dennstaedtiaceae; Pte, Pteridaceae; Asp, Aspleniaceae; The, Thelypteridaceae; Woo, Woodsiaceae; Ble, Blechnaceae; Ono, Onocleaceae; Dry, Dryopteridaceae; Lom, Lomariopsidaceae; Tec, Tectariaceae; Ole, Oleandraceae; Dav, Davalliaceae; Pol, Polypodiaceae.

	Oph	Psi	Equ	Mar	Osm	Hym	Gle	Dip	Mat	Lyg	Ane	Sch	Mrs	Sal	Thy	Lox	Cul	Pla	Cib	Cya	Dic	Met	Lin	Sac	Den	Pte	Asp	The	Woo	Ble	Ono	Dry	Lom	Tec	Ole	Dav	Pol
Bierhorst, 1971	+	+	+	+	+	+	+	+	+	+	+	+	+	+	+	+	+	+	+	+	+	+	+	+	+	+	+	+	+	+	+	+	+	+	+	+	+
Ching, 1978	−	+	+	−	+	+	+	?	−	?	+	+	+	+	?	?	?	+	?	+	−	?	+	+	0	−	+	+	−	+	−	+	+	+	+	−	−
Christensen, 1938	+	?	?	−	+	+	+	0	0	0	0	+	+	+	0	?	?	+	0	+	−	?	0	0	0	−	0	0	0	0	0	0	0	0	0	0	−
Copeland, 1947	+	?	?	+	+	+	+	0	+	0	0	+	−	+	+	+	+	+	0	+	−	?	0	0	0	−	0	+	0	+	0	0	0	0	0	0	−
Crabbe et al., 1975	+	+	+	+	+	+	+	+	+	0	0	+	+	+	+	+	+	+	0	+	+	?	+	+	+	−	+	+	+	+	+	0	+	0	+	+	−
Hennipman, 1996	+	?	?	?	+	+	+	0	?	0	0	−	−	−	0	?	?	0	0	0	0	0	0	0	−	−	−	−	−	−	−	−	−	−	−	−	+
Holttum, 1947, 1949	?	?	?	+	+	+	+	?	+	0	0	+	+	+	?	?	?	+	0	+	0	?	0	0	0	−	0	+	0	+	0	0	−	0	−	0	−
Kubitzki, 1990	+	+	+	+	+	+	+	−	+	0	0	+	+	+	+	+	+	+	0	+	+	0	+	+	+	−	+	+	+	+	+	0	−	0	0	0	−
Nayar, 1970	+	?	?	+	+	+	+	−	−	−	0	+	+	+	+	+	+	+	0	+	−	+	+	+	0	+	+	+	+	+	+	0	−	0	−	−	−
Pichi Sermolli, 1977	−	+	+	+	+	+	+	+	+	+	+	+	+	+	+	+	+	+	+	+	+	+	+	+	+	+	+	+	+	+	+	0	+	+	+	+	+
Smith et al., 2006a	+	+	+	+	+	+	+	+	+	+	+	+	+	+	+	+	+	+	+	+	+	+	+	+	+	+	+	+	+	+	+	+	+	+	+	+	+
Stevenson and Loconte, 1996	−	−	+	+	−	−	+	−	−	?	0	−	−	−	0	0	0	0	0	0	−	−	0	0	−	−	−	−	−	0	−	0	−	−	−	−	+
Tryon and Tryon, 1982	+	+	+	+	+	+	+	+	+	?	?	+	+	+	+	+	0	+	0	+	−	+	+	0	+	−	+	+	+	+	+	−	0	−	−	−	+
% agreement with Smith et al., 2006b	73	75	100	45	92	92	92	42	11	90	33	27	27	90	27	9	80	9	92	0	8	0	45	25	0	8	17	75	67	0	67	17	0	0	0	17	0	25

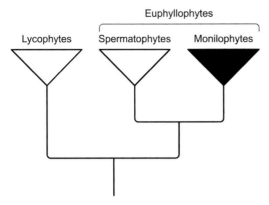

Figure 16.1 Consensus phylogeny depicting relationships of major vascular plant lineages. The topology summarizes the results of previously published phylogenetic studies (e.g., Raubeson and Jansen, 1992; Kenrick and Crane, 1997; Renzaglia *et al.*, 2000; Pryer *et al.*, 2001a; see main text for others). Resolved nodes have received bootstrap support ≥ 70. (From Smith *et al.*, 2006b.)

2003), and ferns (*sensu* Pryer *et al.*, 2004b), with about 9000 species, including horsetails, whisk ferns, and all eusporangiate and leptosporangiate ferns. Based on current evidence, extant ferns appear to be a monophyletic group characterized by lateral root origin in the endodermis, usually mesarch protoxylem in shoots, a pseudoendospore, plasmodial tapetum, and sperm cells with 30–1000 flagella (Renzaglia *et al.*, 2000; Schneider *et al.*, 2002a). Plants in the lycophyte and fern clades are all spore bearing or "seed-free," and because of this common feature their members have been lumped together historically under various terms, such as "pteridophytes" and "ferns and fern allies" – paraphyletic assemblages of plants.

Ideally, morphological data from fossils should also be used in constructing classifications, but this has generally not been done for ferns, except at higher levels in the hierarchy (e.g., by Kenrick and Crane, 1997). Rothwell and Nixon (2006) argued that the addition of fossil data in phylogenetic analyses can alter the topologies of trees obtained using data from only extant species. However, trees produced by adding selected fossil taxa to morphological and/or molecular data sets containing extant species (e.g., Fig. 4B in Rothwell and Nixon, 2006) are often poorly resolved. Although trees based only on morphological data from fossils and extant taxa (e.g., Rothwell, 1999) differ in many respects from phylogenetic hypotheses based on morphological and molecular data, or from hypotheses based only on molecular data, they do provide perspectives that may provide useful alternative hypotheses. Many factors contribute to the lack of resolution in trees containing fossils including: (1) the paucity of extinct taxa that successfully fossilized; (2) our inability to find (or recognize) crucial fossils that aid in filling evolutionary gaps; (3) the low proportion of phylogenetically informative characters, in both gametophytic and sporophytic phases, that are assessable in most fossils; (4) the incompleteness or fractured nature of most fossil taxa; and (5) the inherent difficulty of scoring morphological characters, which in turn may be due to considerable morphological homoplasy in ferns (Ranker et al., 2004). All of these features make the incorporation of fossil data into matrices containing data from extant plants challenging. Fossil data can be used to pinpoint divergence times of lineages (Lovis, 1977; Pryer et al., 2004b; Schneider et al., 2004c), and as the quality and abundance of fossil evidence for ferns improve, fossil data will aid in building even more robust phylogenies. This, in turn, may provide reciprocal illumination for understanding the evolution of characters and character states of extant taxa. Because we consider it impractical at present to include fossil data in taxonomically diverse datasets, in this chapter we develop a classification based solely on morphological and molecular data from extant species.

16.2 Review of critical recent advances

Molecular phylogenetic hypotheses for extant ferns (see Chapter 15) have utilized data from several chloroplast markers (*rbcL*, *atpA*, *atpB*, *accD*, *rps4*, 16S rDNA, ITS), one nuclear gene (18S rDNA), and three mitochondrial genes (*atp1*, *nad2*, *nad5*). A natural outgrowth of these one-gene or few-gene studies on a wide array of ferns has led to still broader, and increasingly robust multiple-gene phylogenetic analyses, e.g., studies by Wolf (1996b), Wolf et al. (1998), Pryer et al. (2001a, 2004b), Schneider et al. (2004c), Wikström and Pryer (2005), Schuettpelz et al. (2006), and Schuettpelz and Pryer (2007). A multiple-gene analysis examining the relationships of all land plants (Qiu et al., 2006) has largely confirmed the basic relationships uncovered in the fern studies. Attempts to generate morphologically based phylogenetic hypotheses for ferns are far fewer (Pryer et al. 1996, 2001a; Schneider, 1996a; Stevenson and Loconte, 1996; Rothwell, 1999; Schneider et al., in preparation), perhaps because of the inherent difficulties in defining characters and assessing characters states, but at least in some cases this approach has added support for molecular-based consensus phylogenies.

All of these studies have given rise to growing confidence in relationships and correspondingly to the composition of taxa at familial and ordinal ranks. This convinced us that it was timely to attempt a reclassification of extant ferns (Smith et al., 2006b). In that paper, the classificatory decisions/innovations reflect a growing reliance on a solid phylogenetic underpinning for the development of useful multipurpose classifications. As in that publication, we focus on the ranks of class, order, and family. We assert that the information available is most appropriate for resolution and understanding of relationships at these levels, and that within most families, especially at the generic level, there is still insufficient evidence to attempt many classificatory decisions. Only one previous fern classification has employed cladistic methodology in a rigorous way; Stevenson and Loconte (1996) superimposed on their tree a hierarchical classification, but the phylogeny that they generated as a basis for their classification was based exclusively on morphological data and differs radically from the most up-to-date phylogenetic hypotheses. Our classification, in contrast, is based on consensus of a variety of morphological and molecular studies.

The impact of molecular phylogenetic studies on classification will likely continue, perhaps at an accelerated rate. Already, since our recent reclassification (Smith *et al.*, 2006b), several additional phylogenetic studies have been published, e.g., for the Cyatheales (Korall *et al.*, 2006a, 2006b, 2007), Dryopteridaceae and related families (Li and Lu, 2006; Lu and Li, 2006; Tsutsumi and Kato, 2006), Hymenophyllaceae (Ebihara *et al.*, 2006; Hennequin *et al.*, 2006a, 2006b), Marsileaceae (Nagalingum *et al.*, 2006, 2007), Pteridaceae (Schuettpelz *et al.*, 2007), and Salviniaceae (Metzgar *et al.*, 2007). These generally reinforce decisions made in our previous paper, and largely confirm the higher level (order and family) structure of the classification (Schuettpelz *et al.*, 2006; Schuettpelz and Pryer, 2007) or suggest the recognition or rejection of taxa at generic level.

16.3 Synthesis of current perspectives: the classification of ferns

In our reclassification (Smith *et al.*, 2006b), we combined the principle of monophyly with a decision to maintain well-established names to update ordinal and familial ranks within ferns so that they are better reconciled with our current best estimates of phylogenetic relationships. We utilized a minimum number of ranks to categorize only the most well-supported splits in the phylogeny, and we treat all classes, orders, and families of extant ferns, which constitute a monophyletic group, sometimes called Infradivision Moniliformopses (Kenrick and Crane, 1997), or monilophytes (Pryer *et al.*, 2001a, 2004a, 2004b; Donoghue in Judd *et al.*, 2002). However, "Infradivision" is not a recognized rank in the International Code of Botanical Nomenclature (Greuter *et al.*, 2000); moreover, the name "Moniliformopses" was never validly published, lacking a Latin diagnosis or description, or a reference to one. Because validly published names for ferns (as we define them) at ranks above class are either not available or have been interpreted in a manner we think might be confusing or incompatible with current hypotheses of relationship, we avoid their use.

For each family, we provide important heterotypic synonyms, approximate numbers of genera and species, names of constituent genera, references to relevant phylogenetic literature, and, where appropriate, discussion of unresolved problems. DNA sequence data are now available for all families we recognize, and for most genera of ferns. A superscript number one (^{1}) denotes those genera for which DNA sequence data are not available; nonetheless, taxonomic placement for most of these is relatively certain, based on morphological evidence. Lack of a superscript indicates that some molecular evidence (either published or unpublished) has been available to us for consideration. The classification presented below is based on the consensus relationships depicted in Chapter 15, which are derived from, and guided by, recent and ongoing phylogenetic studies.

Fern names at family rank used in this classification were obtained from the website of James Reveal, University of Maryland (www.life.umd.edu/emeritus/reveal/PBIO/fam/hightaxaindex.html) and from Hoogland and Reveal (2005); all family names, and cross-references if any, are listed in Appendix A. Most names at family rank are also listed and discussed by Pichi Sermolli (1970, 1982) and summarized in reports by the Subcommittee for Family Names of Pteridophyta (Pichi Sermolli, 1981, 1986); this list was further emended and updated by Pichi Sermolli (1993). Names at ordinal, supraordinal, and subordinal ranks are also available from the Reveal website and from Hoogland and Reveal (2005) (see Appendices 1 and 2 in Smith *et al.*, 2006b); literature citations for all names were given by Smith *et al.* (2006b, Appendix 4). At the end of this chapter, we give an index to commonly accepted genera with family assignments proposed here (Appendix B).

Within ferns, we recognize four classes (Psilotopsida, Equisetopsida, Marattiopsida, Polypodiopsida), 11 orders, and 37 families.

I. Class Psilotopsida

A. Order Ophioglossales

1. Family Ophioglossaceae (incl. Botrychiaceae, Helminthostachyaceae). Four genera: *Botrychium* (grapeferns, moonworts), *Helminthostachys*, *Mankyua*^{1}, *Ophioglossum* (adder tongues); *Botrychium* (incl. *Botrychium s.s.*, *Sceptridium*, *Botrypus*, and *Japanobotrychium*) and *Ophioglossum* (incl. *Cheiroglossa*, *Ophioderma*) are sometimes divided more finely (Kato, 1987; Hauk *et al.*, 2003). Ca. 80 species; monophyletic (Hasebe *et al.*, 1996; Hauk, 1996; Pryer *et al.*, 2001a, 2004b; Hauk *et al.*, 2003). *Mankyua*, from Cheju Island, Korea, has recently been described, but no molecular data are available (Sun *et al.*, 2001). Non-coding chloroplast genes recover similar topologies as coding genes in *Botrychium s.l.* (Small *et al.*, 2005).

Species mostly terrestrial (a few epiphytic), temperate and boreal, but a few pantropical. Characters: vernation nodding (not circinate); rhizomes and petioles fleshy; root hairs lacking; aerophores absent (Davies, 1991); fertile leaves each with a single sporophore arising at the base of, or along, the trophophore stalk, or at the base of the trophophore blade (several sporophores per blade in *Cheiroglossa*); sporangia large, with walls two cells thick, lacking an annulus; spores globose-tetrahedral, trilete, many (>1000) per sporangium; gametophytes subterranean, non-photosynthetic, mycorrhizal; $x = 45$ (46).

B.   Order Psilotales

2. Family Psilotaceae (whisk ferns; incl. Tmesipteridaceae). Two genera (*Psilotum*, *Tmesipteris*), ca. 12 species total (two in *Psilotum*); monophyletic (Hasebe et al., 1996; Pryer et al., 2001a, 2004b). Characters: roots absent; stems bearing reduced, unveined or single-veined euphylls; sporangia large, with walls two cells thick, lacking an annulus; two or three sporangia fused to form a synangium, seemingly borne on the adaxial side of a forked leaf; spores reniform, monolete, many (>1000) per sporangium; gametophytes subterranean (*Psilotum*), non-photosynthetic, mycorrhizal; $x = 52$.

II.   *Class Equisetopsida [= Sphenopsida]*

C.   Order Equisetales

3. Family Equisetaceae (horsetails). A single genus (*Equisetum*), 15 species usually placed in two well-marked subgenera, subg. *Equisetum* and subg. *Hippochaete*; monophyletic (Pryer et al., 2001a, 2004b; Des Marais et al., 2003; Guillon, 2004). The spermatozoids of *Equisetum* share several important features with other ferns that support their inclusion in this clade (Renzaglia et al., 2000). Kato (1983) adduced additional morphological characters, including root characters, supporting a relationship between horsetails and ferns. Characters: stems with whorled branching, lacunate; leaves whorled, connate; sporangia with helical secondary wall thickenings (Bateman, 1991), borne on peltate sporangiophores that collectively comprise strobili; sporangia large, lacking an annulus, many (>1000) per sporangium; spores green, with circular aperture and four paddle-like, coiled elaters; gametophytes green, surficial; $x = 108$.

III.   *Class Marattiopsida*

D.   Order Marattiales (incl. Christenseniales)

4. Family Marattiaceae (marattioids; incl. Angiopteridaceae, Christenseniaceae, Danaeaceae, Kaulfussiaceae). Four genera: *Angiopteris*, *Christensenia*, *Danaea*, *Marattia*. As currently circumscribed, *Marattia* is paraphyletic, and will be subdivided into three elements, one requiring a new generic name (A. G. Murdock, personal communication). *Archangiopteris* has been recognized by some (e.g., Pichi Sermolli, 1977) but appears to nest within *Angiopteris* (Murdock, 2005, personal communication). *Danaea* is sister to the other three genera (Pryer *et al.* 2001a; 2004b; Murdock, 2005) and represents a neotropical radiation (Christenhusz *et al.*, in review). *Angiopteris* and *Christensenia* are restricted to eastern and southeastern Asia, Australasia, and Polynesia, while *Marattia s.l.* is pantropical. Ca. 150 species, but monographic revision is needed at the species level in several genera; monophyletic (Hill and Camus, 1986; Pryer *et al.*, 2001a, 2004b; Murdock, 2005). We see no advantage or good reason for recognizing several of the constituent genera as monogeneric families, as done by Pichi Sermolli (1977), and the paraphyly of *Marattia* vis-à-vis *Angiopteris*, necessitating a recircumscription of marattioid genera (A. G. Murdock, personal communication), reinforces this opinion. The name Danaeaceae has been found to pre-date Marattiaceae; however, Marattiaceae has been proposed for conservation by Murdock *et al.* (2006), and we maintain its usage in the usual broad sense.

Terrestrial (rarely epipetric), pantropical, fossils beginning in Carboniferous (Collinson, 1996). Characters: roots large, fleshy, with polyarch xylem; root hairs septate; roots, stems, and leaves with mucilage canals; rhizomes fleshy, short, upright or creeping, with a polycyclic dictyostele; vernation circinate; leaves large, fleshy, 1–3-pinnate (rarely simple in *Danaea*, or 3–5-foliate in *Christensenia*) with enlarged, fleshy, starchy stipules at the base and swollen pulvinae along petioles and rachises (and sometimes other axes); petiole and stem xylem polycyclic; stems and blades bearing scales; pneumathodes (lenticels) scattered all around petioles and/or rachises; sporangia free or in round or elongate synangia (fused sporangia), lacking an annulus, enclosing 1000–7000 spores; spores usually bilateral or ellipsoid, monolete; gametophytes green, surficial; $x = 40$ (39).

IV.   *Class Polypodiopsida [= Filicopsida]*

E.   *Order Osmundales*

5. Family Osmundaceae. Four genera: *Leptopteris*, *Osmunda*, *Osmundastrum*, *Todea*. Ca. 20 species; monophyletic (Hasebe *et al.*, 1996; Yatabe *et al.*, 1999; Pryer *et al.*, 2001a, 2004b). Evidence from morphology (Miller, 1971) and molecules (Yatabe *et al.*, 1999) suggests that *Osmundastrum cinnamomea* (L.) C. Presl be recognized as an independent, monotypic genus (Yatabe *et al.*, 2005); there is support for three subgenera within *Osmunda s.s.*: subg. *Osmunda*, subg. *Claytosmunda* Y. Yatabe *et al.*, and subg. *Plenasium* (C. Presl) Milde. Fossils from Permian; temperate and tropical. Characters: stem anatomy distinctive, an ectophloic siphonostele, with a ring of discrete xylem strands, these often conduplicate or twice conduplicate in cross-section; stipules at bases of petioles; leaves dimorphic or with fertile portions dissimilar to sterile; sporangia large, with 128–512 spores, opening by an apical slit, annulus lateral; spores green, subglobose, trilete; gametophytes large, green, cordate, surficial; $x = 22$.

F.   Order Hymenophyllales

6. **Family Hymenophyllaceae** (filmy ferns; incl. Trichomanaceae). Nine genera (Ebihara et al., 2006), two major clades (Pryer et al., 2001b), "trichomanoid" and "hymenophylloid," roughly corresponding to the classical genera *Trichomanes* s.l. and *Hymenophyllum* s.l. Ca. 600 species; monophyletic (Hasebe et al., 1996; Dubuisson, 1996, 1997; Pryer et al., 2001b, 2004b; Ebihara et al., 2002, 2006; Dubuisson et al., 2003; Hennequin et al., 2003). Several segregate and monotypic genera are nested within *Hymenophyllum* s.l.: *Cardiomanes, Hymenoglossum, Rosenstockia*, and *Serpyllopsis* (Ebihara et al., 2002, 2006; Hennequin et al., 2003, 2006a, 2006b). Several other classically defined hymenophylloid genera (subgenera) are not monophyletic, e.g., *Mecodium* and *Sphaerocionium* (Hennequin et al., 2003, 2006a, 2006b; Ebihara et al., 2006). *Microtrichomanes* appears to be polyphyletic (Ebihara et al., 2004). *Trichomanes* s.l. comprises eight monophyletic groups that are regarded here as genera: *Abrodictyum, Callistopteris, Cephalomanes, Crepidomanes, Didymoglossum, Polyphlebium, Trichomanes* s.s., and *Vandenboschia*; several of these have been subdivided into putatively monophyletic subgenera and sections (Ebihara et al., 2006). Terrestrial and epiphytic; pantropical and south-temperate, but gametophytes survive in north-temperate regions as far north as Alaska (Farrar, 1993, p. 191). Characters: rhizomes slender, creeping, wiry, or sometimes erect and stouter, protostelic; vernation circinate; blades one cell thick between veins (a few exceptions); stomata lacking; cuticles lacking or highly reduced; scales usually lacking on blades, indument sometimes of hairs; sori marginal, indusia conical (campanulate), tubular, or clam-shaped (bivalvate), with receptacles (at least in trichomanoid genera) usually elongate, protruding from the involucres; sporangia maturing gradately in basipetal fashion, each with an uninterrupted, oblique annulus; spores green, globose, trilete; gametophytes filamentous or ribbon-like, often reproducing by fragmentation or production of gemmae; $x = 11, 12, 18, 28, 32, 33, 34, 36$, and perhaps others.

G.   Order Gleicheniales

(incl. Dipteridales, Matoniales, Stromatopteridales). Monophyletic (Pryer et al., 2004b; Schuettpelz et al., 2006). Characters: root steles with 3–5 protoxylem poles (Schneider, 1996a); antheridia with 6–12 narrow, twisted or curved cells in walls.

7. **Family Gleicheniaceae** (gleichenioids, forking ferns; incl. Dicranopteridaceae, Stromatopteridaceae). Six genera (*Dicranopteris, Diplopterygium, Gleichenella, Gleichenia, Sticherus, Stromatopteris*), ca. 125 species; monophyletic (Hasebe et al., 1996; Pryer et al., 1996, 2001a, 2004b). Hennipman (1996) also suggested inclusion of the next two families in Gleicheniaceae; however, we recognize these as distinct based on their significant morphological disparity. Fossil record beginning in Cretaceous (Jurassic and older fossils may belong to the Gleicheniales or represent ancestors of extant Gleicheniaceae); pantropical. Characters: rhizomes with a "vitalized" protostele, or rarely solenostele; leaves indeterminate, blades pseudodichotomously forked (except *Stromatopteris*); veins free; sori abaxial, not marginal, with 5–15 sporangia, each with a transverse-oblique annulus, exindusiate, round, with 128–800 spores; sporangia maturing simultaneously within sori; spores globose-tetrahedral or bilateral; gametophytes green, surficial, with club-shaped hairs; $x = 22, 34, 39, 43, 56$.

8. **Family Dipteridaceae** (incl. Cheiropleuriaceae). Two genera (*Cheiropleuria, Dipteris*) from India, southeast Asia, eastern and southern China, central and southern Japan, and Malesia, to Melanesia and western Polynesia (Samoa), ca. 11 species; monophyletic (Kato et al., 2001; Pryer et al., 2004b). Fossil record beginning in upper Triassic. Characters: stems long-creeping, solenostelic or protostelic, covered with bristles or articulate hairs; petioles with a single vascular bundle proximally and polystelic distally; blades (sterile ones, at least) cleft into two or often more subequal parts; veins highly reticulate, areoles with included veinlets; sori exindusiate, discrete, compital (served by many veins), scattered over the surface, or leaves dimorphic and the fertile ones covered with sporangia; sporangia maturing simultaneously or maturation mixed, with a four-seriate stalk; annuli almost vertical or slightly oblique; spores ellipsoid and monolete, or tetrahedral and trilete, 64 or 128 per sporangium; gametophytes cordate-thalloid; $x = 33$. *Dipteris* differs from *Cheiropleuria* primarily in having bilateral, monolete spores (tetrahedral and trilete in *Cheiropleuria*) and monomorphic leaves with discrete sori (sporangia acrostichoid in *Cheiropleuria*).

9. **Family Matoniaceae** (matonioids). Two genera (*Matonia, Phanerosorus*), each with two species; monophyletic, sister to Dipteridaceae (Kato and Setoguchi, 1998; Pryer et al., 2004b; Schuettpelz et al., 2006). Malesia–Pacific Basin; fossil record beginning in mid-Mesozoic. Characters: stems solenostelic with at least two concentric vascular cylinders (polycyclic) and a central vascular bundle; blades flabellate (*Matonia*), unevenly dichotomously branched or with dichotomous pinnae; veins free or slightly anastomosing around sori; sori with peltate indusia; sporangia maturing simultaneously, with very short stalks and oblique annuli; spores

globose-tetrahedral, trilete; gametophytes green, thalloid, with ruffled margins; antheridia large, many-celled; $x = 26$ (*Matonia*), 25 (*Phanerosorus*).

H.   Order Schizaeales

Monophyletic (Hasebe et al., 1996; Pryer et al., 2001a, 2004b; Skog et al., 2002; Wikström et al., 2002). The three constituent families are given recognition because of their numerous, we consider significant, morphological differences, differences embracing gametophytes, stelar anatomy, leaf morphology, soral types, spores, and chromosome numbers. Fossil record beginning in the Jurassic (Collinson, 1996). Characters: fertile–sterile leaf blade differentiation; absence of well-defined sori; sporangia each with a transverse, subapical, continuous annulus.

10. Family Lygodiaceae (climbing ferns). A single genus (*Lygodium*), ca. 25 species; monophyletic (Skog et al., 2002; Wikström et al., 2002). Terrestrial, pantropical. Characters: rhizomes creeping, slender, protostelic, bearing hairs; leaves indeterminate, climbing, alternately pinnate; primary blade divisions (pinnae) pseudodichotomously forking with a dormant bud in the axils; veins free or anastomosing; sori on lobes of the ultimate segments; sporangia abaxial, solitary, one per sorus, each sporangium covered by an antrorse indusium-like subtending flange; spores 128–256 per sporangium, tetrahedral and trilete; gametophytes green, cordate, surficial; $x = 29, 30$.

11. Family Anemiaceae (incl. Mohriaceae). One genus (*Anemia*, incl. *Mohria*), ca. 100+ species; monophyletic (Skog et al., 2002; Wikström et al., 2002). Terrestrial; primarily New World, but a few species in Africa, India, and islands in the Indian Ocean. Characters: rhizomes creeping to suberect, bearing hairs; leaves determinate, mostly hemidimorphic or dimorphic; veins free, dichotomous, occasionally casually anastomosing; sporangia usually on a basal pair (sometimes more than two pinnae, or all pinnae modified and fertile) of skeletonized, highly modified, often erect pinnae; spores 128–256 per sporangium, tetrahedral, with strongly parallel ridges (Tryon and Lugardon, 1991); gametophytes green, cordate, surficial; $x = 38$.

12. Family Schizaeaceae. Two genera (*Actinostachys*, *Schizaea*), ca. 30 species; monophyletic (Skog et al., 2002; Wikström et al., 2002). The Cretaceous *Schizaeopsis* is the oldest fossil assigned to this lineage (Wikström et al., 2002). Terrestrial, pantropical. Characters: blades simple (linear) or fan-shaped, variously cleft and with dichotomous free veins; sporangia on marginal, elaminate, branched or unbranched projections at blade tips, not in discrete sori, exindusiate; spores bilateral, monolete, 128–256 per sporangium; gametophytes green and filamentous (*Schizaea*), or subterranean and non-green, tuberous (*Actinostachys*); a puzzling array of base chromosome numbers: $x = 77, 94, 103$.

I. Order Salviniales

(water ferns, heterosporous ferns; incl. "Hydropteridales," Marsileales, Pilulariales). Monophyletic (Hasebe *et al.*, 1996; Pryer, 1999; Pryer *et al.*, 2001a, 2004b). The fossil *Hydropteris pinnata* provides evidence linking the two families of this order (Rothwell and Stockey, 1994; Pryer, 1999), although hypotheses differ about the exact relationships of *Hydropteris* with extant genera. Characters: fertile–sterile leaf blade differentiation; veins anastomosing; aerenchyma tissue often present in roots, shoots, and petioles; annulus absent; plants heterosporous, spores with endosporous germination; monomegaspory; gametophytes reduced.

13. **Family Marsileaceae** (clover ferns) (incl. Pilulariaceae). Three genera (*Marsilea*, *Pilularia*, *Regnellidium*), ca. 75 species total; monophyletic (Hasebe *et al.*, 1996; Pryer, 1999; Nagalingum *et al.*, 2007). Hennipman (1996) included both Salviniaceae and Azollaceae within Marsileaceae, but the spores of Marsileaceae differ markedly from those of Salviniaceae and Azollaceae (Schneider and Pryer, 2002). Rooted aquatics, in ponds, shallow water, or vernal pools, with floating or emergent leaf blades; subcosmopolitan. Characters: stems usually long-creeping, slender, often bearing hairs; leaflets 4, 2 or 0 per leaf; veins dichotomously branched but often fusing toward their tips; sori borne in stalked bean-shaped sporocarps (Nagalingum *et al.*, 2006), these arising from the rhizomes or from the base of the petioles, one to many per plant; heterosporous, microspores globose, trilete, megaspores globose, each with an acrolamella positioned over the exine aperture (Schneider and Pryer, 2002); perine gelatinous; $x = 10$ (*Pilularia*), 20 (*Marsilea*).

14. **Family Salviniaceae** (floating ferns, mosquito ferns; incl. Azollaceae). Two genera (*Salvinia*, *Azolla*), ca. 16 species; monophyletic (Pryer *et al.*, 1996, 2004b; Reid *et al.*, 2006; Metzgar *et al.*, 2007). Some authors separate the genera into two families (Schneller in Kubitzki, 1990), a perfectly acceptable alternative, given the significant differences between the two genera. Plants free-floating, subcosmopolitan; fossil record beginning in Cretaceous (Collinson, 1996). Characters: roots present (*Azolla*) or lacking (*Salvinia*); stems protostelic, dichotomously branched; leaves sessile, alternate, small (ca. 1–25 mm long), round to oblong, entire; veins free (*Azolla*) or anastomosing (*Salvinia*); spores of two kinds (plants heterosporous), large megaspores and small microspores, these

globose, trilete; spore germination endoscopic; $x = 9$ (*Salvinia*), the lowest base chromosome number known in ferns, 22 (*Azolla*).

J.   Order Cyatheales

(tree ferns; incl. Dicksoniales, Hymenophyllopsidales, Loxomatales, Metaxyales, Plagiogyriales) (Hasebe et al., 1996; Wolf et al., 1999; Pryer et al., 2004b, Korall et al., 2006b). Existing molecular evidence indicates a close relationship among the included families. The order is without obvious defining morphological characters: some of the species have trunk-like stems but others have creeping rhizomes; some have only hairs on the stems and blades, others have scales; sori are abaxial or marginal, either indusiate or exindusiate; spores are globose or tetrahedral-globose, each with a trilete scar; gametophytes green, cordate.

15. Family Thyrsopteridaceae. One genus (*Thyrsopteris*) with a single species, *T. elegans*, endemic to the Juan Fernández Islands; clearly related to tree ferns, but of uncertain phylogenetic position within this group (Korall et al., 2006b). Characters: rhizomes ascending to erect, solenostelic, bearing runners, clothed with stiff, pluricellular hairs; leaves large, 2–3.5 m long; blades 3–5-pinnate, partially dimorphic (sori often restricted to proximal segments); blade axes adaxially grooved; veins free; sori terminal on the veins, the outer and inner indusia fused to form asymmetric cup-like structures, each sorus with a columnar, clavate receptacle; sporangia with oblique annuli; spores globose-tetrahedral, with prominent angles; $x = $ ca. 78.

16. Family Loxomataceae (often spelled Loxsomataceae). Two genera (*Loxoma*, *Loxsomopsis*), each with a single species; monophyletic (Lehnert et al., 2001; Pryer et al., 2001a, 2004b; Korall et al., 2006b). South American Andes, southern Central America, and New Zealand. Characters: rhizomes long-creeping, solenostelic, bearing hairs with a circular, multicellular base; blades bipinnate or more divided; veins free, forked; indument of uniseriate (*Loxsomopsis*) to pluriseriate (*Loxoma*) bristles; sori marginal, terminal on veins, each with an urceolate indusium and elongate, often exserted receptacle; sporangia on thick, short stalks, with a slightly oblique annulus; spores tetrahedral, trilete; gametophytes with scale-like hairs (occurring also in some Cyatheaceae); $x = 46$ (*Loxsomopsis*), 50 (*Loxoma*).

17. Family Culcitaceae. One genus (*Culcita*) with two species; monophyletic (Korall et al., 2006b). Sister to Plagiogyriaceae, and not closely related to *Calochlaena*, with which *Culcita* has historically been associated. This separation is supported by anatomical characters (White and Turner, 1988; Schneider, 1996a). Terrestrial; Azores, Madeira, Tenerife, southwestern Europe, and the Neotropics. Characters: rhizomes creeping or ascending, solenostelic, bearing articulate hairs; petioles in cross-section each with gutter-shaped vascular bundles; blades large, 4–5-pinnate-pinnatifid, sparingly hairy; veins free, often forked; sori to 3 mm wide, terminal on veins, paraphysate; outer indusia scarcely differentiated from the laminar tissue, inner noticeably modified; spores tetrahedral-globose, trilete; $x = 66$.

18. Family Plagiogyriaceae. A single genus (*Plagiogyria*), with ca. 15 species (Zhang and Nooteboom, 1998); monophyletic (Korall et al., 2006b). Characters: stems creeping to usually erect, lacking hairs or scales; leaves dimorphic; blades pectinate to 1-pinnate; veins simple to 1-forked, free, or in fertile blades somewhat anastomosing at their ends; young leaves densely covered with pluricellular, glandular, mucilage-secreting hairs; sori exindusiate; sporangia borne on distal parts of veins, seemingly acrostichoid; sporangial stalks 6-rowed; annuli slightly oblique, continuous; spores tetrahedral, trilete; gametophytes green, cordate; $x = 66$?

19. Family Cibotiaceae. One genus (*Cibotium*), ca. 11 species; monophyletic, with some affinity to Dicksoniaceae, as circumscribed here (Korall et al., 2006b). Terrestrial, amphipacific (eastern Asia, Malesia, Hawaii, southern Mexico, and Central America). Characters: rhizomes massive, creeping to ascending or erect (to 6 m), solenostelic or dictyostelic, bearing soft yellowish hairs at apices and persistent petiolar bases; fronds monomorphic, mostly 2–4 m long; petioles hairy at bases, with three corrugated vascular bundles arranged in an omega-shape; blades large, bipinnate to bipinnate-pinnatifid or more divided; secondary and tertiary blade axes adaxially ridged; veins free, simple or forked to pinnate; stomata with three subsidiary cells; sori marginal at vein ends, indusia bivalvate, each with a strongly differentiated, non-green outer indusium and a similarly modified tongue-like inner indusium, paraphyses filiform; spores globose-tetrahedral, with prominent angles and a well-developed equatorial flange; antheridial walls 5-celled; $x = 68$.

The spores of Cibotiaceae are unlike those of all other families in Cyatheales (Gastony, 1982; Tryon and Lugardon, 1991), with a prominent equatorial flange, and with usually thick, bold, ± parallel, sometimes anastomosing ridges on the distal face; these ridges are the result of a coarsely ridged exospore, which is overlain by a thin, sometimes granulate perispore. The spores of *Lophosoria* (Dicksoniaceae) also have a prominent equatorial flange but lack distal ridges (Tryon and Lugardon, 1991; Tryon and Tryon, 1982). As far as is known, the chromosome base number of $x = 68$ for *Cibotium* is also unique in the tree fern clade. The Hawaiian species have been extensively studied by Palmer (1994).

20. Family Cyatheaceae (cyatheoids, scaly tree ferns; incl. Alsophilaceae, Hymenophyllopsidaceae). Ca. five genera: *Alsophila* (incl. *Nephelea*), *Cyathea* (incl. *Cnemidaria*, *Hemitelia*, *Trichipteris*), *Gymnosphaera*, *Hymenophyllopsis*, *Sphaeropteris* (incl. *Fourniera*). *Alsophila*, as often construed, may be paraphyletic (Korall et al., 2007); 600+ species; monophyletic, together with Dicksoniaceae, Metaxyaceae, and Cibotiaceae constituting the "core tree ferns" (Korall et al., 2006b). Several studies have addressed relationships within Cyatheaceae (Conant et al., 1996a, 1996b; Hasebe et al., 1996; Stein et al., 1996; Lantz et al., 1999; Conant and Stein, 2001; Korall et al., 2007), and circumscriptions of genera have varied widely (compare, e.g., Tryon, 1970; Holttum and Edwards, 1983). Several studies show convincingly that *Cnemidaria* nests within *Cyathea* (Conant et al., 1996a, 1996b; Conant and Stein, 2001; Korall et al., 2006b, 2007), and this close relationship is supported by the existence of sterile hybrids (see, e.g., Conant, 1975). Hennipman (1996) included all other families here placed in Cyatheales (excepting Hymenophyllopsidaceae, unplaced in his "consensus" classification) in his Cyatheaceae. *Hymenophyllopsis* (ca. eight species) has thin leaves lacking stomates, and is confined to the sandstone tepuis of the Guayana Shield (eastern Venezuela, Guyana, northern Brazil). It has nearly always been regarded as an isolated genus in its own family, or order (e.g., by Copeland, 1947; Pichi Sermolli, 1977; Tryon and Tryon, 1982). In the analysis by Wolf et al. (1999), a close relationship of *Hymenophyllopsis* to Cyatheaceae was suggested, based on a small taxonomic sampling. A larger sampling by Korall et al. (2006b) indicates that *Hymenophyllopsis*, as well as *Cnemidaria* and *Trichipteris*, all nest within *Cyathea*, and together form a well-supported neotropical clade. The spores of *Hymenophyllopsis* are remarkably similar to those of some species of *Cyathea* (compare, e.g., Figs. 14.8–14.11 with 26.15–26.18 in Tryon and Tryon, 1982). Characters associating *Hymenophyllopsis* with Cyatheaceae include the presence of true scales on the rhizomes, petiole bases, and sometimes on the blades. Tree ferns are mostly arborescent (but many exceptions known), and pantropical; fossils beginning in Jurassic or early Cretaceous.

Characters: stems with polycyclic dictyosteles, apices (and usually petiole bases) covered with large scales, sometimes also with trichomidia (scurf = small scales) or hairs; leaves usually large (to 5 m); petioles with obvious, usually discontinuous pneumathodes in two lines; blades 1–3-pinnate (rarely simple); veins simple to forked, free, rarely anastomosing (mostly in some *Cyathea*); sori superficial (abaxial) or terminal on the veins and marginal or submarginal (*Hymenophyllopsis*), round, exindusiate, or indusia saucer-like, cup-like, or globose and completely surrounding sporangia, or bivalvate (*Hymenophyllopsis*); sporangia maturing gradately, with oblique annuli; receptacles raised; paraphyses usually present; spores tetrahedral, trilete, variously ornamented; gametophytes green, cordate; $x = 69$ (*Hymenophyllopsis* not yet counted).

21. Family Dicksoniaceae, *nom. cons.* (dicksonioids; incl. Lophosoriaceae). Three genera (*Calochlaena, Dicksonia, Lophosoria*). Ca. 30 species; monophyletic (Korall et al., 2006b). Terrestrial; eastern Asia, Australasia, Neotropics, St. Helena. Characters: mostly arborescent or with erect or ascending rhizomes; rhizomes with polycyclic dictyosteles, or solenostelic (*Calochlaena*); stem apices and usually petiolar bases covered with uniseriate hairs; blades large, 2–3 pinnate; veins simple to forked, free; sori abaxial and exindusiate (*Lophosoria*) or marginal (*Calochlaena, Dicksonia*) and each with a bivalvate or cup-like indusium, the adaxial (outer) valve formed by the reflexed segment margin and often differently colored; sporangia with oblique annuli; receptacles raised; paraphyses often present, filiform; spores globose or tetrahedral, trilete; $x = 56$ (*Calochlaena*), 65 (*Dicksonia, Lophosoria*).

Lophosoria (three species) is distinctive in having spores with a prominent subequatorial flange, with the proximal face coarsely tuberculate, the distal face perforate. It has often been placed in its own family, Lophosoriaceae (Pichi Sermolli, 1977; Tryon and Tryon, 1982; Kramer in Kubitzki, 1990).

22. Family Metaxyaceae. A single genus (*Metaxya*), two species; monophyletic (Smith et al., 2001). Terrestrial, Neotropics. Characters: rhizomes short-creeping to ascending, dorsiventral, solenostelic, apices covered with pluricellular hairs; petioles each with an omega-shaped, corrugated, vascular bundle; blades simply pinnate; veins free, simple or forked at the base, ± parallel; sori abaxial, round, scattered in several ill-defined rows, often with several sori on the same vein, with numerous filiform paraphyses, exindusiate; sporangia maturing simultaneously; sporangial stalks 4-rowed; annuli vertical or slightly oblique; spores 64 per sporangium, globose, trilete; $x = 95, 96$.

K. Order Polypodiales

(incl. "Aspidiales," Aspleniales, Athyriales, Blechnales, "Davalliales," Dennstaedtiales, Dryopteridales, Lindsaeales, Lonchitidales, Monachosorales, Negripteridales, Parkeriales, Platyzomatales, Pteridales, Saccolomatales, Thelypteridales). Monophyletic (Hasebe et al., 1996; Pryer et al., 1996, 2001a, 2004b; Schneider et al. 2004d). Characters: indusia laterally or centrally attached (indusia lost in many lineages); sporangial stalks 1–3 cells thick, often long; sporangial maturation mixed; sporangia each with a vertical annulus interrupted by the stalk and stomium; gametophytes green, usually cordate (sometimes ribbon-shaped in some epiphytes), surficial.

23. **Family Lindsaeaceae** (lindsaeoids; incl. Cystodiaceae, Lonchitidaceae). Ca. eight genera (*Cystodium, Lindsaea, Lonchitis, Odontosoria, Ormoloma*[1], *Sphenomeris, Tapeinidium, Xyropteris*[1]); in an unpublished thesis, Barcelona (2000) advocated the establishment of three additional genera allied to *Odontosoria* and *Sphenomeris*. Ca. 200 species; most likely monophyletic (Wolf *et al.*, 1994; Pryer *et al.*, 2004b; Korall *et al.*, 2006a; Schuettpelz *et al.*, 2006). The inclusion of *Lonchitis* (traditionally associated with dennstaedtioid ferns) in Lindsaeaceae is puzzling on morphological grounds, but molecular evidence strongly suggests it belongs near the lindsaeoid ferns (Schuettpelz and Pryer, 2007). Terrestrial, or infrequently epipetric or epiphytic, pantropical. Characters: roots with sclerenchymatous outer cortex combined with an innermost cortical layer six cells wide (Schneider, 1996a) (excepting *Lonchitis* and *Cystodium*); rhizomes short- to long-creeping, protostelic with internal phloem, or in a few taxa solenostelic, bearing generally narrow, basally attached, non-clathrate scales or uniseriate hairs; blades 1–3-pinnate or more divided, generally glabrous; veins usually free, forking, occasionally anastomosing, without included veinlets; sori marginal or submarginal, indusiate, indusia opening towards the margin (extrorse), sometimes also attached at the sides, or sori covered by the reflexed segment margin (*Lonchitis*); spores tetrahedral, trilete, infrequently bilateral, monolete; gametophytes green, cordate; $x = 34, 38, 39, 44, 47, 48, 49, 50, 51$, perhaps others.

The position of *Cystodium* is clearly among Polypodiales, and not Dicksoniaceae (Cyatheales), where it has historically been placed, e.g., by Kramer in Kubitzki (1990) and Stevenson and Loconte (1996). Croft (1986) discussed its differences from dicksonioids and elevated it to family rank. A relationship to other lindsaeoids is suggested by molecular evidence, and this is reflected in our classification. However, expanded taxon sampling within early-diverging lineages of Polypodiales is necessary to confirm this or to determine whether recognition of a monotypic family Cystodiaceae is warranted (Korall *et al.*, 2006a; Schuettpelz and Pryer, 2007).

24. **Family Saccolomataceae.** One genus, ca. 12 species; apparently monophyletic, but more sampling is needed to determine whether the Old World species are congeneric with those from the New World. The relationships of *Saccoloma* (incl. *Orthiopteris*) have been contentious. Kramer (in Kubitzki, 1990) treated *Saccoloma* and Lindsaeoideae as subfamilies within Dennstaedtiaceae. Molecular data suggest that it lies at or near the base of the polypodialean radiation, just below *Cystodium* and *Lonchitis* (Schuettpelz and Pryer, 2007). Terrestrial,

pantropical. Characters: rhizomes short-creeping to erect and somewhat trunk-like (long-creeping in most Lindsaeaceae and Dennstaedtiaceae) and dictyostelic (usually solenostelic in Dennstaedtiaceae, protostelic with internal phloem in Lindsaeaceae); petioles each with an omega-shaped vascular strand (open end adaxial); blades pinnate to decompound, lacking articulate hairs (as found in Dennstaedtiaceae); veins free; sori terminal on the veins, indusia pouch- or cup-shaped; spores globose-tetrahedral, surface with distinctive ± parallel, branched ridges; $x =$ ca. 63.

25. Family Dennstaedtiaceae (dennstaedtioids; incl. Hypolepidaceae, Monachosoraceae, Pteridiaceae). Ca. 11 genera: *Blotiella, Coptodipteris, Dennstaedtia* (incl. *Costaricia*[1]), *Histiopteris, Hypolepis, Leptolepia, Microlepia, Monachosorum, Oenotrichia s.s.*[1], *Paesia, Pteridium* (bracken). The north-temperate *Dennstaedtia punctilobula* (Michx.) T. Moore is aberrant in *Dennstaedtia*, probably rendering that genus polyphyletic (Schuettpelz and Pryer, 2007). Ca. 170 species; monophyletic, if lindsaeoid ferns are excluded (Pryer *et al.*, 2004b; Schuettpelz *et al.*, 2006). Monachosoraceae nests within Dennstaedtiaceae (Wolf *et al.*, 1994; Wolf, 1996a, 1997; Pryer *et al.*, 2004b; Schuettpelz *et al.*, 2006). Terrestrial, sometimes scandent; pantropical. Characters: rhizomes mostly long-creeping, often siphonostelic or polystelic, bearing jointed hairs; petioles often with epipetiolar buds, usually with a gutter-shaped vascular strand (adaxial face open); blades often large, 2–3-pinnate or more divided; indument of hairs; veins free, forked or pinnate, rarely anastomosing and then without included veinlets; sori marginal or submarginal, linear or discrete, indusia linear or cup-like at blade margins, or reflexed over sori; spores tetrahedral and trilete, or reniform and monolete; gametophytes green, cordate; $x =$ 26, 29, 30, 31, 33, 34, 38, 46, 47, 48, and probably others.

26. Family Pteridaceae. (pteroids or pteridoids); incl. Acrostichaceae, Actiniopteridaceae, Adiantaceae (adiantoids, maidenhairs), Antrophyaceae, Ceratopteridaceae, Cheilanthaceae (cheilanthoids), Cryptogrammaceae, Hemionitidaceae, Negripteridaceae, Parkeriaceae, Platyzomataceae, Sinopteridaceae, Taenitidaceae (taenitidoids), Vittariaceae (vittarioids, shoestring ferns). Ca. 50 genera, 950 species; monophyletic (Prado *et al.*, 2007; Schuettpelz *et al.*, 2007). Constituent genera, some of them notoriously polyphyletic or paraphyletic and in need of redefinition (e.g., *Cheilanthes*), include *Acrostichum, Actiniopteris, Adiantopsis, Adiantum, Aleuritopteris, Ananthacorus, Anetium, Anogramma, Antrophyum, Argyrochosma, Aspidotis, Astrolepis, Austrogramme, Bommeria, Cassebeera, Ceratopteris, Cerosora*[1], *Cheilanthes, Cheiloplecton, Coniogramme, Cosentinia* (Nakazato and Gastony, 2001), *Cryptogramma, Doryopteris, Eriosorus, Haplopteris, Hecistopteris, Hemionitis, Holcochlaena, Jamesonia, Llavea, Mildella, Monogramma, Nephopteris*[1], *Neurocallis* (probably nested within a portion of *Pteris s.l.*; Schuettpelz and Pryer, 2007), *Notholaena, Ochropteris, Onychium, Paraceterach, Parahemionitis, Pellaea* (Kirkpatrick, 2007), *Pentagramma, Pityrogramma, Platyloma, Platyzoma, Polytaenium, Pteris* (incl. *Afropteris, Anopteris*), *Pterozonium, Radiovittaria, Rheopteris, Scoliosorus, Syngramma, Taenitis, Trachypteris,* and *Vittaria*. The family thus defined is monophyletic (Gastony and Rollo, 1996, 1998; Hasebe et al., 1996; Pryer et al., 1996; Gastony and Johnson, 2001; Schneider et al. 2004c; Zhang et al., 2005; Prado et al., 2007). Pteridaceae comprises five monophyletic groups (Prado et al., 2007; Schuettpelz and Pryer, 2007; Schuettpelz et al., 2007), and if it were to be formally subdivided to reflect this, at either family or subfamily rank, the following taxa could be recognized: (1) Parkeriaceae, or Parkerioideae (*Acrostichum* and *Ceratopteris*); (2) Adiantaceae, or Adiantoideae, but Vittarioideae and Antrophyoideae have priority at subfamily rank (*Adiantum* and the ten vittarioid genera; Crane et al., 1996; Hasebe et al., 1996; Hennipman, 1996; Crane, 1997; Huiet and Smith, unpublished data); (3) Cryptogrammaceae (comprising *Coniogramme, Cryptogramma,* and *Llavea*; Zhang et al., 2005; Schuettpelz et al., 2007), no subfamily name available; (4) Sinopteridaceae, or Cheilanthoideae, but Notholaenoideae has priority); and (5) Pteridaceae s.s., or Pteridoideae, containing *Pteris* (probably not monophyletic; Schuettpelz and Pryer, 2007; Schuettpelz et al., 2007), its immediate allies, and the taenitoid ferns (*Taenitis* and allies; Sánchez-Baracaldo, 2004a, 2004b).

Terrestrial, epipetric, or epiphytic, subcosmopolitan, but most numerous in tropics and arid regions. Characters: rhizomes long- to short-creeping, ascending, suberect, or erect, bearing scales (less often, only hairs); blades monomorphic, hemidimorphic, or dimorphic in a few genera, simple (mostly vittarioids), pinnate, or sometimes pedate, sometimes decompound; veins free and forking, or variously anastomosing and forming a reticulate pattern without included veinlets; sori marginal or intramarginal, lacking a true indusium, often protected by the reflexed segment margin, or sporangia along the veins; sporangia each with a vertical, interrupted annulus, receptacles not or only obscurely raised; spores globose or tetrahedral, trilete, variously ornamented; mostly $x = 29, 30$.

Platyzoma, sometimes recognized as an isolated monotypic family, is aberrant in having numerous tiny, pouchlike pinnae (100+ pairs per frond), distinctive finely ridged and reticulate spore ornamentation (Tryon and Lugardon, 1991), an unusual (in the family) chromosome base number ($x = 38$; Tindale and Roy, 2002), and dimorphic spores (so-called "incipient heterospory"; A. Tryon, 1964), but *Platyzoma* nests with other genera of Pteridaceae, subfamily Pteridoideae (Hasebe et al., 1996; Pryer et al., 1996), perhaps near the base of the subfamily (Hasebe et al., 1996; Schuettpelz et al., 2007).

Ceratopteris (three species, monophyletic) nests within Pteridaceae in all molecular analyses, and it appears to be sister to *Acrostichum* (Hasebe et al., 1996; Pryer et al., 1996). It has a number of strong autapomorphies that separate it from other Pteridaceae: coarsely ridged spores with parallel striations; spores 32 or fewer per sporangium; sporangia with ill-defined annuli; aquatic habitat; $x = 38$. Consequently, many taxonomists have placed it in its own family, Parkeriaceae (e.g., Copeland, 1947; Pichi Sermolli, 1977). Many of these autapomorphies (reduced spore number, loss of annulus) are probably a consequence of the shift to aquatic habitats.

The vittarioid genera include: *Ananthacorus*, *Anetium*, *Antrophyum*, *Haplopteris*, *Hecistopteris*, *Monogramma*, *Polytaenium*, *Radiovittaria*, *Rheopteris*, *Scoliosorus*, and *Vittaria*. The presence of *Rheopteris* in this clade is now clear (Schuettpelz et al., 2007; Huiet and Smith, unpublished data). Characters include linear, mostly simple blades, sori along veins or in linear grooves, clathrate stem scales; presence of "spicular" cells in blade epidermis (shared with a few genera of Pteridaceae, e.g., *Adiantum*). Spores in the vittarioid ferns are predominantly trilete, but reversals to the monolete condition have occurred in *Vittaria*.

27. Family Aspleniaceae (asplenioids, spleenworts). From one to ten genera (generic delimitation in doubt, in light of all recent molecular data, e.g., van den Heede et al., 2003; Schneider et al., 2004b, 2005; Perrie and Brownsey, 2005). Regardless of the classification adopted, a huge proportion of the species are in *Asplenium*, even if that genus is construed in a fairly strict sense; the segregate genera *Camptosorus*, *Loxoscaphe*, *Phyllitis*, and *Pleurosorus* clearly nest within *Asplenium* s.l., or appear related to species heretofore generally placed in *Asplenium* (Murakami et al., 1999; Gastony and Johnson, 2001; Schneider et al. 2004b). So also are the generic segregates *Diellia* (endemic to Hawaii), *Pleurosorus*, *Phyllitis*, *Ceterach*, *Thamnopteris*, and several others little used in recent years (Murakami et al., 1999; Pinter et al., 2002; van den Heede et al., 2003; Schneider et al., 2005). One expects that the oft-recognized, but still unsampled, genera *Antigramma*[1], *Holodictyum*[1], *Schaffneria*[1], and *Sinephropteris*[1] also nest in *Asplenium*. *Hymenasplenium*, however, with a different chromosome base number than nearly all of the other segregates, as well as distinctive root characters (Schneider, 1996b; Schneider et al., 2004b, 2005), appears to represent the sister clade to the rest of the species in the family, and this name could be adopted for a well-supported segregate genus. Ca. 700+ species; monophyletic (Murakami and Schaal, 1994; Hasebe et al., 1996; Murakami et al., 1999; Gastony and Johnson, 2001; van den Heede et al., 2003; Perrie and Brownsey, 2005; Schneider et al., 2005).

Terrestrial, epipetric, or epiphytic, subcosmopolitan, but most numerous in the tropics. Characters: rhizomes creeping, ascending, or suberect, bearing clathrate scales at apices and petiole bases (and sometimes other axes); petioles with back-to-back C-shaped vascular strands, these fusing distally into an X-shape; blades monomorphic, usually lacking acicular hairs on axes and/or lamina, often with microscopic clavate hairs; veins pinnate or forking, usually free, infrequently reticulate and then without included veinlets; sori elongate (linear) along the veins, not usually back-to-back on the same vein, usually with laterally attached, linear indusia; sporangial stalks long, 1-rowed; spores reniform, monolete, with a decidedly winged perine; $x = 36$ (mostly), but $x = 38, 39$ in *Hymenasplenium* (Murakami, 1995), 38 in *Boniniella*.

28. **Family Thelypteridaceae** (thelypteroids or thelypteridoids; incl. "Sphaerostephanaceae"). Ca. 5–30 genera, depending on taxonomic viewpoint: commonly accepted segregates are *Cyclosorus* (incl. *Ampelopteris*[1], *Amphineuron*, *Chingia*, *Christella*, *Cyclogramma*[1], *Cyclosorus s.s.*, *Glaphyropteridopsis*, *Goniopteris*, *Meniscium*, *Menisorus*[1], *Mesophlebion*, *Pelazoneuron*, *Plesioneuron*, *Pneumatopteris*, *Pronephrium*, *Pseudocyclosorus*, *Sphaerostephanos*, *Stegnogramma*, *Steiropteris*, *Trigonospora*), *Macrothelypteris*, *Phegopteris*, *Pseudophegopteris*, and *Thelypteris* (incl. *Amauropelta*, *Coryphopteris*, *Metathelypteris*, *Oreopteris*, *Parathelypteris*, and *Thelypteris s.s.*) (see Holttum, 1971; Smith and Cranfill, 2002). Smith (in Kubitzki, 1990) advocated recognition of five genera. Existing studies suggest that the phegopteroid genera (*Macrothelypteris*, *Phegopteris*, *Pseudophegopteris*) constitute a monophyletic clade at the base of the family, sister to all other "genera" (Smith and Cranfill, 2002; Schuettpelz and Pryer, 2007). These same studies also indicate the paraphyly or polyphyly of certain other segregates (e.g., *Christella*, *Pronephrium*), as well as the monophyly of other segregates (e.g., *Amauropelta*, *Goniopteris*; Schuettpelz and Pryer, 2007). Clearly, much more sampling is needed to understand the complex relationships within this species-rich family. Ca. 950 species; monophyletic (Hasebe et al., 1996; Smith and Cranfill, 2002; Yatabe et al., 2002; Schuettpelz and Pryer, 2007). Hennipman (1996) also included Blechnaceae and the athyrioid ferns in this family, a definition that would make Thelypteridaceae difficult or impossible to define morphologically. It is unclear whether the numerous genera recognized by Holttum (1971 and later publications) will hold up when additional molecular sampling has been done.

Terrestrial, rarely epipetric, pantropical, a few temperate. Characters: rhizomes creeping, ascending, or erect, bearing scales at apices, these non-clathrate, usually bearing acicular hairs; petioles in cross-section with two elongate or crescent-shaped vascular bundles facing one another, these uniting distally into a gutter-shape; blades monomorphic or occasionally dimorphic, usually pinnate or pinnate-pinnatifid; veins pinnate, free to variously and usually very regularly anastomosing, with or without included veinlets; indument of acicular hyaline hairs on blades and rhizome scales; sori abaxial, round to oblong, rarely elongate along veins, with reniform indusia or exindusiate; sporangia with 3-rowed, short to long stalks; spores ellipsoid, monolete, perine winged to spinulose; $x = 27$–36. Indusia have been lost independently in many lineages within the family.

29. Family Woodsiaceae (athyrioids, lady ferns; incl. Athyriaceae, Cystopteridaceae). Ca. 15 genera as defined here, ca. 700 species, nearly 85% of them in the two main genera, *Athyrium* and *Diplazium* (incl. *Callipteris, Monomelangium*), which are both probably paraphyletic (Wang et al., 2003). Other widely recognized genera include *Acystopteris, Cheilanthopsis, Cornopteris, Cystopteris, Deparia* (incl. *Lunathyrium, Dryoathyrium, Athyriopsis,* and *Dictyodroma*; Sano et al., 2000b), *Diplaziopsis, Gymnocarpium* (incl. *Currania*), *Hemidictyum, Homalosorus, Protowoodsia*[1], *Pseudocystopteris, Rhachidosorus,* and *Woodsia* (incl. *Hymenocystis*[1]; see Shmakov, 2003). This family has been variously circumscribed, and its limits are still uncertain (Hasebe et al., 1996; Sano et al., 2000a, 2000b; Schuettpelz and Pryer, 2007). Wang et al. (2004) divided Athyriaceae (excluding woodsioid ferns, in their circumscription), by far the largest component in our concept of Woodsiaceae, into five subfamilies: Cystopteroideae, Athyrioideae, Deparioideae, Diplazioideae, and Rhachidosoroideae. As delimited here, the Woodsiaceae s.l. may be paraphyletic with respect to the Aspleniaceae, Blechnaceae + Onocleaceae, and Thelypteridaceae clades, but support for this paraphyly, or alternatively for the monophyly of the family as here defined, is lacking in broad analyses (Hasebe et al., 1996; Sano et al., 2000a). Because of this uncertainty, combined with the morphological grounds for the recognition of the Woodsiaceae as here circumscribed, we believe it is premature to adopt the alternative of erecting (or resurrecting) numerous small families to house its constituent genera. The most aberrant genera, within the Woodsiaceae as circumscribed here, appear to be *Woodsia* itself, *Cystopteris, Gymnocarpium,* and *Hemidictyum* (Schuettpelz and Pryer, 2007). Further sampling may support the recognition of several additional families.

Mostly terrestrial, subcosmopolitan. Characters: rhizomes creeping, ascending, or erect; scales at apices, these usually non-clathrate, glabrous, glandular, or ciliate; petioles with two elongate or crescent-shaped vascular bundles facing one another, these uniting distally into a gutter-shape; blades monomorphic, rarely dimorphic; veins pinnate or forking, free, uncommonly anastomosing and then without included veinlets; sori abaxial, round, J-shaped, or linear with reniform to linear indusia, or exindusiate; spores reniform, monolete, perine winged, ridged, or spiny; mostly $x = 40, 41$; also 31 (*Hemidictyum*), 33, 38, 39 (*Woodsia*), and 42 (*Cystopteris*).

30. Family Blechnaceae (blechnoids; incl. Stenochlaenaceae). Currently ca. nine genera recognized (*Blechnum s.l.*, *Brainea*, *Doodia*, *Pteridoblechnum*, *Sadleria*, *Salpichlaena*, *Steenisioblechnum*, *Stenochlaena*, *Woodwardia*). Most of the existing recognized genera nest within *Blechnum s.l.*, and their acceptance is dependent on a revised recircumscription of *Blechnum s.l.*, which is manifestly paraphyletic in its current usage (Nakahira, 2000; Cranfill, 2001; Shepherd et al., 2007; Schuettpelz and Pryer, 2007). Ca. 200 species; monophyletic, sister to Onocleaceae (Hasebe et al., 1996; Cranfill, 2001; Cranfill and Kato, 2003). *Woodwardia* (incl. *Anchistea*, *Chieniopteris*, *Lorinseria*) appears to be an early-branching member of the Blechnaceae (Cranfill and Kato, 2003). Characters: rhizomes creeping, ascending, or erect, sometimes trunk-like, often bearing stolons, scaly at apex (and on blades), scales non-clathrate; petioles with numerous, round, vascular bundles arranged in a ring; leaves monomorphic or often dimorphic; veins pinnate or forking, free to variously anastomosing, areoles without included veinlets, on fertile leaves forming costular areoles bearing the sori; sori in chains or linear, often parallel and adjacent to midribs, indusiate, with linear indusia opening inwardly (toward midribs); sporangia with 3-rowed, short to long stalks; spores reniform, monolete, perine winged; gametophytes green, cordate; $x = 27, 28, 31–37$ (*Blechnum* and segregates, *Woodwardia*); 40 (*Salpichlaena*).

31. Family Onocleaceae (onocleoids). Four genera (*Matteuccia*, *Onoclea*, *Onocleopsis*, *Pentarhizidium*), five species; monophyletic, sister to Blechnaceae (Hasebe et al., 1996; Gastony and Ungerer, 1997). Family circumscription follows Pichi Sermolli (1977) and Gastony and Ungerer (1997, their tribe Onocleeae of Dryopteridaceae). Terrestrial, largely in north-temperate regions. Characters: rhizomes long- to short-creeping to ascending, sometimes stoloniferous (*Matteuccia* and *Onocleopsis*); leaves strongly dimorphic; petioles with two vascular bundles uniting distally into a gutter-shape; blades pinnatifid or pinnate-pinnatifid; veins free or anastomosing, lacking included veinlets; spores reniform, brownish to green; sori enclosed (sometimes tightly) by reflexed laminar margins, also with membranous, often fugacious true indusia; $x = 37$ (*Onoclea*), 39 (*Matteuccia*), 40 (*Onocleopsis*, *Pentarhizidium*).

32. Family Dryopteridaceae (dryopteroids or dryopteridoids; incl. "Aspidiaceae," Bolbitidaceae, Elaphoglossaceae, Hypodematiaceae, Peranemataceae). Ca. 30–35 genera, 1700 species, of which 70% are in four genera (*Ctenitis*, *Dryopteris*, *Elaphoglossum*, and *Polystichum*) (Li and Lu, 2006). Genera include *Acrophorus*, *Adenoderris*, *Arachniodes* (incl. *Lithostegia*, *Phanerophlebiopsis*; Li and Lu, 2006), *Ataxipteris*[1], *Bolbitis* (incl. *Egenolfia*), *Coveniella*[1], *Ctenitis*, *Cyclodium*, *Cyrtogonellum* (incl. *Cyrtomidictyum*), *Cyrtomium* (Lu et al., 2005, 2007), *Didymochlaena*, *Dryopolystichum*[1], *Dryopsis*,

Dryopteris (incl. *Acrorumohra*, *Nothoperanema*, and probably other segregates tentatively accepted in this paper; for *Nothoperanema*, see Geiger and Ranker, 2005; for *Acrorumohra*, see Li and Lu, 2006), *Elaphoglossum* (incl. *Microstaphyla*[1], *Peltapteris*; Rouhan *et al.*, 2004; Skog *et al.*, 2004), *Hypodematium*, *Lastreopsis*, *Leucostegia*, *Lomagramma*, *Maxonia*, *Megalastrum*, *Oenotrichia* p.p. (Tindale and Roy, 2002), *Olfersia*, *Peranema*, *Phanerophlebia*, *Polystichum* (incl. *Papuapteris*, *Plecosorus*, *Sorolepidium*; Little and Barrington, 2003; Driscoll and Barrington, 2007; Lu *et al.*, 2007), *Polybotrya*, *Polystichopsis*, *Revwattsia*[1] (Tindale and Roy, 2002), *Rumohra*, *Stenolepia*[1], *Stigmatopteris*, and *Teratophyllum*.

Terrestrial, epipetric, hemiepiphytic, or epiphytic, pantropical, also with many temperate representatives. Characters: rhizomes creeping, ascending, or erect, sometimes scandent or climbing, with non-clathrate scales at apices; petioles with numerous round, vascular bundles arranged in a ring; blades monomorphic, less often dimorphic, sometimes scaly or glandular, uncommonly hairy; veins pinnate or forking, free to variously anastomosing, with or without included veinlets; sori usually round, indusia round-reniform or peltate (lost in several lineages), or sori exindusiate, acrostichoid in a few lineages; sporangia with 3-rowed, short to long stalks; spores reniform, monolete, perine winged; $x = 41$ (nearly all genera counted), rarely 40 (presumably derived).

Dryopteridaceae, as defined here, is almost certainly monophyletic, if *Didymochlaena*, *Hypodematium*, and *Leucostegia* are excluded (Hasebe *et al.*, 1996; Tsutsumi and Kato, 2006; Schuettpelz and Pryer, 2007). The inclusion of these three genera may render this family paraphyletic, but they are tentatively included here pending further studies to address their precise phylogenetic affinities. *Didymochlaena*, with a single species, has generally been associated with other members of the Dryopteridaceae (as here defined). The three closely related species of *Hypodematium*, on the other hand, have been variously treated: as composing a monogeneric family Hypodematiaceae; as allied to the athyrioid ferns (e.g., by Kramer in Kubitzki, 1990, presumably based on the presence of two vascular bundles in the petiole bases); or close to *Dryopteris* (e.g., Tryon and Lugardon, 1991, using evidence from spore morphology). *Leucostegia* is nearly always placed in Davalliaceae (e.g., by Kramer, in Kubitzki, 1990), because of its similar indusia and sori terminal on the veins, but it differs from members of Davalliaceae in the terrestrial habit, the more strongly verrucate spores with rugulate perispore (Tryon and Lugardon, 1991), and $x = 41$ (versus $x = 40$ in Davalliaceae). In a molecular phylogenetic analysis by Schneider *et al.* (2004c), *Didymochlaena* and *Hypodematium* were resolved as sister to one another, and together sister to the remainder of the eupolypods I clade (their Figure 2), but support for these relationships was lacking. Tsutsumi and Kato (2006) found support for a sister relationship between *Hypodematium* and *Leucostegia*, and also support for these as sister to the remaining eupolypods I, but *Didymochlaena* was unsampled in their analysis. In the analysis by Schuettpelz and Pryer (2007), these three genera form a weakly supported clade at the base of Dryopteridaceae. Based on these results, we therefore consider it premature to segregate these genera from the Dryopteridaceae.

The indusium, either reniform or peltate and superior in most members of Dryopteridaceae, has undergone remarkable transformation in some genera, e.g., *Peranema*, which has inferior, initially globose indusia, and *Acrophorus*, with shallow, ± semicircular, cuplike indusia. The remarkably different indusia in *Peranema* are the basis for the segregate family Peranemataceae, recognized by some authorities, but it seems likely that this genus is very closely related to *Dryopteris*, and may not be separable even generically. Soral position in Dryopteridaceae is also remarkably variable, sori sometimes being borne on the tips of marginal teeth in *Dryopteris deparioides* (T. Moore) Kuntze subsp. *deparioides* (Fraser-Jenkins, 1989), or on elevated receptacles in *Stenolepia*. Indusia have been lost independently along many evolutionary lines in Dryopteridaceae, even within genera, e.g., in *Ctenitis*, *Dryopteris*, *Lastreopsis*, *Megalastrum*, *Phanerophlebia*, *Polystichum*, and *Stigmatopteris*, as well as in a suite of dimorphic genera, e.g., *Elaphoglossum*, *Maxonia*, *Olfersia*, and *Polybotrya*.

Within Dryopteridaceae, as defined here, nests *Elaphoglossum* (Hasebe et al., 1996; Sano et al., 2000a). Sometimes it is included in its own family Elaphoglossaceae (e.g., by Pichi Sermolli, 1977), with 600–800 species, many still undescribed. Elaphoglossaceae was regarded as comprising three genera by Pichi Sermolli (1977), but *Microstaphyla* and *Peltapteris* nest within *Elaphoglossum* (www.nybg.org/bsci/res/moran/elaphoglossum.html; Mickel and Atehortúa, 1980; Rouhan et al., 2004; Skog et al., 2004). Relationships of *Elaphoglossum* are often considered to be with *Lomariopsis* (Kramer in Kubitzki, 1990), but this is refuted by two unpublished topologies. Elaphoglossaceae, narrowly defined, is monophyletic (Skog et al., 2004), but to exclude it from Dryopteridaceae *s.s.*, as delimited above, renders the latter paraphyletic. Characters of *Elaphoglossum* include simple blades (usually) and dimorphic leaves with acrostichoid sporangia.

Several authors have treated most of the genera within our concept of Dryopteridaceae, as well as Tectariaceae, Woodsiaceae, and Onocleaceae, as comprising a much larger family Dryopteridaceae *s.l.*, with slightly varying circumscriptions (e.g., Tryon and Tryon, 1982; Kramer in Kubitzki, 1990; Wagner and Smith, 1993). With such a broad circumscription, and unless several other well-circumscribed families (e.g., Aspleniaceae, Blechnaceae, Polypodiaceae, Thelypteridaceae) are included, Dryopteridaceae is certainly paraphyletic.

33. Lomariopsidaceae (lomariopsids; incl. Nephrolepidaceae, sword ferns). Four genera (*Cyclopeltis*, *Lomariopsis*, *Nephrolepis*, *Thysanosoria*[1]); ca. 70 species. Characters: rhizomes creeping or sometimes climbing (plants hemi-epiphytic); petioles with round vascular bundles arranged in a gutter-shape; blades 1-pinnate, pinnae entire or crenate, often articulate, auriculate in some genera; veins free, ± parallel or pinnate; sori discrete, round, and with round-reniform to reniform indusia, or exindusiate, or sporangia acrostichoid and the fronds dimorphic; spores bilateral, monolete, variously winged or ornamented; $x = 41$ (lower numbers known in some *Lomariopsis* species).

Based on published and unpublished results (especially Schuettpelz and Pryer, 2007), it appears likely that these genera form a monophyletic group, despite the fact that such an assemblage has never been proposed. Lomariopsidaceae (sensu Kramer in Kubitzki, 1990; Moran in Davidse et al., 1995) was construed to comprise six genera (containing ca. 800+ species): *Bolbitis* (and segregates *Edanyoa*, *Egenolfia*), *Elaphoglossum*, *Lomagramma*, *Lomariopsis*, *Teratophyllum*, and *Thysanosoria*[1]. We place all of the aforementioned genera except *Lomariopsis* (and *Thysanosoria*, which lacks molecular data, but appears to be closely related to *Lomariopsis*) in the Dryopteridaceae (see above). *Nephrolepis*, with ca. 20 species, has sometimes been included in a monogeneric family Nephrolepidaceae (Kramer in Kubitzki, 1990). There is support in some analyses for *Nephrolepis* as sister to a large clade comprising the Tectariaceae, Oleandraceae, Polypodiaceae, and Davalliaceae (Hasebe et al., 1996; Schneider et al., 2004c); however, *Lomariopsis* was not included in these analyses. When *Lomariopsis* is included, *Nephrolepis* is resolved as sister to it, and these two genera, in turn, are strongly supported as sister to the aforementioned larger clade (Tsutsumi and Kato, 2006) and therefore to be expunged from the Dryopteridaceae. Although we have here decided tentatively to include *Nephrolepis* in the Lomariopsidaceae, the monophyly of this clade requires additional scrutiny, and thus Nephrolepidaceae may eventually require recognition.

Lu and Li (2006a) attempted to show that *Cyclopeltis*, based on a sample from one species, and using only *rbcL* data, should be placed in Tectariaceae. However, their own molecular trees do not support this placement, and we also consider their sampling of genera to be inadequate for resolving affinities of this genus.

34. Family Tectariaceae (tectarioids; incl. "Dictyoxiphiaceae," "Hypoderriaceae"). 8–15 genera: *Aenigmopteris*[1], *Arthropteris*, *Heterogonium*, *Hypoderris*[1], *Pleocnemia*, *Psammiosorus*, *Psomiocarpa*[1], *Pteridrys*, *Tectaria* s.l. (incl. *Amphiblestra*[1], *Camptodium*[1], *Chlamydogramme*[1], *Cionidium*, *Ctenitopsis*, *Dictyoxiphium*, *Fadyenia*, *Hemigramma*, *Pleuroderris*[1], *Pseudotectaria*[1], *Quercifilix*, and perhaps other genera mentioned above), and *Triplophyllum* (Holttum, 1986); ca. 230 species, most in *Tectaria* s.l.

Generic limits, especially within *Tectaria s.l.*, are still very much in doubt. With the definition given here, Tectariaceae appears monophyletic, with moderate support (Schuettpelz and Pryer, 2007). Including Tectariaceae within an expanded Dryopteridaceae renders the latter polyphyletic. *Ctenitis, Lastreopsis*, and several other genera here included in Dryopteridaceae have often been considered closely related to tectarioid ferns (Pichi Sermolli, 1977; Holttum, 1986; Moran in Davidse *et al.*, 1995), but molecular data suggest otherwise (Hasebe *et al.*, 1996; Schuettpelz and Pryer, 2007). Terrestrial, pantropical. Characters: rhizomes usually short-creeping to ascending, dictyostelic, bearing scales; petioles not abscising, with a ring of vascular bundles in cross-section; blades simple, pinnate, or bipinnate, sometimes decompound; indument of jointed, usually short stubby hairs on the axes, veins, and sometimes laminar tissue, especially on rachises and costae adaxially; veins free or often highly anastomosing, sometimes with included veinlets; indusia reniform or peltate (lost in several lineages); spores brownish, reniform, monolete, variously ornamented; $x = 40$ (a few genera with $x = 41$, some dyploids with $x = 39$).

Arthropteris is apparently not closely related to *Oleandra*, as previously suggested (Kramer in Kubitzki, 1990), nor to *Nephrolepis*, as suggested by Pichi Sermolli (1977). Analyses that have included it show it to be sister to tectarioid ferns (Hasebe *et al.*, 1996; Tsutsumi and Kato, 2006; Schuettpelz and Pryer, 2007). *Psammiosorus*, a monotypic genus endemic to Madagascar, has in turn been placed close to *Arthropteris* (Kramer, in Kubitzki, 1990) or even within *Arthropteris* (Tryon and Lugardon, 1991, on the basis of spore ornamentation). Therefore, both *Arthropteris* and *Psammiosorus* are tentatively assigned to Tectariaceae, although a Tectariaceae that includes them is more difficult to define morphologically.

35. Oleandraceae. Monogeneric, ca. 40 species, sister to Davalliaceae + Polypodiaceae (Hasebe *et al.*, 1996; Schneider *et al.*, 2004c, 2004d; Tsutsumi and Kato, 2006). Kramer (in Kubitzki, 1990), included two genera in addition to *Oleandra*: *Arthropteris* (ca. 12 species), and *Psammiosorus* (monotypic), but with this broader circumscription, the family is clearly polyphyletic; we include both of these genera in Tectariaceae. Species are terrestrial, epilithic or often secondary hemiepiphytes. Characters: blades simple; leaves articulate, abscising cleanly upon senescence from pronounced phyllopodia; sori indusiate, indusia round-reniform; spores reniform, monolete; $x = 41$.

36. Family Davalliaceae (davallioids; excl. Gymnogrammitidaceae). Four or five genera: *Araiostegia, Davallia* (incl. *Humata, Parasorus, Scyphularia*), *Davallodes, Pachypleuria*; ca. 65 species. Monophyletic, sister to Polypodiaceae (Hasebe *et al.*, 1996; Ranker *et al.*, 2004; Schneider *et al.*, 2004c, 2004d; Tsutsumi and Kato, 2005), but more information needed. *Gymnogrammitis* and *Leucostegia* are often included in Davalliaceae but the former belongs in Polypodiaceae (Schneider *et al.*, 2002b), while the latter is seemingly allied to *Hypodematium* (Dryopteridaceae; Tsutsumi and Kato, 2005; Schuettpelz and Pryer, 2007). Generic limits of *Araiostegia*, *Davallia*, and *Pachypleuria* relative to each other are ill defined, and all of these genera appear to be paraphyletic or polyphyletic (Tsutsumi and Kato, 2005). Paleotropics and subtropics, Pacific Basin. Characters: plants epiphytic (most genera) or epipetric; rhizomes long-creeping, dictyostelic, dorsiventral, bearing scales; old leaves cleanly abscising at petiole bases; blades usually 1–4-pinnate (rarely simple), monomorphic (rarely dimorphic); veins free, forking or pinnate; indument generally lacking on blades and axes, but sometimes of articulate hairs; sori abaxial, inframarginal to well back from the margin, ± round, with cup-shaped to reniform or lunate indusia (rarely forming a submarginal coenosorus in *Parasorus*); sporangia with 3-rowed, usually long stalks; annuli vertical; spores ellipsoid, monolete, yellowish to tan, perine various, but usually not strongly winged or cristate; gametophytes green, cordate; $x = 40$.

37. Family Polypodiaceae (polygrams; incl. Drynariaceae, Grammitidaceae (grammitids), Gymnogrammitidaceae, Loxogrammaceae, Platyceriaceae, Pleurisoriopsidaceae). Ca. 56 genera, ca. 1200 species Pantropical, a few temperate. Genera include *Acrosorus*, *Adenophorus* (Ranker *et al.*, 2003), *Aglaomorpha* (incl. *Photinopteris, Merinthosorus, Pseudodrynaria, Holostachyum*; Janssen and Schneider, 2005), *Arthromeris*, *Belvisia*, *Calymmodon*, *Campyloneurum* (incl. *Hyalotrichopteris*; Kreier *et al.*, 2007), *Ceradenia*, *Christiopteris* (Kreier and Schneider, unpublished data), *Chrysogrammitis*, *Cochlidium*, *Colysis*, *Ctenopteris*, *Dicranoglossum*, *Dictymia*, *Drynaria* (Janssen and Schneider, 2005), *Enterosora*, *Goniophlebium* s.l., *Grammitis*, *Lecanopteris* (Haufler *et al.*, 2003), *Lellingeria*, *Lemmaphyllum*, *Lepisorus* (incl. *Platygyria*), *Leptochilus*, *Loxogramme* (including *Anarthropteris*, a monotype from New Zealand; Kreier and Schneider, 2006b), *Melpomene*, *Microgramma* (incl. *Solanopteris*), *Micropolypodium*, *Microsorum*, *Neocheiropteris* (incl. *Neolepisorus*), *Neurodium*, *Niphidium*, *Pecluma*, *Phlebodium*, *Phymatosorus*, *Platycerium* (Kreier and Schneider, 2006a), *Pleopeltis*, *Polypodiodes* (incl. *Metapolypodium*; Lu and Li, 2006b), *Polypodium*, *Prosaptia*, *Pyrrosia* (incl. *Drymoglossum*), *Scleroglossum*, *Selliguea* (incl. *Crypsinus*, *Polypodiopteris*), *Serpocaulon* (Smith *et al.*, 2006a), *Synammia* (Schneider *et al.*, 2006a), *Terpsichore*, *Themelium*, *Thylacopteris* (Schneider *et al.*, 2004a), and *Zygophlebia*[1]. Additional monotypic genera, include *Caobangia*[1], *Drymotaenium*, *Gymnogrammitis* (Schneider *et al.*, 2002b), *Kontumia*[1] (Wu *et al.*, 2005), *Luisma*[1], *Pleurosoriopsis*, and *Podosorus*[1].

Polypodiaceae *s.s.*, as often recognized (e.g., by Kramer in Kubitzki, 1990), is paraphyletic, because it excludes the grammitids, often segregated as Grammitidaceae (Ranker *et al.*, 2004; Schneider *et al.*, 2004d). Generic boundaries need clarification, and, in particular, *Polypodium* and *Microsorum*, two of the largest assemblages, are known to be polyphyletic (Schneider *et al.*, 2004d; 2006b). Certain previously misplaced genera are now shown to be nested within Polypodiaceae, e.g., *Pleurosoriopsis* (Hasebe *et al.*, 1996, Schneider *et al.*, 2004d) and *Gymnogrammitis* (Schneider *et al.*, 2002b). Polypodiaceae contains large wholly Neotropical and wholly Paleotropical clades (Schneider *et al.*, 2004d; Haufler, 2007).

Mostly epiphytic and epipetric, a few terrestrial; pantropical. Characters: rhizomes long-creeping to short-creeping, dictyostelic, bearing scales; petioles cleanly abscising near their bases or not (most grammitids), leaving short phyllopodia; blades monomorphic or dimorphic, mostly simple to pinnatifid or 1-pinnate (uncommonly more divided); indument lacking or of hairs and/or scales on the blade; veins often anastomosing or reticulate, sometimes with included veinlets, or veins free (most grammitids); indument various, of scales, hairs, or glands; sori abaxial (rarely marginal), round to oblong or elliptic, occasionally elongate, or the sporangia acrostichoid, sometimes deeply embedded; sori exindusiate, sometimes covered by caducous scales when young (e.g., *Lepisorus*, *Pleopeltis*); sporangia with 1–3-rowed, usually long stalks, frequently with paraphyses on sporangia or on receptacle; spores hyaline to yellowish, reniform, and monolete (non-grammitids), or greenish and globose-tetrahedral, trilete (nearly all grammitids); perine various, usually thin, not strongly winged or cristate; mostly $x = 35, 36, 37$ (25 and other numbers also known).

The grammitid ferns clearly nest within Polypodiaceae (Ranker *et al.*, 2004; Schneider *et al.*, 2004d). Tryon and Tryon (1982) and Hennipman (1996) have previously subsumed the grammitids in Polypodiaceae, as we now do here. Grammitids (ca. 20 genera, 600 species, pantropical) share a large number of morphological synapomorphies: veins free (mostly); scales lacking on blades; setiform, often dark red-brown hairs on leaves; sporangial stalks 1-rowed; spores green, trilete; gametophytes ribbon-shaped. Some genera of grammitids have been shown to be polyphyletic and their limits are the subject of re-interpretation, e.g., *Ctenopteris*, *Grammitis*, *Micropolypodium*, and *Terpsichore*, while others are likely monophyletic, e.g., *Ceradenia, Melpomene, Prosaptia* s.l. (Ranker *et al.*, 2004). Schuettpelz and Pryer (2007) support the newly described genus *Serpocaulon* (Smith *et al.*, 2006a) as sister to the grammitid ferns.

16.4 Synthesis: lessons learned from morphology and molecular systematics, and unexpected surprises

With the benefit of hindsight, it is instructive to note how classifications based on morphology have fared with the advent of molecular data. Further, we explore where more information is needed in order to better circumscribe natural lineages.

Many of the fern families recognized during the past one hundred years still have strong support, and clear evidence of monophyly, in the most recent large-scale molecular analysis (Schuettpelz and Pryer, 2007). Bitypic or polytypic families (two or more genera) with essentially the same, or only minor, changes in circumscription from that utilized in most recent classifications include the eusporangiate families Marattiaceae, Ophioglossaceae, and Psilotaceae; early-diverging leptosporangiate families Cyatheaceae (with the addition of Hymenophyllopsidaceae), Dipteridaceae (including Cheiropleuriaceae), Gleicheniaceae (including *Stromatopteris*), Hymenophyllaceae, Loxomataceae, Marsileaceae, Matoniaceae, Osmundaceae, Salviniaceae *s.l.*, and Schizaeaceae *s.s.*; and more derived leptosporangiate families Aspleniaceae, Blechnaceae, and Thelypteridaceae (Table 16.1). Many monotypic families in older classifications are still recognized, e.g., Equisetaceae, Plagiogyriaceae, Metaxyaceae, and several new ones have been added by virtue of their seeming isolation (as judged from molecular and morphological analyses) from existing families: Thyrsopteridaceae, Culcitaceae, Cibotiaceae (all members of the order Cyatheales), and Saccolomataceae.

The biggest surprises have come from demonstration of relationships between taxa previously considered to be only remotely related. Among these, we mention: (1) the unanticipated sister relationship between Ophioglossaceae and Psilotaceae, which is shown by nearly all molecular analyses (e.g., Hasebe *et al.*, 1996; Manhart, 1996; Pryer *et al.*, 2001a, 2004b); (2) the intimate relationship of Equisetaceae with ferns, rather than with a grade of so-called fern allies (Pryer *et al.*, 2001a; but excluded from the ferns by Rothwell and Nixon, 2006, who include fossil data); and (3) the recognition of genera of previously uncertain placement (sometimes placed in monotypic families) in existing often diverse families, e.g., *Stromatopteris* (Stromatopteridaceae) in Gleicheniaceae (Pryer *et al.*, 2004b; Schuettpelz and Pryer, 2007); *Cystodium* allied to lindsaeoid (rather than dicksonioid) ferns (Korall *et al.*, 2006a); *Hymenophyllopsis* in tree ferns (Wolf *et al.*, 1999), probably even nested in *Cyathea* itself (Korall *et al.*, 2006b); *Rheopteris* in Pteridaceae, among the vittarioid ferns (Schuettpelz *et al.*, 2007); *Leucostegia* among the dryopteroid ferns (rather than with Davalliaceae; Tsutsumi and Kato, 2006; Schuettpelz and Pryer, 2007); *Gymnogrammitis* in Polypodiaceae (Schneider *et al.*, 2002b); and *Pleurosoriopsis* (Pleurosoriopsidaceae) in Polypodiaceae (Hasebe *et al.*, 1996; Schneider *et al.*, 2004d). Moreover, several relatively large, diverse families have been shown to nest within families considered distinct by many: Vittariaceae, Parkeriaceae, and Platyzomataceae within Pteridaceae (Crane *et al.*, 1996; Gastony and Rollo, 1996, 1998; Hasebe *et al.*, 1996; Schuettpelz and Pryer,

2007; Schuettpelz et al., 2007); Grammitidaceae in Polypodiaceae (Schneider et al., 2004d).

Other surprises have resulted from evidence that suggests that morphologically "odd," small or monotypic genera nest within much larger, highly diverse genera, e.g., *Serpyllopsis*, *Rosenstockia*, and *Hymenoglossum* within *Hymenophyllum* (Ebihara et al., 2002, 2006; Hennequin et al., 2003, 2006a, 2006b); *Diellia* within *Asplenium* (Schneider et al., 2005). Other monotypic or small genera now recognized as probably better placed in larger genera include *Anarthropteris* (in *Loxogramme*, Polypodiaceae; Kreier and Schneider, 2006b); *Neurodium*, *Dicranoglossum*, and *Microphlebodium* (in *Pleopeltis*, Polypodiaceae; Schneider et al., 2004d); *Ochropteris* and *Neurocallis* (in *Pteris*, Pteridaceae; Schuettpelz et al., 2007); and almost all *Asplenium* segregates, e.g., *Camptosorus*, *Ceterach*, *Loxoscaphe*, *Phyllitis*, and *Pleurosorus* (in *Asplenium*, Aspleniaceae; Schneider et al., 2004b). Unsampled monotypes likely to disappear once they can be scrutinized include *Schaffneria* and *Holodictyum* (in *Asplenium*); *Costaricia* (in *Dennstaedtia*); and *Amphiblestra*, *Cionidium*, *Fadyenia*, *Quercifilix*, and *Dictyoxiphium* (in *Tectaria*, Tectariaceae). However, a significant number of monotypic genera in a wide variety of families, are supported in molecular analyses, including *Stromatopteris* (Gleicheniaceae; Pryer et al., 2004b, *Onoclea* (Onocleaceae; Gastony and Ungerer, 1997); *Regnellidium* (Marsileaceae; Pryer, 1999); *Helminthostachys* (Ophioglossaceae; Hasebe et al., 1996); *Thylacopteris* (Polypodiaceae; Schneider et al., 2004a); *Llavea* (Pteridaceae; Gastony and Rollo, 1998; Zhang et al., 2005; Schuettpelz et al., 2007); *Anetium* and *Ananthacorus* (Pteridaceae; Crane et al., 1996; Crane, 1997).

Still other major generic-level recircumscriptions have been suggested or seem likely in Polypodiaceae, with a redefinition of *Polypodium*, *Pleopeltis*, and allied genera (Schneider et al., 2004d; Smith et al., 2006a), and in Pteridaceae, with a redefinition of *Pellaea* (Kirkpatrick, 2007).

It is now clear that many morphological characters traditionally utilized in fern classification, at family and higher ranks, are still extremely useful in characterizing monophyletic groups at these ranks. Among the more import and useful characters are rhizome anatomy, venation pattern, indument type, indusial presence and type, spore type and ornamentation, eusporangiate versus leptosporangiate development, sporangial capacity, annulus position and form, chromosome base number, gametophyte morphology, and antheridial and archegonial characters. It is equally clear that certain characters usually given high importance in recognition of genera are highly homoplastic, and hence often of dubious importance, e.g., dimorphism and blade dissection.

16.5 Future goals and directions

Although many questions have been satisfactorily resolved, at least in a preliminary way, other questions remain, with regard to relationships and circumscription of fern families and genera. At higher taxonomic levels, unresolved questions include: (1) circumscription of some families, in a monophyletic way, particularly Dryopteridaceae, Woodsiaceae, and Lomariopsidaceae *sensu* Smith *et al.* (2006b); (2) relationships of some genera and families, relative to other genera and families, e.g., *Saccoloma* (Saccolomataceae) and *Cystodium* (Lindsaeaceae?); and (3) circumscription of genera in some of the larger, more diverse families, particularly in Pteridaceae, Thelypteridaceae (Smith and Cranfill, 2002), Blechnaceae (Cranfill, 2001), Woodsiaceae, Dryopteridaceae, Davalliaceae (Tsutsumi and Kato, 2006), and Polypodiaceae (Ranker *et al.*, 2004; Schneider *et al.*, 2004d).

References

Barcelona, J. F. (2000). Systematics of the fern genus *Odontosoria* sensu lato (Lindsaeaceae). Unpublished Ph.D. Thesis, Miami University, Oxford, OH.

Bateman, R. M. (1991). Paleobiological and phylogenetic implications of anatomically-preserved *Archeocalamites* from the Dinantian of Oxroad Bay and Loch Humphrey Burn, southern Scotland. *Palaeontographica*, **B223**, 1–59.

Bierhorst, D. W. (1971). *Morphology of Vascular Plants*. New York: Macmillan.

Bower, F. O. (1926). *The Ferns (Filicales)*, Vol. 2, *The Eusporangiate and Other Relatively Primitive Ferns*. London: Cambridge University Press.

Ching, R. C. (1940). On natural classification of the family "Polypodiaceae." *Sunyatsenia*, **5**, 201–268.

Ching, R. C. (1978). The Chinese fern families and genera: systematic arrangement and historical origin. *Acta Phytotaxonomica Sinica*, **16** (3), 1–19; **16** (4), 16–37.

Christenhusz, M. J. M., Tuomisto, H., Metzgar, J., and Pryer, K. M. (in review). Evolutionary relationships within the neotropical, eusporangiate fern genus *Danaea* (Marattiaceae). *Molecular Phylogenetics and Evolution*.

Christensen, C. (1938). Filicinae. In *Manual of Pteridology*, ed. F. Verdoorn. The Hague: Martinus Nijhoff, pp. 522–550.

Collinson, M. E. (1996). "What use are fossil ferns?" – 20 years on: with a review of the fossil history of extant pteridophyte families and genera. In *Pteridology in Perspective*, ed. J. M. Camus, M. Gibby, and R. J. Johns. Kew: Royal Botanic Gardens, pp. 349–394.

Conant, D. S. (1975). Hybrids in American Cyatheaceae. *Rhodora*, **77**, 441–455.

Conant, D. S. and Stein, D. B. (2001). Phylogenetic and geographic relationships of the tree ferns (Cyatheaceae) on Mount Kinabalu. *Sabah Parks Nature Journal*, **4**, 25–43.

Conant, D. S., Raubeson, L. A., Attwood, D. K., Perera, S., Zimmer, E. A., Sweere, J. A., and Stein, D. B. (1996a). Phylogenetic and evolutionary implications of combined analysis of DNA and morphology in the Cyatheaceae. In *Pteridology in Perspective*, ed. J. M. Camus, M. Gibby, and R. J. Johns. Kew: Royal Botanic Gardens, pp. 231–248.

Conant, D. S., Raubeson, L. A., Attwood, D. K., and Stein, D. B. (1996b) ["1995"]. The relationships of Papuasian Cyatheaceae to New World tree ferns. *American Fern Journal*, **85**, 328–340.

Copeland, E. B. (1947). *Genera Filicum*. Waltham, MA: Chronica Botanica.

Crabbe, J. A., Jermy, A. C., and Mickel, J. T. (1975). A new generic sequence for the pteridophyte herbarium. *Fern Gazette*, **11**, 141–162.

Crane, E. H. (1997). A revised circumscription of the genera of the fern family Vittariaceae. *Systematic Botany*, **22**, 509–517.

Crane, E. H., Farrar, D. R., and Wendel, J. F. (1996) ["1995"]. Phylogeny of the Vittariaceae: convergent simplification leads to a polyphyletic *Vittaria*. *American Fern Journal*, **85**, 283–305.

Cranfill, R. B. (2001). Phylogenetic studies in the Polypodiales (Pteridophyta) with an emphasis on the family Blechnaceae. Unpublished Ph.D. Thesis, University of California, Berkeley, CA.

Cranfill, R. B. and Kato, M. (2003). Phylogenetics, biogeography and classification of the woodwardioid ferns (Blechnaceae). In *Pteridology in the New Millennium*, ed. S. Chandra and M. Srivastava. Dordrecht: Kluwer, pp. 25–48.

Croft, J. R. (1986). The stipe and rachis vasculature of the dicksonioid fern, *Cystodium sorbifolium* (Cystodiaceae). *Kew Bulletin*, **41**, 789–803.

Davidse, G., Sousa S., M., and Knapp, S. (eds.) (1995). *Flora Mesoamericana*, Vol. 1, *Psilotaceae a Salviniaceae*, ed. R. C. Moran and R. Riba. México: Universidad Nacional Autónoma de México.

Davies, K. L. (1991). A brief comparative survey of aerophore structure within the Filicopsida. *Botanical Journal of the Linnean Society*, **197**, 115–137.

Des Marais, D. L., Smith, A. R., Britton, D. M., and Pryer, K. M. (2003). Phylogenetic relationships and evolution of extant horsetails, *Equisetum*, based on chloroplast DNA sequence data (*rbcL* and *trnL-F*). *International Journal of Plant Sciences*, **164**, 737–751.

Dickason, F. G. (1946). The ferns of Burma. *Ohio Journal of Science*, **46**, 109–141.

Driscoll, H. E. and Barrington, D. S. (2007). Origin of Hawaiian *Polystichum* (Dryopteridaceae) in the context of a world phylogeny. *American Journal of Botany*, **94**, 1413–1424.

Dubuisson, J.-Y. (1996). Evolutionary relationships within the genus *Trichomanes* sensu lato (Hymenophyllaceae) based on anatomical and morphological characters and a comparison with *rbcL* nucleotide sequences; preliminary results. In *Pteridology in Perspective*, ed. J. M. Camus, M. Gibby, and R J. Johns. Kew: Royal Botanic Gardens, pp. 285–287.

Dubuisson, J.-Y. (1997). *rbcL* sequences: a promising tool for the molecular systematics of the fern genus *Trichomanes* (Hymenophyllaceae)? *Molecular Phylogenetics and Evolution*, **8**, 128–138.

Dubuisson, J.-Y., Hennequin, S., Douzery, E. J. P., Cranfill, R. B., Smith, A. R., and Pryer, K. M. (2003). rbcL phylogeny of the fern genus *Trichomanes* (Hymenophyllaceae), with special reference to neotropical taxa. *International Journal of Plant Sciences*, **164**, 753–761.

Ebihara, A., Iwatsuki, K., Kurita, S., and Ito, M. (2002). Systematic position of *Hymenophyllum rolandi-principis* Rosenst. or a monotypic genus *Rosenstockia* Copel. (Hymenophyllaceae) endemic to New Caledonia. *Acta Phytotaxonomica et Geobotanica*, **53**, 35–49.

Ebihara, A., Hennequin, S., Iwatsuki, K., Bostock, P. D., Matsumoto, S., Jaman, R., Dubuisson, J.-Y., and Ito, M. (2004). Polyphyletic origin of *Microtrichomanes* (Prantl) Copel. (Hymenophyllaceae), with a revision of the species. *Taxon*, **53**, 935–948.

Ebihara, A., Dubuisson, J.-Y., Iwatsuki, K., Hennequin, S., and Ito, M. (2006). A taxonomic revision of Hymenophyllaceae. *Blumea*, **51**, 221–280.

Farrar, D. R. (1993). Hymenophyllaceae. In *Flora of North America North of Mexico*, Vol. 2, ed. Flora of North America Editorial Committee. New York: Oxford University Press, pp. 190–197.

Fraser-Jenkins, C. R. (1989). A monograph of *Dryopteris* (Pteridophyta: Dryopteridaceae) in the Indian subcontinent. *Bulletin of the British Museum (Natural History), Botany*, **18**, 323–477.

Gastony, G. J. (1982). Spore morphology of the Dicksoniaceae. 2. The genus *Cibotium*. *Canadian Journal of Botany*, **60**, 955–972.

Gastony, G. J. and Johnson, W. P. (2001). Phylogenetic placements of *Loxoscaphe thecifera* (Aspleniaceae) and *Actiniopteris radiata* (Pteridaceae) based on analysis of rbcL nucleotide sequences. *American Fern Journal*, **91**, 197–213.

Gastony, G. J. and Rollo, D. R. (1996) ["1995"]. Phylogeny and generic circumscriptions of cheilanthoid ferns (Pteridaceae: Cheilanthoideae) inferred from rbcL nucleotide sequences. *American Fern Journal*, **85**, 341–360.

Gastony, G. J. and Rollo, D. R. (1998). Cheilanthoid ferns (Pteridaceae: Cheilanthoideae) in the southwestern United States and adjacent Mexico – a molecular phylogenetic reassessment of generic lines. *Aliso*, **17**, 131–144.

Gastony, G. J. and Ungerer, M. C. (1997). Molecular systematics and a revised taxonomy of the onocleoid ferns (Dryopteridaceae: Onocleeae). *American Journal of Botany*, **84**, 840–849.

Geiger, J. M. O. and Ranker, T. A. (2005). Molecular phylogenetics and historical biogeography of Hawaiian *Dryopteris* (Dryopteridaceae). *Molecular Phylogenetics and Evolution*, **34**, 392–407.

Greuter, W., McNeill, J., Barrie, F. R., Burdet, H.-M., Demoulin, V., Filgueiras, T. S., Nicolson, D. H., Silva, P. C., Skog, J. E., Trehane, P., Turland, N. J., and Hawksworth, D. L. (eds.). (2000). *International Code of Botanical Nomenclature (Saint Louis Code)* [Regnum Vegetabile 138]. Königstein: Koeltz Scientific Books.

Guillon, J. M. (2004). Phylogeny of horsetails (*Equisetum*) based on the chloroplast rps4 gene and adjacent noncoding sequences. *Systematic Botany*, **29**, 251–259.

Hasebe, M., Wolf, P. G., Pryer, K. M., Ueda, K., Ito, M., Sano, R., Gastony, G. J., Yokoyama, J., Manhart, J. R., Murakami, N., Crane, E. H., Haufler, C. H., and

Hauk, W. D. (1996) ["1995"]. Fern phylogeny based on *rbcL* nucleotide sequences. *American Fern Journal*, **85**, 134–181.

Haufler, C. H. (2007). Genetics, phylogenetics, and biogeography: considering how shifting paradigms and continents influence fern diversity. *Brittonia*, **59**, 108–114.

Haufler, C. H., Grammer, W. A., Hennipman, E., Ranker, T. A., Smith, A. R., and Schneider, H. (2003). Systematics of the ant-fern genus *Lecanopteris* (Polypodiaceae): testing phylogenetic hypotheses with DNA sequences. *Systematic Botany*, **28**, 217–227.

Hauk, W. D. (1996) ["1995"]. A molecular assessment of relationships among cryptic species of *Botrychium* subgenus *Botrychium* (Ophioglossaceae). *American Fern Journal*, **85**, 375–394.

Hauk, W. D., Parks, C. R., and Chase, M. W. (2003). Phylogenetic studies of Ophioglossaceae: evidence from *rbcL* and *trnL-F* plastid DNA sequences and morphology. *Molecular Phylogenetics and Evolution*, **28**, 131–51.

Hennequin, S., Ebihara, A., Ito, M., Iwatsuki, K., and Dubuisson, J.-Y. (2003). Molecular systematics of the fern genus *Hymenophyllum* s.l. (Hymenophyllaceae) based on chloroplastic coding and noncoding regions. *Molecular Phylogenetics and Evolution*, **27**, 283–301.

Hennequin, S., Ebihara, A., Ito, M., Iwatsuki, K., and Dubuisson, J.-Y. (2006a). Phylogenetic systematics and evolution of the genus *Hymenophyllum* (Hymenophyllaceae: Pteridophyta). *Fern Gazette*, **17**, 247–257.

Hennequin, S., Ebihara, A., Ito, M., Iwatsuki, K., and Dubuisson, J.-Y. (2006b). New insights into the phylogeny of the genus *Hymenophyllum* s.l. (Hymenophyllaceae): revealing the polyphyly of *Mecodium*. *Systematic Botany*, **31**, 271–284.

Hennipman, E. (1996). Scientific consensus classification of Pteridophyta. In *Pteridology in Perspective*, ed. J. M. Camus, M. Gibby, and R. J. Johns. Kew: Royal Botanic Gardens, pp. 191–202.

Hill, C. R. and Camus, J. M. (1986). Evolutionary cladistics of marattialean ferns. *Bulletin of the British Museum (Natural History), Botany*, **14**, 219–300.

Holttum, R. E. (1947). A revised classification of leptosporangiate ferns. *Journal of the Linnean Society, Botany*, **53**, 123–158.

Holttum, R. E. (1949). The classification of ferns. *Biological Review*, **24**, 267–296.

Holttum, R. E. (1971). Studies in the family Thelypteridaceae. III. A new system of genera in the Old World. *Blumea*, **19**, 17–52.

Holttum, R. E. (1973). Posing the problems. In *The Phylogeny and Classification of the Ferns*, ed. A. C. Jermy, J. A. Crabbe, and B. A. Thomas, *Botanical Journal of the Linnean Society*, **67** (Suppl. 1), 1–10.

Holttum, R. E. (1986). Studies in the genera allied to *Tectaria* Cav., V. *Triplophyllum*, a new genus of Africa and America. *Kew Bulletin*, **41**, 237–260.

Holttum, R. E. and Edwards, P. (1983). The tree ferns of Mt. Roraima and neighboring areas of the Guayana Highlands with comments on the family Cyatheaceae. *Kew Bulletin*, **38**, 155–188.

Hoogland, R. D. and Reveal, J. L. (2005). Index nominum familiarum plantarum vascularium. *The Botanical Review*, **71**, 1–291.

Janssen, T. and Schneider, H. (2005). Exploring the evolution of humus collecting leaves in drynarioid ferns (Polypodiaceae, Polypodiidae) based on phylogenetic evidence. *Plant Systematics and Evolution*, **252**, 175–197.

Judd, W., Campbell, C. S., Kellogg, E. A., Stevens, P. F., and Donoghue, M. J. (2002). *Plant Systematics: A Phylogenetic Approach*, 2nd edn. Sunderland, MA: Sinauer.

Kato, M. (1983). The classification of major groups of pteridophytes. *Journal of the Faculty of Science, University of Tokyo, Section III, Botany*, **13**, 263–283.

Kato, M. (1987). A phylogenetic classification of Ophioglossaceae. *Gardens' Bulletin, Singapore*, **40**, 1–14.

Kato, M. and Setoguchi, H. (1998). An *rbcL*-based phylogeny and heteroblastic leaf morphology of Matoniaceae. *Systematic Botany*, **23**, 391–400.

Kato, M., Yatabe, Y., Sahashi, N., and Murakami, N. (2001). Taxonomic studies of *Cheiropleuria* (Dipteridaceae). *Blumea*, **46**, 513–525.

Kenrick, P. and Crane, P. R. (1997). *The Origin and Early Diversification of Land Plants*. Washington, DC: Smithsonian Institution Press.

Kirkpatrick, R. (2007). Investigating the monophyly of *Pellaea* (Pteridaceae) in the context of a phylogenetic analysis of cheilanthoid ferns. *Systematic Botany*, **32**, 504–518.

Korall, P., Conant, D. S., Schneider, H., Ueda, K., Nishida, H., and Pryer, K. M. (2006a). On the phylogenetic position of *Cystodium*: it's not a tree fern – it's a polypod! *American Fern Journal*, **96**, 45–53.

Korall, P., Pryer, K. M., Metzgar, J. S., Schneider, H., and Conant, D. S. (2006b). Tree ferns: monophyletic groups and their relationships as revealed by four protein-coding plastid loci. *Molecular Phylogenetics and Evolution*, **39**, 830–845.

Korall, P., Conant, D. S., Metzgar, J. S., Schneider, H., and Pryer, K. M. (2007). A molecular phylogeny of scaly tree ferns (Cyatheaceae). *American Journal of Botany*, **94**, 873–886.

Kreier, H.-P. and Schneider, H. (2006a). Phylogeny and biogeography of the staghorn fern genus *Platycerium* (Polypodiaceae, Polypodiidae). *American Journal of Botany*, **93**, 217–225.

Kreier, H.-P. and Schneider, H. (2006b). Reinstatement of *Loxogramme dictyopteris*, based on phylogenetic evidence, for the New Zealand endemic fern, *Anarthropteris lanceolata* (Polypodiaceae, Polypodiidae). *Australian Systematic Botany*, **19**, 309–314.

Kreier, H.-P., Rojas-Alvarado, A. F., Smith, A. R., and Schneider, H. (2007). *Hyalotrichopteris* is indeed a *Campyloneurum* (Polypodiaceae). *American Fern Journal*, **97**, 127–135.

Kubitzki, K. (ed.). (1990). *The Families and Genera of Vascular Plants*, Vol. 1, *Pteridophytes and Gymnosperms*, ed. K. U. Kramer and P. S. Green. Berlin: Springer-Verlag.

Lantz, T. C., Rothwell, G. W., and Stockey, R. A. (1999). *Conantiopteris schuchmanii*, gen. et sp. nov., and the role of fossils in resolving the phylogeny of Cyatheaceae s.l. *Journal of Plant Research*, **112**, 361–381.

Lehnert, M., Mönnich, M., Pleines, T., Schmidt-Lebuhn, A., and Kessler, M. (2001). The relictual fern genus *Loxsomopsis*. *American Fern Journal*, **91**, 13–24.

Li, C.-X. and Lu, C. G. (2006). Phylogenetic analysis of Dryopteridaceae based on chloroplast *rbcL* sequences. *Acta Phytotaxonomica Sinica*, **44**, 503–515.

Little, D. P. and Barrington, D. S. (2003). Major evolutionary events in the origin and diversification of the fern genus *Polystichum* (Dryopteridaceae). *American Journal of Botany*, **90**, 508–514.

Lovis, J. D. (1977). Evolutionary patterns and processes in ferns. In *Advances in Botanical Research*, Vol. 4, ed. R. D. Preston and H. W. Woolhouse. London: Academic Press, pp. 229–415.

Lu, J.-M. and Li, D.-Z. (2006a). The study on systematic position of *Cyclopeltis*. *Acta Botanica Yunnanica*, **28**, 337–340.

Lu, S.-G. and Li, C.-X. (2006b). Phylogenetic position of the monotypic genus *Metapolypodium* Ching endemic to Asia: evidence from chloroplast DNA sequences of *rbcL* gene and *rps4-trnS* region. *Acta Phytotaxonomica Sinica*, **44**, 494–502.

Lu, J.-M., Li, D.-Z., Gao, L.-M., and Cheng, X. (2005). Paraphyly of *Cyrtomium* (Dryopteridaceae): evidence from *rbcL* and *trnL-F* sequence data. *Journal of Plant Research*, **118**, 129–135.

Lu, J.-M., Barrington, D. S., and Li, D.-Z. (2007). Molecular phylogeny of the polystichoid ferns in Asia based on *rbcL* sequences. *Systematic Botany*, **32**, 26–33.

Manhart, J. R. (1996) ["1995"]. Chloroplast 16S rDNA sequences and phylogenetic relationships of fern allies and ferns. *American Fern Journal*, **85**, 182–192.

Metzgar, J. S., Schneider, H., and Pryer, K. M. (2007). Phylogeny and divergence time estimates for the fern genus *Azolla* (Salviniaceae). *International Journal of Plant Sciences*, **168**, 1045–1053.

Mickel, J. T. (1974). Phyletic lines in the modern ferns. *Annals of the Missouri Botanical Garden*, **61**, 474–482.

Mickel, J. T. and Atehortúa, L. G. (1980). Subdivision of the genus *Elaphoglossum*. *American Fern Journal*, **70**, 47–68.

Miller, C. N., Jr. (1971). Evolution of the fern family Osmundaceae based on anatomical studies. *Contributions from the Museum of Paleontology, Ann Arbor, Michigan*, **23**, 105–169.

Murakami, N. (1995). Systematics and evolutionary biology of the fern genus *Hymenasplenium* (Aspleniaceae). *Journal of Plant Research*, **108**, 257–268.

Murakami, N. and Schaal, B. A. (1994). Chloroplast DNA variation and the phylogeny of *Asplenium* sect. *Hymenasplenium* (Aspleniaceae) in the New World tropics. *Journal of Plant Research*, **107**, 245–251.

Murakami, N., Nogami, S., Watanabe, M., and Iwatsuki, K. (1999). Phylogeny of Aspleniaceae inferred from *rbcL* nucleotide sequences. *American Fern Journal*, **89**, 232–243.

Murdock, A. (2005). Molecular evolution and phylogeny of marattioid ferns, an ancient lineage of land plants. www.2005.botanyconference.org/engine/search/index.php.

Murdock, A. G., Reveal, J. L., and Doweld, A. (2006). (1746) Proposal to conserve the name Marattiaceae against Danaeaceae (Pteridophyta). *Taxon*, **55**, 1040–1042.

Nagalingum, N. S., Schneider, H., and Pryer, K. M. (2006). Comparative morphology of reproductive structures in heterosporous water ferns and a re-evaluation of the sporocarp. *International Journal of Plant Sciences*, **167**, 805–815.

Nagalingum, N. S., Schneider, H., and Pryer, K. M. (2007). Molecular phylogenetic relationships and morphological evolution in the heterosporous fern genus *Marsilea*. *Systematic Botany*, **32**, 16–25.

Nakahira, Y. (2000). A molecular phylogenetic analysis of the family Blechnaceae, using the chloroplast gene *rbcL*. M. S. Thesis, Graduate School of Science, University of Tokyo, Tokyo.

Nakazato, T. and Gastony, G. J. (2001). Molecular phylogenetics of *Anogramma* species and related genera (Pteridaceae: Taenitidoideae). In *Botany 2001, Plants and People, Albuquerque, NM,* Abstract.

Nayar, B. K. (1970). A phylogenetic classification of the homosporous ferns. *Taxon*, **19**, 229–236.

Palmer, D. D. (1994). The Hawaiian species of *Cibotium*. *American Fern Journal*, **84**, 73–85.

Perrie, L. R. and Brownsey, P. J. (2005). Insights into the biogeography and polyploid evolution of New Zealand *Asplenium* from chloroplast DNA sequence data. *American Fern Journal*, **95**, 1–21.

Pichi Sermolli, R. E. G. (1970). A provisional catalogue of the family names of living pteridophytes. *Webbia*, **25**, 219–297.

Pichi Sermolli, R. E. G. (1973). Historical review of the higher classification of the Filicopsida. In *The Phylogeny and Classification of the Ferns*, ed. A. C. Jermy, J. A. Crabbe, and B. A. Thomas. *Botanical Journal of the Linnean Society*, **67** (Suppl. 1), 11–40.

Pichi Sermolli, R. E. G. (1977). Tentamen pteridophytorum genera in taxonomicum ordinem redigendi. *Webbia*, **31**, 313–512.

Pichi Sermolli, R. E. G. (1981). Report of the subcommittee for family names of Pteridophyta. *Taxon*, **30**, 163–168.

Pichi Sermolli, R. E. G. (1982). A further contribution to the nomenclature of the families of Pteridophyta. *Webbia*, **35**, 223–237.

Pichi Sermolli, R. E. G. (1986). Report of the subcommittee for family names of Pteridophyta. *Taxon*, **35**, 686–691.

Pichi Sermolli, R. E. G. (1993). New studies on some family names of Pteridophyta. *Webbia*, **47**, 121–143.

Pinter, I., Bakker, F., Barrett, J., Cox, C., Gibby, M., Henderson, S., Morgan-Richards, M., Rumsey, F., Russell, S., Trewick, S., Schneider, H., and Vogel, J. (2002). Phylogenetic and biosystematic relationships in four highly disjunct polyploid complexes in the subgenera *Ceterach* and *Phyllitis* in *Asplenium* (Aspleniaceae). *Organisms, Diversity, and Evolution*, **2**, 299–311.

Prado, J., Rodrigues, C. D. N., Salatino, A., and Salatino, M. L. F. (2007). Phylogenetic relationships among Pteridaceae, including Brazilian species, inferred from *rbcL* sequences. *Taxon*, **56**, 355–368.

Pryer, K. M. (1999). Phylogeny of marsileaceous ferns and relationships of the fossil *Hydropteris pinnata* reconsidered. *International Journal of Plant Sciences*, **160**, 931–954.

Pryer, K. M., Smith, A. R., and Skog, J. E. (1996) ["1995"]. Phylogenetic relationships of extant ferns based on evidence from morphology and *rbcL* sequences. *American Fern Journal*, **85**, 205–282.

Pryer, K. M., Schneider, H., Smith, A. R., Cranfill, R., Wolf, P. G., Hunt, J. S., and Sipes, S. D. (2001a). Horsetails and ferns are a monophyletic group and the closest living relatives to seed plants. *Nature*, **409**, 618–622.

Pryer, K. M., Smith, A. R., Hunt, J. S., and Dubuisson, J.-Y. (2001b). *rbcL* data reveal two monophyletic groups of filmy ferns (Filicopsida: Hymenophyllaceae). *American Journal of Botany*, **88**, 1118–1130.

Pryer, K. M., Schneider, H., and Magallón, S. (2004a). The radiation of vascular plants. In *Assembling the Tree of Life*, ed. J. Cracraft and M. J. Donoghue. New York: Oxford University Press, pp. 138–153.

Pryer, K. M., Schuettpelz, E., Wolf, P. G., Schneider, H., Smith, A. R., and Cranfill, R. (2004b). Phylogeny and evolution of ferns (monilophytes) with a focus on the early leptosporangiate divergences. *American Journal of Botany*, **91**, 1582–1598.

Qiu, Y.-L., et al. (2006). The deepest divergences in land plants inferred from phylogenomic evidence. *Proceedings of the National Academy of Sciences of the United States of America*, **103**, 15511–15516.

Ranker, T. A., Geiger, J. M. O., Kennedy, S. C., Smith, A. R., Haufler, C. H., and Parris, B. S. (2003). Molecular phylogenetics and evolution of the endemic Hawaiian genus *Adenophorus* (Grammitidaceae). *Molecular Phylogenetics and Evolution*, **26**, 337–347.

Ranker, T. A., Smith, A. R., Parris, B. S., Geiger, J. M. O., Haufler, C. H., Straub, S. C. K., and Schneider, H. (2004). Phylogeny and evolution of grammitid ferns (Grammitidaceae): a case of rampant morphological homoplasy. *Taxon*, **53**, 415–428.

Raubeson, L. A. and Jansen, R. K. (1992). Chloroplast DNA evidence on the ancient evolutionary split in vascular land plants. *Science*, **255**, 1697–1699.

Reid, J. D., Plunkett, G. M., and Peters, G. A. (2006). Phylogenetic relationships in the heterosporous fern genus *Azolla* (Azollaceae) based on DNA sequence data from three noncoding regions. *International Journal of Plant Sciences*, **167**, 529–538.

Renzaglia, K. S., Duff, R. J., Nickrent, D. L., and Garbary, D. J. (2000). Vegetative and reproductive innovations of early land plants; implications for a unified phylogeny. *Philosophical Transactions of the Royal Society of London, Series B*, **355**, 769–793.

Rothwell, G. W. (1999). Fossils and ferns in the resolution of land plant phylogeny. *The Botanical Review*, **65**, 188–218.

Rothwell, G. W. and Nixon, K. C. (2006). How does the inclusion of fossil data change our conclusions about the phylogenetic history of euphyllophytes? *International Journal of Plant Sciences*, **167**, 737–749.

Rothwell, G. W. and Stockey, R. A. (1994). The role of *Hydropteris pinnata* gen. et sp. nov. in reconstructing the cladistics of heterosporous ferns. *American Journal of Botany*, **81**, 479–492.

Rouhan, G., Dubuisson, J.-Y., Rakotondrainibe, F., Motley, T. J., Mickel, J. T., Labat, J.-N., and Moran, R. C. (2004). Molecular phylogeny of the fern genus *Elaphoglossum* (Elaphoglossaceae) based on chloroplast non-coding DNA sequences: contributions of species from the Indian Ocean area. *Molecular Phylogenetics and Evolution*, **33**, 745–763.

Sánchez-Baracaldo, P. (2004a). Phylogenetics and biogeography of the neotropical fern genera *Jamesonia* and *Eriosorus* (Pteridaceae). *American Journal of Botany*, **91**, 274–284.

Sánchez-Baracaldo, P. (2004b). Phylogenetic relationships of the subfamily Taenitoideae, Pteridaceae. *American Fern Journal*, **94**, 126–142.

Sano, R., Takamiya, M., Ito, M., Kurita, S., and Hasebe, M. (2000a). Phylogeny of the lady fern group, tribe Physematieae (Dryopteridaceae), based on chloroplast *rbcL* gene sequences. *Molecular Phylogenetics and Evolution*, **15**, 403–413.

Sano, R., Takamiya, M., Kurita, S., Ito, M., and Hasebe, M. (2000b). *Diplazium subsinuatum* and *Di. tomitaroanum* should be moved to *Deparia* according to molecular, morphological, and cytological characters. *Journal of Plant Research*, **113**, 157–163.

Schneider, H. (1996a.) Vergleichende Wurzelanatomie der Farne. Unpublished Ph.D. Thesis, Universität Zürich, Shaker, Aachen.

Schneider, H. (1996b). Root anatomy of Aspleniaceae and the implications for systematics of the fern family. *Fern Gazette*, **12**, 160–168.

Schneider, H. and Pryer, K. M. (2002). Structure and function of spores in the aquatic heterosporous fern family Marsileaceae. *International Journal of Plant Sciences*, **163**, 485–505.

Schneider, H., Pryer, K. M., Cranfill, R., Smith, A. R., and Wolf, P. G. (2002a). Evolution of vascular plant body plans: a phylogenetic perspective. In *Developmental Genetics and Plant Evolution*, ed. Q. C. B. Cronk, R. M. Bateman, and J. A. Hawkins. London: Taylor and Francis, pp. 330–364.

Schneider, H., Smith, A. R., Cranfill, R., Haufler, C. H., Ranker, T. A., and Hildebrand, T. (2002b). *Gymnogrammitis dareiformis* is a polygrammoid fern (Polypodiaceae) – resolving an apparent conflict between morphological and molecular data. *Plant Systematics and Evolution*, **234**, 121–136.

Schneider, H., Janssen, T., Hovenkamp, P., Smith, A. R., Cranfill, R., Haufler, C. H., and Ranker, T. A. (2004a). Phylogenetic relationships of the enigmatic Malesian fern *Thylacopteris* (Polypodiaceae, Polypodiidae). *International Journal of Plant Sciences*, **165**, 1077–1087.

Schneider, H., Russell, S. J., Cox, C. J., Bakker, F., Henderson, S., Gibby, M., and Vogel, J. C. (2004b). Chloroplast phylogeny of asplenioid ferns based on *rbcL* and *trnL-F* spacer sequences (Polypodiidae, Aspleniaceae) and its implications for the biogeography. *Systematic Botany*, **29**, 260–274.

Schneider, H., Schuettpelz, E., Pryer, K. M., Cranfill, R., Magallón, S., and Lupia, R. (2004c). Ferns diversified in the shadow of angiosperms. *Nature*, **428**, 553–557.

Schneider, H., Smith, A. R., Cranfill, R., Hildebrand, T. E., Haufler, C. H., and Ranker, T. A. (2004d). Unraveling the phylogeny of polygrammoid ferns (Polypodiaceae and Grammitidaceae): exploring aspects of the diversification of epiphytic plants. *Molecular Phylogenetics and Evolution*, **31**, 1041–1063.

Schneider, H., Ranker, T. A., Russell, S. J., Cranfill, R., Geiger, J. M. O., Aguraiuja, R., Wood, K. R., Grundmann, M., Kloberdanz, K., and Vogel, J. C. (2005). Origin of the endemic fern genus *Diellia* coincides with the renewal of Hawaiian terrestrial life in the Miocene. *Proceedings of the Royal Society of London, Series B, Biological Sciences*, **272**, 455–460.

Schneider, H., Kreier, H.-P., Wilson, R., and Smith, A. R. (2006a). The *Synammia* enigma: evidence for a temperate lineage of polygrammoid ferns (Polypodiaceae, Polypodiidae) in southern South America. *Systematic Botany*, **31**, 31–41.

Schneider, H., Kreier, H.-P., Perrie, L. R., and Brownsey, P. J. (2006b). The relationships of *Microsorum* (Polypodiaceae) species occurring in New Zealand. *New Zealand Journal of Botany*, **44**, 121–127.

Schuettpelz, E. and Pryer, K. M. (2007). Fern phylogeny inferred from 400 leptosporangiate species and three plastid genes. *Taxon*, **56**, 1037–1050.

Schuettpelz, E., Korall, P., and Pryer, K. M. (2006). Plastid *atpA* data provide improved support for deep relationships among ferns. *Taxon*, **55**, 897–906.

Schuettpelz, E., Schneider, H., Huiet, L., Windham, M. D., and Pryer, K. M. (2007). A molecular phylogeny of the fern family Pteridaceae: assessing overall relationships and the affinities of previously unsampled genera. *Molecular Phylogenetics and Evolution*, **44**, 1172–1185.

Scotland, R. W. and Wortley, A. H. (2003). How many species of seed plants are there? *Taxon* **52**, 101–104.

Shepherd, L. D., Perrie, L. R., Parris, B. S., and Brownsey, P. J. (2007). A molecular phylogeny for the New Zealand Blechnaceae ferns from analysis of chloroplast *trnL-trnF* DNA sequences. *New Zealand Journal of Botany*, **45**, 67–80.

Shmakov, A. I. (2003). Review of the family Woodsiaceae (Diels) Herter of Eurasia. In *Pteridology in the New Millennium*, ed. S. Chandra and M. Srivastava. Dordrecht: Kluwer, pp. 49–64.

Skog, J. E., Zimmer, E., and Mickel, J. T. (2002). Additional support for two subgenera of *Anemia* (Schizaeaceae) from data for the chloroplast intergenic spacer region *trnL-F* and morphology. *American Fern Journal*, **92**, 119–130.

Skog, J. E., Mickel, J. T., Moran, R. C., Volovsek, M., and Zimmer, E. A. (2004). Molecular studies of representative species in the fern genus *Elaphoglossum* (Dryopteridaceae) based on cpDNA sequences *rbcL*, *trnL-F*, and *rps4-TRNS*. *International Journal of Plant Sciences*, **165**, 1063–1075.

Small, R. L., Lickey, E. B., Shaw, J., and Hauk, W. D. (2005). Amplification of noncoding chloroplast DNA for phylogenetic studies in lycophytes and monilophytes with a comparative example of relative phylogenetic utility from Ophioglossaceae. *Molecular Phylogenetics and Evolution*, **36**, 509–522.

Smith, A. R. (1996) ["1995"]. Non-molecular phylogenetic hypotheses for ferns. *American Fern Journal*, **85**, 104–122.

Smith, A. R. and Cranfill, R. B. (2002). Intrafamilial relationships of the thelypteroid ferns (Thelypteridaceae). *American Fern Journal*, **92**, 131–149.

Smith, A. R., Tuomisto, H., Pryer, K. M., Hunt, J. S., and Wolf, P. G. (2001). *Metaxya lanosa*, a second species in the genus and fern family Metaxyaceae. *Systematic Botany*, **26**, 480–486.

Smith, A. R., Kreier, H.-P., Haufler, C. H., Ranker, T. A., and Schneider, H. (2006a). *Serpocaulon* (Polypodiaceae), a new genus segregated from *Polypodium*. *Taxon*, **55**, 919–930.

Smith, A. R., Pryer, K. M., Schuettpelz, E., Korall, P., Schneider, H., and Wolf, P. G. (2006b). A classification for extant ferns. *Taxon*, **55**, 705–731.

Stein, D. B., Conant, D. S., and Valinski, A. E. C. (1996). The implications of chloroplast DNA restriction site variation on the classification and phylogeny of the Cyatheaceae. In *Holttum Memorial Volume*, ed. R. J. Johns. Kew: Royal Botanic Gardens, pp. 235–254.

Stevenson, D. W. and Loconte, H. (1996). Ordinal and familial relationships of pteridophyte genera. In *Pteridology in Perspective*, ed. J. M. Camus, M. Gibby, and R. J. Johns. Kew: Royal Botanic Gardens, pp. 435–467.

Sun, B.-Y., Kim, M. H., Kim, C. H., and Park, C.-W. (2001). *Mankyua* (Ophioglossaceae): a new fern genus from Cheju Island, Korea. *Taxon*, **50**, 1019–1024.

Tagawa, M. and Iwatsuki, K. (1972). Families and genera of the pteridophytes known from Thailand. *Memoirs of the Faculty of Science, Kyoto University, Series of Biology*, **5**, 67–88.

Thorne, R. F. (2002). How many species of seed plants are there? *Taxon*, **51**, 511–522.

Tindale, M. D. and Roy, S. K. (2002). A cytotaxonomic survey of the Pteridophyta of Australia. *Australian Systematic Botany*, **15**, 839–937.

Tryon, A. F. (1964). *Platyzoma* – a Queensland fern with incipient heterospory. *American Journal of Botany*, **51**, 939–942.

Tryon, A. F. and Lugardon, B. (1991). *Spores of the Pteridophyta*. New York: Springer-Verlag.

Tryon, R. M. (1952). A sketch of the history of fern classification. *Annals of the Missouri Botanical Garden*, **39**, 255–262.

Tryon, R. (1970). The classification of the Cyatheaceae. *Contributions of the Gray Herbarium of Harvard University*, **200**, 1–53.

Tryon, R. M. and Tryon, A. F. (1982). *Ferns and Allied Plants, with Special Reference to Tropical America*. Berlin: Springer-Verlag.

Tsutsumi, C. and Kato, M. (2005). Molecular phylogenetic study on Davalliaceae. *Fern Gazette*, **17**, 147–162.

Tsutsumi, C. and Kato, M. (2006). Evolution of epiphytes in Davalliaceae and related ferns. *Botanical Journal of the Linnean Society*, **151**, 495–510.

van den Heede, C. J., Viane, R. L. L., and Chase, M. W. (2003). Phylogenetic analysis of *Asplenium* subgenus *Ceterach* (Pteridophyta: Aspleniaceae) based on plastid and nuclear ribosomal ITS DNA sequences. *American Journal of Botany*, **90**, 481–493.

Wagner, W. H., Jr. (1969). The construction of a classification. In *Systematic Biology*, United States National Academy of Sciences, Science Publication 1692. Washington, DC: National Academy Press, pp. 67–90.

Wagner, W. H., Jr. and Smith, A. R. (1993). Pteridophytes of North America. In *Flora of North America North of Mexico*, Vol. 1, ed. Flora of North America Editorial Committee. New York: Oxford University Press, pp. 247–266.

Wang, M.-L., Chen, Z.-D., Zhang, X.-C, Lu, S.-G., and Zhao, G.-F. (2003). Phylogeny of the Athyriaceae: evidence from chloroplast *trnL-F* region sequences. *Acta Phytotaxonomica Sinica*, **41**, 416–426.

Wang, M.-L., Hsieh, Y.-T., and Zhao, G.-F. (2004). A revised subdivision of the Athyriaceae. *Acta Phytotaxonomica Sinica*, **42**, 524–527.

White, R. A. and Turner, M. D. (1988). *Calochlaena*, a new genus of dicksonioid ferns. *American Fern Journal*, **78**, 86–95.

Wikström, N. and Pryer, K. M. (2005). Incongruence between primary sequence data and the distribution of a mitochondrial *atp1* group II intron among ferns and horsetails. *Molecular Phylogenetics and Evolution*, **36**, 484–493.

Wikström, N., Kenrick, P., and Vogel, J. C. (2002). Schizaeaceae: a phylogenetic approach. *Review of Palaeobotany and Palynology*, **119**, 35–50.

Wolf, P. G. (1996a) ["1995"]. Phylogenetic analyses of *rbcL* and nuclear ribosomal RNA gene sequences in Dennstaedtiaceae. *American Fern Journal*, **85**, 306–327.

Wolf, P. G. (1996b). Pteridophyte phylogenies based on analysis of DNA sequences: a multiple gene approach. In *Pteridology in Perspective*, ed. J. M. Camus, M. Gibby, and R. J. Johns. Kew: Royal Botanic Gardens, pp. 203–215.

Wolf, P. G. (1997). Evaluation of *atpB* nucleotide sequences for phylogenetic studies of ferns and other pteridophytes. *American Journal of Botany*, **84**, 1429–1440.

Wolf, P. G., Soltis, P. S., and Soltis, D. E. (1994). Phylogenetic relationships of dennstaedtioid ferns: evidence from *rbcL* sequences. *Molecular Phylogenetics and Evolution*, **3**, 383–392.

Wolf, P. G., Pryer, K. M., Smith, A. R., and Hasebe, M. (1998). Phylogenetic studies of extant pteridophytes. In *Molecular Systematics of Plants II. DNA Sequencing*, ed. D. E. Soltis, P. S. Soltis, and J. J. Doyle. Boston, MA: Kluwer, pp. 541–556.

Wolf, P. G., Sipes, S. D., White, M. R., Martines, M. L., Pryer, K. M., Smith, A. R., and Ueda, K. (1999). Phylogenetic relationships of the enigmatic fern families Hymenophyllopsidaceae and Lophosoriaceae: evidence from *rbcL* nucleotide sequences. *Plant Systematics and Evolution*, **219**, 263–270.

Wu, S., Phan, K. L., and Xiang, J. (2005). A new genus and two new species of ferns from Vietnam. *Novon*, **15**, 245–249.

Yatabe, Y., Nishida, H., and Murakami, N. (1999). Phylogeny of Osmundaceae inferred from *rbcL* nucleotide sequences and comparison to the fossil evidence. *Journal of Plant Research*, **112**, 397–404.

Yatabe, Y., Watkins, J. E., Farrar, D. R., and Murakami, N. (2002). Genetic variation in populations of the morphologically and ecologically variable fern *Stegnogramma* pozoi subsp. *mollissima* (Thelypteridaceae) in Japan. *Journal of Plant Research*, **115**, 29–38.

Yatabe, Y., Murakami, N., and Iwatsuki, K. (2005). *Claytosmunda*; a new subgenus of *Osmunda* (Osmundaceae). *Acta Phytotaxonomica et Geobotanica*, **56**, 127–128.

Zhang, X.-C. and Nooteboom, H. P. (1998). A taxonomic revision of Plagiogyriaceae (Pteridophyta). *Blumea*, **43**, 401–469.

Zhang, G., Zhang, X., and Chen, Z. (2005). Phylogeny of cryptogrammoid ferns and related taxa based on *rbcL* sequences. *Nordic Journal of Botany*, **23**, 485–493.

Appendix A: Familial names applied to extant ferns

Familial names applied to extant ferns, and their taxonomic disposition. Family names accepted by us are in **boldface**. Synonyms are in *italics*. Unpublished or otherwise illegitimate names are in quotation marks. An equal sign (=) is intended to indicate that we regard the first name as a heterotypic synonym of the family name we adopt.

Acrostichaceae Mett. ex A. B. Frank = Pteridaceae

Actiniopteridaceae Pic. Serm. = Pteridaceae

Adiantaceae Newman, *nom. cons.* over Parkeriaceae = Pteridaceae

Alsophilaceae C. Presl = Cyatheaceae

Anemiaceae Link; here included in Schizaeales

Angiopteridaceae Fée ex J. Bommer = Marattiaceae

Antrophyaceae Ching = Pteridaceae

"*Aspidiaceae*" Burnett, *nom. illeg.* = Dryopteridaceae

Aspleniaceae Newman; here included in Polypodiales

Athyriaceae Alston = Woodsiaceae

Azollaceae Wettst. = Salviniaceae

Blechnaceae Newman; here included in Polypodiales

Bolbitidaceae Ching = Dryopteridaceae

Botrychiaceae Horan. = Ophioglossaceae

Ceratopteridaceae Underw. = Parkeriaceae = Pteridaceae

Cheilanthaceae B. K. Nayar = Pteridaceae

Cheiropleuriaceae Nakai = Dipteridaceae

Christenseniaceae Ching = Marattiaceae

Cibotiaceae Korall; here included in Cyatheales

Cryptogrammaceae Pic. Serm. = Pteridaceae

Culcitaceae Pic. Serm.; here included in Cyatheales

Cyatheaceae Kaulf.; here included in Cyatheales

Cystodiaceae J. R. Croft = Lindsaeaceae

Cystopteridaceae Schmakov

Danaeaceae C. Agardh = Marattiaceae

Davalliaceae M. R. Schomb.; here included in Polypodiales

Dennstaedtiaceae Lotsy; here included in Polypodiales

Dicksoniaceae M. R. Schomb., *nom. cons.* over Thyrsopteridaceae; here included in Cyatheales

Dicranopteridaceae Ching ex Doweld = Gleicheniaceae

"*Dictyoxiphiaceae*" Ching, *nom. nud.* = Tectariaceae

"*Didymochlaenaceae*" Ching, *nom. nud.* = Dryopteridaceae, tentatively

Dipteridaceae Seward and E. Dale; here included in Gleicheniales

Drynariaceae Ching = Polypodiaceae

Dryopteridaceae Herter, *nom. cons.* over Peranemataceae; here included in Polypodiales

Elaphoglossaceae Pic. Serm. = Dryopteridaceae

Equisetaceae Michx. ex DC.; here included in Equisetales

"*Filicaceae*" Juss., *nom. illeg.*

Gleicheniaceae C. Presl; here included in Gleicheniales

Grammitidaceae Newman [often misspelled Grammitaceae] = Polypodiaceae

Gymnogrammitidaceae Ching (incl. Gymnogrammaceae, spelling variant used by some authors) = Polypodiaceae

Helminthostachyaceae Ching = Ophioglossaceae

Hemionitidaceae Pic. Serm. = Pteridaceae

Hymenophyllaceae Mart.; here included in Hymenophyllales

Hymenophyllopsidaceae Pic. Serm. = Cyatheaceae

Hypodematiaceae Ching = Dryopteridaceae, tentatively

"*Hypoderriaceae*" Ching, *nom. nud.*, used by various authors, incl. Dickason (1946) = Tectariaceae

Hypolepidaceae Pic. Serm. = Dennstaedtiaceae

"*Kaulfussiaceae*" Campb., *nom. illeg.* = Marattiaceae

Lindsaeaceae C. Presl; here included in Polypodiales

Lomariopsidaceae Alston; here included in Polypodiales

Lonchitidaceae Doweld = Lindsaeaceae

Lophosoriaceae Pic. Serm.; here included in Dicksoniaceae

Loxogrammaceae Ching ex Pic. Serm. = Polypodiaceae

Loxomataceae C. Presl [often misspelled "Loxsomaceae"]; here included in Cyatheales

Lygodiaceae M. Roem.; here included in Schizaeales

Marattiaceae Kaulf., *nom. cons. prop.*; here included in Marattiales; antedated by Danaeaceae (Murdock *et al.*, 2006)

Marsileaceae Mirb.; here included in Salviniales

Matoniaceae C. Presl; here included in Gleicheniales

Metaxyaceae Pic. Serm.; here included in Cyatheales

Mohriaceae C. F. Reed. = Anemiaceae

Monachosoraceae Ching = Dennstaedtiaceae

Negripteridaceae Pic. Serm. = Pteridaceae

Nephrolepidaceae Pic. Serm. = Lomariopsidaceae, tentatively

Oleandraceae Ching ex Pic. Serm.; here included in Polypodiales

Onocleaceae Pic. Serm.; here included in Polypodiales

Ophioglossaceae Martynov; here included in Ophioglossales

Osmundaceae Martynov; here included in Osmundales

Parkeriaceae Hook. = Pteridaceae

Peranemataceae (C. Presl) Ching = Dryopteridaceae

Pilulariaceae Mirb. ex DC. (Pilulariae) = Marsileaceae

Plagiogyriaceae Bower; here included in Cyatheales

Platyceriaceae Ching = Polypodiaceae

Platyzomataceae Nakai = Pteridaceae

Pleurosoriopsidaceae Kurita and Ikebe ex Ching = Polypodiaceae

Polypodiaceae J. Presl; here included in Polypodiales

Psilotaceae J. W. Griff. and Henfr.; here included in Psilotales

Pteridaceae E. D. M. Kirchn.; here included in Polypodiales

Pteridiaceae Ching = Dennstaedtiaceae

Saccolomataceae Doweld; here included in Polypodiales

Salviniaceae Martynov; here included in Salviniales

Schizaeaceae Kaulf.; here included in Schizaeales

Sinopteridaceae Koidz., *nom. rej.* in favor of Adiantaceae = Pteridaceae

"*Sphaerostephanaceae*" Ching, *nom. nud.* = Thelypteridaceae

Stenochlaenaceae Ching = Blechnaceae

Stromatopteridaceae Bierh. = Gleicheniaceae

Taenitidaceae Pic. Serm. = Pteridaceae

Tectariaceae Panigrahi; here included in Polypodiales

Thelypteridaceae Pic. Serm.; here included in Polypodiales

Thyrsopteridaceae C. Presl; here included in Cyatheales

Tmesipteridaceae Nakai = Psilotaceae

Trichomanaceae Burmeist. = Hymenophyllaceae

Vittariaceae Ching = Pteridaceae

Woodsiaceae Herter; here included in Polypodiales

Appendix B: Index to genera

Index to genera with family assignments proposed in this classification. All accepted genera (but not all synonyms) in Kramer in Kubitzki (1990) are accounted for here. Genera newly described or recircumscribed since 1990 are also included. Accepted names are in roman, synonyms are in *italics*. Families are listed below with numbers in parentheses corresponding to the family numbers assigned in the text.

Ophioglossaceae (1)
Psilotaceae (2)
Equisetaceae (3)
Marattiaceae (4)
Osmundaceae (5)
Hymenophyllaceae (6)
Gleicheniaceae (7)
Dipteridaceae (8)
Matoniaceae (9)
Lygodiaceae (10)
Anemiaceae (11)
Schizaeaceae (12)
Marsileaceae (13)
Salviniaceae (14)
Thyrsopteridaceae (15)
Loxomataceae (16)
Culcitaceae (17)
Plagiogyriaceae (18)
Cibotiaceae (19)
Cyatheaceae (20)
Dicksoniaceae (21)
Metaxyaceae (22)
Lindsaeaceae (23)
Saccolomataceae (24)

Dennstaedtiaceae (25)
Pteridaceae (26)
Aspleniaceae (27)
Thelypteridaceae (28)
Woodsiaceae (29)
Blechnaceae (30)
Onocleaceae (31)

Dryopteridaceae (32)
Lomariopsidaceae (33)
Tectariaceae (34)
Oleandraceae (35)
Davalliaceae (36)
Polypodiaceae (37)

Abacopteris = Cyclosorus
Abrodictyum (6)
Acrophorus (32)
Acrorumohra = Dryopteris
Acrosorus (37)
Acrostichum (26)
Actiniopteris (26)
Actinostachys (12)
Acystopteris (29)
Adenoderris (32)
Adenophorus (37)
Adiantopsis (26)
Adiantum (26)
Aenigmopteris (34)
Afropteris = Pteris
Aglaomorpha (37)
Aleuritopteris (26)
Allantodia = Diplazium
Alsophila (20)
Amauropelta = Thelypteris
Ampelopteris = Cyclosorus
Amphiblestra = Tectaria
Amphineuron = Cyclosorus
Ananthacorus (26)
Anarthropteris = Loxogramme
Anchistea = Woodwardia
Anemia (11)
Anetium (26)
Angiopteris (4)
Anogramma (26)
Anopteris = Pteris
Antigramma = Asplenium
Antrophyum (26)
Arachniodes (32)
Araiostegia (36)
Archangiopteris = Angiopteris

Argyrochosma (26)
Arthromeris (37)
Arthropteris (34)
Aspidotis (26)
Aspleniopsis = Austrogramme
Asplenium (27)
Astrolepis (26)
Ataxipteris (32)
Athyriopsis = Deparia
Athyrium (29)
Austrogramme (26)
Azolla (14)
Belvisia (37)
Blechnum (30)
Blotiella (25)
Bolbitis (32)
Bommeria (26)
Botrychium (1)
Botrypus = Botrychium
Brainea (30)
Callipteris = Diplazium
Callistopteris (6)
Calochlaena (21)
Calymmodon (37)
Camptodium = Tectaria
Camptosorus = Asplenium
Campyloneurum (37)
Caobangia (37)
Cardiomanes = Hymenophyllum
Cassebeera (26)
Cephalomanes (6)
Ceradenia (37)
Ceratopteris (26)
Cerosora (26)
Ceterach = Asplenium
Ceterachopsis = Asplenium

Cheilanthes (26)
Cheilanthopsis (29)
Cheiloplecton (26)
Cheiroglossa = Ophioglossum
Cheiropleuria (8)
Chieniopteris = Woodwardia
Chingia = Cyclosorus
Chlamydogramme = Tectaria
Christella = Cyclosorus
Christensenia (4)
Christiopteris (37)
Chrysochosma = Notholaena
Chrysogrammitis (37)
Cibotium (19)
Cionidium = Tectaria
Cnemidaria = Cyathea
Cochlidium (37)
Colysis (37)
Coniogramme (26)
Coptodipteris (25)
Cornopteris (29)
Coryphopteris = Thelypteris
Cosentinia (26)
Costaricia = Dennstaedtia
Coveniella (32)
Crepidomanes (6)
Crypsinus = Selliguea
Cryptogramma (26)
Ctenitis (32)
Ctenitopsis = Tectaria
Ctenopteris (37)
Culcita (17)
Currania = Gymnocarpium
Cyathea (20)
Cyclodium (32)
Cyclogramma = Cyclosorus
Cyclopeltis (33)
Cyclosorus (28)
Cyrtogonellum (32)
Cyrtomidictyum = Cyrtogonellum
Cyrtomium (32)
Cystodium (23)
Cystopteris (29)
Danaea (4)

Davallia (36)
Davalliopsis = Trichomanes
Davallodes (36)
Dennstaedtia (25)
Deparia (29)
Diacalpe = Peranema
Dicksonia (21)
Dicranoglossum (37)
Dicranopteris (7)
Dictymia (37)
Dictyocline = Cyclosorus
Dictyodroma = Deparia
Dictyoxiphium = Tectaria
Didymochlaena (32)
Didymoglossum (6)
Diellia = Asplenium
Diplaziopsis (29)
Diplazium (29)
Diplopterygium (7)
Dipteris (8)
Doodia (30)
Doryopteris (26)
Drymoglossum = Pyrrosia
Drymotaenium (37)
Drynaria (37)
Dryoathyrium = Deparia
Dryopolystichum (32)
Dryopsis (32)
Dryopteris (32)
Edanyoa = Bolbitis
Egenolfia = Bolbitis
Elaphoglossum (32)
Enterosora (37)
Equisetum (3)
Eriosorus (26)
Fadyenia = Tectaria
Feea = Trichomanes
Fourniera = Sphaeropteris
Glaphyropteridopsis = Cyclosorus
Glaphyropteris = Cyclosorus
Gleichenella (7)
Gleichenia (7)
Goniophlebium (37)
Goniopteris = Cyclosorus

Gonocormus = Crepidomanes
Grammitis (37)
Gymnocarpium (29)
Gymnogramma = Hemionitis
Gymnopteris = Hemionitis
Gymnogrammitis (37)
Gymnosphaera (20)
Haplopteris (26)
Hecistopteris (26)
Helminthostachys (1)
Hemidictyum (29)
Hemigramma = Tectaria
Hemionitis (26)
Hemitelia = Cyathea
Heterogonium (34)
Hippochaete = Equisetum
Histiopteris (25)
Holcochlaena (26)
Holodictyum = Asplenium
Holostachyum = Aglaomorpha
Homalosorus (29)
Humata = Davallia
Hyalotricha = Campyloneurum
Hyalotrichopteris = Campyloneurum
Hymenasplenium (27)
Hymenocystis = Woodsia
Hymenoglossum = Hymenophyllum
Hymenophyllopsis (20)
Hymenophyllum (6)
Hypodematium (32)
Hypoderris (34)
Hypolepis (25)
Idiopteris = Pteris
Jamesonia (26)
Japanobotrychium = Botrychium
Kontumia (37)
Kuniwatsukia = Athyrium
Lacostea = Trichomanes
Lacosteopsis = Vandenboschia
Lastrea = Thelypteris
Lastreopsis (32)
Lecanium = Didymoglossum
Lecanopteris (37)
Lellingeria (37)

Lemmaphyllum (37)
Lepisorus (37)
Leptochilus (37)
Leptogramma = Cyclosorus
Leptolepia (25)
Leptopteris (5)
Leptorumohra = Arachniodes
Leucostegia (32)
Lindsaea (23)
Lindsayoides = Nephrolepis
Lithostegia = Arachniodes
Litobrochia = Pteris
Llavea (26)
Lomagramma (32)
Lomaphlebia = Grammitis?
Lomaria = Blechnum
Lomariopsis (33)
Lonchitis (23)
Lophosoria (21)
Lorinseria = Woodwardia
Loxogramme (37)
Loxoma (16)
Loxoscaphe = Asplenium
Loxsomopsis (16)
Luisma (37)
Lunathyrium = Deparia
Lygodium (10)
Macroglena = Abrodictyum
Macrothelypteris (28)
Mankyua (1)
Marattia (4)
Marginariopsis = Pleopeltis
Marsilea (13)
Matonia (9)
Matteuccia (31)
Maxonia (32)
Mecodium = Hymenophyllum
Megalastrum (32)
Melpomene (37)
Meniscium = Cyclosorus
Menisorus = Cyclosorus
Merinthosorus = Aglaomorpha
Meringium = Hymenophyllum
Mesophlebion = Cyclosorus

Metapolypodium = Polypodiodes
Metathelypteris = Thelypteris
Metaxya (22)
Microgonium = Didymoglossum
Microgramma (37)
Microlepia (25)
Microphlebodium = Pleopeltis
Micropolypodium (37)
Microsorum (37)
Microstaphyla = Elaphoglossum
Microtrichomanes = Hymenophyllum
Mildella (26)
Mohria = Anemia
Monachosorum (25)
Monogramma (26)
Monomelangium = Diplazium
Neocheiropteris (37)
Nephelea = Alsophila
Nephopteris (26)
Nephrolepis (33)
Neurocallis (26)
Neurodium (37)
Neuromanes = Trichomanes
Niphidium (37)
Notholaena (26)
Nothoperanema = Dryopteris
Ochropteris (26)
Odontosoria (23)
Oenotrichia (25)
Oenotrichia p.p (32)
Oleandra (35)
Olfersia (32)
Onoclea (31)
Onocleopsis (31)
Onychium (26)
Ophioderma = Ophioglossum
Ophioglossum (1)
Oreopteris = Thelypteris
Ormoloma (23)
Orthiopteris = Saccoloma
Osmunda (5)
Osmundastrum (5)
Pachypleuria (36)
Paesia (25)
Paltonium = Neurodium

Papuapteris = Polystichum
Paraceterach (26)
Parahemionitis (26)
Parasorus = Davallia
Parathelypteris = Thelypteris
Pecluma (37)
Pelazoneuron = Cyclosorus
Pellaea (26)
Peltapteris = Elaphoglossum
Pentagramma (26)
Pentarhizidium (31)
Peranema (32)
Phanerophlebia (32)
Phanerosorus (9)
Phegopteris (28)
Phlebodium (37)
Photinopteris = Aglaomorpha
Phyllitis = Asplenium
Phymatosorus (37)
Pilularia (13)
Pityrogramma (26)
Plagiogyria (18)
Platycerium (37)
Platygyria = Lepisorus
Platyloma (26)
Platyzoma (26)
Plecosorus = Polystichum
Pleocnemia (34)
Pleopeltis (37)
Plesioneuron = Cyclosorus
Pleuroderris = Tectaria
Pleurosoriopsis (37)
Pleurosorus = Asplenium
Pneumatopteris = Cyclosorus
Podosorus (37)
Polybotrya (32)
Polyphlebium (6)
Polypodiodes (37)
Polypodiopteris = Selliguea
Polypodium (37)
Polystichopsis (32)
Polystichum (32)
Polytaenium (26)
Pronephrium = Cyclosorus
Prosaptia (37)

Protowoodsia (29)
Psammiosorus (34)
Pseudocolysis = Pleopeltis
Pseudocyclosorus = Cyclosorus
Pseudocystopteris (29)
Pseudodrynaria = Aglaomorpha
Pseudophegopteris (28)
Pseudotectaria = Tectaria
Psilotum (2)
Psomiocarpa (34)
Pteridium (25)
Pteridoblechnum (30)
Pteridrys (34)
Pteris (26)
Pterozonium (26)
Ptilopteris = Monachorosum
Pycnodoria = Pteris
Pyrrosia (37)
Quercifilix = Tectaria
Radiovittaria (26)
Regnellidium (13)
Revwattsia (32)
Rhachidosorus (29)
Rheopteris (26)
Rosenstockia = Hymenophyllum
Rumohra (32)
Saccoloma (24)
Sadleria (30)
Saffordia = Trachypteris
Sagenia = Tectaria
Salpichlaena (30)
Salvinia (14)
Sceptridium = Botrychium
Schaffneria = Asplenium
Schizaea (12)
Scleroglossum (37)
Scoliosorus (26)
Scyphularia = Davallia
Selliguea (37)
Serpocaulon (37)
Serpyllopsis = Hymenophyllum
Sinephropteris = Asplenium
Sinopteris = Aleuritopteris
Solanopteris = Microgramma

Sorolepidium = Polystichum
Sphaerocionium = Hymenophyllum
Sphaeropteris (20)
Sphaerostephanos = Cyclosorus
Sphenomeris (23)
Steenisioblechnum (30)
Stegnogramma = Cyclosorus
Steiropteris = Cyclosorus
Stenochlaena (30)
Stenolepia (32)
Sticherus (7)
Stigmatopteris (32)
Stromatopteris (7)
Synammia (37)
Syngramma (26)
Taenitis (26)
Tapeinidium (23)
Tectaria (34)
Teratophyllum (32)
Terpsichore (37)
Thamnopteris = Asplenium
Thelypteris (28)
Themelium (37)
Thylacopteris (37)
Thyrsopteris (15)
Thysanosoria (33)
Tmesipteris (2)
Todea (5)
Trachypteris (26)
Trichoneuron = Lastreopsis
Trichipteris = Cyathea
Trichomanes (6)
Trigonospora = Cyclosorus
Triplophyllum (34)
Trismeria = Pityrogramma
Vaginularia = Monogramma
Vandenboschia (6)
Vittaria (26)
Weatherbya = Lemmaphyllum
Woodsia (29)
Woodwardia (30)
Xiphopteris = Cochlidium
Xyropteris (23)
Zygophlebia (37)

Index

Abrodictyum 403, 425
Acrophorus 439
Acrorumohra 440
Acrosorus 444
Acrostichaceae 434
Acrostichum 210, 434, 435
  A. aureum 210
  A. danaeifolium 203, 204, 210
  A. speciosum 210
actin 30
Actiniopteridaceae 434
Actiniopteris 434
Actinostachys 427
Acystopteris 438
Adenoderris 439
Adenophorus 444
  A. periens 115
Adiantaceae 354, 434, 435
Adiantoideae 435
Adiantopsis 434
Adiantopteris 354
Adiantum 161–162, 165, 311, 434, 435
  plastid genome of 163
  A. capillus-veneris 6, 7, 8, 9, 10, 11, 12, 15, 17, 18, 19, 21, 22, 22, 25, 27, 29, 30, 31, 32, 33, 34–35, 37, 38, 120, 162, 176, 187, 203
  A. caudatum 21

A. diaphanum 21
A. latifolium 237, 376
A. pedatum 308, 378
A. philippense 203
A. reniforme 203
A. tenerum 292
Aenigmopteris 442
Afropteris 435
agamospory 307
Aglaomorpha 212, 444
  A. cornucopia 268
Aleuritopteris 434
alleles
  deleterious 110
  recessive 110
allohomoploidy
  lycophytes and 319
  secondary speciation through (*see also* speciation, secondary) 318–320
  tree ferns and 318–319
Alloiopteris 341
allopolyploidy
  secondary speciation through (*see also* speciation, secondary) 320–321
Alsophila 405, 431
  A. auneae (*see also Cyathea pubescens*) 202

A. firma 206, 207, 215
A. salvinii 202, 206, 380
A. setosa 208
A. spinulosa 115
Alsophilaceae 431
Amauropelta 437
Amazonia 370
Ampelopteris 437
Amphiblestra 442
Amphineuron 437
Anachoropteris clavata see Kaplanopteris clavata
Ananthacorus 227, 434
Anarthropteris 444
Anchistea 439
Andes 371
Anemia 9, 135, 138, 140, 142, 143, 351, 427
  A. fremontii 352, 353
  A. phyllitidis 138, 139, 150
Anemiaceae 404, 427
Anetium 434
Angiopteridaceae 423
Angiopteris 82, 86, 163, 165, 423
  A. lygodiifolia 76
Ankyropteris 349
  A. brongniartii 344, 348
Anogramma 434
Anopteris 435

Antarctica 368
antheridiogen 31, 68, 138, 143, 144, 148
  chemical structure of 135, 139
  dark germination and 143–145
  general effect of 135–138
  in nature 145–149
  laboratory conditions and 140–143
  response in polyploids 149
Anthoceros 163
Antigramma 436
Antrophyaceae 434
Antrophyoideae 435
Antrophyum 434
  A. williamsii 268
apical cell 76, 80–81
  bulging of 19–20
apogamy 58, 56–59
  and antheridiogen response 142, 144
  and sporogenesis 58–59
  facultative 56–57
  obligate 57–59
apospory 59–60
Arabidopsis 20, 29, 33, 34, 35, 37
  A. thaliana 33, 80, 183
Arachniodes 406, 439
Arachnoxylon kopfii 339
Araiostegia 443
Archangiopteris 424
Argyrochosma 211, 310, 434
Arthromeris 444
Arthropteris 407, 442
Asia 369
Aspidiaceae 439
Aspidiales 432
Aspidotis 434
  A. densa 110

Aspleniaceae 124, 212, 354, 378, 436
Aspleniales 432
asplenioid ferns 408
Asplenium 115, 135, 149, 211, 212, 288, 305, 379, 381, 408, 436
  A. adiantum-nigrum 373
  A. adulterinum 211
  A. alatum 210
  A. cimmeriorum 268
  A. csikii 111
  A. dalhousiae 376
  A. harpeodes 213
  A. heterosiliens 290
  A. nidus 308
  A. pimpinellifolium 145
  A. platyneuron 124, 237, 375
  A. praemorsum 209, 210
  A. rhizophyllum 237, 378
  A. ruprechtii 378
  A. ruta-muraria 135, 140, 148, 149–150, 288
  A. septentrionale 111, 115, 149
  A. sessilifolium 210
  A. subglandulosum 375
  A. trichomanes 138, 140, 149, 211
  A. trichomanes ssp. *quadrivalens* 111
  A. viride 211
Astrolepis 310, 434
A-substance 134, 135
Ataxipteris 439
Athyriaceae 353, 438
Athyriales 432
athyrioid ferns 408
Athyrioideae 438
Athyrium 263, 306, 310, 408, 438

  A. filix-femina 135, 144, 143–144, 145, 146, 148, 145–149
  A. filix-femina complex 314–315
  A. filix-femina var. *angustum* 314–315
  A. filix-femina var. *asplenioides* 314–315
  A. oblitescens 308
  A. yokoscense 211, 260
Athyropsis 438
Australia 378
Austrogramme 434
autopolyploidy 317
Azolla 357, 404, 428
  A. microphylla 264
  A. standleyi 356, 357
Azollaceae 428

bacterial artificial chromosomes (BAC) 176
Belvisia 444
biogeographical disjunctions 375
Biscalitheca 342
  B. musata 341
Blechnaceae 353, 354, 355, 356, 357, 439
Blechnales 432
Blechnum 145, 241, 408, 439
  B. fragile 213, 215
  B. orientale 377
  B. spicant 120, 124, 142
Blotiella 434
Bolbitidaceae 439
Bolbitis 209, 407, 439
Bolivia 270
Bommeria 135, 142, 144, 434
  B. hispida 179
  B. pedata 57, 142
Boniniella 437
Borneo 369, 371
Botrychiaceae 422

Botrychium 82, 268, 343–345, 422
  B. australe 263
  B. dissectum 121
  B. lanceolatum 121
  B. multifidum var. *robustum* 120, 121
  B. nipponicum 121
  B. simplex 121
  B. ternatum 120, 121
  B. triangularifolium 121
  B. virginianum 120, 121, 123, 345
  B. wightonii 343, 344
  subg. *Botrychium* 308
Botryopteridaceae 347, 348
Botryopteris 347–349
  B. forensis 213, 348, 349
  B. tridentata 344, 348, 349
Botryopus 422
bracken fern *see Pteridium*
Brainea 439
Brazil 270, 371
bryophyte 81, 82

Caobangia 444
Callipteris 438
Callistopteris 403, 425
Calochlaena 432
Calymmodon 444
Camptodium 442
Camptosorus 408, 436
Campyloneurum 444
  C. angustifolium 111
  C. brevifolium 235, 241
Carboniferous 96, 213, 342
  Lower 339, 341, 343, 347
  Upper 343
Cardiomanes 425
Cassebeera 434
cell division 16–19
Cenozoic 347

Centres of Plant Diversity (WWF and IUCN) 270
Cephalomanes 403, 425
Ceradenia 444
Ceratopteridaceae 434
Ceratopteris 9, 59, 135, 138, 138, 140, 142, 143, 434, 435
  C. richardii 57, 140, 176, 180, 181, 184, 185, 183–185, 187
  C. thalictroides 111, 112, 178, 180, 308
Cerosora 434
Ceterach 436
Cheilanthaceae 434
Cheilanthes 144, 203, 211
  C. acrostica 111
  C. gracillima 120
  C. tinaei 111
Cheilanthoideae 435
Cheilanthopsis 438
Cheiloplecton 434
Cheiroglossa 422
Cheiropleuria 351, 426
Cheiropleuriaceae 426
Chieniopteris 439
China 268
Chingia 437
Chlamydogramme 442
chloroplast 25
chloroplast DNA
  PCR mapping of 167
chloroplast movement 20–25, 26, 27, 29, 30
  mechanism of 30
  photorelocation 22, 24, 20–25, 30
  speed of 29–30
Christella 437
Christensenia 423
Christenseniaceae 423
Christenseniales 423
Christiopteris 444

chromosomes
  counts/numbers 177
  homeologous pairing of 112, 178
  size of 186–187
Chrysogrammitis 444
Cibotiaceae 262, 404, 430
Cibotium 263, 430
  C. barometz 262
  C. schiedei 290
  C. taiwanense 204
Cionidium 442
cladoxylopsids 338–341
climate change (*see also* conservation) 263–264
Cnemidaria 431
Cochlidium 444
Colombia 271
colonization
  genetics of 123–124
Colysis 444
commercial collection of ferns 262–263
Coniogramme 434, 435
conservation
  climate change and 263–264
  ex situ banking of spores and 285–290
  ex situ cryostorage of gametophytes and 291–292
  ex situ cryostorage of sporophytes and 292–293
  ex situ cultivation and 285
  ex situ propagation and 265
  gametophyte generation and (*see also* gametophytes) 265
  genetics and 269–270
  in vitro cultures and collections and 290

life cycles and 264–265
protected areas and 270–271
regional and ecosystem level 265–268
restoration and 271–272
spore banks and 264–265
Cooksonia 53
Coptodipteris 434
Cornopteris 438
cortical microfibrils 21
cortical microtubules 21
Corynepteris 343
Coryphopteris 437
Cosentinia 434
　C. vellea 111
Costa Rica 370
Costaricia 434
Coveniella 439
Crepidomanes 403, 425
　C. venosum 203
Cretaceous 37, 343, 345, 347, 350, 351, 353, 354, 356, 358
　Lower 349, 351, 352, 354, 358
　Upper 349, 352, 353, 355, 356, 357
Crypsinus 444
cryptochrome 32, 38–39
Cryptogramma 434, 435
　C. crispa 135, 289
Cryptogrammaceae 434, 435
Ctenitis 407, 439, 443
　C. bigarellae 258
　C. humilis 268
　C. maritima 211
Ctenitopsis 442
Ctenopteris 444
Culcita 429
　C. macrocarpa 110
Culcitaceae 404, 429
Currania 438
C-value 185–186, 192

Cyathea 162, 353, 405, 431
　C. australis 259, 261
　C. bicrenata 207, 208
　C. caracasana 62, 271
　C. cranhamii 353
　C. delgadii 287
　C. pubescens (see also Alsophila auneae) 202
　C. spinulosa 288, 295
Cyatheaceae 350, 353, 379, 404, 431
Cyatheales 232, 350, 429
Cyclodium 439
Cyclogramma 437
Cyclopeltis 407, 442
　C. semicordata 237
Cyclosorus 408, 437
Cyrtogonellum 439
Cyrtomidictyum 439
Cyrtomium 268, 406, 439
　C. falcatum 142, 287
　C. fortunei 142
　C. macrophyllum 142
Cystodiaceae 433
Cystodium 405, 433
Cystopteridaceae 438
Cystopteridoideae 438
Cystopteris 306, 308, 311, 320, 407, 438
　C. tennesseensis 149

Danaea 423
　D. wendlandii 203, 206, 235
Danaeaceae 423
Davallia 87, 212, 443
　D. parvula 203
Davalliaceae 354, 443
Davalliales 432
davallioid ferns 407
Davallodes 443
Dennstaedtia 405, 434
　D. bipinnata 237
　D. cicutaris 66

D. macgregori 268
D. punctilobula 67
Dennstaedtiaceae 352, 353, 405, 434
Dennstaedtiales 432
dennstaedtioid clade 405
Dennstaedtiopsis aerenchymata 352
Deparia 438
　D. acrostichoides 378
Deparioideae 438
detrivore 65
Devonian 81, 232, 335, 337, 339, 342, 343
　Late 342
　Upper 342
Dicksonia 86, 262, 271, 432
　D. antarctica 259, 261, 262
　D. sellowiana 215
Dicksoniaceae 262, 404, 405, 432
Dicksoniales 429
Dickwhitea allenbyensis 355
Dicranoglossum 444
Dicranopteridaceae 426
Dicranopteris 88, 426
　D. dichotoma 77
　D. nitida 88
Dictymia 444
Dictyodroma 438
Dictyoxiphiaceae 442
Dictyoxiphium 442
Didymochlaena 406, 439
Didymoglossum 403, 425
Diellia 124, 269, 408, 436
Diphasiastrum 319
Diplazioideae 438
Diplaziopsis 438
Diplazium 408, 438
　D. subsilvaticum 237, 239
Diplopterygium 87, 426
Dipteridaceae 350, 351, 352, 403, 426
Dipteridales 425

Dipteris 351, 426
  D. lobbiana 203, 208
dispersal
  genetics of 123–124
  long-distance 372–378
divergence times
  estimates of 378
diversification
  patterns of in the fossil record 346–358
diversity 368–372
DNA interference (DNAi) 187
Doodia 439
Doryopteris 434
Drosophila 259
Drymoglossum 444
Drymotaenium 444
Drynaria 212, 444
  D. fortunei 202
  D. quercifolia 265
Drynariaceae 444
Dryoathyrium 438
Dryopolystichum 439
Dryopsis 439
Dryopteridaceae 354, 356, 439, 443
Dryopteridales 432
Dryopteris 13, 35, 145, 145, 226, 263, 306, 310, 320, 406, 439
  D. affinis 142, 144
  D. affinis ssp. *affinis* 57
  D. carthusiana 124
  D. dilitata 138, 147
  D. expansa 120
  D. filix-mas 135, 138, 144, 143–144, 147, 148, 149
  D. filix-mas complex 320
  D. goldiana 262
  D. intermedia 262
  D. nipponensis 112
dryopteroid ferns 406–407

Ecuador 369
Egenolfia 439
Elaphoglossaceae 230, 231, 439
Elaphoglossum 111, 114, 212, 213, 377, 407, 439, 440
  E. lonchophyllum 213, 215
  E. peltatum 213, 215
endemism 371, 373
endopolyploidy 80
Enterosora 444
Eocene 352
  Middle 354, 355
Equisetaceae 402, 423
Equisetales 232, 423
equisetophytes 335
Equisetopsida 423
Equisetum 57, 163, 286, 289, 423
  E. arvense 120, 288
  E. hyemale 287, 289
  E. telmateia 287
  subg. *Equisetum* 423
  subg. *Hippochaete* 423
Eriosorus 434
ethnobotanical studies 259
euphyllophytes 333
eupolypod clade 405
eupolypods I 354
  divergences within 406–407
eupolypods II 354, 355
  divergences within 407–408
Eutracheophyta 54
expressed sequence tags (EST) 176, 190

Fadyenia 442
filicalean radiation 334
Filicopsida 424
filmy ferns (*see also* Hymenophyllaceae) 351
fixation index 116

floras
  species diversity on continent versus island 231
Florida 369
floristics 379–381
fluorescent in situ hybridization (FISH) 180
fossil record, nature of 335
Fourniera 431
F_{ST} 121

Galápagos Islands 263
gametophyte
  autecology of 234–236
  carbon relations of 240–242
  cell growth 9–12
  cell growth cessation 10
  conservation and (*see also* conservation) 265
  desiccation tolerance of 239, 236–240
  ecological adaptations of 228–229
  ecomorphology of 225–233
  ecophysiology of 233–242
  ecovalidation of laboratory results 242–250
  evolution of growth from of 232–233
  gemmae-forming (*see also* gemmae) 213, 231
  growth forms of 226
  light stress relations of 240–242
  longevities of 234
  photosynthetic rates of 241
  subterranean 121
  survivorship of 235
  water relations of 236–240

gemmae (see also gametophyte, gemmae-forming) 50, 142, 150, 377
gene
  chlorophyll a/b binding protein (CAB) 179
  copy number and restriction fragment length polymorphisms (RFLPs) 179–180
  copy number, nuclear 179–180
  duplications, plastid 182
  flow 61, 122–123
  feminization (FEM1) 143
  HER 143
  transformer (TRA) 143
genetic
  divergence (see also populations) 122–123
  diversity 112–115
  drift 122
  linkage map 175, 180–183
  load 110–112, 124
genome
  inversions, plastid 165, 167, 170
  reorganization, polyploids and 321
  size 176, 185–187
Gillespiea randolphensis 341, 342
Ginkgo 161
Glaphyropteridopsis 437
Gleichenella 426
Gleichenia 162, 426
  G. appianensis 351, 352
Gleicheniaceae 88, 216, 350, 351, 352, 378, 403, 426
Gleicheniales 232, 351, 425
gleichenioid ferns 351, 403
Gondwana 345, 368
Goniophlebium 444

Goniopteris 437
Grammitidaceae 407, 444
grammitid ferns 213, 216, 229, 231, 377, 407
Grammitis 444
Greenland 369
Guatemala 369
Gymnocarpium 306, 311, 320, 369, 407, 438
  G. dryopteris 308
  G. dryopteris ssp. *disjunctum* 120, 140
  G. robertianum 135
Gymnogrammitidaceae 443, 444
Gymnogrammitis 444
Gymnosphaera 431

habitat
  fragmentation 260–262
  preference see habitat specificity
  specificity 203, 209–215
  specificity and epiphytes 212–215
  specificity and lithophytes 211
  specificity and mangrove species 210–211
  specificity and terrestrial species 209
Haplopteris 434
Hardy–Weinberg equilibrium 116
Hausmannia 351
  H. morinii 352
Hawaiian Islands 64, 65, 115, 123, 124, 259, 263, 264, 269, 271, 373
heavy metals
  contaminated soils 211
  hyper-accumulation of 211
Hecistopteris 434

Helminthostachyaceae 422
Helminthostachys 87, 422
Hemidictyum 407, 438
Hemigramma 442
Hemionitidaceae 434
Hemionitis 434
  H. palmata 120, 123, 140
  H. pinnatifida 111
Hemitelia 431
herbivore 65
Heterogonium 442
heteromorphy 51
heterosporous leptosporangiates (see also water ferns) 357–358, 404
heterospory 177, 350
Histiopteris 87, 434
Hokkaido Island 369
Holcochlaena 434
Holodictyum 436
Holostachyum 444
Homalosorus 438
Honshu Island 369
Hopetedia praetermissa 351
Humata 443
Huperzia 379
  H. lucidula 162, 176
  H. miyoshiana 120
  H. saururus 376
Hyalotrichopteris 444
hybridization
  secondary speciation through (see also speciation, secondary) 318
hydropterid radiation 334
Hydropteridales 350, 356, 357–358, 428
Hydropteris pinnata 356, 357, 428
Hymenasplenium 408, 436
Hymenocystis 438
Hymenoglossum 425

Hymenophyllaceae 87, 212, 213, 216, 227, 228, 230, 231, 232, 350, 351, 377, 379, 403, 425
Hymenophyllales 425
hymenophylloid clade 403
Hymenophyllopsidaceae 431
Hymenophyllopsidales 429
Hymenophyllopsis 405, 431
Hymenophyllum 227, 403, 404, 425
 H. tunbrigense 269, 291
Hypodematiaceae 439
Hypodematium 406, 440
Hypoderriaceae 442
 Hypoderris 442
Hypolepidaceae 434
Hypolepis 88, 434
 H. punctata 78, 91

Iberia 270
Ibyka 335–336
inbreeding coefficient 116
intergametophytic
 crossing 108
 selfing 108, 121
intragametophytic selfing 108, 124, 178
 rates of 120–121
invasive species 263
iridopterids 338–341
islands
 oceanic 263, 373
Isoëtaceae 77, 79, 82, 83, 90, 92, 96–97, 177
Isoëtes 65, 83, 90, 163, 268, 306, 308, 320, 379
 I. asiatica 93
 I. engelmannii 293
 I. georgiana 268
 I. hypsophila 268
 I. louisianensis 293, 294
 I. olympica 268
 I. setacea 271

 I. sinensis 65, 115, 268
 I. taiwanensis 268
 I. yunguiensis 268
isomorphy 50–51
isozymes
 ploidy and 178–179
IUCN 270
 Red List 268

Jamesonia 434
Japan 270, 369
Japanobotrychium 422
Jurassic 37, 343, 345, 347, 351, 358
 Upper 358

Kamchatka Peninsula 369
Kaplanopteridaceae 347, 348, 349–350
Kaplanopteris clavata 349, 349
Kaulfussiaceae 423
Killarney fern *see Trichomanes speciosum*
Klukia 351
KNOX gene 86
Kontumia 444
Korea
 Republic of 269
Krakatau 377

Lastreopsis 407, 440, 443
leaf 82
 apical meristem (LAM) 83
 life span 202
 marginal blastozone 83
 marginal meristem 83
Lecanopteris 212, 444
Lellingeria 444
Lemmaphyllum 444
Lepisorus 444
 L. thunbergianus 111
Leptochilus 444
Leptolepia 434

Leptopteris 403, 424
leptosporangiate ferns 334, 402
 early divergences in 403–405
 oldest evidence for 347
Leucostegia 406, 440, 444
life cycle
 biphasic 51, 53
light
 fluence rates of 25, 26
 sensitivity to 22
Lindsaea 433
Lindsaeaceae 405, 433
Lindsaeales 432
lindsaeoid ferns 405
Lithostegia 439
Llavea 434, 435
Lomagramma 87, 440
Lomariopsidaceae 354, 442
Lomariopsis 209, 227, 379, 407, 442
 L. guineensis 377
 L. palustris 377
 L. vestita 235
Lonchitidaceae 433
Lonchitidales 432
Lonchitis 405, 433
long-term studies 216–217
Lophosoria 271, 432
Lophosoriaceae 432
Lorinseria 439
Loxogrammaceae 444
Loxogramme 444
Loxoma 429
Loxomataceae 404, 429
Loxomatales 429
Loxoscaphe 408, 436
Loxsomataceae (*see also* Loxomataceae) 429
Loxsomopsis 429
Luisma 444
Lunathyrium 438

Lycopodiaceae 77, 79, 82, 83, 86, 89–90, 92, 177
Lycopodiella 319
  L. cernua 78, 377
Lycopodium 56, 83, 209, 306, 319
  L. annotinum 120, 264
  L. clavatum 78, 85, 93, 120
  L. complanatum 89, 89
  L. tristachyum 89
Lygodiaceae 404, 427
Lygodium 9, 142, 143, 351, 427
  L. heterodoxum 145, 288
  L. japonicum 82, 83, 124, 350
  L. microphyllum 124, 376
  L. palmatum 257
  L. venustum 203, 208–209

Macrothelypteris 408, 437
  M. torresiana 205, 376
Makotopteris princetonensis 353, 354
Malesia 266
Mankyua 422
  M. chejuense 269
Marattia 423
Marattiaceae 91, 402, 423
Marattiales 232, 344, 423
Marattioid 345–346
Marattiopsida 423
Marchantia 160, 161
Marsilea 39, 286, 357, 404, 428
  M. strigosa 121
  M. vestita 187
  M. villosa 271
Marsileaceae 357–358, 404, 428
Marsileaceaephyllum 358
Marsileales 428
mating systems 116–121
Matonia 426

Matoniaceae 350, 403, 426
Matoniales 425
Matonidium 351, 353
Matteuccia 439
  M. struthiopteris 135, 287
Maxonia 440
Mecodium 425
Medicago trunculata 183
Megalastrum 407, 440
megaphyll 82
Melpomene 444
  M. flabelliformis 376
Mendelian inheritance 107
Meniscium 437
Menisorus 437
Merinthosorus 444
merophytes 76, 94
Mesophlebion 437
Mesozoic 347, 350, 351
Metaclepsydropsis duplex 341
Metapolypodium 444
Metathelypteris 437
Metaxya 432
Metaxyaceae 405
Metaxyales 429
microbeam irradiation/irradiator 6, 27
microfibrils
  cortical 21
Microgramma 87, 444
  M. reptans 237, 239, 241
Microlepia 434
microphyll 82
Micropolypodium 444
Microsorum 444
Microstaphyla 440
Microtrichomanes 425
microtubules
  cortical 21
mid-domain effect 371
Midlandia nishidae 353, 355
Mildella 434
Miocene 356

Mississippian 347, 351
mixed-mating system 120
moa-nalo *see Thambetochen chauliodous*
Mohria 427
  M. caffrorum 143
Mohriaceae 427
Monachosoraceae 434
Monachosorales 432
Monachosorum 434
Moniliformopsis (*see also* moniliforms) 336–337
moniliforms 334, 339, 341, 337–343
Monogramma 434
Monomelangium 438
morphotaxa 335
mycorrhizae 209
myosin 30

natural selection 122
Negripteridaceae 434
Negripteridales 432
Nei's unbiased genetic identity (*I*) 121
Neocheiropteris 444
neochrome 35–37
Neolepisorus 444
neopolyploidy 189
Nephelea 431
Nephopteris 434
Nephrolepidaceae 442
Nephrolepis 214, 407, 442
  N. biserrata 237
  N. exaltata 377
Neurocallis 434
Neurodium 444
New England 369
New Guinea 266, 371
New York state fern law 258
New Zealand 263, 268, 378
Nicotiana 160, 161
Niphidium 212, 444
Notholaena 203, 211, 310, 435

Notholaena (cont.)
  N. distans 135
  N. grayii 307
  N. sinuata 135
  N. vellea 135
Notholaenoideae 435
Nothoperanema 440
nuclear movement 30

Ocropteris 435
Odontosoria (see also
    Sphenomeris chinensis)
    433
Oenotrichia 434, 440
Oleandra 407, 443
  O. pistillaris 77
Oleandraceae 354, 443
oleandroid ferns 407
Olfersia 440
Oligocene 351
Onoclea 355, 408, 439
  O. sensibilis 139, 140,
    141–142, 287, 288, 355,
    356, 378
Onocleaceae 354, 355, 357,
    403
onocleoid ferns 408
Onocleopsis 439
Onychium 435
Ophioderma 422
Ophioglossaceae 87, 121,
    400, 422
Ophioglossales 232, 344, 422
ophioglossid ferns 343–345
Ophioglossum 209, 422
  O. reticulatum 177
Oreopteris 437
Ormoloma 405, 433
Orthiopteris 433
Osmunda 9, 35, 82, 161, 162,
    288, 307, 403, 424
  O. cinnamomea 62, 349,
    350, 355, 403
  O. claytoniana 7–9, 83

O. regalis 57, 111, 259, 288
O. vancouverensis 349
subg. *Claytosmunda* 424
subg. *Osmunda* 424
subg. *Plenasium* 424
Osmundaceae 91, 348,
    350–351, 357, 424
Osmundales 232, 424
Osmundastrum 403, 424

Pachypleuria 443
Paesia 434
Paleocene 344, 355, 356
Paleogene 343, 351, 354
paleopolyploidy 188–189
Paleozoic 341, 346, 347, 351
Pandanus 203
Papuapteris 440
Paraceterach 435
Parahemionitis 435
Paralygodium vancouverensis
    352
Parasorus 443
parastichy 85
Parathelypteris 437
Parazolla 357
Parkeriaceae 434, 435
Parkeriales 432
Parkerioideae 435
pathogens
  fungal 263
PD networks *see*
  plasmodesmatal
  networks
Pecluma 444
Pecopteris 344
Pelazoneuron 437
Pellaea 211, 286, 306, 310,
    435
  P. andromedifolia 179,
    308
  P. rufa 179, 191
  P. viridis 135
Peltapteris 440

Pennsylvanian 342, 345,
    347, 349, 351
  Late 346
  Lower 345
  Middle 344, 348, 349
  Upper 341, 344, 348
Pentagramma 435
Pentarhizidium 439
Peranema 440
Peranemataceae 439
Permian 337, 342, 344,
    345, 346, 347, 349, 350,
    351
  Early 342
  Late 347
  Lower 347
Peru
  Andean highlands of 268
Phanerophlebia 406, 440
Phanerophlebiopsis 439
Phanerosorus 426
Phegopteris 369, 408, 437
  P. decursive-pinnata 111
phenology 202
  seasonality and 204–209
Philippine Islands 369
Phlebodium 214, 444
  P. pseudoaureum 237
Photinopteris 444
photoreceptive sites 8
photoreceptor 24, 28–29, 32,
    31–39
photosynthetically available
  radiation (PAR) 62–63,
  64
phototropic response 36
phototropin 32, 32–33,
    34–35
phototropism 13, 14, 12–16
Phyllitis 436
  P. scolopendrium var.
    americana 264
phyllotaxis 87
Phymatosorus 444

phytochrome 13–16, 32, 33–34
phytoremediators 211
pigs
    feral 263
Pilularia 357, 404, 428
Pilulariaceae 428
Pilulariales 428
Pisum 161
Pityrogramma 435
    P. calomelanos 57, 376, 377
    P. tartarea 235, 237
Plagiogyria 430
Plagiogyriaceae 404, 430
Plagiogyriales 429
plasmodesmatal (PD)
    networks 79, 79–80, 81–82
plastid genome sequences 162–164
Platyceriaceae 444
Platycerium 212, 233, 444
    P. andinum 212, 378
    P. bifurcatum 59, 289
    P. coronarium 202
    P. grande 202
    P. stemaria 262
Platygyria 444
Platyloma 435
Platyzoma 435
Platyzomataceae 434
Platyzomatales 432
Plecosorus 440
Pleocnemia 442
Pleopeltis 120, 315, 444
    P. furfuraceum 209
    P. macrocarpa 376
Plesioneuron 437
Pleurisoriopsidaceae 444
Pleurisoriopsis 444
Pleuroderris 442
Pleurosorus 436
Pneumatopteris 437
Podosorus 444

polarotropism 14, 12–16
Polybotrya 209, 440
polygrammoid ferns 407
Polyphlebium 403, 425
    P. capillaceum (see also *Trichomanes capillaceum*) 213
polyploidization 177, 179, 180–193
polyploidy
    origins of 321
polypod ferns
    early divergences of 405–406
Polypodiaceae 111, 212, 213, 216, 230, 231, 354, 407, 444
polypodiaceous ferns 353, 377
Polypodiales 226, 232, 353, 432
Polypodiodes 444
Polypodiopsida 424
Polypodiopteris 444
Polypodium 62, 212, 213, 307, 308–310, 320, 444
    P. amorphum 309, 313
    P. appalachianum 309, 313, 378
    P. australe 309
    P. crassifolium 144
    P. feei 144
    P. furfuraceum 210
    P. glycyrrhiza 309
    P. hesperium 309
    P. interjectum 309
    P. pseudoaureum 241
    P. rhodopleuron 209
    P. saximontanum 310
    P. sibiricum 309, 378
    P. sibiricum group 313–314
    P. virginianum 308
    P. vulgare 62, 63, 115

P. vulgare complex 308, 309, 310, 320
polysporangiophytes 335
Polystichopsis 406, 440
Polystichum 62, 162, 306, 311, 320, 378, 379, 406, 439, 440
    P. acrostichoides 62, 146, 262
    P. aleuticum 295
    P. drepanum 290, 295
    P. imbricans 120, 318
    P. munitum 120, 144, 179, 318
    P. otomasui 120
    P. tsus-sinense 287
Polytaenium 435
populations
    genetic divergence of 122–123
    genetic structure of 121–122
predators 263
Premnornis guttuligera 259
pre-prophase band 4, 5
Pronephrium 437
Prosaptia 444
Protowoodsia 438
Psalixochlaenaceae 347
Psammiosorus 407, 442
Psaronius 213, 344, 345–346, 349
    P. brasiliensis 344
Pseudocyclosorus 437
Pseudocystopteris 438
Pseudodrynaria 444
Pseudophegopteris 408, 437
Pseudosporochnus 335–336
    P. hueberi 339
    P. nodosus 339, 341
Pseudotectaria 442
Psilotaceae 400, 423
Psilotaceae, rhizomes of 94–95

Index

Psilotales 232, 423
psilotophytes 333
Psilotopsida 422
Psilotum 94–95, 209, 423
  P. nudum 94, 176
  P. nudum var. *gasa* 186
Psomiocarpa 442
Pteridaceae 149, 212, 213, 230, 353, 354, 434
Pteridales 432
Pteridiaceae 434, 435
Pteridium 52, 54, 55, 58, 60, 60–61, 68, 88, 138, 263, 434
  P. aquilinum 76, 120, 135, 139, 141, 142, 144, 245, 246, 247, 249, 377
Pteridoblechnum 439
Pteridoideae 435
Pteridrys 442
Pteris 21, 226, 230, 287, 318, 435
  P. altissima 237
  P. cretica 21, 56, 58, 187
  P. dispar 111
  P. ensiformis 57
  P. multifida 124
  P. vittata 7, 31, 139, 211, 289, 377
Pterisorus radiata 353, 354
pteroid clade 405
Pterozonium 354, 435
Pyrrhula murina 259
Pyrrosia 212, 444
  P. piloselloides 237

Quaternary 345, 351
Quercifilix 442
Quercus 215
quiescent center (QC) 92
quillwort *see Isoëtes*

Radiovittaria 435

RAM *see* root apical meristem
Regnellidium 357, 404, 428
Regnellites nagashimae 358
remote sensing 271
reproductive organs 31
restriction fragment length polymorphisms (RFLPs) and gene copy number 179–180
Revwattsia 440
Rhachidosoroideae 438
Rhachidosorus 438
Rhacophyton 335–336, 342
Rheopteris 435
rhizomes
  branching of (*see also* shoots, branching of) 94
rhizomorph 95–97
  Lepidodendrid 96
rhizophore 96, 95–97
Rhynia gwynn-vaughanii 81
Rhynie chert 53, 232
rhyniophyte 54
Rickwoodopteris hirsuta 353
RNA interference (RNAi) 187
root apical meristem (RAM) 91, 93, 90–94
roots 90–94
  branching of 91, 93
  evolution of 93–94
  initiation of 91
Rosenstockia 425
Ruffordia 351
Rumohra 407, 440
  R. adiantiformis 263
Russia 369

Saccoloma 405, 433
Saccolomataceae 405, 433
Saccolomatales 432
Sadleria 245, 439

S. cyatheoides 110, 120, 124, 135, 140
S. pallida 110, 120
Saint Helena 114
Salpichlaena 439
Salvinia 357, 404, 428
Salviniaceae 356, 357–358, 404, 428
Salviniales 428
SAM *see* shoot apical meristem
Sceptridium 422
  S. multifidum var. *robustum* 120, 121
  S. nipponicum 121
  S. ternatum 120, 121
  S. triangularifolium 121
Schaffneria 436
Schizaea 351, 427
  S. pusilla 380
Schizaeaceae 142, 350, 351, 352, 404, 427
Schizaeales 232, 350, 427
schizaeoid ferns 404
Schizaeopsis 427
Scleroglossum 444
Scolecopteris 344
Scoliosorus 435
Scolopendrium vulgare 107, 258
Scyphularia 443
seasonality (*see also* phenology) 204–209
  climbing ferns and 208–209
  epiphytes and 208–209
  rheophytes and 208
  terrestrial ferns and 206–208
Selaginella 80, 83, 86, 163, 268, 314
  S. caudata 96
  S. densa 313
  S. kraussiana 96, 96, 186

S. lepidophylla 176
S. martensii 78, 84
S. moellendorffii 176, 193
S. rupestris 313
S. sylvestris 292, 293
S. uncinata 96, 292
S. underwoodii 313
Selaginellaceae 77, 79, 83, 90, 92, 95–97, 177
Selliguea 444
Senftenbergia 351
Sermayaceae 347
Serpocaulon 407, 444
Serpyllopsis 425
shoot apical meristem (SAM) 76, 77,78, 79, 76–82, 84, 85, 91, 95
shoots, branching of
　axillary 87
　dichotomous 87, 88, 86–90
　epipetiolar 87
　equal 89
　extra-axillary 87
　lateral 88
　monopodial 86–90
　unequal 89
signal transduction 19, 26–28
Sinephropteris 436
Sinopteridaceae 434, 435
Solanopteris 212, 444
Sorolepidium 440
Southern Hemisphere 376
speciation 311–322
　allopatric 312–315
　primary 312–316
　primary, adaptation and 315–316
　primary, ecology and 315–316
　secondary 316–321
　secondary, allohomoploidy and 318–320

secondary, allopolyploidy and 320–321
secondary, hybridization and 318
tertiary 321–322
species
　asexual 307–308
　biological 306
　concepts 304–307
　cryptic 307–308
　definitions 304–307
　introduced 376
　morphological 305
　number of (see also diversity) 368
　taxonomic 305
　boundaries of 308–311
species concept
　evolutionary 304
　genetic 307
Speirseopteris orbiculata 353, 355
Sphaerocionium 425
Sphaeropteris 405, 431
　S. cooperi 206, 263
Sphaerostephanaceae 437
Sphaerostephanos 437
Sphenomeris 433
　S. chinensis 120, 123
Sphenophyllum 346
Sphenopsida 423
Spinacia 161
spores
　bank 67–68
　bank conservation and (see also conservation) 264–265
　chlorophyllous 7, 286
　germination 7–9
　non-chlorophyllous 7–9, 286
　release and dispersal 65–68
sporogenesis 61–68

Stauropteridales 341
stauropterids 334, 341–342
Stauropteris burntislandica 342
Steenisioblechnum 439
Stegnogramma 437
Steiropteris 437
Stenochlaena 439
　S. areolaris 203
　S. palustris 211
Stenochlaenaceae 439
Stenolepia 440
Sticherus 426
　S. flabellatus 120
Stigmatopteris 407, 440
stress tolerance 50
Stromatopteridaceae 426
Stromatopteridales 425
Stromatopteris 87, 426
Stylites 90
Synammia 369, 444
Syngramma 435
Syria 268
Sumatra 272

Taenitidaceae 434
Taenitis 435
Taiwan 369
Tapeinidium 405, 433
Tectaria 442
　T. semibipinnata 203
Tectariaceae 354, 442
Tedelea 349
　T. glabra 348
Tedeleaceae 347, 348, 349, 351
telome 84–86
Tempskya 349, 353
Tempskyaceae 348, 353
Teratophyllum 440
Terpsichore 444
　T. asplenifolia 213, 215
Tertiary 345, 350, 351, 352, 353, 356

Thambetochen chauliodous 65, 259
Thamnopteris 436
Thelypteridaceae 353, 354, 355, 408, 437
Thelypteridales 432
Thelypteris 230, 408, 437
   T. angustifolia 64, 203, 208
   T. balbisii 237, 241
   T. curta 237
   T. dentata 376
   T. nicaraguensis 237, 241
   T. opulenta 376
   T. palustris var. *pubescens* 378
   T. torresiana 376
thelypteroid ferns 408
Themelium 444
Thylacopteris 444
   T. capillaceum (see also *Polyphlebium capillaceum*) 213, 215
   T. ferrugineum 203
   T. osmundoides 230
   T. rigidum 230
   T. speciosum 65
Thyrsopteridaceae 404, 429
Thysanosoria 442

Tmesipteridaceae 423
Tmesipteris 209, 423
Todea 403, 424
   T. barbara 290
Trachypteris 435
transposable element (TE) 187
Trawetsia princetonensis 353, 355
Triassic 307, 345, 350, 351, 353
   Late 351
Trichipteris 431
Trichomanaceae 425
Trichomanes 377, 379, 403, 425
trichomanoid clade 403
Trigonospora 437
trimerophytes 333
Triplophyllum 442
tropistic response 14, 15, 16

Uredinales 263

Vandenboschia 403
vicariance 378–379
Vittaria 50, 135, 142, 150, 435

V. isoetifolia 213
V. lineata 235
Vittariaceae 434
vittarioid ferns 212, 213, 227, 228, 230, 231, 377
Vittarioideae 435

water ferns 404
Woodsia 407, 438
   W. ilvensis 285
Woodsiaceae 354, 408, 438
Woodwardia 439
   W. radicans 110
   W. virginica 355, 356
woodwardioid ferns 379
World Conservation Union 270
World Wildlife Fund (WWF) 270

Xyropteris 405, 433

Zygophlebia 444
Zygopteridales 341
zygopterids 334, 342–343
Zygopteris illinoiensis 341